Sequoia Wellingtonia.

THE

VEGETABLE WORLD:

BEING

A History of Plants,

WITH

THEIR STRUCTURE AND PECULIAR PROPERTIES.

ADAPTED FROM THE WORK OF

LOUIS FIGUIER.

WITH A GLOSSARY OF BOTANICAL TERMS.

New and Revised Edition.

WITH 473 ILLUSTRATIONS.

D. APPLETON AND CO.,

NEW YORK.

PREFACE.

THIS History of Plants, founded on the well-known work of M. Figuier, is divided into four parts:—

I. The Organography and Physiology of Plants, comprehending under these heads the description of the different parts or organs of vegetables, and some explanation of the various functions performed by means of them.

II. The principles upon which the Classification of Plants into particular groups rests; with brief sketches of the lives of the more eminent Botanists who have devoted themselves to its study.

III. The systematic arrangement of Plants. In this section the Editor has departed from the original work, being desirous of giving as complete a view of the Vegetable World as his limited space would permit. Although not following exactly M. Figuier's plan, the subjects which that writer had selected for more special illustration have been adhered to. His idea was to make a selection of the more important orders, describe one or two prominent and well-known plants, which were taken as types of the group, and give a slighter notice of the less prominent species belonging to it, with their properties. This idea has been adopted and considerably enlarged upon, the arrangement of families adopted by Dr. Lindley in his "Vegetable Kingdom" being, with some few exceptions, employed as the framework with which these descriptions are interwoven.

IV. The Geographical distribution of Plants on the surface of the globe.

A Glossary of Botanical terms has been added.

With respect to the illustrations, it may be mentioned that they are nearly all drawn from Nature. Those which belong to Cryptogams are borrowed from the original memoirs which have appeared in the "Annales des Sciences Naturelles." M. Faguet has in these designs happily united the sentiment of the artist with the precision of the naturalist.

CONTENTS.

PART I.

ORGANOGRAPHY AND PHYSIOLOGY.

PART II.

PART III.

SYSTEMATIC ARRANGEMENT OF PLANTS.

PART IV.

FULL PAGE ILLUSTRATIONS.

FRONTISPIECE—SEQUOIA WELLINGTONIA.

Fig. 1.—Definite root-stock of the Iris.*

THE VEGETABLE WORLD.

ORGANOGRAPHY, OR THE STRUCTURE OF PLANTS.

COMMIT a seed to the earth; plant, for example, a Haricot bean (Fig. 2) at the depth of two inches in moist vegetable soil, and if the temperature is between 40° and 100° F., the seed will not be slow to germinate, first swelling, and then bursting its outer skin. By this

* Fig. 1 is an example of Rhizome, or underground stem or root-stock; *a*, the parent root-stock, which has produced *b* from a lateral bud, which in its turn has produced *c c c;* the rootlets *d* consist of thick adventitious fibres; *e*, young root fibres from *c*, which have just begun to sprout. This kind of rhizome is said to be definite, because the growth of each individual branch is limited; compound, because it consists of separate branches.

admirable arrangement of which Nature permits us to contemplate the wonderful results, but without as yet enabling us to comprehend the strange mystery, a plant in miniature, eventually the counterpart of its parent, will, after a time, reveal itself to the observer. In the meantime two parts, very distinct, make their appearance: one, yellowish in colour, usually branched, sinks into the soil—this is the *radicle* or root; the other, of a pale greenish colour, takes the opposite direction, ascends to the surface, and rises above the ground—this is the stem.

Let us consider at first, in a general manner, this root and stem, with their functions. They are the essential organs of vegetation, which

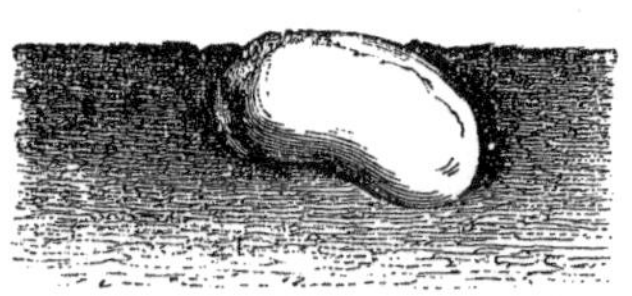

Fig. 2.—Haricot bean (*Phaseolus vulgaris*).

Fig. 3.—Haricot bean germinating.

exist in all plants, or at least when we have excepted certain vegetables of an inferior order, in all plants provided with leaves and flowers.

I.—The Root.

The design of the Creator of the world seems to have been to embellish and make beautiful all which was to be exposed to our eyes, while that which was to be hidden was left destitute of grace or beauty. Leaves suspended from their branches balance themselves gracefully with every movement of the air; the stems, branches, and flowers are the ornament of the landscape, and satisfy the eye with their beauty; but the root is without colours or brilliancy, and is usually of a dull uniform brown, yet performs in obscurity functions as important as those of stem, branches, leaves, or flowers. Yet how vast the difference between the verdant top of a tree, which rises graceful and elegant into mid air—not to speak of the flowers it bears—and the coarse mass of its roots, divided into tortuous branches, without harmony, without symmetry, and forming a tangled, disordered mass! These organs, so little favoured in their appearance, have, however, very important functions in the order of vegetable action.

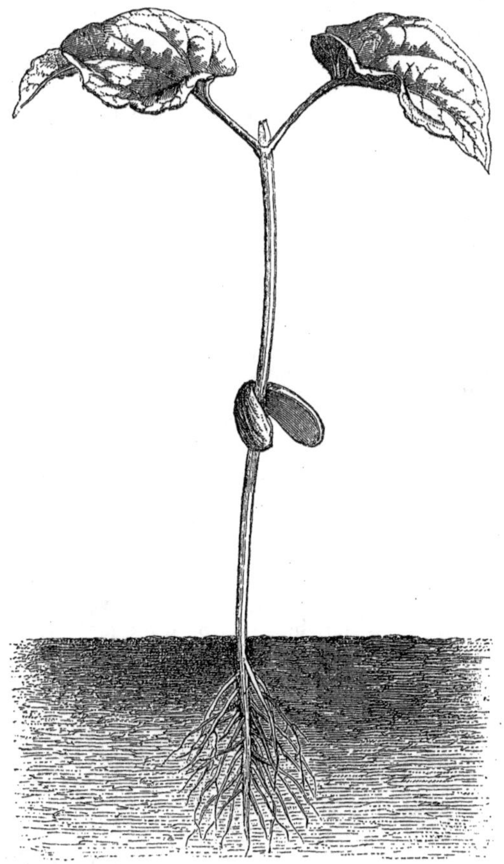

Fig. 4.—Young Haricot.

All plants which germinate with two seed leaves or *cotyledons*, have, at first, a single descending root, the *tap-root.* From this central tap-root, lateral roots branch out more or less regularly, and these lateral roots subdivide again and again. In many cases, especially at first, the lateral roots issue from the tap-root with great order and regularity, as much as in the arrangement of the branches of a young fir-tree; in older plants this order is lost. The tap-root is conspicuous in the dock (*Rumex*) and in seedling fruit-trees; its upper portion in many cultivated plants, such as the beet and carrot, expands under cultivation, and becomes abnormally fleshy (Fig. 6).

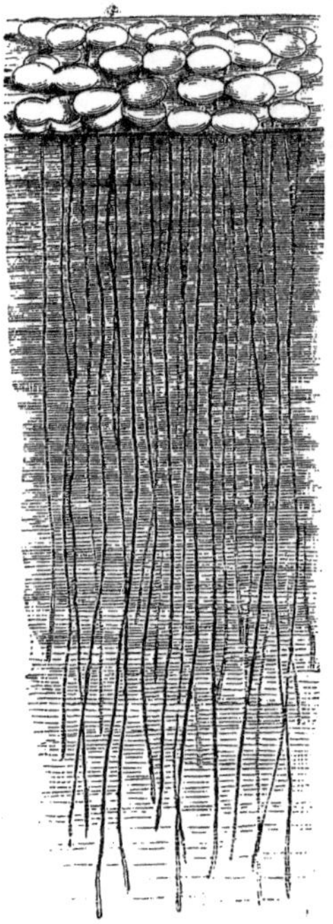

Fig. 5.—Duckweed (*Lemna*).

But all roots are not planted in the soil. There are some plants which develop roots in water, as the Duckweed (*Lemna*), Fig. 5, which never touches the earth. Others nourish themselves on the tissues of other plants, as the Mistletoe, a singular parasitic plant, which forms tufts or branches of a delicate pale green, attaching itself to apple-trees, poplars, and a number of other trees. Some roots appear, moreover, to have no other function than to fix the plants to the soil: they seem to contribute nothing to their nourishment. In the Museum of Natural History of Paris there has been for some years a magnificent Peruvian cactus, of an extraordinary height, which has been growing vigorously, throwing out enormous branches with great rapidity. Its roots are shut up in a box of three feet square, filled with earth, which has never been renewed and never watered. It is therefore evident that in this case the roots have little to do with the nourishment of the plant. Other instances confirm these inferences. "In a country where many months pass without a drop of rain falling," says Auguste de St.-Hilaire, "I have seen, during the dry season, cactuses covered with flowers, maintaining themselves on the burning rocks by the aid of a few weak slender roots, which sink into the dried-up humus which has found its way into the narrow clefts of the rock." Nevertheless, most plants are nourished, to a large extent, through their roots. Other plants, like the Screw-pine (*Pandanus*), and the Mangrove (*Rhizophora*), from their habit of emitting roots from the stem, which descend until they reach the ground, when they bury themselves in the soil, are sometimes said—not very correctly, however—to possess *aërial* roots. The roots, of course, as long as they are aërial in these plants are useless.

Fig. 6.—Tap-root of Tankard Turnip. Fig. 7.—Fibrous root of Melon.

The growth of roots takes place sometimes by their elongation and thickening, as in the Carrot, Turnip (Fig. 6), and Beet; short and slender rootlets are emitted from the sides of the tap-root. Sometimes the root is entirely composed of fibres more or less thick, more or less numerous, and nearly of the same size, which unite at the base of the stem. This produces the fibrous root, of which the Melon (Fig. 7), the Wheat-plant, the Lily, and the Palms afford examples.

This difference in the structure and constitution of the root must be taken into consideration under a great number of circumstances. The old Fir-tree, firmly anchored in the ground by its deep and spreading roots, braves the most violent storms, and even on the mountain-top resists the most terrible tempests. But the Fan-palm, whose cord-like roots spread themselves horizontally in the sand, is overthrown, beaten down by the wind, when it has reached the height of five or six feet. If the stem of this palm be artificially supported, it may attain, even in our climate, to a height of fifty or sixty feet. In front of the great amphitheatre in the Museum of Natural History at Paris, two Fan-palms thus supported rear their lofty heads, surmounted with their tuft of fan-like leaves (Fig. 8).

Some acquaintance with the form of roots will soon find its practical uses. In watering a plant, it is recommended to pour in the water at the foot of the stem, if it is tap-rooted; on the contrary, if the root is fibrous, it should be poured out at some distance from the stem, in order that the spreading roots may receive the benefit of the water. In the cultivation of plants we manure the surface of the soil, or of the deeper beds, according as the plant has tap-roots or fibrous roots. In scientific farming a plant with fibrous roots which exhausts the soil on the surface, is succeeded by a plant with a tap-root, which seeks its nourishment at a greater depth in the soil.

This diversity in the structure of roots is not the work of chance, but the result of design. The composition of the soil varies singularly in different parts of the globe. In order that every point of the surface of the earth should be covered with vegetation, and that no part of it should be without that incomparable adornment, roots must take very varying shapes in order to accommodate themselves to these varieties in the composition of the soil. In one place the soil is hard and stony, heavy or light, formed of sand or clay; in another it is dry or moist; elsewhere it is exposed to the heat of a burning sun, or swept, on the heights, by the violence of the winds and atmospheric currents; sometimes it is sheltered from these movements of the wind in the depth of some warm valley. Roots, hard and woody, separated into strong ramifications, yet finely divided at their terminations, are requisite for mountain plants, whose roots are to live in the midst of rocks or between the stones, in order that they may penetrate between the chinks of the rocks, and cling to them with sufficient force to resist the violence of hurricanes and other aërial tempests. Straight tap-

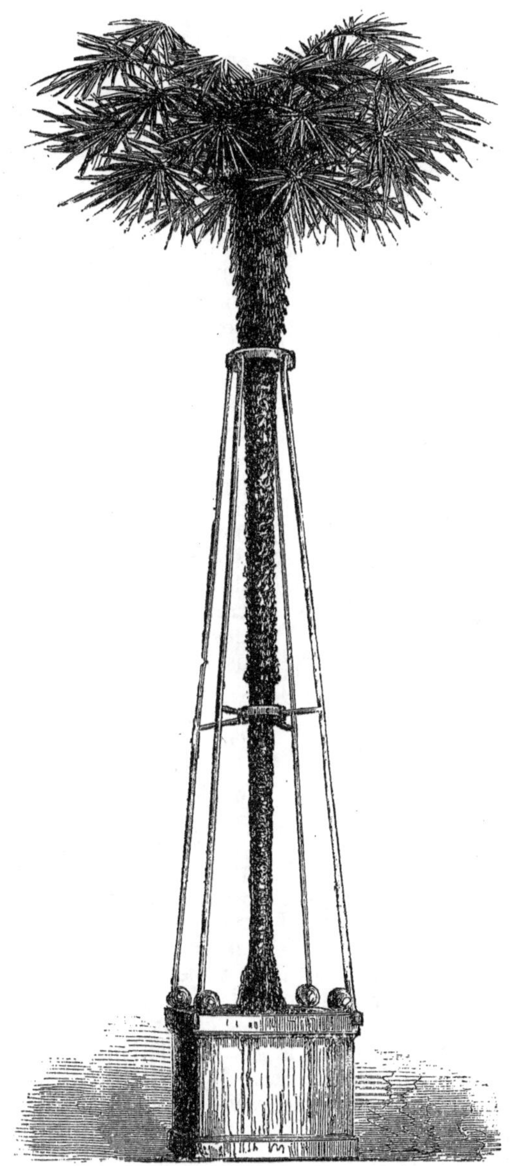

Fig. 8.—Fan-palm at the entrance of the Amphitheatre at the Jardin des Plantes.

roots, slightly branching, are fit for light and permeable soils. They would not suit close, clayey, and shallow soils. Such districts are suitable for plants whose roots spread themselves horizontally just under the surface of the soil.

These considerations are of great importance to the cultivator, who, if he would propagate plants successfully, must carefully study the nature of the soil, and choose for his experiments plants having roots adapted to it.

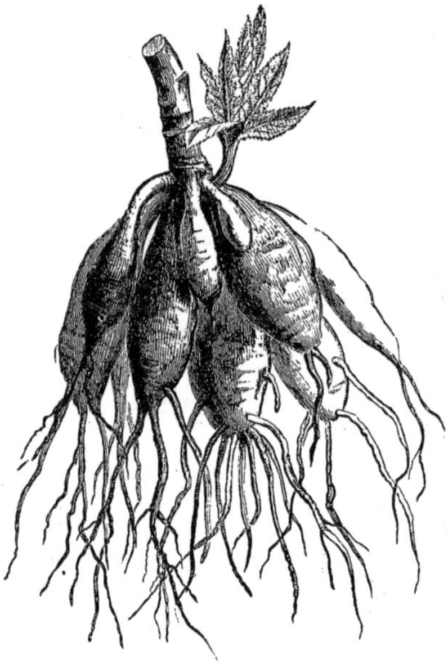

Fig. 9.—Tuberous root of the Dahlia.

Two modifications may be found in the two classes of roots of which we have been speaking. It sometimes happens that these roots develop into masses more or less voluminous, full of nutritive matter, which is destined to nourish the plant or to favour its increase. Common examples of this structure are presented to us by the *Orchis mascula* of our meadows and woods, the Anemone, Ranunculus, and Dahlia of our flower gardens. These roots are called *tuberous* when they take the form of the roots of the Dahlia (Fig. 9), or *tubero-fibrous* when they take the form of those of the Orchis (Fig. 10), where some of the rootlets enlarge while others do not.

These enlargements of the root have a special use in the life of the plant. It is their function to accumulate, in the lower part of the vegetable, supplies of nutritive substances, consisting chiefly of starchy matter, whose purpose is to aid in the development of the plant during a certain period of its existence.

Plants derive their principal nourishment from their roots. We should, then, naturally be led to think that the bulk of the roots would be always in proportion to the size of the stem and branches of a plant. This is generally true for the *same* species; we know, for instance, that the more numerous the branches of an Oak are, the more abundant are its roots; more than this, it is known that the strongest roots in the Oak correspond in direction with its strongest branches. But if we turn from one species of plant to another, we find, not without surprise, that the roots of the Palms and Firs bear little proportion to their height; whilst some plants, such as Lucerne and *Ononis* (Rest-Harrow), are provided with enormously long roots in proportion to the small dimensions of their stems.

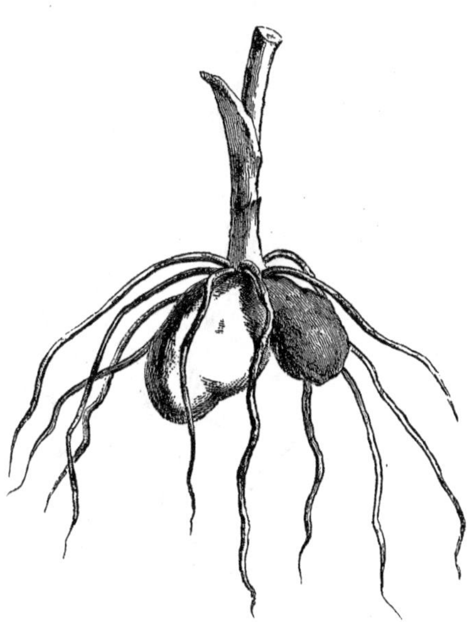

Fig. 10.—Tubero-fibrous root of *Orchis mascula.*

If roots do not show throughout their growth in their ramifications the same regular and unvarying arrangement that we see in leaves and boughs, the cause is not difficult to understand. In the bosom of the earth they meet with obstacles which leaves and branches never meet with in the air. The latter consequently spread freely in every direction, whilst roots are incessantly stopped by all sorts of obstacles. They are constantly cramped in their lengthening or thickening, and are forced to turn aside from the course which they ought naturally to follow, and obliged to twist round to surmount the impediments opposed to them by the unequal hardness of the soil, the presence of walls, rocks, or of other roots. From these causes arise the deformities which we notice in their outward structure, and the numerous windings observable in their branches.

The manner in which roots succeed in overcoming obstacles has always been a subject of surprise to the observer. The roots of trees and shrubs, when cramped or hindered in their progress, have

Fig. 11.—Adventitious roots of the Couch-grass (*Triticum repens*).

been observed to exhibit considerable mechanical force, throwing down walls or splitting rocks; and in other cases clinging together in bunches, or spreading out their fibres over a prodigious space, in order to follow the course of a rivulet with its friendly moisture. Who has not seen with admiration how roots will adapt themselves to the special circumstances of the soil, dividing their filaments, in a soil fit for them, almost to infinity, elsewhere abandoning a sterile

Fig. 12.—Adventitious roots of the Primrose.

soil to seek one farther off, which is favourable to them; and as the ground was more or less hard, wet or dry, heavy or light, sandy or stony, varying their shapes accordingly? We might be disposed to think that there is in these selections made by roots a true manifestation of vital instinct. A sufficient explanation is, however, found, if we remember that roots will almost always grow most readily in that direction which offers least obstacle and gives most encouragement.

Duhamel, a botanist of the last century, relates, that, wishing to preserve a field of rich soil from the roots of a row of elms which would soon have exhausted it, he had a ditch dug between the field and the trees, in order to cut the roots off from it. But he saw, with surprise, that those roots which had not been severed in the operation had made their way down the slope, had passed under the ditch, and were again spreading themselves over the field. It was in reference to an occurrence of this kind that Bonnet, the Swiss naturalist, said that it was sometimes difficult to distinguish "a cat from a rose-tree;" a quaint, if not a witty remark.

Hitherto we have occupied ourselves in considering the roots constituting the descending and normal system of vegetation. There are, however, some roots which are developed along the stem itself. Organs, supplementary in some sort, they come as helps to the roots properly so called, and replace them when by any cause they have been destroyed. In the Wheat-plant, the Couch-grass, and in general in all plants of the grass family, the lower part of the stem gives rise to supplementary roots, to which these common field plants owe a portion of their vigour and their resistance to the causes which would destroy them (Fig. 11).

In the Primrose (*Primula*) both the principal and the secondary roots which spring from it perish after some years of growth. But the *adventitious* roots (Fig. 12) springing from the lower part of the stalk prevent the plant from dying.

In the tropical forests of America and Asia, the Vanilla, whose fruit is so sought after for its sweet aroma, twines its slender stem round the neighbouring trees, forming an elegant, flexible, and aërial garland, at once a grateful and pleasing ornament in these vast solitudes. The underground roots of the Vanilla would not be sufficient for the nutriment of the plant, and the rising of the nourishing sap would take place too slowly. But Nature has provided for this inconvenience by the adventitious roots which the plant throws out at intervals along its stem. Living in the warm and humid atmosphere of tropical forests, they eventually reach the ground, and root themselves in the soil. Others float freely in the atmosphere, absorbing the humidity which trickles down them, and conveying it to the parent stem. All these processes may be observed in full operation in many well-ordered conservatories.

A grand tree, the Banyan (*Ficus indica*), adorns the landscape of India, and presents the most remarkable development of adventitious roots. It rarely vegetates on the ground, but usually in the crown of Palms where birds have deposited the seed. This sends down roots to the ground which embrace and finally kill the foster-plant. When

Fig. 13.—Conservatory of the Jardin des Plantes, Paris, with the adventitious roots of the Vanilla.

the parent stem has attained the height of some fifty or sixty feet, it throws out its lateral branches in every direction, and each branch in its turn throws out adventitious roots, which descend perpendicularly in long slender shoots till they reach the ground. (PLATE I.) When they have rooted themselves in the soil, they increase rapidly in

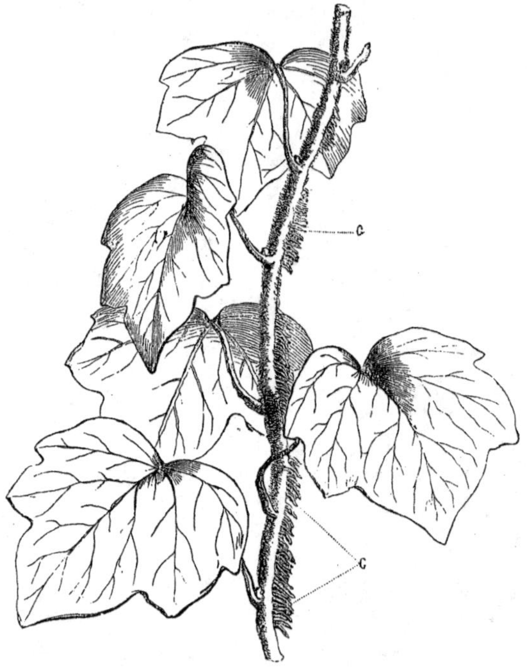

Fig. 14.—Adventitious roots of the Ivy.

diameter, and soon form around the parent stem thousands of columns, which extend their ramifications, each throwing out new lateral branches and new adventitious roots. The famous Banyan-tree on the Nerbuddah, is said, by Forbes, to have 300 large and 3,000 smaller stems. It is capable of sheltering 3,000 men; and forms one of the marvels of the vegetable world. It is, in short, a forest within a forest.

I.—The Banyan of India (*Ficus indica*).

The stem of the Ivy (*Hedera helix*) is furnished with root-like processes or suckers, which seem to have no other function than that of mechanically supporting the plant (Fig. 14). By insinuating their spurs into the bark of trees or on the surface of a wall, they sustain the plant, but without nourishing it.*

There is one genus of plants, that of *Cuscuta*, which is, above all

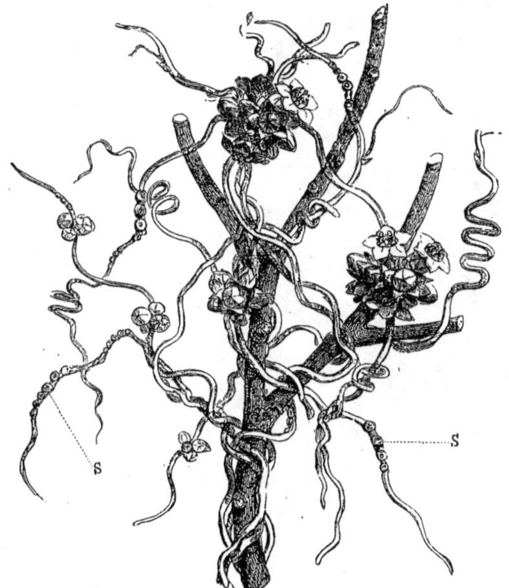

Fig. 15.—Dodder Plant (*Cuscuta*).

others parasitic. Different species have a partiality for particular kinds of plants. They produce in the autumn abundance of sweet-scented flowers, but the plants to which they have attached themselves find their sap resistlessly drawn from them. The European Dodder (*Cuscuta*), Fig. 15, illustrates the mode in which these plants form true nourishing roots.

* This assertion may be doubted: it is no unusual thing to plant ivy on a damp wall, and the invariable result is to dry up the moisture.

The fundamental property of roots, in a physiological point of view, is their constant endeavour to bury themselves in the earth. They seem to shun the light of day; and this tendency is to be remarked from the very first moment of the root showing itself in the germinating seed. It is a tendency so decided and appears so inherent in the life of all vegetables, that if we try to go contrary to it —if, for example, we reverse a germinating seed, placing it with the root upwards—the root and the stem will twist round of themselves; the stem will stretch upwards, and the root will bury itself in the ground.

Fig. 16.—Tube experiment with germinating seeds.

We can convince ourselves, by a very simple experiment, of the natural inclination which stems have to seek the light of day, and which roots have to avoid it. In a room lighted by a single window, place a few germinating mustard seeds on a piece of cotton, and let it float on water in a vessel. It will soon be seen that the small roots point towards the dark part of the room, while the stalklets bend over to meet the rays of light coming from the window.

What can be the cause which determines this natural and invincible tendency of roots towards the interior of the earth? Is it that they would avoid the light because its action might be injurious to them? Do they seek for moisture? The two following experiments will assist the reader to answer these questions.

Place a few seeds upon a wet sponge contained in a glass tube, and light the apparatus from below. When the plant shall have germinated, and pushed out roots and rootlets, they will appear as represented in Fig. 16; the small fibres descending towards the lower part of the tube, and consequently towards the light, in obedience to their natural tendency. Therefore roots do not bury themselves in the ground to

Fig. 17.—Box experiment with germinating seeds.

avoid the light, for in this experiment it is precisely towards the light that they take their course.

Take a box whose bottom is pierced with holes, as represented in Fig. 17, and fill it with mould; place a few kidney beans in these holes, and suspend the apparatus in the open air. The roots will not ascend in order to seek the humid earth. Obedient to the inflexible law which guides them, they will be found to descend through the holes in the box into the dry air, in which they will soon be dried up. It is not moisture therefore that roots seek after.

It has been suggested that the action of gravitation would take some part in the guidance of the roots. This is, in fact, the apparent tendency of the following experiments.

Beans have been made to germinate when placed on the circumference of an iron or wooden wheel surrounded with moss, so as to maintain the moisture of the seeds, and holding little troughs full of

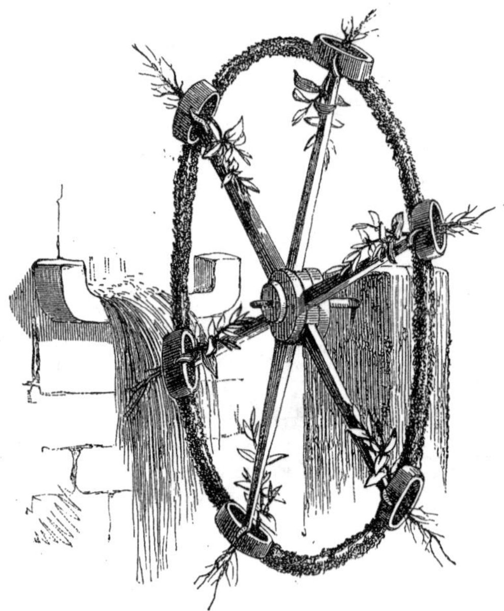

Fig. 18.—Knight's Vertical Wheel experiment.

mould, open on two sides (Fig. 18); the wheel being put in motion in a vertical direction by a current of water, and made to describe many revolutions in a minute. In consequence of this rotary movement, producing the particular force known in mechanics as *centrifugal force*, the action of gravitation is as it were annihilated, and the sprouting seed removed from its influence, is subjected to *centrifugal force* only. See what occurs: the small stems, which, in ordinary circumstances, would be directed upwards, that is to say, in a direction opposite to the action of gravitation, now turn themselves in the direction opposite to the direction of the centrifugal force, or towards the centre of the wheel. The rootlets, which, under ordinary circumstances, would bury themselves in the earth, and in the direction required by the laws of gravitation, in reality now point in the direction of the force which has taken the place of gravitation.

This curious experiment, carried out for the first time by Mr. J. A. Knight, a former president of the Royal Horticultural Society, has been repeated and modified in France by the ingenious naturalist, Dutrochet. He replaced the vertical wheel by a horizontal one, the force of gravitation acting constantly on the same points of the ger-

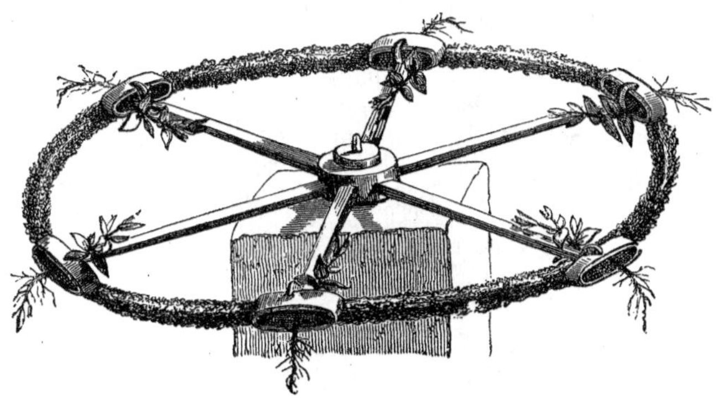

Fig. 19.—Knight's Wheel turning horizontally.

minating seed; but as this seed is exposed at the same time to the action of centrifugal force, produced by the movement of the wheel, the rootlets follow an intermediate direction between a vertical one, which would be determined by the power of gravitation, and a horizontal one, resulting from centrifugal force. As the movement communicated to the wheel is increased in rapidity, the angle made by the root with the plane of the wheel becomes more acute also. When this angle becomes nothing, the root is horizontal.

The influence of gravitation in directing the course of the root is put beyond doubt by these curious experiments.

It must, however, be acknowledged that all is not mechanical in this tendency of roots to bury themselves in the earth. There exists beyond any doubt a real organic tendency belonging to the living plant.

If we compare a transverse section of the stem with one cut from the root of one of our forest trees, the difference between the two parts of the vegetable amounts to very little. The exterior of the root is covered with a bark, very similar to that on the trunk of the tree, only the *parenchyma*, or cellular tissue, is never green in roots. The interior is a woody cylinder, composed of fibres, vessels, and medullary rays. Wood, therefore, forms the central portion of the root, but the

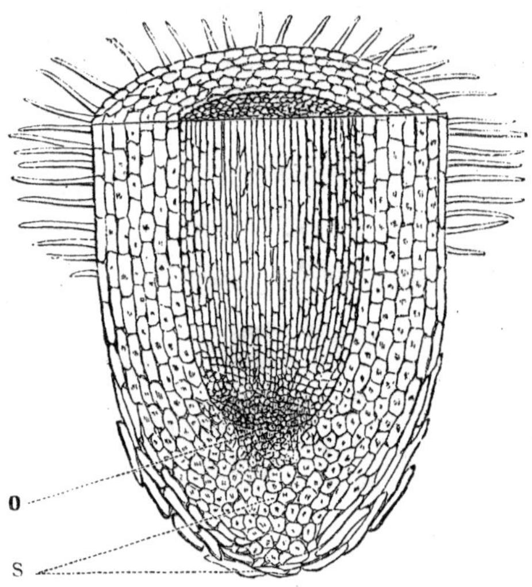

Fig. 20.—Vertical Section of the Extremity of a Root seen under the Microscope.

pith, which is a characteristic feature in the structure of stems, is very rarely well marked in roots.

Roots increase their growth at their extremities only. These extremities are always fresh, and always furnished with permeable and soft cellular tissues. It is in the neighbourhood of these tender extremities that the absorption takes place of the liquid matters which are destined to penetrate into the interior of the plant. This absorption is facilitated and increased by means of the fine elongated hair-like fibres attached to the younger portions of all roots. Fig. 20 represents the terminal part of a root, as seen under the microscope. The true seat of new growth is not situated, as one might suppose, at the extremity of the radicle, that is to say, at the point S, but rather at a certain distance from the end, in the part marked in the engraving by the letter O.

The material which rootlets and their root-hairs take up from the soil in order to pass into the system of the plant must be liquid. Solid bodies, however attenuated, or however subdivided, even when held in suspension in water, cannot penetrate into the infinitely narrow channel which the extremities of the root-fibres present. All substances so absorbed must therefore be in a state of chemical solution in water. The more important of these substances for the purpose of vegetation are the salts of potash, of soda, and of lime, ammoniacal compounds, and carbonic acid gas in solution.

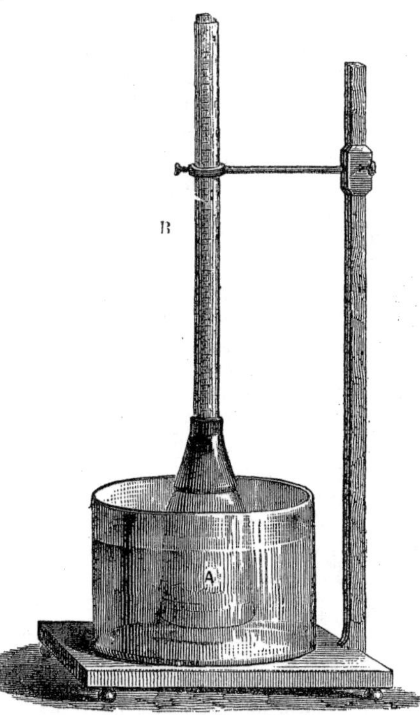

Fig. 21.—Endosmometer.

But what is the mysterious power which produces the operation of absorption in plants, this operation, by which a liquid from the exterior enters and traverses an organ already gorged with liquids? Botanists have now agreed that this result is due to the triple influence, in successive or combined action of *endosmose*, of *capillary attraction*, and of *exhalation* of watery vapour from the leaves. Let us explain ourselves.

Take a small vessel, A, Fig. 21, formed of animal or vegetable membrane, containing a solution of sugar or water, and plunge this vessel into another containing pure water. The syrupy liquid contained in the small vessel will be heavier than the pure water which surrounds it. This unequal density creates immediately a double current through the walls of the vessel; the pure water flows towards the denser syrup, while the other flows in the opposite direction, the least dense liquid passing, however, more rapidly than the other; and if we adapt a vertical tube to the vessel B, we see the liquid gradually rising in the tube. This curious result is called *endosmose.* It is a phenomenon which has been carefully studied by chemists and botanists, and is visibly in action in the vital functions of plants. The extremities of the rootlets of plants are filled with liquids and nutritive matters denser than the water which surrounds them in the soil. By the phenomenon of endosmose the infiltration, or passage of water through the thin cellular exterior tissues takes place; thence the water rises up through the interior vessels of the plant, as we have seen it rise in the tube of the *endosmometer.* In this manner the first movement of ascension is produced.

But the mere power of *endosmose* would not force the foreign fluids very far up into the vessels of the plant. A second force which here intervenes singularly accelerates their upward progress. When a liquid has begun to penetrate, by means of *endosmose*, the extremities of the roots, and the density of the liquid contained in these radicular extremities has been thus diminished by dilution, a current from them is formed in the interior of the root. After that, the force known in physics by the name of *capillarity* promotes the ascent of the liquid in the more elevated parts of the root. The internal walls of each cell of the root exercise on the liquid which it contains the force of capillarity; in other words, an attraction which partly counteracts the effect of gravitation, and determines the ascent of the liquid to a much higher level than it would attain in a larger tube. The phenomenon of capillarity is, then, added to the action of endosmose to favour the absorption of liquids by the radicular extremities.

When the plant is furnished with leaves, there is a third force which unites with the two others in accelerating the absorption. The leaves are the seat of a considerable evaporation. The water dispersed into the atmosphere in the form of vapour leaves the vessels partially empty; its place is immediately supplied by the afflux of the liquids flowing from the roots. In this manner leaf action is produced; it is a sort of suction which draws towards the leaves an afflux of liquid, which the radicular absorption is constantly compelled to supply.

Thus *endosmose*, *capillarity*, and *suction* in the upper part of the plant, are the physical forces which appear to play the principal parts in the absorption carried on by roots.

II.—The Stem of Plants.

The stem is the axis of the ascending system of the vegetable. It is furnished at intervals with *nodes*, or knots, from which spring the leaves, buds, and branches, arranged in a perfectly regular order. The *root* presents nothing of a similar nature. This characteristic enables us easily to distinguish in the vegetable axis between that which belongs to the stem, and that which is peculiar to the root. The stem is that part of the plant which, rising into the air, produces and supports the branches, boughs, leaves, and flowers. Through its tissues the liquids absorbed by the roots penetrate into the interior of the vegetable, for the purpose of supplying it with nourishing juices, increasing its growth, and maintaining its vital functions.

Fig. 22.—Bindweed.

The form, size, and direction of the stem, depend on the part which each plant has to take amongst the vast vegetable population which covers and adorns the globe. Plants which require to live in a pure and often-renewed air have a straight stem, either robust or slender, according to their individual habit. When they only require a moist and denser atmosphere, when they have to creep along the ground or to glide among the brambles, the stems are usually long, flexible, and trailing. If they have to float in

Fig. 23.—Trunk of an Oak.

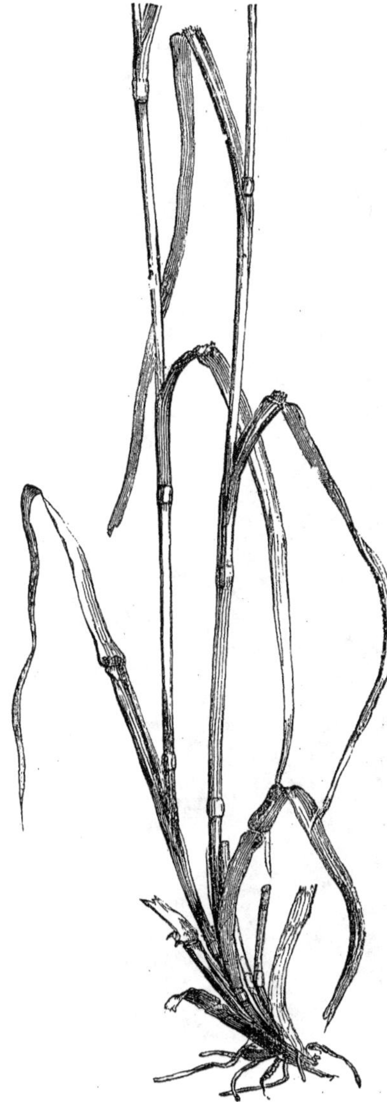
Fig. 24.—Culms of Rye.

the air, supporting themselves on plants of more robust growth, or to hang suspended from forest trees in graceful festoons and light garlands, they are provided with flexible, slender, and pliant stems, which enable them to embrace with their tendrils the trunks of trees or shrubs. Thus, Nature fashions the outward forms of plants according to the part which they are intended to fill, and according to the functions which have been allotted to them.

Nothing is more variable than the appearance of the stem in vegetables and trees; in their infinite variety they sometimes present to us perfect types of beauty and elegance. Sculpture and painting have borrowed from the trunks of certain trees models of architecture at once elegant and majestic—types which have been handed down to us from the most remote antiquity, and which are still the models we use. Man has discovered in vegetable forms his first designs for adornment, construction, and grace. The stem of the Palm-tree and the Date formed copies for majestic columns; the leaves of the Acanthus supplied the stately capitals of the Corinthian order with their ornament; the leaves of the vine and the natural garlands of

Fig. 25.—Palm-trees

young climbing plants furnished ancient art with types of ornamental design, which are still preserved in modern architecture.

In botanical language, the stems of plants are not always called by the same name. The stems of trees, as the Oak (*Quercus*), in our climate (Fig. 23), or of Palms (Fig. 25), bear the name of *trunk*. The stems of the *Graminaceæ* (grass family), commonly cylindrical, and nearly always indented by annular knots from which the leaves spring are called the *culm* (Fig. 24).

The thickness and height of stems vary very much among plants. Whilst the trunk of certain exotic trees, as the monstrous Baobab, attains gigantic dimensions, the stems of many of our spring plants, as those of the Saxifrage, and the early Whitlow grass (*Draba verna*), scarcely attains the thickness of a thread. "While crossing the Rio Claro, a river in the province of Goyas, in Brazil," says A. de Saint-Hilaire, "I perceived growing on a rock a plant not more than a quarter of an inch in size, which I took at first for a moss. It was, however, a species of a superior order, and provided with a reproductive apparatus like that of our oaks and beeches. By the side of it gigantic trees reared their majestic heads to the height of a hundred feet."

Accordingly as stems last one year, two years, or more, they are called *annual*, *biennial*, or *perennial*. Arborescent stems which live a greater or less number of years, and form solid wood, are said to be *ligneous*, or woody. The soft stems of annual, biennial, or perennial plants are called *herbaceous;* and the stems of the house-leek, the cactus, and some of the euphorbias, are called *succulent*. Fig. 26 represents the stem of a cactus in flower.

In a great number of plants the stem rises firm and straight into the air. It is then called an *upright stem*. There are some, on the contrary, which have not rigidity enough to keep themselves upright; they stretch along the ground, only lifting up their heads, so to speak: these are *procumbent stems;* or, being quite prostrate, they are fixed by adventitious roots, and are called *creeping stems*.

Fig. 27 represents the procumbent stem of the *Veronica officinalis*. Other plants, like the *ivy*, hang on neighbouring bodies by the aid of their suckers or adventitious roots; or, like the bindweed, they entwine themselves spirally round trees. The first are called *scandent stems*, the latter *volubile*.

Volubile stems do not all twine in the same manner; but the direction of twining of stems is invariable in the same species, and even resists efforts made to change it. Some, like the Bindweed (Fig. 22), if we suppose that they are twining round our own body, go from right to left; others, as the Hop (Fig. 28), go from left to right. "The Lianes, which in the primitive forests produce the most marked varied effects," says A. de Saint-Hilaire, "and which impart to these forests their most picturesque beauties, are ligneous plants, some of them *scandent*, others *volubile.* These are the species of

Fig. 26.—Stem and Flowers of a Cactus.

Bignonia, *Bauhinia*, *Cissus*, &c., and though they all need a support, yet each plant has a bearing which is peculiar to itself. Some Lianes resemble wavy ribbons; others are twisted, and describe large spirals. They hang in festoons, wind about among the trees, leap from one tree to another, and entangling themselves together, form masses of leaves and flowers, so that the observer finds it difficult to distinguish to which tree they belong." These Lianes of the American forests are very imperfectly represented in our climate by the Ivy, the Honeysuckle and the Clematis, the Bindweed and the Hop-plant.

The stems of which we have hitherto spoken are *aërial;* but there are subterranean stems. The Solomon's Seal (Fig. 29) presents a subterranean stem, thick, fleshy, whitish, and indented on its upper surface with scars corresponding with the bases of old aërial stems (thence the name of *seal* which this plant has preserved). This sub-

Fig. 27.—Procumbent stem of *Veronica officinalis.*

terranean stem terminates at its foremost extremity with a leafy and flower-bearing axis, placed behind a terminal bud, which will develop itself the following year. Many plants, as the iris, flowering rush, water trefoil, and sedge, alike present subterranean stems. Fig. 30 represents the subterranean stem of the *Iris Germanica.* These stems have received from botanists the name of *rhizomes*, from ῥίζωμα, a root or root-stock. They creep obliquely or horizontally under the surface of the soil, and vegetate at their most advanced point, whilst the hinder part is gradually destroyed by the decay of age. This mode of existence in subterranean stems is well exemplified in

Fig. 31, which represents the growth of a *Carex*, or sedge. In this engraving is shown the horizontal and creeping axis, which bears at once scales or modified leaves, and root-fibres, and which sends out leafy shoots at intervals. The shoot 1 is only one year old; in the next spring it will assume the form of the shoot 2; the following year it will bear flowers and fruit, as in 3; the production of fruit will mark the term of its existence, as shown in 4.

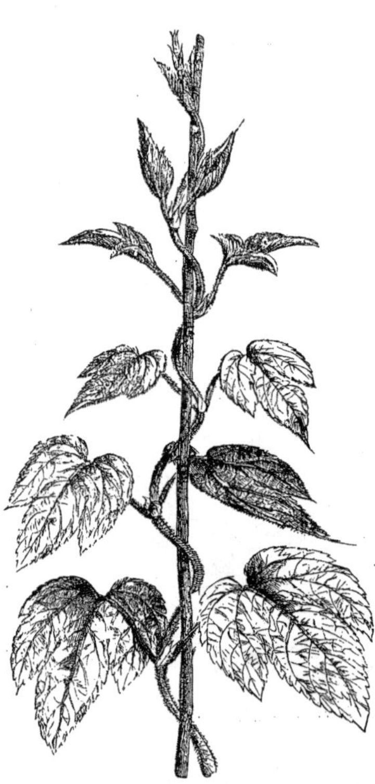

Fig. 28.—The Hop (*Humulus lupulus*).

Another very remarkable kind of subterranean stem is that which forms the central or essential part of *bulbous* plants. Cut the bulb of a hyacinth or lily longitudinally; it will be observed that it is composed of a fleshy mass, more or less conical in the upper part, and truncated below, constituting a short stem, with *internodes* or knots placed very close together. This fleshy mass gives rise, at its upper face, to fleshy scales, which are modified leaves pressing one against the other, and to a central bud formed of leaves and rudimentary flowers, whilst from its lower face spring the root-fibres. In the Hyacinth (Figs. 32 and 33) the scales form nearly complete sheaths, which grow one over and around the other, and its bulb is said to be *tunicated*. In the Lily (Figs. 34 and 35), the scales are smaller, and overlap one another like tiles on the roof of a house; its bulb is said to be *scaly*. In the Crocus (Figs. 36 and 37), the base of the stem is extremely broad, of a globular or depressed shape, and only produces a few thin and membranous scales; this kind of underground stem is called a *corm*.

The *rhizome* and the *corm* are only distinguished from each other by the length of the axis (as shown in the vertical section); a corm is annual, while a rhizome is perennial. A rhizome may be regarded,

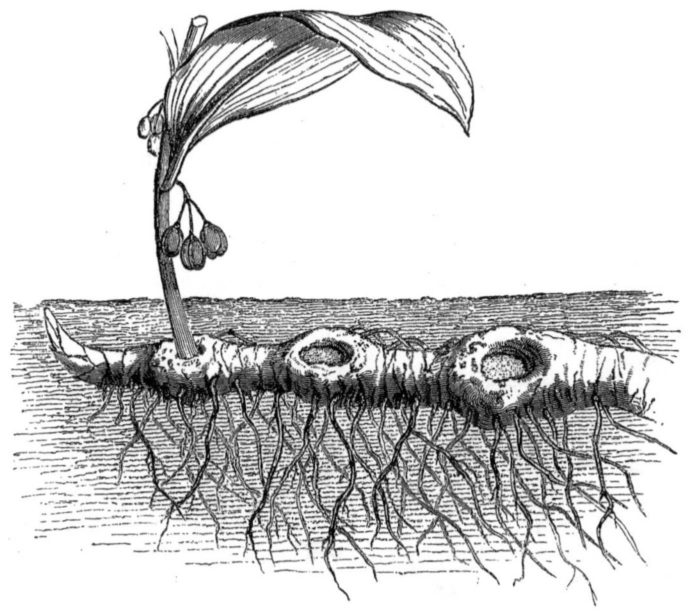

Fig. 29.—Common Solomon's Seal (*Convallaria multinora*).

therefore, as consisting of a number of corms attached together, and more or less persistent.

We now have to consider the structure of stems in different kinds of plants. In order to arrive at a correct idea of their structure, let us consider, first, the stem of *forest* trees; secondly, that of *palm* trees; thirdly, that of *arborescent* or tree *ferns.*

An acquaintance with the ligneous stems of forest trees is interesting in more than one respect. Nature has united all her powers to give to trees the strength necessary to resist the dangers and the causes of destruction which threaten them. Their wide-spread and branching summits, the immense mass of foliage which they support, and the

Fig. 30.—Rhizome or Subterranean stem of *Iris Germanica.*

great height to which they attain at the end of their growth, expose them to the fury of tempests. Their trunks must be solid, yet to some extent yielding and elastic, in order to brave all the violence of the

Fig. 31.—Subterranean stem of Carex, with shoots of four years.

winds. Nature has constructed them with the particular aim of resistance. Year after year she accumulates on their exterior successive layers of solid woody substance. In proportion as the plant increases in size and needs a more powerful support, the interior concentric rings, which by their combinations form the strong and compact tissues of our forest trees, are more and more consolidated by woody matter deposited in their texture. In its origin—that is to say, at the

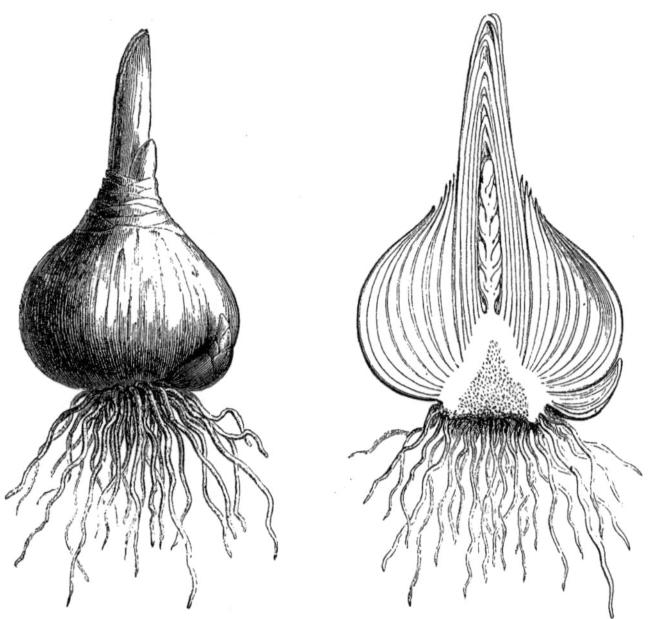

Fig. 32.
Bulb of the Hyacinth.

Fig. 33.
Vertical section of the Bulb of the Hyacinth.

moment when the young stem, just sprung up out of the ground, begins to rear itself in the air—nothing is observable in its interior except an abundance of pith, surrounded by a sheath of spiral vessels. But as the plant increases, new elements interpose between the pith and the bark; and when the trunk has lengthened and strengthened, it presents a complicated internal structure, and one well calculated for resistance to all outward forces. A mere glance at the section of a log of firewood informs us that the stems of forest trees present three essential parts, namely, pith at the centre, surrounded by woody fibre, and exterior bark. Let us examine more closely each of these parts in an exogenous tree.

The *pith* forms a sort of column in the centre of the woody axis,

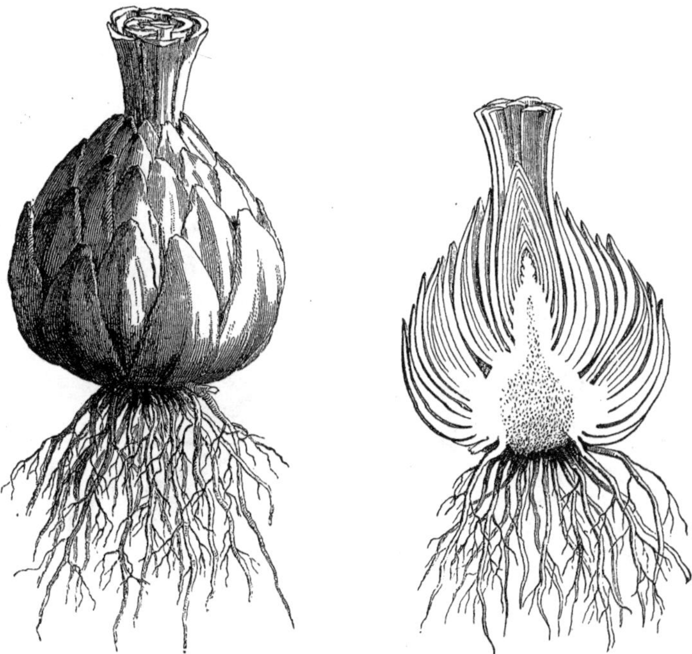

Fig. 34.—Bulb of the Lily. Fig. 35.—Vertical section of the Bulb of the Lily.

as in Fig. 38, which represents a horizontal section of the trunk of the Maple. In very thick and solid stems, the diameter of the pith appears very small, and for a long time it was even supposed that in the trunk of very old trees pith completely disappeared. But it is not so. It is asserted that, according to exact measurement and observation, the diameter retains perceptibly and invariably the same proportion from the time when the young ligneous axis has begun to solidify, up to the period of its maturity. The fact is that the size of the pith varies in different shoots of the same tree, according as their first growth has or has not been rapid.

The pith is formed by a combination of *cells*, to use the scientific term. Cells are simple, primitive organs, which are present in every

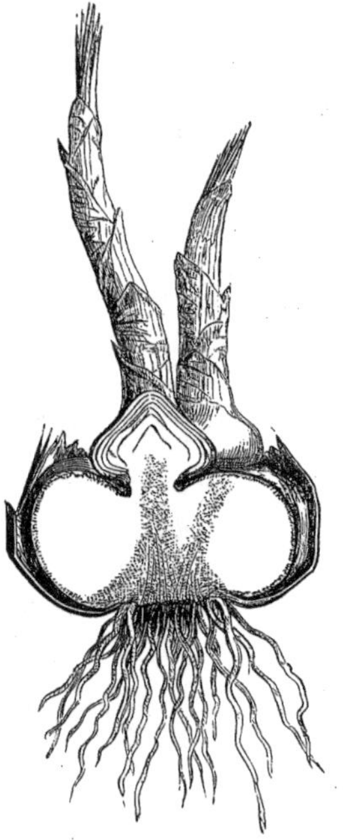

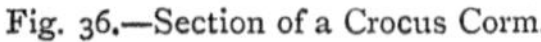

Fig. 36.—Section of a Crocus Corm.

Fig. 37.—Corm of the Crocus.

vegetable structure. Each cell is a sort of sac or cavity, surrounded by walls of transparent membrane; a vegetable cell, in short, is well represented by a soap bubble. A cellular mass without intercellular spaces may be compared to an aggregation of soap bubbles, pressed against each other. The cavity is completely closed; sometimes it is empty, sometimes it is filled with more or less fluid matter.

Fig. 39 represents the transverse section of a cluster of young vegetable cells; they are, as we see, nearly circular in section.

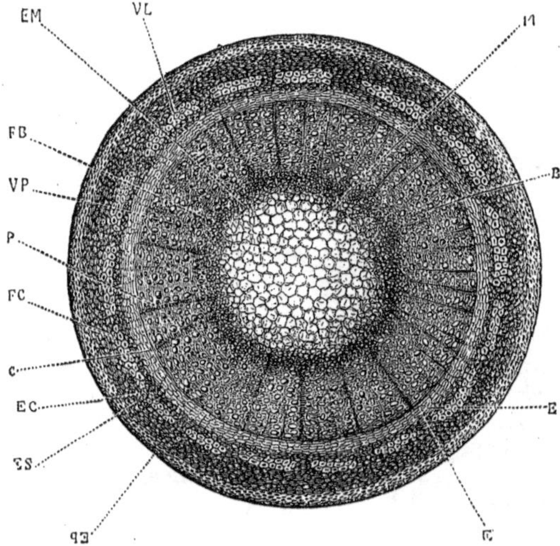

Fig. 38.—Horizontal section of a young Maple-stem.

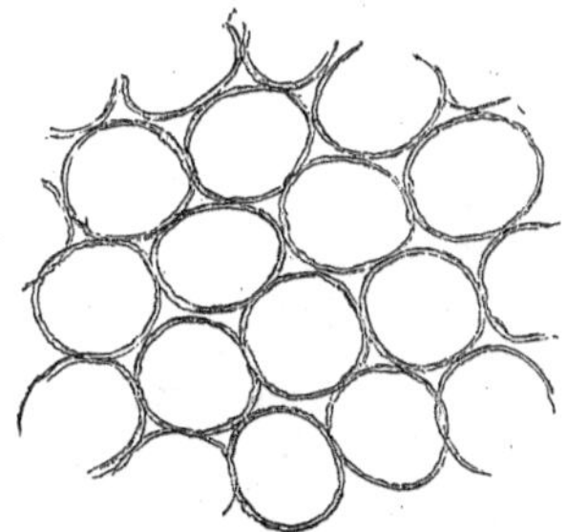

Fig. 39.—Spherical cells of Pith.

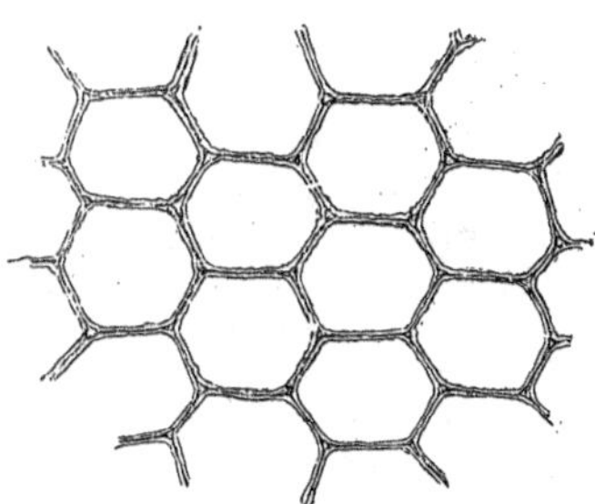

Fig. 40.—Polyhedral tissues of Pith.

When they have become larger they mutually compress each other, so that their form, at first nearly spherical, becomes polyhedral, as represented in Fig. 40.

The pith of young trees, as represented in Fig. 39, is, as we have said, an aggregation of cells, at first nearly spherical, which become polyhedral from compression as the stem increases in size, and this is the *medulla* or medullary tissue, which occupies the central position in the trunks of most forest trees. This medullary tissue is totally deficient in vital energy in stems of more than a year old.

Between the pith and the bark we find concentric zones, which bear the name of *ligneous* layers, the aggregation of which forms what is commonly called *wood.* If we examine the trunk of the oak, the apple, or the cherry-tree, a very sensible difference is observable between the innermost woody layers, which are of a darker colour and denser texture than the exterior ones, which are, on the contrary, of a paler hue and softer. In Fig. 41, which represents a vertical section of an Oak of eighteen years' growth; the *sap-wood* (*alburnum*) is represented by the letter A, the *heart wood* by the letter B, the bark by the letter E. The pith is in the centre, with the stellate appearance which, in the oak, it often presents. The *medullary rays*, to which we shall presently return, are very apparent in this section. The name of sap-wood (*alburnum*) is given to the outside layer of wood, and that of heart-wood (*duramen*) to the innermost ones. In some trees, and notably in those which are not hard-grained, as the poplar, willow, and chestnut, the line of demarcation between the wood and the sap-wood is slightly marked. In hard woods, on the contrary, it is strongly defined. Thus, in ebony the heart-wood is of an intense black, whilst the sap-wood is white; in the Judas-tree the heart is yellow and the sap-wood white; in the *Phillyrea* (mock privet) the heart is red and the sap-wood white.

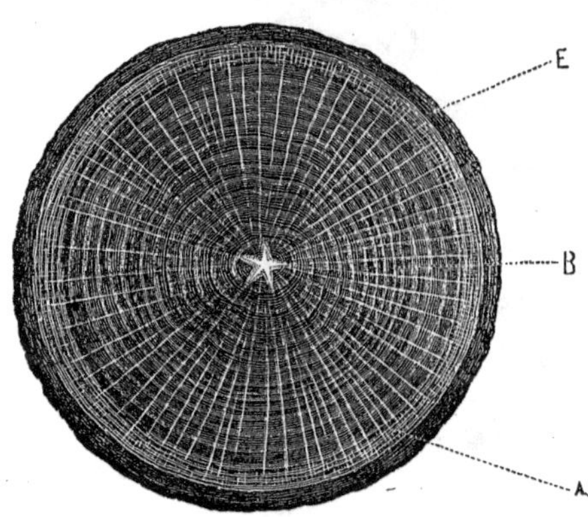

Fig. 41.
Vertical section of Oak of eighteen years' growth.

Workmen who work in wood are well aware that the sap-wood is much less solid than the heart, and that the latter only ought to be employed for wood-work. Examined in masses, the *ligneous* layers are harder the nearer they are to the centre; but, individually, each layer is more compact towards its exterior. Finally, all the layers are not of an equal thickness, whether compared with each other, or in their several parts.

The substance which prevails in the wood, and which gives it its hardness, is the *woody fibre*, represented in Fig. 42. This is an elongated cellular body, as there represented, terminating in a point at the two extremities. Its walls are very thick, generally so thick that their interior cavity is much reduced. This thickness, as well as the colouring of the fibres, varies with the different parts of wood, with the age of the stem, and even with the nature of the tree that is under examination. The woody fibres press end to end, one against the other, and become so entangled as to constitute what is called a *fibrous* tissue, very difficult to pierce when cut across, but, on the contrary, easy enough to divide when cut longitudinally.

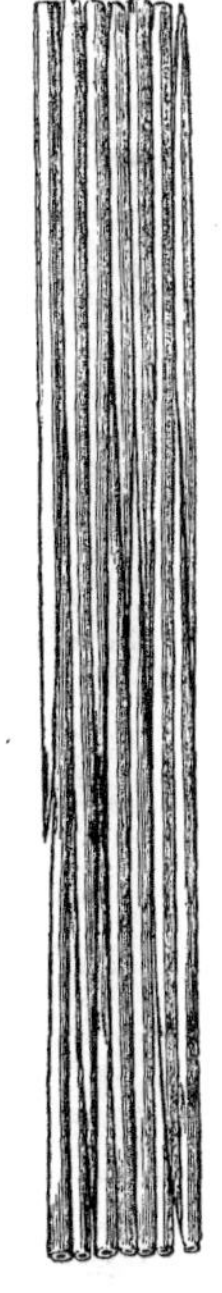

Fig. 42. Woody fibres magnified.

This woody *fibre* is not the only element composing wood. Cut transversely a branch of the Vine (a plant in which the elements we are going to speak of acquire a considerable volume), and apply the eye to one end; if the branch is straight, you will see the light at the other end. Examine the surface of the section of the branch, either with the naked eye or with a magnifying glass, and it will be observed that it is perforated with a considerable number of small holes, of unequal size. If you introduce a hair or a very fine thread into one of these openings, you will succeed in passing the thread to the other end of the branch. Continuous canals, therefore, exist in the interior of the vine branch. These canals, formed of a membrane peculiar to them, are the *vessels*.

If a portion of the transverse surface of a log of Oak or Elm is neatly cut, it will be observed that the inside edge of each ligneous zone presents a certain number of small holes, clearly perceptible to the naked eye, or at least with the help of a common lens; these are the orifices of vessels, rather large in size. In the middle of the ligneous zone the vessels are much smaller, and sometimes almost imperceptible. Examine the wood of the Hornbeam, Lime, or Maple in the same manner, and it will be observed that the internal edge of the zone is not now occupied by large vessels, but is almost entirely riddled by the orifices of smaller and more evenly-sized vessels, which become indistinct towards the external edge of each zone.

What is the structure and function of these different vessels? They resemble a cylinder, with constrictions at intervals, more or

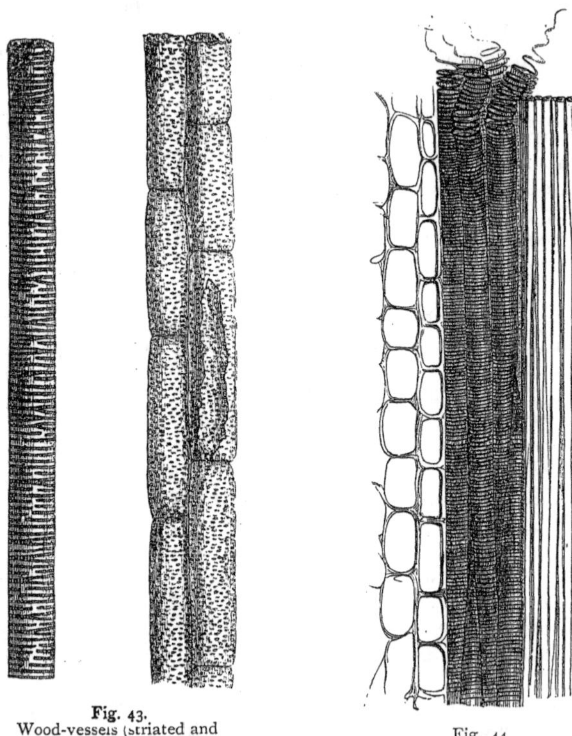

Fig. 43.
Wood-vessels (striated and punctated vessels of the Melon).

Fig. 44.
Spiral vessels between pith and woody fibres.

less marked, having also transverse lines dividing them into superposed joints. Sometimes the remains of diaphragms correspond in the interior of the vessel to these contractions and lines. In one word, these cylinders appear formed of cells, placed end to end, the partitions of which have been destroyed. Their exterior walls show punctations, striæ, and reticulations, presenting very beautiful patterns, which result from the unequal thickening of the walls, forming inequalities which are the result of certain laws of symmetrical deposition. Fig. 34 represents these *vessels* in the *melon*. From the peculiar appearance offered by their external walls, which are marked with small dots, furrows, and streaks, they are called *dotted* and *striated vessels*.

There is a particular part of the wood, however, in which the vessels are of a very different nature from that just indicated. We find these round the pith, in the innermost portion of the woody circle, and never anywhere else in the wood. These vessels, with the slender fibres which accompany them, have received, and most improperly retained, the name of the *medullary sheath*. We say improperly, for here there is only a combination of vessels, and no *sheath*. The image which this word recalls is not of a nature to enable us to understand the important modification of structure which belongs to the innermost vessels of the woody circle.

Fig. 44 represents the central part of a piece of tree as seen under the microscope, with a very strong magnifying power. In this central portion, the vessels, of which we are speaking, on one side touch the pith at the centre of the stem, and are on the other side in contact with the woody fibres.

The structure of these vessels is very singular. They form somewhat elongated fibres, still more slender at the extremities. Any one would believe at the first glance that these vessels were very finely streaked transversely, and that their external coat was continuous; but if subjected to the slightest pulling, they unroll like a spiral spring. These vessels are, then, formed of a spiral thread, twisted into a coil with contiguous spiral turns, which are joined together by a membrane, which is so extremely thin that it is difficult to find the traces of it when the spiral tube has been unrolled.

There is one final peculiarity which marks the section of the stem of one of our forest trees. It is that assemblage of diverging lines which bear the name of *medullary rays*. In a transverse section of the stem of a tree, the mass of wood is traversed by a great number of radiating lines, all of which start from the bark and converge towards the pith, or medulla. But they do not all reach it; there is a certain number which stop short in some of the layers, more or less deep in the trunk, without reaching the pith at all. These radiating lines result from the transverse section of cellular laminæ, the edge of which we see, and the length and thickness of which are variable.

Fig. 45 represents the medullary rays of a trunk of Cork-tree (*Quercus Suber*), in a transverse section. Fig. 46 shows the same organs in a similar section of the stem of the Maple, magnified by means of the microscope. In this last figure, M R are the medullary rays, which go from the centre to the circumference of the stem. The spiral vessels, the woody fibres, and the dotted vessels are

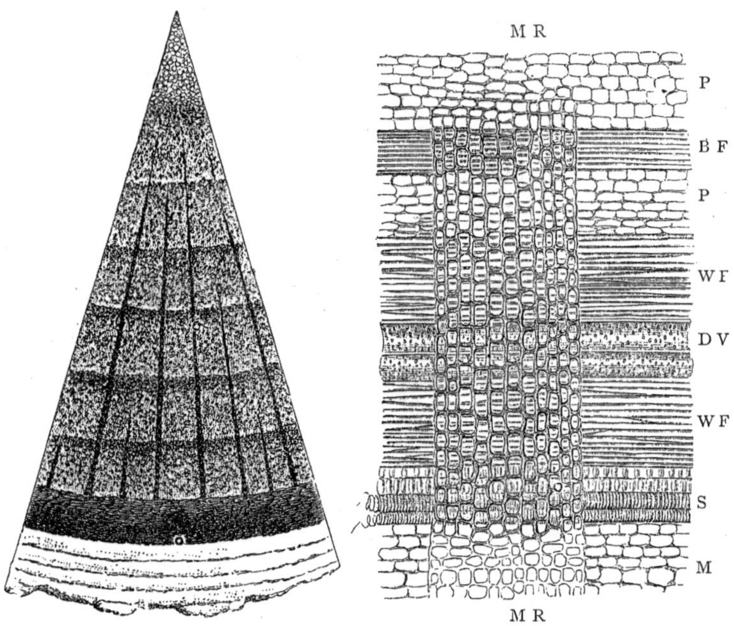

Fig. 45.—Medullary rays of the Cork-tree (horizontal section).

Fig. 46.—Medullary rays of the Maple. (radial section).

represented by the letters S, W F, D V, and the pith by the letter M. To make this more clear to the reader, we avail ourselves of an illustration borrowed from Mr. Christopher Dresser's "Rudiments of Botany" (Fig. 47), in which A is a horizontal section of an exogenous stem—that is, of a stem consisting of pith, wood, and bark, which is enlarged by external additions. *a b c* are the cellular mass of the stem, in which bundles of woody fibre, *d*, have been developed. By the disposition of these bundles, the cellular matter has become divided into a central portion, *a*, which is the pith, an outer portion, *c*, which is the bark, and the plates, *b*, intervening between the bundles of wood fibre which connect the bark and the pith; these are the medullary rays. B is the section of an older stem, in which a bundle of woody fibre has been deposited between each of the older bundles. *a*, pith; *b*, medullary rays; *c*, cortex, or bark; *d*, bundles of wood-fibre.

The vertical section, C, represents a portion of a stem with the bark removed, showing that the bundles of fibre, *d*, do not descend vertically through the cellular tissue, *b*, but in waved lines, dividing the cellular tissue into lenticular masses, which form the medullary rays, *b*.

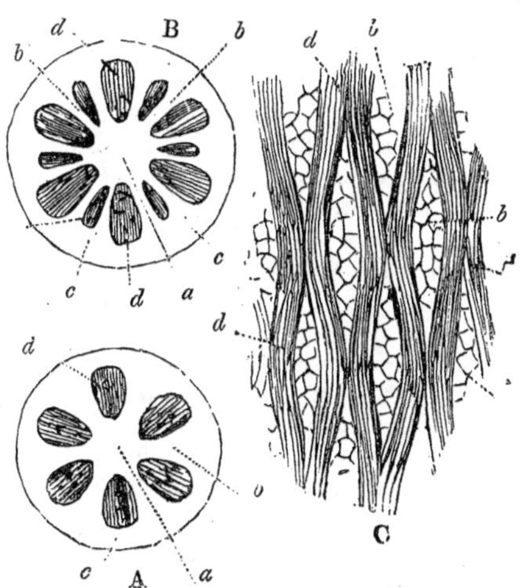

Fig. 47.—Sections of an Exogenous Stem.

Most trees are provided with medullary rays of only one kind. A few only present thick and thin rays together. Thus in the Oak or the Hornbeam we find both thick and thin rays, whilst in Willow and Maple the rays are visibly equal.

During the summer and autumn, the wood cells, which are as yet unindurated, are the seat of disposition of starchy matter in small granules. This is dissolved in the sap during the following spring, and supplies the plant with the materials for the development of new foliage.

The *bark* of trees is essentially composed of fibrous and cellular tissue; but it is easy to understand how varied are the forms, disposition, and structure of these elements, when we consider the extraordinary variety in the appearance of the bark of trees, and the diversity of their products. To explain everything which relates to the structure of the bark would lead us into details which our space does not permit; we must therefore limit our remarks, and content ourselves with pointing out the principal characteristics of bark, considering in most cases trees of our own climate. Briefly, the young stem is invariably covered with a thin cuticle, the *epidermis*. As the stem increases, new bundles of woody fibres are deposited in regular annular layers, one in each year, the new layers being deposited outside those already formed. The new layers of bark and wood are thus formed almost in contact, being juxtaposed. The epidermis covers the bark, as it does every other part of the vegetable, but its existence is altogether ephemeral. It is destroyed at an early stage as much by the growth of the vegetable as by the action of external agents. It is otherwise with the *cork* or *suber*, which forms the next layer, the cells of which are of a cubical form, and are closely united to each other with thin walls or partitions, without colour at first, but afterwards acquiring a brownish tint.

In many trees the *cork* is very slightly developed. But this is not the case with the Cork-oak (*Quercus Suber*). In this beautiful tree, which furnishes man with one of his most useful commercial products, the *suberous layer* acquires an extraordinary thickness; it is, in short, the substance known commercially as cork, in Latin *suber*, whence the specific name of the tree. When about five years old the corky layer of the bark which constitutes the greater part of the bark in the Cork-tree, begins to make a remarkably quick growth; then all the energy of its vegetation seems to concentrate itself on this part of the tree. New cells appear on the internal face of the primitive zone, pushing outwards the exterior cells which preceded them. Independently of these cells, the successive accumulation of which constitutes the mass of cork, others are formed which are shorter, darker in colour, of a flat or plate-like form, and which divide the mass of cork into successive zones of growth. This mass attains by degrees to a considerable thickness. If left to itself, it would crack so deeply as to become unfit for the uses to which cork is destined. It is necessary, therefore, to strip it off before it acquires this hard and fissured appearance.

Barking or peeling off the suber of the Cork-oak does no injury to the tree; it is so managed as to avoid injury to the newly formed *suberous layer*, and consequently to the living and under-lying layers of the bark. The operation is usually performed when the trees have attained the circumference of ten or twelve inches. The process is performed during the summer months, by cutting a longitudinal notch in the trunk of the tree, intersecting it with several transverse incisions distant about forty inches from one another. The bark is beaten in order to break away the adhesion of the cork to the living layers, and separate the under-lying tissues, and it is then detached

II.—Gathering the Bark of the Cork Tree.

in the shape of cylindrical pieces, by means of the handle of an axe, made crooked and thin at the end, as represented in PLATE II.

The Cork-tree is peculiar to hot countries. Algeria possesses several forests of this tree in course of working. Spain has long been celebrated for its produce. The crops of cork are generally gathered in each forest once every eight years.

The cork lies immediately over a cellular mass of a very different nature. The cells which constitute this layer are polyhedral, they are thicker and more closely joined, and of a greenish colour. This colouring is owing to the presence of *chlorophyll* or *leaf-green*, a matter peculiar to all the green organs of vegetables, which is applied in granules to the internal face of the cellular walls. *Chlorophyll* presents itself, in a mature state, under the form of very small rounded masses, formed of albuminous and fatty matter, sometimes enclosing small kernels of starch in their interior, and appearing to be superficially penetrated by the green colouring matter.

To these three cortical formations a fourth must be added, which bears the name of *bast* or *liber*, and generally appears formed of layers composed of elongated cells with alternately thick and thin walls. The first are fibres of a brilliant white, longer and more slender than the woody fibres; their walls are very thick, and are often dotted and extremely tough.

The fibres of bast render an important service to human industry, since they furnish the materials for ropes, threads, and the strongest as well as the most delicate cloths. In Fig. 48 the fibres of *Hemp* (*Cannabis sativa*) are shown as a common example of the vegetable tissue known under the name of bast. These fibres are joined in bundles. The bundles are arranged in concentric circles, frequently joined to one another by anastomosis, and constitute very thin super-imposed layers, which appear like a sort of tissue of a more or less loose texture. The whole of these layers together were formerly compared to a book, every leaf of which would represent a layer. Hence the rather unsuitable name of *liber*.

The cells with thin walls interposed between the layers of bast in spring enclose starch, and their very thin walls are extensively punctured, the punctations being filled in by wonderfully delicate network, with interstices often not more than $\frac{1}{1000}$th part of a line in diameter. These cells, whose physiological functions appear to be important, are called *sieve-cells*.

We cannot conclude our examination of bark without noticing the existence of a product which latterly has occupied the attention of botanists. We mean the *latex* and *laticiferous vessels*.

In the bark and in the pith of some trees, vessels are noticed very different from those we have hitherto spoken of. They are remarkable at once for their structure and their contents. They are tubes, simple or ramified, sometimes completely independent, sometimes attaching themselves one to the other in a continuous length. While the vessels traceable in the woody fibres are formed

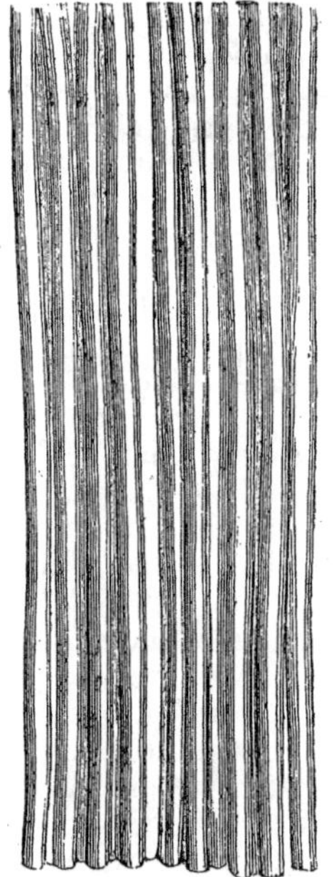

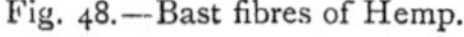
Fig. 48.—Bast fibres of Hemp.

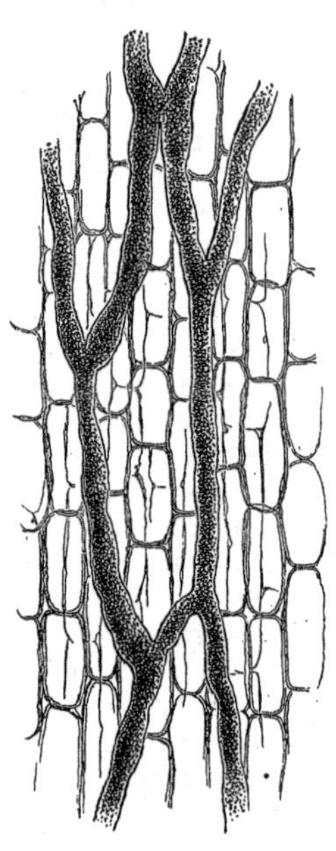

Fig 49.—Laticiferous vessels in Celandine.

of cells which can be separated one from the other by the use of proper means, the cells constituting the *laticiferous vessels* are, on the contrary, so intimately blended together that neither mechanical nor chemical action can separate them.

The laticiferous vessels contain a juice, generally coloured. It is easily proved, under the microscope, that this liquid is composed of an uncoloured serum, holding in suspension numerous and very small globules, to which it owes its colouration. This liquid is called *latex.* But what strikes the observer as above all remarkable, is the circulating movement which is the property of latex. The transparency of the vascular walls, and the presence of the granules, render this movement very perceptible.

The *latex* is very abundant in some plants. Place on the stage of a microscope, and on a thin plate of glass, a young leaf, for instance, of *Chelidonium majus*—better known as Celandine (Fig. 49)—still attached to the branch, or a sepal of the same plant, of which the latex is an orange yellow; or a petal of the *Poppy*, of which the latex is white; or a stipule of *Ficus elastica* (one of the *caoutchouc trees*); we shall see in all these cases the latex descending in one branch of the network of the laticiferous vessels, and ascending in another, returning sometimes to its point of departure, and, in fact, circulating with a rapidity greater in proportion as the temperature is warmer and vegetation more active. Gutta-percha, caoutchouc, and opium are the products of the latex of certain plants.

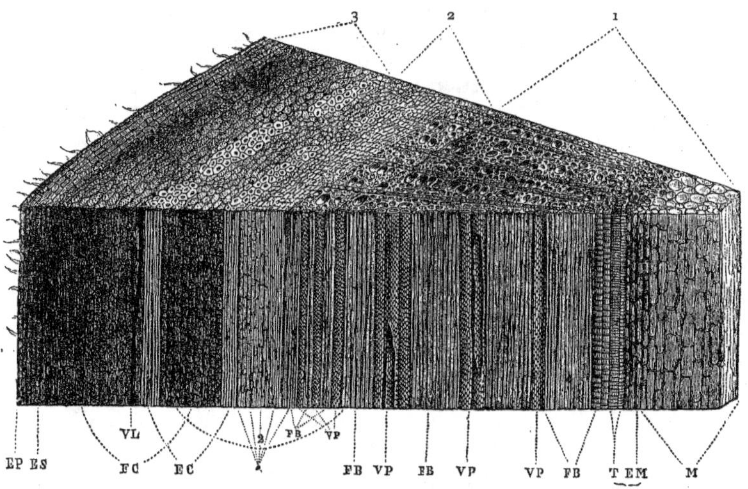

Fig. 50.—Transverse section of the trunk of the Maple.

The elements entering into the composition of the trunks of forest trees are, as we see, rather complex. Having described each in its turn, it will be instructive to bring the whole of them under the eyes of the reader.

Fig. 50 represents a section, both horizontal and vertical, of a trunk of the Maple (*Acer campestre*). The structures embraced by the lines marked 1 represent the wood of the first year; those in 2 the wood formed during the second year; and those in 3 the tissues of the first year forming the bark. In the centre of the stem, M represents the pith, the cells of which are polyhedral. The spiral vessels of the *medullary sheath*, coming next to the pith, and enveloping it on all sides, are represented by the letters T E M. Then follow three groups of *woody fibres*, F B; the dotted vessels, V P, being placed alternately with three groups of woody fibres; *c* is the cambium layer. The bark, enclosed by the line marked 3, succeeds these substances, the fibres of the bast lying at E C, the cork at the letters E S, the laticiferous vessels at the letters V L, and the herbaceous layer at the letters F C, the epidermis, E P, bristling with hairs, forming the external surface of the trunk. The medullary rays are plainly enough observable on the horizontal section; they commence with the pith, and stop with that part of the wood which belongs to the second year's growth.

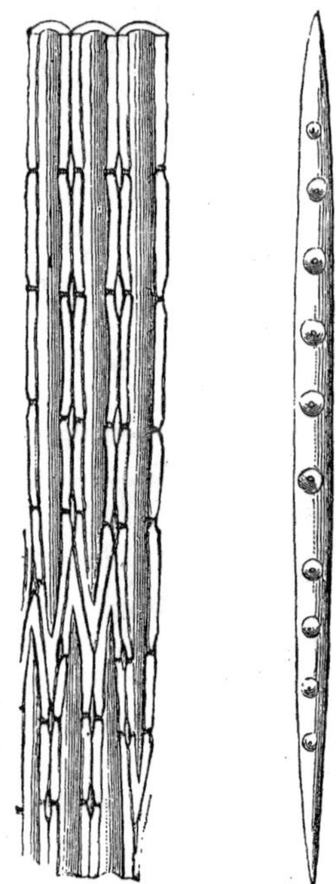

Fig. 51.
Woody fibre of the Fir.

Having reviewed all the constituent elements belonging to the trunks of forest trees with membranous leaves, we have now to speak of the structure of the stems of Firs and Pines—cone-bearing trees. These are at once easily distinguished from the trees we have been considering by the structure of their wood, which is exclusively formed of large fibres, without any appearance of vessels, except the ring of spiral vessels forming the medullary sheath. These woody fibres (Fig. 51) present besides the singular peculiarity of exhibiting on each of their lateral faces—namely, those which look towards the medullary rays—a row of dots or punctations, each surrounded with a disk-like ring. The wood is traversed by *resiniferous ducts*, which are interstices in which the resin produced by the cells surrounding it is deposited and accumulates. Fig. 52 represents the transverse section of the stem of a Fir-tree. We see that conifers, like other trees, present a central medullary canal, concentric woody and cortical layers. But vessels have no existence in these stems, and the medullary rays are scarcely visible.

Fig. 52.—Transverse section of a Fir.

Stems which increase in the way which has been described by external circumferential additions are said to be exogenous. Their seedlings usually have two seed-leaves or cotyledons. Hence they are also called Dicotyledons. All British trees, and a vast number of our smaller shrubs and plants, have this form of stem.

Fig. 53.—Section of the stem of a Palm.

The general appearance of Palm-trees is very different from that of our indigenous trees—their long and drawn-out stem approximately equal in thickness from the base to the summit, and completely bare, that is, not divided by boughs and branches, making them like some tall column surmounted with a thick tuft of leaves. What is the interior structure of such a stem? To give an idea of its formation, we must first understand that the growth of the Palm-tree differs from that of any group of trees we have hitherto considered. Palm-trees do not, like our evergreen and forest trees, increase their growth by concentric external layers, deposited between the wood and the bark. The interior structure, therefore, must show arrangements very different from those we have been describing. Here there is no single central pith, no concentric layers distinctly separating the pith, the wood, and the bark; no medullary rays diverging from the centre to the circumference. If we cut the stem of a Palm across, we shall immediately see that it differs as much from the trunk of our trees in its inmost organisation as it does in its outward appearance. We look in vain for the central pith, the concentric zones, the radiating lines, which so plainly characterise the wood of our indigenous trees. We see, on a groundwork of palish colour, little spots of a deeper tint, formed by a more solid tissue. These little rounded or half-moon shaped spots are more numerous, more crowded, darker coloured, and in general larger towards the circumference of the stem than they are in the central part. This stem, therefore, appears at first sight formed of two descriptions of tissue, one rather soft and pulpy, forming the principal mass, the other very solid, forming little islets scattered through the former (Fig. 53). It is called *endogenous;* it is characteristic of plants with a single seed-leaf to the seedling.

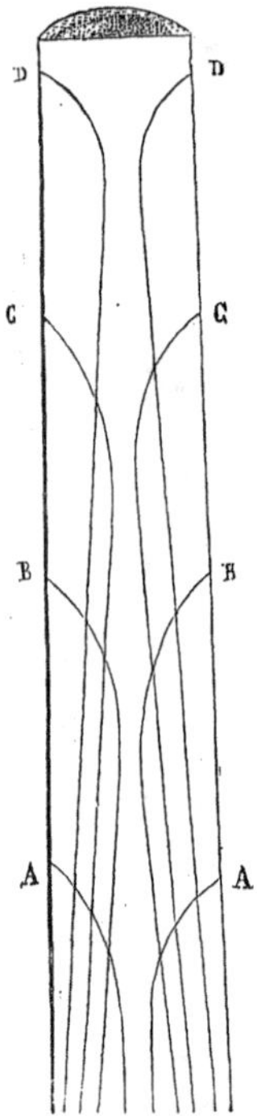

Fig. 54.
Theoretical figure showing the internal structure of the stem of the Palm.

Microscopic examination has shown us that the first of these tissues is exclusively formed of cells, and may be compared to the pith of our indigenous trees. It is traversed also by vascular bundles of very tough fibres, the tortuous course of which may be traced theoretically by the help of Fig. 54, in which letters A B C D, represent the different interlacings of these bundles in the middle of the pith. The fibrous bundles which traverse the stem of the Palm and other trees belonging to the same natural group, present arrangements which are very interesting to examine. The anatomical structure of each of them does not appear to be alike through their whole length; it seems to become more simple the farther they are distant from the point where they leave the stem to pass into the leaves. In its

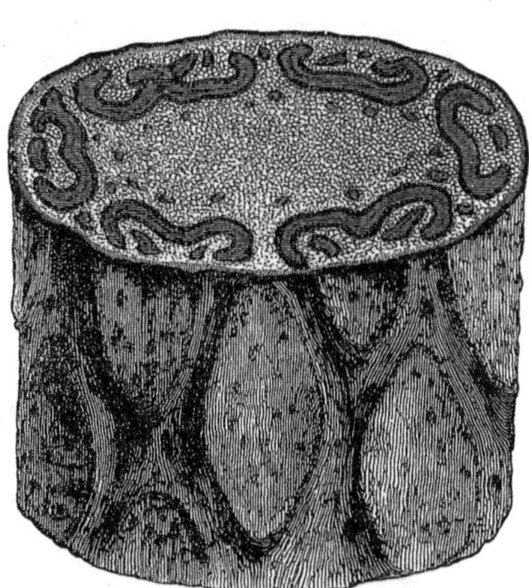

Fig. 55.—Section of the stem of a Tree Fern.

higher part, at the end of its course, the fibrous bundle possesses the characteristic structure which it has in the stems of our indigenous trees, including the medullary sheath; it presents, in fact, spiral dotted and striated vessels of a greater or less size, and woody fibres.

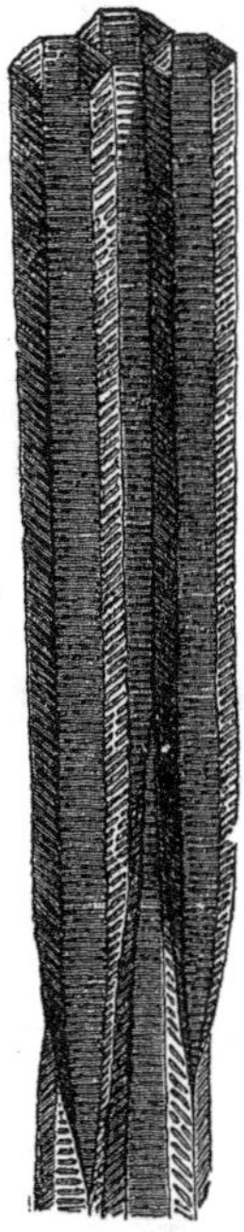

Fig. 56. Scalariform vessels from a Tree Fern.

The *Tree Ferns* of warm climates approach in their appearance much nearer to Palms than to our indigenous trees. Their slender trunks, simple and branchless, and of nearly equal thickness from the base to the summit, support at a great height a tuft of leaves. Nevertheless, Ferns differ much from Palms in their internal structure. Round an abundant pith the fibro-vascular bundles are arranged in plates, showing in the transverse section of the stem winding forms, more or less irregular, and grouped in circles towards the circumference of the trunk. This is shown in Fig. 55, representing the horizontal section of the stem of a Tree Fern.

The bundles traverse the stem of the Tree Fern from top to bottom; they form, in fact, a perforated cylinder, the perforations corresponding to the bases of the fronds. The constituent fibres are very tough, impregnated with a dark brown colour, and containing cellular tissue and vessels. Among these vessels we particularly notice prismatic tubes, which show on each of their faces horizontal clefts very close to each other, and at equal distances—called *scalariform vessels.* Fig. 56 shows the structure and relative arrangement of the scalariform vessels from the trunk of a Tree Fern. These vessels are represented as they appear under the microscope. All ferns have stems of this form.

The rind or bark marked by the cicatrices of leaves or fronds, shows that its leaves are produced at the summit only.

Fig. 57.—Pollard Willows.

III.—Of Buds.

We have studied the tortuous and deformed roots, and denuded trunks of trees; before considering the branches, the boughs, the leaves, and the flowers which decorate them, we pause at the organ from which emanate all these elements. We speak of the *bud*, which hides under its delicate green envelope the source of these brilliant ornaments of Nature, of which every year witnesses the birth and death. The bud is, in fact, the cradle of the young plant. This organ alone is capable of reproducing a new individual, and the horticulturist is familiar with many wonderful multiplications of species through its means. In ordinary circumstances, however, the bud is not intended to be separated from the mother plant: the function of the mother plant is to nourish, strengthen, and increase its growth until it becomes an organ taking part with the others in the life of the plant.

The bud may therefore be considered as a fundamental element in the plant, which, without it, would soon perish. It is the bud which year by year repairs the losses, supplies the flowers, the leaves and the branches which have disappeared. Through its means the plant increases in growth. Through it, its existence is prolonged. The bud is the true *renovator* of the vegetable world. It may be said, in fact, that a plant is all bud; there is scarcely any part that does not produce them; the roots, the leaves, the flowers even, may accidentally give birth to buds, for Nature never loses sight of the phenomena essential to organic life—namely, the production of new beings.

Buds are of two kinds—namely, buds which produce leaves and branches, and buds which contain both leaves and flowers.

The leaf-bud is a scaly conical organ placed in the axil of a leaf—in fact, a rudimentary branch, so to speak, formed as the growing season is about to close. The growing point is composed of cellular tissue, in direct communication with the pith of the stem.

The arrangements of the scales of buds are very various: the scales being rudimentary leaves, the arrangements of scales are also the arrangements of leaves. Some of their forms are familiar to us. There are no vascular structures at the growing point of the bud itself, but spiral vessels and woody fibres make their appearance near to the base of the cone.

The flower-bud, on the other hand, is a stationary axis surrounded by rudimentary leaves; its growing point has become quasi-paralysed, and has no power of elongating itself; it is, in short, a stunted branch, from which the power of growth has been withdrawn. It is well known that any violent check to the vegetative growth of a plant is favourable to the production of flower-buds. A conifer, for example, whose roots have got down into bad soil, will suddenly produce cones all over. This is often the precursor of its decease.

It is necessary to avoid confounding the leaf-bud with the flower-buds, which contain only the flower ready to burst, to astonish with its beauty, and disappear. The *leaf-bud*, in its serried and complex mass, includes all the elements necessary to the production of a young plant, and as we shall see, it often suffices to produce a new individual.

Buds are the first stage and the earliest form of the vegetable axis; they occupy the summit of the stem which they are destined to prolong, or the arm-pit or axil of the leaves upon the stem of which they are destined to form the branches. In the case of herbaceous plants in general, and with a great many trees of equinoctial countries, whose vegetation, so to speak, enjoys no annual repose, the buds are *naked*, that is to say, all the young leaves resemble each other, and produce the true leaves as they enlarge. But in countries where the winter, more or less rigorous, would destroy the delicate organs, the external leaves which cover the others are subjected to modifications which transform them into protecting organs. They are changed into *scales*—coriaceous membranes, frequently furnished on the interior with an abundant down, or thick hair, or with a coating of resinous juice, insoluble in water, and preserving considerable warmth. Under this

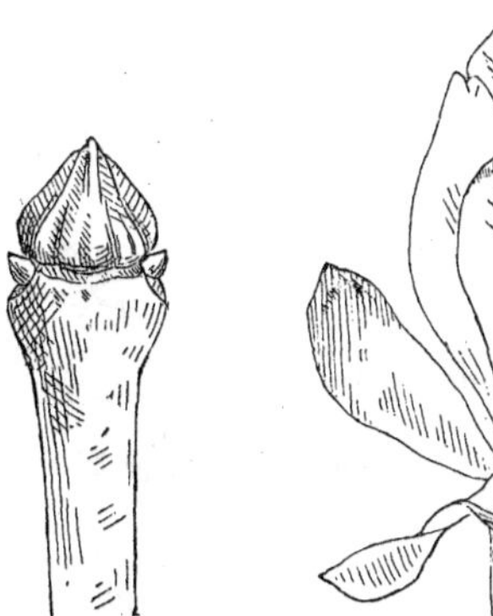

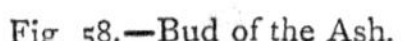

Fig 58.—Bud of the Ash.

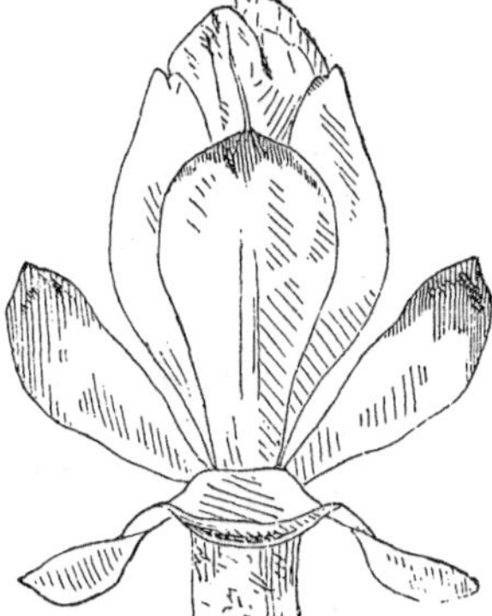

Fig. 59.—Opening leaf-bud of Horse Chestnut.

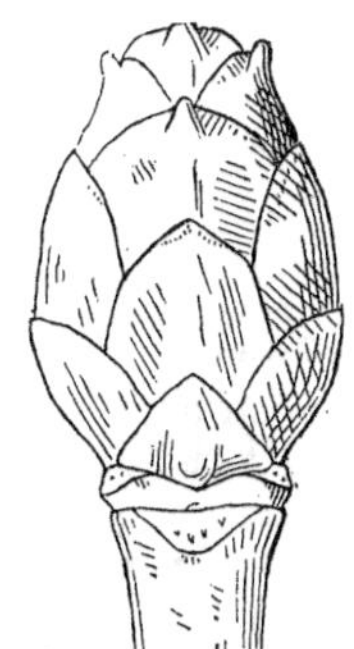

Fig. 60.—Bud of the Horse Chestnut.

shelter the rudiments of the young plant are so effectually swaddled up, so to speak, as to be thoroughly protected from the external air. Experiment proves that where the buds are detached from the tree, and the wound covered over with a varnish, they have remained for a long time under water without experiencing the least change.

The scales are modified leaves, then, but it is not always the same part of the leaf which constitutes them. Nature employs divers processes for transforming a leaf into a scale. Again, between the scales of a *bud* and the leaf which they enclose, we frequently find a series of intermediate forms which throw considerable light on the metamorphoses of which the leaf is the seat when it passes insensibly from one state to the other. Fig. 61, which represents the leaf of the Currant (*Ribes*) gradually passing from the leafy state to the scaly state, shows sufficiently the transition from the one organ to the other, to render any further detail unnecessary.

The leaves are not always disposed in the same manner in the bud, whether we consider them in their isolated state, or in the

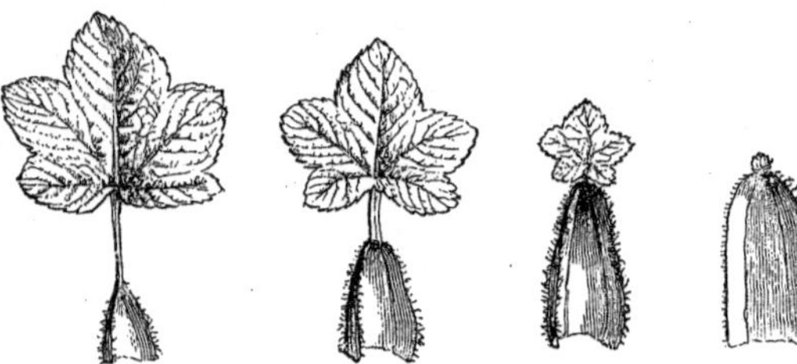

Fig. 61.—Gradual transformation of the leaves of the Currant (*Ribes*) into scales and leaves.

Fig 62.
Bud of the Birch.

positions they occupy in reference to the others. The structure of buds sometimes becomes a characteristic very useful to the forester when he wishes to acquaint himself with the names of trees during winter.

Let us consider each leaf independently of the others. These are the different situations that the leaf may assume in the interior of the bud:—It may be folded up transversely in such a manner that the upper part rests over the lower, as in the Tulip-tree (*Liriodendron tulipiferum*), Fig. 63. It may fold in its length in such a manner that one half of the leaf may lie over the other half, as in the Almond-tree (*Amygdalus communis*), Fig. 64. It may be folded several times in fan shape, as in the graceful Birch-tree (Fig. 62); rolled round itself, as it were, as in the Indian Shot (*Canna indica*), Fig. 66; rolled on both edges, outwardly, as in the Dock (*Rumex*), Fig. 71, or inwardly, as in the Poplar, Fig. 65.

We need not enter into more minute details on this subject. Figs. 67, 68, and 69, which represent vertical sections of the buds of Sage (*Salvia officinalis*), Lilac (*Syringa vulgaris*), and the well-known

Iris, will suffice to prove the mutual connection of the young leaves in certain vegetables, while they are yet shut up in the bud.

In most trees of temperate regions the buds make their appearance in spring, stopping at an early stage of their development, and only elongating themselves in the following spring. They ramify slowly, and it is only once a year that branches are produced. Nevertheless, in the case of the Peach-tree and the Vine, two generations of branches are produced. The cause of this is, that their scaly buds

Fig. 63.– Expanding Bud of the Tulip-tree.

Fig. 64.—Expanding Bud of the Almond.

have remained stationary during the autumn and winter of the preceding year, have elongated themselves in the spring, and given birth at the axils of their leaves to buds, which in place of remaining stationary, and developing themselves only at the commencement of the approaching season, grow without interruption, and produce new branches. French horticulturists have given these the name of *precocious buds.* The branches which these produce, on the other hand, only carry scale buds, developing themselves the year after, and therefore called *dormant buds.*

We have described normal buds, which are borne in the axils of the leaves, or which terminate the axis. There are others which present themselves without any order, and the exact spot where they may present themselves cannot be foreseen. These are *adventitious*

Fig. 65.—Expanding Buds of the Poplar.

Fig. 66.—Expanding Leaves of the Indian Shot.

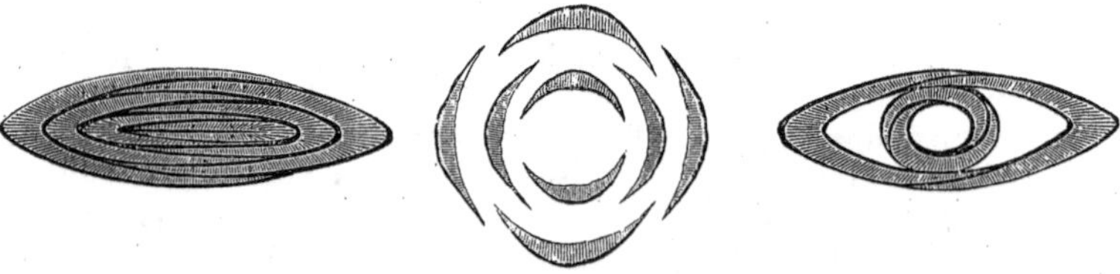

Fig. 67.—Transverse section of Sage Bud.

Fig. 68.—Section of Bud of Iris.

Fig. 69.—Section of a Bud of Lilac.

buds: they present themselves on all parts of the vegetable—upon the stem, the leaves, the flowers, and the roots. The root of the Sumach (*Rhus*), for example; of the Poplar or Pollard Willow (*Salix alba*); of the common Acacia (*Robinia pseud-acacia*), and many others, run horizontally in the soil very near the surface, producing adventitious buds, which root themselves quickly, and rapidly multiply the plant, so that in a few years they become a considerable nuisance.

The formation of adventitious buds is frequently produced by accidental irritation. The wheel of a cart, for instance, grazes the trunk, or the root of a tree is wounded by the passing ploughshare, and an adventitious bud results. If we cut down the head of a group of forest trees, the plants which, left to themselves, would have become stately trees, are transformed into stumps, which cover themselves afterwards with branches, all of the same age and of the same strength; they have been transformed from trees of stately growth to pollarded dwarfs. In the case of the Willow, this principle of adventitious budding has been largely utilised. Willows of enormous trunks, but short and deformed, surmounted by a thick tuft of branches, as in the engraving at the head of this chapter, are commonly known as pollards, and owe their singular appearance to the regular and periodical cutting to which they are subjected. In consequence of this mutilation, a great number of adventitious shoots are formed, which subsequently produce so many branches of like size. These branches are cut to make supports for young trees, for pea-sticks in agricultural districts, and as props for the vine in wine-producing countries. In Epping Forest, in the neighbourhood of London, it has been the custom from time immemorial to have annual sales of these cuttings, at which the neighbouring inhabitants are supplied with wood both for firing and horticultural purposes.

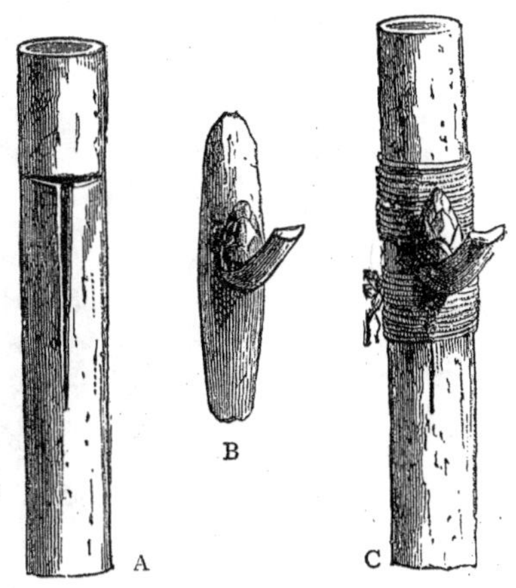

Fig 70.—Budding.

When the Lombardy Poplar-tree (*Populus dilatata*) reaches twenty-five or thirty years, it is cut down, when it forms planks of some value; but it is also lopped every five years, the result of lopping being numerous adventitious buds, which produce branches much used for fences and for firewood.

Fig. 71.—Dock.

Buds are placed upon the stem at regular intervals, where they develop themselves in the form of branches, and extend the tree, nourishment being carried through them to every leaf and fibre. It is also one of their peculiarities, that, without injury to these organs, they may be separated from the parent plant and placed upon another, which, so to speak, becomes its nurse. Horticulturists profit by this circumstance to produce upon a wild and worthless stock some of their finest flowers and fruits. This process, known to gardeners as budding, is practised in many different ways, but in all the principle is the same; the bud without any of the wood is carefully removed from the parent tree and applied to a corresponding cut in the nursing one, covering the wound so as to keep out the air. Fig. 70 shows the manner in which this is performed; B represents the bud after it has been removed from the parent branch; A the nursing stem in which an incision in the form of a T has been made to receive it; C the bud secured in its place by means of wool or cotton thread, wound lightly, but closely round both. The bud continues to grow on its new nurse, and in course of time it forms a branch or head of a tree, producing the same flowers and fruit for which the parent may have been celebrated. We need not enlarge here on the importance of this principle; it is applied most successfully in horticulture, where some delicate species of fruit or flowers is produced on a stem destitute of the vigour necessary to nourish and bring it to maturity.

The leaf-bud is thus a conical organ placed in the axil of a leaf; in short, a rudimentary branch formed as the growing season is closing, and is the *nidus* in which the foliage will be formed in the coming spring, rather than the leaves themselves, in a rudimentary condition. The central growing point (Fig. 59) is composed of cellular tissue possessing special powers of vitality and growth, and closely connected with the pith of the stem. From this point all the future leaves have their development. This growing point has a certain analogy with the embryo in the seed, inasmuch as both are provisions for growth and reproduction; but they differ in this, that the leaf-bud needs no fertilisation for its development, and propagates the individual as well as the species, while the embryo imperatively needs fertilisation, and continues the species, but not necessarily the individual.

Fig. 72.—Italian or Stone Pine (*Pinus pinea*)

Fig. 73.—Branch of Butcher's Broom (*Ruscus aculeatus*).

IV.—Boughs and Branches.

The branch is formed by the development of the bud, and this bud, as we have said, originates in the axil of the leaf.

The branch being only a secondary stem emanating from the principal trunk, necessarily presents the same modifications of form, of structure, and of disposition of the leaf, which we observe in the trunk, properly so called; but the resemblance between the stem and its branches is not always complete; thus, in the Butcher's Broom (*Ruscus aculeatus*), Fig. 73, the branches are short, and take so immediately the form of leaves, that the early botanists considered them to be such. But an attentive observer will not be deceived if he considers that these flattened organs with their foliaceous appearance spring from the axils of *scales* which are the true leaves, and carry flowers; these are the exclusive characteristics of branches.

In some plants the branches expand considerably, but in most others they remain slender; their terminal bud becomes abortive, they become pointed and hardened at the extremities; in short, they are changed into spines, as in the Hawthorn (*Cratægus oxycantha*).

A modification, extremely curious and interesting, in the form and consistence of branches, occurs in the Potato (*Solanum tuberosum*), Fig. 74, which is developed under ground. The subterranean part of the stem is not green, and the leaves, if we can call them so, are only small rudimentary scales, in whose axils branches, which extend more or less in a horizontal direction, develop themselves, and are supplied with abortive leaves. These branches, which are thin and slender at their origin, swell at their extremities, are filled with a starchy matter, and finally become the tuber, which we recognise as the potato. In short, if we examine a potato, we see that it is covered at intervals with eyes, or scales. At the axils of these scales a bud is found; every one knows that these buds, when the potato is stored in dark cellars, push out long slender shoots at the return of spring. The fact that the parts here described are really portions of the stem, is curiously proved by the following instance, recorded in the *Gardeners' Chronicle*, vol. ii., p. 85. A potato plant had grown underneath an inverted flower-pot in a dark cellar, where it had formed itself into a perfect miniature plant. Being only surrounded by air, it had thrown out its branches, and meeting with no resistance, it had grown with the same regularity as an ordinary plant would have done above ground. The set, or old tuber, was shrivelled up, and formed a wrinkled knob, out of which grew many branches and branchlets. Of the latter, some had become thickened at the point, resembling young potatoes; others, having no power of extending themselves, had swelled close to the parent tuber. All were covered with scales, the rudiments of leaves.

At first sight the plant appeared as if it had been unable to form roots, but a more minute inspection showed that they were really

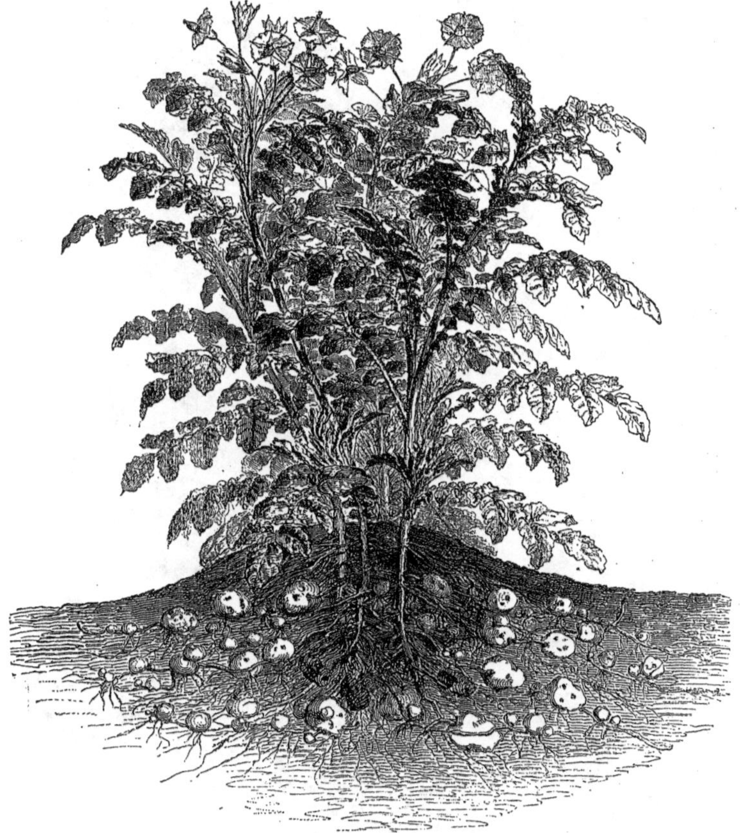

Fig. 74.—Subterranean branches of the Potato.

beginning to form here and there in many places upon the surface of the branches.

There is a great difference between *tubers* and tuberous roots—between the *Potato* and the *Dahlia*, for instance. The tuber-like

Fig. 75.—Spruce Fir.

root of the Dahlia may be called a true root; it has no *nodes*, or

Fig. 76.—Horse Chestnut Tree.

joints. On the other hand, the stem of the Potato bears many of these nodes.

The length and direction of the branches, as compared with the parent stem, are extremely varied, and this variety tends to give to each plant its special appearance, its peculiar physiognomy. If the lower, and consequently first formed, branches continue to extend themselves in the same proportion, and the upper ones are shorter as they approach the summit, the form of the tree is conical or pyramidal, as in the Firs (Fig. 75). If the central branches extend beyond the lower ones, the form is round or oval, as is the case with the Horse Chestnut (*Æsculus hippocastanum*), Fig. 76. If the upper branches take a fuller development, or rather, if the leading shoot ceases after a time to develop itself, as in the Italian Pine, Fig. 72, the summit of the tree expands in an umbrella form, and has been aptly compared to the spreading of volcanic smoke, expanding over the mouth of a crater before it descends again to the earth. In all these instances the direction and growth of the branch to the tree give its particular appearance. Branches issue from the stem at all imaginable angles—sometimes at right angles, sometimes at angles so acute as to seem at a little distance to rise with the bole of the tree. The tapering branches of the funereal Cypress (*Cupressus*), as compared with a kind of dome formed by the branches of the Oak (*Quercus*), or the Cedar, give us some idea of the contrast which the two kinds of ramifications present (Fig. 77). The Lombardy Poplar (*Populus dilatata*) carries the contrast still further. The aspect of each of these trees places in bold relief the influence which the different modes of the ramification exercise upon the form of the tree.

In some trees the branches take a direction which seems inverse to the usual habit of trees. In place of rising towards the skies, the branches appear to incline towards the earth with a drooping aspect. The Weeping Willow, represented in Fig. 78, presents a striking and well-known example of this habit of growth. The long slender branches of this elegant tree fall by their own weight. But the habit is not confined to this tree. The Ash and other trees occasionally produce seedling varieties with a drooping habit, and gardeners avail themselves largely of these, planting the banks of water basins, ponds, and brooks with the one, and forming shady places on lawns by means of the others. The Willow is at once elegant and sad in its aspect.

The branches of the Sophora of Japan (*Sophora japonica*), or the Weeping Sophora, resemble the Weeping Willow in many respects, but they possess a certain rigidity, and turn down abruptly at their origin towards the soil. (Fig. 79).

We have said that a branch may be considered as a secondary stem, emanating from the principal one, from which it draws its

Fig. 77.—The Cypress.

nourishment. But if we give this secondary stem or branch another source of nourishment, it may be separated from the principal axis

Fig. 78.—Weeping Willow.

which carries it, and become a free and distinct individual of the same species. Upon this natural fact has been founded the processes of layering, budding, and grafting, well known in horticulture. Bending

Fig. 79.—The Sophora of Japan.

a flexible branch towards the humid soil, the gardener maintains it in its position by pegging down, as represented in Fig. 80, until it has thrown out roots. These roots being developed, it contains within itself all the elements of life. The branch may be separated from the stem, from which it no longer requires support. This process is known as *layering*. But every branch which it is desired to layer may neither be within reach of the soil nor sufficiently flexible to bend to the extent

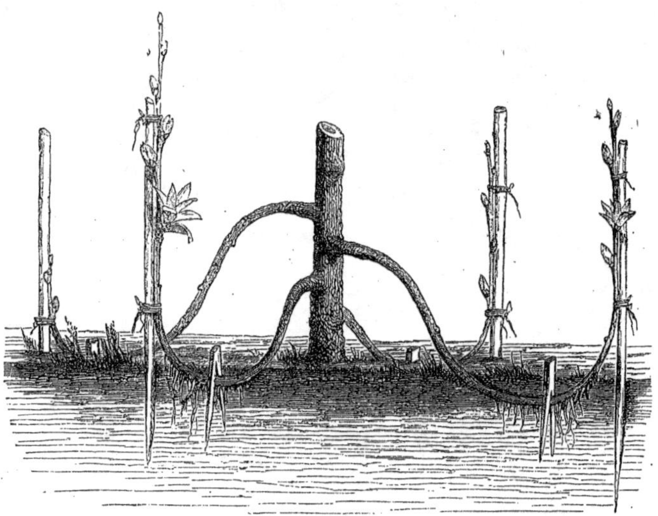

Fig. 80.—Propagation by layers.

requisite. In such cases the soil must be raised to the branch. To effect this, vases or flower-pots of various forms are employed, which are filled with earth, and maintained at the necessary height, the branch being placed in it traverses the vase in contact with the soil, as in Fig. 81. The soil being maintained in a humid state, the portion of the branch in contact with it is not slow to push forth its adventitious roots, which are soon present in sufficient numbers. In due time the branches may be separated from the parent stem, and transplanted elsewhere. This is called *layering by elevation*.

Propagating by *slips* or *cuttings* differs from layering only in this, that the part of the plant employed in the process of multiplication is detached *at once* from the parent plant, and completely abandoned to

Fig. 81.—Layering by elevation.

the resources of Nature. Cut a branch even of considerable size from a willow or poplar; give it a clean sloping cut across a joint or node, and bury it in humid soil, it will immediately push out adventitious roots, and soon begin to grow a new and independent willow or poplar. But all plants will not so readily accommodate themselves to this easy mode of multiplication. There are some plants which will not *take*, to use

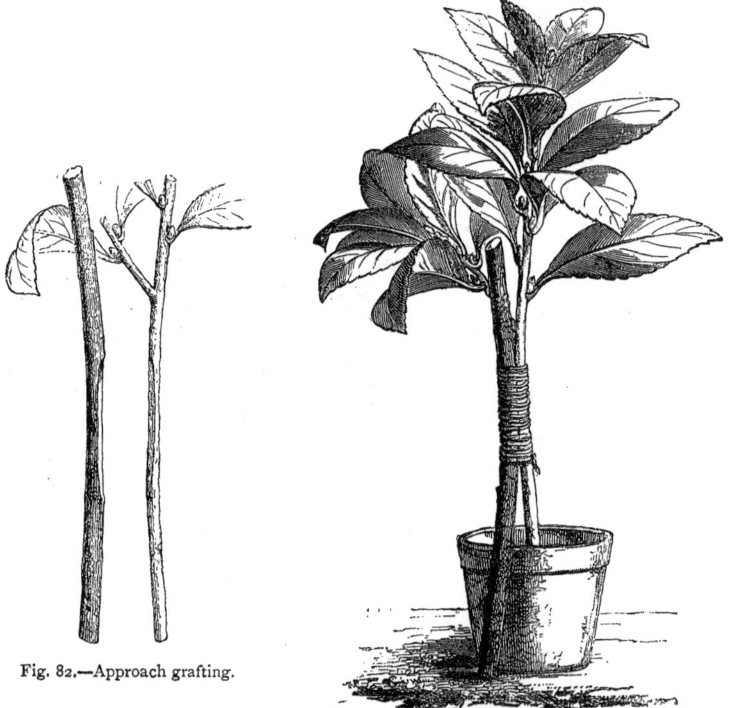

Fig. 82.—Approach grafting.

Fig. 83.—Approach grafting.

the conventional term, without the aid of many complicated artifices; there are even some which resist all means known of propagation by slips; in short, the slip often finds itself subject to these alternatives—to die of inanition, for want of sufficient moisture, or to rot from over-much liquid. The problem for the operator to solve, in order to favour the production of roots, is to establish a proper equilibrium between the aqueous losses to which the slip will submit, and the quantity of water which it absorbs; and to do so is not without its difficulties. But this is not the place to explain the processes by which these operations are successfully performed. We must content ourselves with remarking that the process is not confined to slips or cuttings of branches, as cited above. Cuttings may be made from rhizomes, from leaves, and even from parts of a leaf.

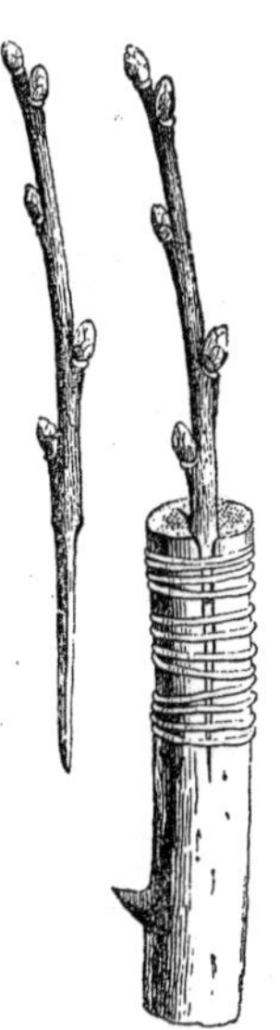

Fig. 84.
Cleft grafting.

Layers and slips are not the only operations in which branches are employed for the purpose of multiplication. There is another, the most important of all in garden operations—namely, *grafting*, or *budding*—of which we have already said a few words in connection with the bud. Its object is to attach one vegetable to another, which is to sustain and furnish matter for its sustenance; to nurse it, in short. We sometimes see in forests certain trees, particularly the Hornbeam, in which a branch of one is firmly united to a neighbouring tree of the same species. This process, which in this case is a natural occurrence, is practised artificially to a great extent in gardening. The operator cuts a corresponding slice of bark from two trees, brings the two equal places into contact, and lashes them firmly together with cord, which is again covered with some sort of clay, to keep the wound moist until a junction has taken place. This is *approach grafting*. Fig. 82 shows the manner of preparing the two subjects intended for approach grafting. Fig. 83 exhibits two subjects firmly attached by means of ligatures.

In *cleft grafting* the trunk of a tree is cut through horizontally, and a vertical cleft is made in it some inches deep. Into this cleft the branch of a graft with several buds, and cut to the shape of the cleft, is inserted, which is closely in contact with the sides of the cleft. The cleft is then covered with mortar of some kind, and bound firmly together by means of cord. In Fig. 84 we have these successive operations represented. Cleft grafting is operated successfully both on the trunks and roots of trees. By its means the horticulturist changes with advantage the products of trees of the same species, making the head bear fruit and flowers other than those belonging to the principal stem. In fact, they restore the vigour and sweetness of youth to a tree already aged.

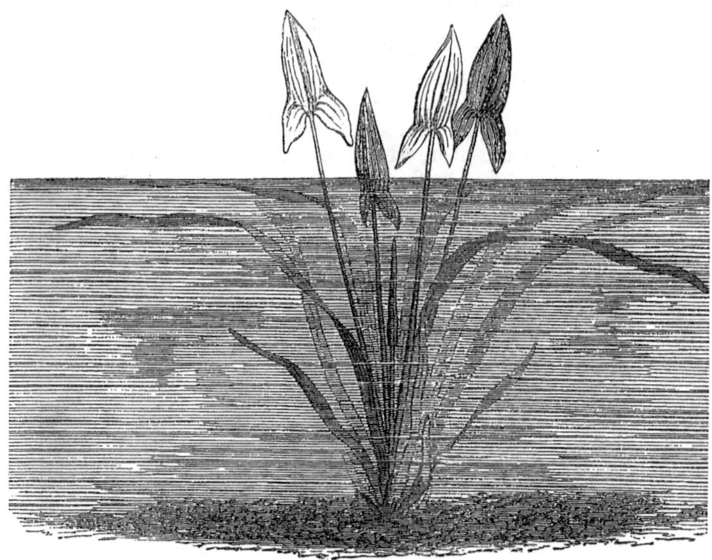

Fig. 85.—Aërial and submerged leaves of the Arrow-head (*Sagittaria*).

V.—Of Leaves.

We have considered buds, which enclose in their green envelopes the promises of spring. At the hour marked by awakening Nature this cradle of the leafy organs will open itself step by step, and in a short time the gardens, the fields, and the woods will be clothed with a dazzling mantle of verdure.

The season of the renaissance of leaves is that which exercises the softest influence on the human soul, when the new vegetation begins to decorate the fields, and gives to the boughs and branches, long denuded by the frost, that tint of vernal green so vivid and dazzling, which brings with it that delicious impression which no animated beings can refuse to feel. The reviving verdure is the forerunner of fine days, the first adornments of the fields announcing a brilliant assembly of flowers, a plentiful tribute of savoury fruits. Renovated Nature offers at once to the eyes and the mind a most seducing picture; what pleasure do we not enjoy in the shade and shelter of the forest in the burning days of summer!

If the leaves have not the dazzling and variegated colours of the flowers of our fields or parterres, their green surface and variegated shades serve to relieve the eye. The movement of the leaves, as they gracefully wave to every breath of wind, serves also to animate the landscape, and gives it a kind of living existence.

But the functions of the leaves are not limited to mere ornament

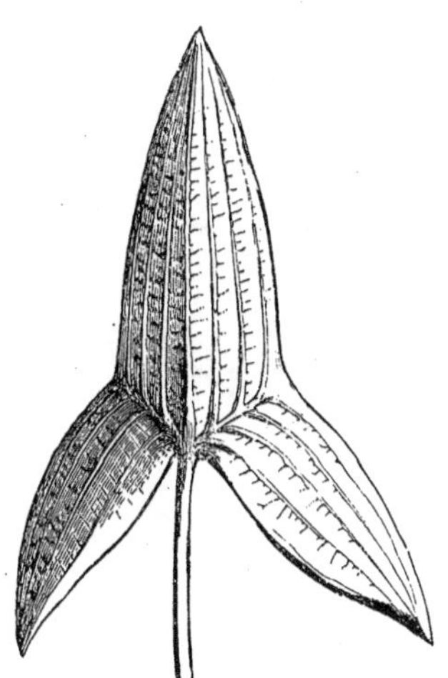

Fig. 86.—Arrow-head.

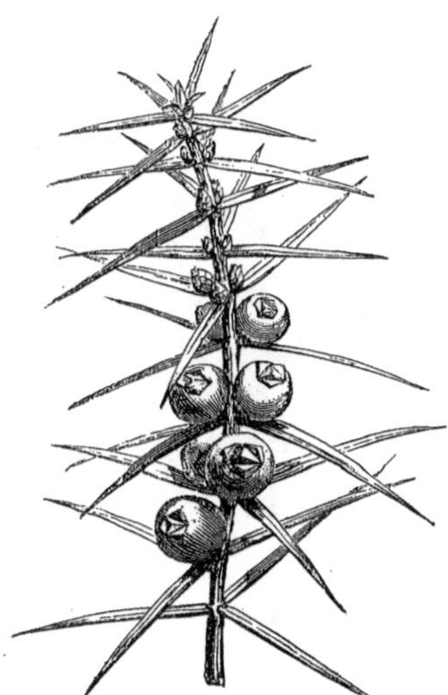

Fig. 87.—Juniper.

and shade. Nature, as we shall see, assigns to them infinitely more important offices, both to surrounding Nature and to the tree of which they form a part. They purify the surrounding atmosphere, restoring it to its normal condition, rendering it healthy and salubrious when vitiated by the breath of animals. Nature has in this, as in all its works, united decorative elegance and beauty of form with direct and immediate utility.

Leaves are borne upon the stem and branch, and nothing is more varied than the forms they assume. In *Sagittaria*, Fig. 86, they resemble an arrow or spear head, whence its name. In the Juniper bush (*Juniperus communis*), Fig. 87, the leaves are like so many needles. Others have scythe-like leaves, as in the *Gladiolus;* others sword-like, as in the *Iris.* Leaves may take the form of a disc, as in the *Tropæolum*, Fig. 89, or the form of a spatula, as in the Daisy (*Bellis perennis*), Fig. 88.

Some leaves have forms so strange that botanists have been puzzled to describe them. For example, in *Nepenthes khasyana*,

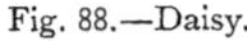

Fig. 88.—Daisy.

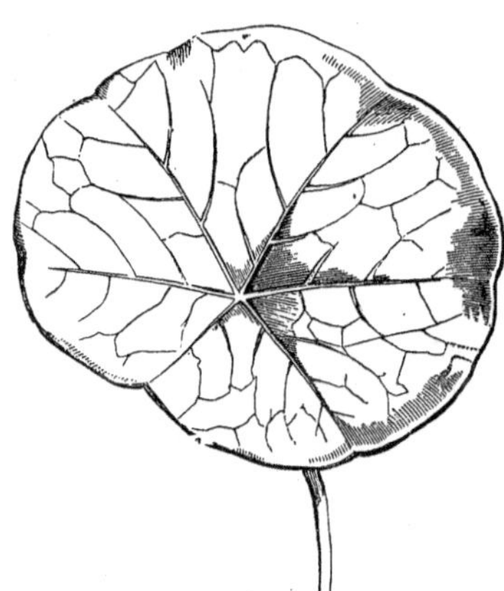

Fig. 89.—Tropæolum.

Fig. 90, the leaves terminate in a most singular manner, forming a sort of urn or vase, surmounted by a cover. This vessel is suspended at the extremity of a thread-like appendage to the end of the leaf, which would seem to be altogether unfit to support it. In a recent work we find the following facts recorded in reference to the leaves of the *Nepenthes*. An officer of marines writes :—" Three days after my arrival at Madagascar, I lost my way during a short excursion into the interior, and was overtaken with an excessive lassitude, accompanied with a devouring thirst. After a long walk I was upon the point of yielding to despair, when I perceived close to me, suspended to leaves, some small vases, somewhat like those used on board to preserve fresh water. I began to think I was under one of those hallucinations by which the sick are visited in fever, when the refreshing draught seems to fly from their parched lips. I approached, however, with some hesitation; I threw a rapid glance into the pitchers: judge of my happiness when I found them filled with a pure and transparent liquid! The draught I partook gave me the

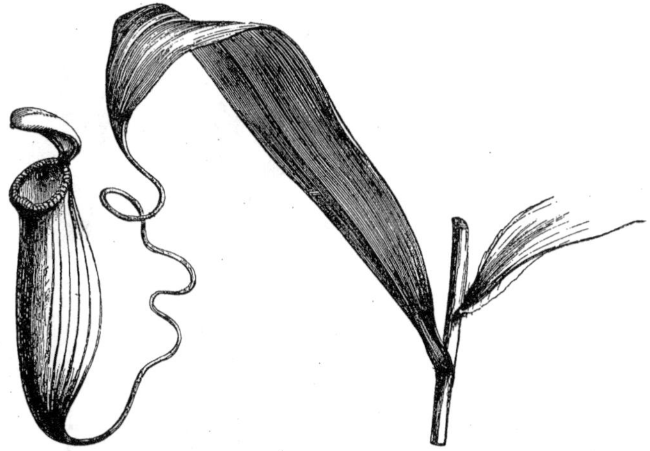

Fig. 90.—*Nepenthes khasyana*

best idea I have ever realised of the nectar served at the table of the gods."

To pursue our subject, however. In the *Sarracenias* a number of the leaves are long and funnel-shaped, somewhat like a long horn or trumpet, as in Fig. 92; while in the Venus's Catchfly (*Dionæa muscipula*), Fig. 91, the leaves are terminated by two rounded plates or leaves, furnished with hairs on the outer edge. When touched, these leaves close upon their victim, and become, as it were, a charnel-house, when thus united.

Among the many species of plants which have been described, there are scarcely two whose leaves can be said to be perfectly alike. "These contrasts surprise the traveller," says Auguste de Saint-Hilaire, "when in traversing equinoctial countries, he finds himself surrounded by thousands of forms which have among them all only one trait of resemblance—their elegance and grace—when he sees the delicate foliage of the *Mimosa*, so sensitive to the touch, hanging over the gigantic leaf of the *Scitamineæ*, and the Ferns with their thousands of finely-cut fronds growing upon the trunk of the *Eugenia*, and mingling with *Bromeliaceæ* and Tillandsias, with their rigid and inflexible leaves.

But more than this: we do not find in Nature any two leaves

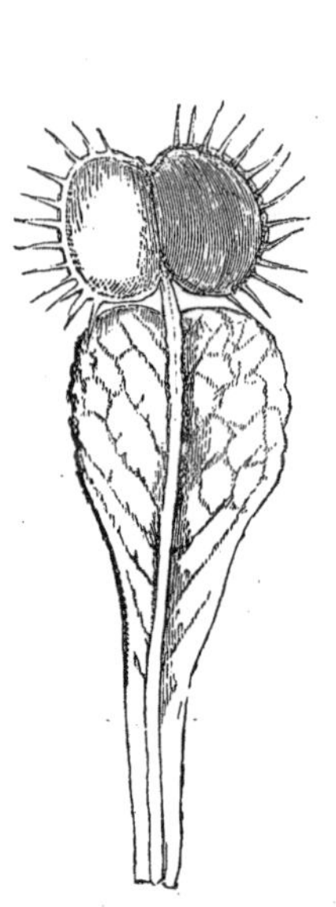

Fig. 91.—Leaf of Venus's Catchfly.

Fig 92.—Leaves of *Sarracenia.*

exactly alike. Sometimes the same plant possesses leaves having less resemblance to each other than those of two different species. The Paper Mulberry (*Broussonetia papyrifera*), Fig. 93, has at the same time heart-shaped and lobed leaves. In the garden Valerian the lower leaves are entire, and those at the summit are deeply notched. In the *Ranunculus aquatilis*, Fig. 94, the leaves which vegetate in the water are divided into thread-like expansions, so narrow that they seem to be leaves reduced to their nerves, or skeleton, while those leaves which grow in the air are entire, and disc-like in form, and more or less notched. When the common Arrow-head (*Sagittaria*) grows in brooks, its submerged leaves form long ribbons; when it

Fig. 93.—Branch of the Paper Mulberry (*Broussonetia papyrifera.*)

grows on the banks of great ponds or tanks, the leaves resemble arrow-heads, Fig. 85.

There is not less diversity in the length and breadth of leaves than in their form. While some leaves are only half a line in length, others attain the dimensions of five or six yards. Nor is their size always proportioned to the thickness of the stem which carries them. The leaves of a small plant, the Dock (*Rumex*), would cover many hundred times the space occupied by the fascicled leaves of the Larch, an imposing mountain tree; and there is a thousand times less vegetable matter in the leaf of the Fir-tree or Cedar than in that of the Plantain, or Banana-tree.

The leaf usually consists of two parts—a stalk, or *petiole*, and the

Fig. 94.—*Ranunculus aquatilis.*

blade, or *lamina*. The petiole connects the leaf with the branch or stem, and is composed of a bundle of unexpanded fibres covered by an epidermis. When the petiole fails, as in the common Flax (*Linum*), Fig. 95, the leaf is said to be *sessile.* In such cases the leaves often partially or entirely surround the stem, when they are said to be *amplexicaul*, or *semiamplexicaul.* The leaf is *simple* when the limb consists of one piece either quite entire or variously indented, cleft, or divided on the edge; and compound when it consists of one or more leaflets, each of which is jointed to the common petiole by intermediate *petiolules*, sometimes very short. The leaf of the Lime-tree (*Tilia*), Fig. 96, is simple; that of the Robinia, or False Acacia, Fig. 97, is compound.

It happens sometimes that the petiole is branching, and bears petioles of the second or third order, upon which are inserted *petiolules* with their leaflets. This occurs in *Gleditschia triacantha*, Fig. 98, which

Fig. 95.
Sessile leaves of the Flax-plant.

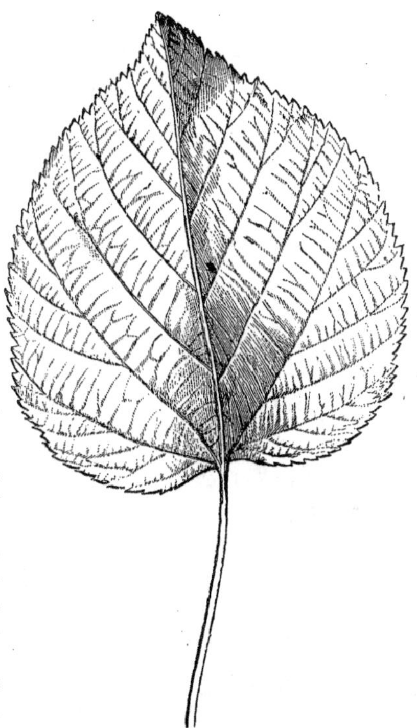

Fig. 96.—Leaf of the Lime-tree.

belongs to a highly ornamental group of trees, with branching thorns in some of the species. When the division of leaves is carried further, the term *decompound* is made use of; the Hemlock (*Conium maculatum*) is said to be *supra-decompound*. The outline of the limb of the leaf is often continuous all round, or *entire*, as we see in the Box-tree, the Iris, and other plants. But it is not so in the majority of cases. The edge of the limb of the leaf is generally more or less serrated; according to the form and depth of these inequalities round the margin, the leaf is said to be *dentate*, *crenate*, *serrate*, *lobed*, or *partite*.

Leaves are dentate when the edge is notched with acute-pointed

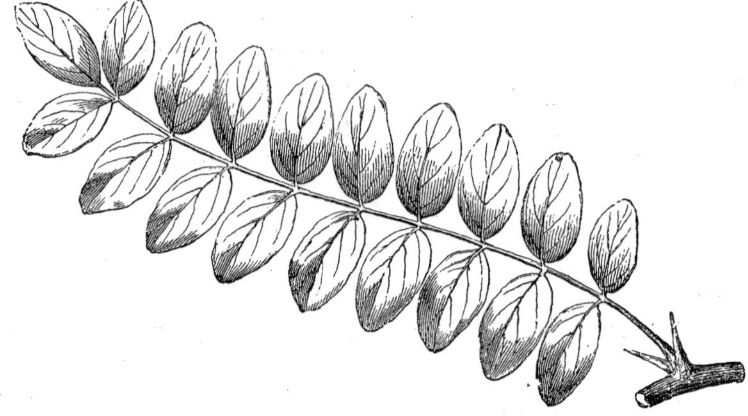

Fig. 97.—Leaf of Robinia.

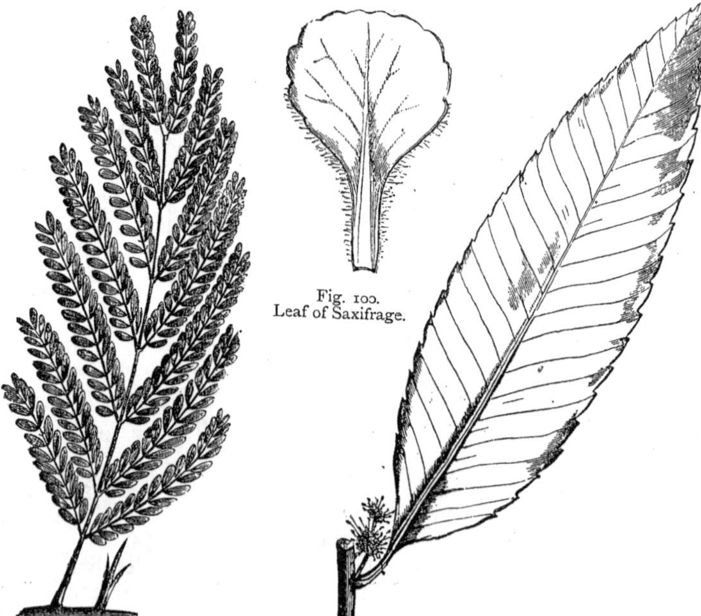

Fig. 100.
Leaf of Saxifrage.

Fig. 98.—Leaf of *Gleditschia*.

Fig. 99.—Chestnut.

teeth, as in the Chestnut, Fig. 99; crenate when the margin is cut into bulging projections, as in the Saxifrage, Fig. 100. They are lobed when the leaf is more deeply indented, as in the Gingko, Fig. 101. They are cleft when their division extends through one-half of the leaf, as in the *Bauhinia*, Fig. 102, the leaf of which gives a very good idea of the cleft leaf. The leaf of Castor-oil tree (*Ricinus communis*), Fig. 103, is cleft into eight sections. Finally, leaves are partite when the separation penetrates nearly to the petiole, or to the mid-rib of the leaf, as in Fig. 104, a bipartite leaf; Fig. 105, which

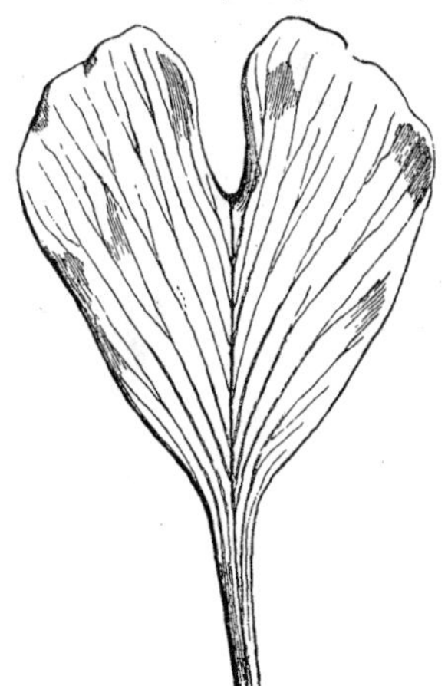

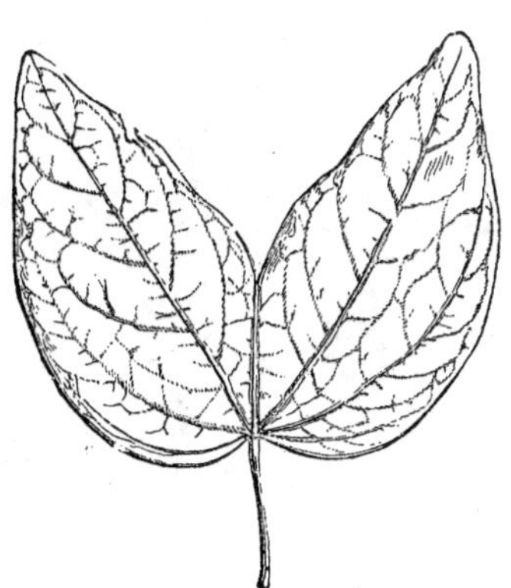

Fig. 101.—Leaf of Gingko.

Fig. 102.—Leaf of *Bauhinia*

represents the leaf of *Cannabis sativa*, the Hemp; Fig. 106, *Echinops sphærocephalus;* and Fig. 107, the leaf of *Scolymus hispanicus*, in which the divisions are more numerous still.

The leaf is, as we have said, a flattened organ having two surfaces, and border, the whole of which constitutes the *lamina*, or blade. The blade of every leaf is traversed by prominent lines or ridges, which are, however, more prominent on the lower face than on the upper. These are formed of woody tissues with spiral vessels. They form the nerves or veins of the leaf, and are retained in their positions, and the intervening space is filled up by cellular tissue. The tissues of the veins are brought into closer proximity in the petiole, which resembles a small branch. Having passed into the stem, one part enters the bark, while the other traverses the wood and penetrates to the medullary sheath at the centre of the stem. Every leaf is thus in communication with the stem; its skeleton is, in fact, a prolongation of the spiral vessels, bast, and wood of the stem. The disposition of the veins differs very little in the three principal types. In the Chestnut, the leaf of which

Fig. 103.—Leaf of the Castor-oil tree. Fig. 104.—A bipartite leaf.

is represented in Fig. 99, the mid-rib runs from the base to the summit of the blade, sending out to the right and left a secondary set of ribs or veins parallel to each other, disposed like the barbs of a feather. In the Mallow (*Malva silvestris*), Fig. 108, five principal nerves run from the base of the leaf to the apex, and radiate in the blade like the foot of a web-footed bird. In the Iris, of which leaves are represented in Fig. 30, p. 34, a great number of delicate veins run from the base of the leaf towards the summit, all being parallel to each other.

The petiole, sometimes long, sometimes short, and sometimes absent altogether, is often cylindrical, sometimes arched and inflated, as in the Water Caltrops (*Trapa natans*), or compressed, as in the Birch-tree (*Carpinus betulus*); and in many Poplars, in which the surface is large, in place of being a continuation of the blade, it abuts upon it at right angles. In such cases the petiole gives little support to the leaf, but presents its two largest sides to the wind, which causes

Fig. 105.—Leaf of the Hemp.

it to oscillate and tremble, producing the rustling sound which distinguishes the Aspen-tree (*Populus tremula*).

In some cases the petiole is wanting; the blade even may be defective; the leaf is then reduced to its petiole. But in such cases, in obedience to the laws of compensation, which intervene when any organ proves abortive, the neighbouring parts take a greater development; the petiole is enlarged and assumes a ribbon-like form, which was long taken for a leaf. It is distinguished by the position of its veins, and also by the fact that in place of being compressed in such a manner as to present the usual upper and under surface, it is set edgeways, and its two faces are lateral. This form of petiole is termed a *phyllode*, and is observed in the case of plants which are really leafless, where the petiole performs the functions of a leaf. *Acacia heterophylla*, Fig. 109, is full of instruction in this respect. We find in it all the intermediate steps between a perfect compound

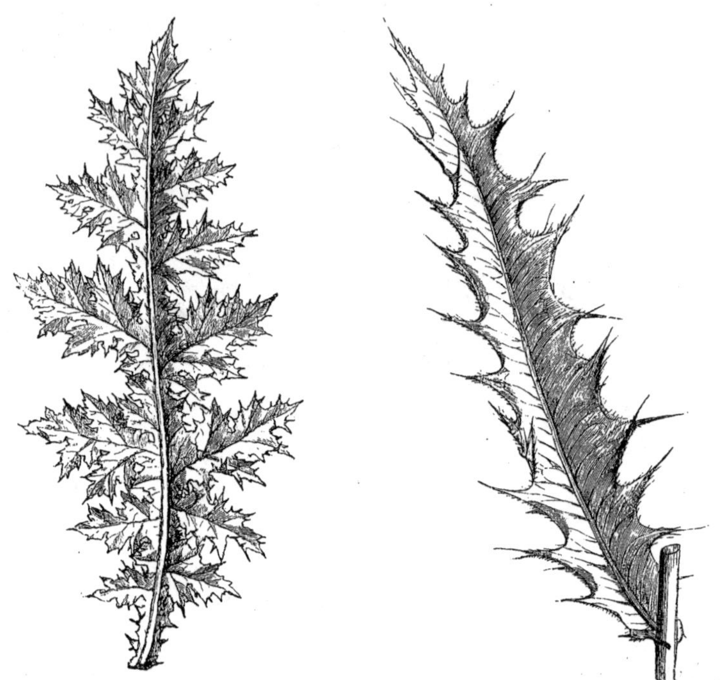

Fig. 106.—Leaf of *Echinops*.

Fig. 107.—Leaf of *Scolymus*.

leaf and a *phyllode*. The petiole is seen to flatten and enlarge in exact proportion as the lamina decreases: the petioles bear leaflets at the earliest stage of their development; they have parallel veins, although occurring in exogenous plants. This transformation of the petiole, which is frequent in the Acacias of Australia, occurs also in many other plants, as in *Dionæa muscipula*, Venus's Fly-trap (Fig. 91), which the petioles extend laterally, and resemble true leaves; in the tendrils of the Pea, and in some others belonging to the leguminous plants, the petiole extends itself longitudinally.

Leaves transform themselves into other organs with wonderful facility. It is, in fact, by modifications of the leaves that Nature produces many organs essential to the life of plants. They are changed into *scales*, a transformation of which we have an example in the young shoots of the Asparagus plant, Fig. 112; into tendrils, as in the Pea, Fig. 111; and into spines, as in the Barberry (*Berberis vulgaris*), Fig. 114.

Fig. 108.—Leaf of the Mallow.

What is the disposition of leaves upon the stems and branches which carry them? Are they placed at random upon the axes of the vegetable? The most superficial examination suffices to satisfy us that the leaves are always placed in the same manner and in a fixed order for every species of plant; in other words, that their relative distance and direction are rigorously fixed by Nature. A more profound examination leads to the conviction that this order is subject to certain laws, and may be expressed by an arithmetical formula. If we examine a branch of the Elm (Fig. 115), of the Willow, or the Cherry-tree, we observe at once that the leaves are all inserted at different heights. In this case they are said to be *alternate.* In the Willow, the Nettle (*Urtica*) Fig. 110, on the contrary, the leaves are grouped in pairs at the same height. These leaves are said to be *opposite.* In the Loosestrife (*Lysimachia vulgaris*), and the Oleander, Fig. 113, three leaves are grouped at the same height round the stem. In this case, as also in cases where many leaves are grouped in the same manner, we say the leaves are *verticillate,* or *whorled.* What gives to these plants their peculiar physiognomy and appearance is, that the leaves which constitute the whorl correspond with the intervals which separate the leaves in the pair or group placed immediately above or below, and it may be added that the leaves of a similar pair or verticil are always equidistant.

To return, however, to the alternate leaves. Let us take a

branch of the Peach or the Plum-tree, Fig. 116, and examine any leaf whatever. We shall find that, higher up, the branch carries another leaf immediately above it, and that in the interval between these two leaves there are four others diversely placed. All these leaves are placed upon the line of an ideal spiral, which commences in one leaf

Fig. 109.—*Acacia heterophylla.*

and terminates in the leaf exactly above it. This spiral, which forms what is called a cycle, may consist of one or more revolutions round the axis. Fixing on any leaf, and ascending the branch till the leaf immediately over it is found, by counting all the intermediate ones the number of leaves contained in the cycle is arrived at. In the Peach and Plum-tree the cycle embraces five leaves, and the spiral goes twice round the branch. This is expressed in botanical language by the fractional formula $\frac{2}{5}$, the numerator indicating the number of turns the spiral makes on the cycle, the denominator the

Fig. 110.—The Dead Nettle. Fig. 111.—The Pea.

number of leaves which constitute the cycle, as in Fig. 117. In the Alder (*Alnus glutinosa*), Fig. 119, three leaves constitute the cycle; and the spiral only describing a single turn on the stem, the disposition of its leaves is represented by the fraction $\frac{1}{3}$, as exhibited in Fig. 119.

In the Elm (*Ulmus*), as will be seen in Fig. 115, and in the Lime-tree, two leaves only constitute the cycle, in which case it is

Fig. 112.—Asparagus.

Fig. 113.—Oleander.

represented by the fraction $\frac{1}{2}$. Let us write these three fractions in a single line, $\frac{1}{2}$, $\frac{1}{3}$, $\frac{2}{5}$, and we find that the numerator and denominator of the fraction $\frac{2}{5}$ are respectively the sum of those of the first two fractions. Let us combine in a similar way the fractions $\frac{1}{3}$ and $\frac{2}{5}$, we obtain $\frac{3}{8}$. Do the same for $\frac{2}{5}$ and $\frac{3}{8}$, and we obtain $\frac{5}{13}$. In this manner the terms $\frac{1}{2}$, $\frac{1}{3}$, $\frac{2}{5}$ $\frac{3}{8}$, $\frac{5}{13}$, $\frac{8}{21}$, etc., are obtained. Singularly enough these fractional formulas express precisely the disposition of leaves which Nature realises. The denominators of these fractions, in giving the leaves of each cycle, indicate at the same time the number of vertical lines along which the leaves are ranged. Thus in the Elm and in the Lime-tree the leaves are disposed in two rows, and are said to be *distichous*, in which they agree with the Ivy

Fig. 114.—The Barberry. Fig. 115.—Branch of the Elm.

(*Hedera helix*), and the Yew (*Taxus baccata*). In the Alder, Sedges, and the Tulip, the leaves are disposed in three rows, and said to be *tristichous*. In the Peach-tree, where the leaves are disposed in five rows, they are said to be *quincuncial*. "The distribution of leaves upon the branches," says De Candolle, "is in accord with their functions, which are almost exclusively determined by the action of the sun. In order that this action should exercise itself properly, it is necessary that the leaves be wide apart and overlap each other as little as possible. We have seen that in all the systems by which the position of the leaves are arranged, it results that the leaves which spring immediately above those below are never covered.

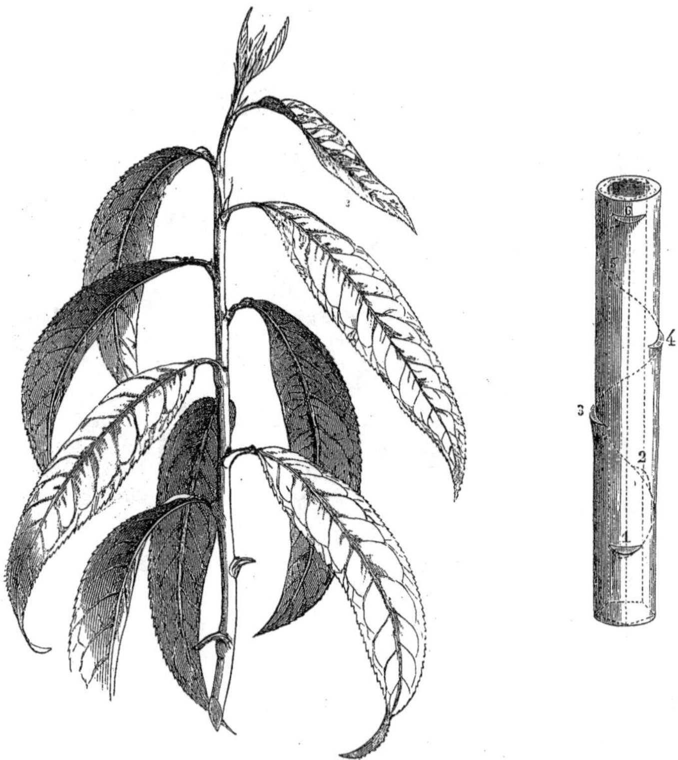

Fig. 116.—Branch of the Peach.

Fig. 117.—Insertion of Peach leaves.

In cases the least favourable, the third covers the first, and the fourth the second. In another case it is only the sixth which overlaps the first. Thus, in all these combinations, whether it be from the distance of the cycles, or of the leaves contained in them, or from the size of the leaves, which diminish as they ascend, we find that all leaves are so arranged as to enjoy the free action of the solar light."

In most plants, when the leaves have accomplished their physiological functions, they fall, even in the year which witnesses their

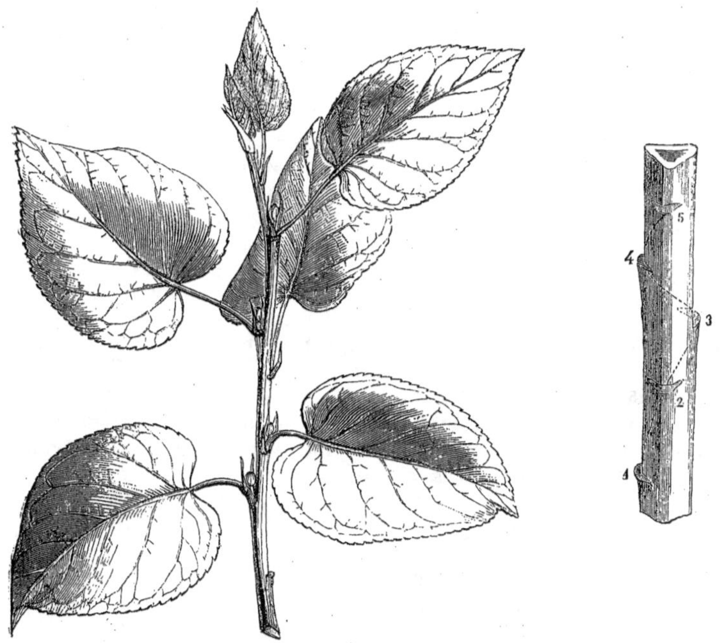

Fig. 118.—Branch of the Alder.

Fig. 119.—Insertion of leaves of the Alder.

birth. But there are others which are not detached till the following year, while others still remain for many years attached to the stem. The leaves of most of the Conifers, those of the Box (*Buxus*), the Holly (*Ilex*), of the Orange trees (*Citrus*), do not fall in the year in which they are developed, but are met by the new ones. These plants are never seen naked; they are therefore commonly known as evergreens. When a plant sheds its leaves before the next spring, it is said to be *deciduous*. Leaves are *persistent*, when they remain longer than a year.

In some plants, as the Cactus, the leaves are shed almost as soon as they appear. These are said to be *caducous*, from *cado*, "I fall." The leaf thus dies, like all created things, when the purpose for which it was created is accomplished. The immediate cause of death seems to be this: the cells of which it is composed have become encrusted with foreign matter, deposited during the processes of assimilation and evaporation carried on by the organ, and it then becomes incapable of further action.

On the subject of the distribution of evergreen trees, Auguste de Saint-Hilaire makes the following remarks:—"As we retire from the tropics, the number of evergreen trees goes on diminishing in rapid succession. At Porto Allegre, near 30° south latitude, I found in the coldest season the trees nearly all covered with leaves. At San Francisco de Paula, near the Rio Grande, in 34°, nearly one-third of the ligneous vegetation had lost their leaves; and finally, at two degrees farther south, a tenth of the trees only preserved their foliage. At Montpellier, the country in winter is not wholly deprived of verdure; and Lisbon, Madeira, and Teneriffe present a still more considerable number of trees always green. It must not be supposed, however, that in the tropics all the trees are evergreens. Even in the vast forests which occupy the Brazilian coast, and where vegetation is maintained in continual activity by its two principal agents, heat and moisture, there exist trees, such as some of the *Bignoniaceæ*, which lose every year, like our trees, all their leaves at once, but immediately after they are covered with flowers, and in a very short time these are succeeded by new foliage. I speak here of woods growing in equinoctial regions, where, as with us, rain and drought have no determinate period. In countries, on the other hand, where six months' continual rain is succeeded by uninterrupted dry weather, there are woods which every year remain for a considerable time destitute of verdure, and the traveller who traverses them is scorched by the ardent blaze of the equinoctial zone, while he has before his eyes the leafless image of European forests during winter. We have even seen this excessive drought continue during two years, and the trees remain, accordingly, for two years without their foliage."

But evergreen trees are only exceptional in the vegetable world. Most trees, shrubs, and herbaceous plants are without their leaves during one half of the year. When the leaves have performed their functions, when the fruits have appeared, matured, and ripened, vegetation has entered into a new phase; the leaves lose their brilliant green, and assume their autumnal tint, sometimes clothed, however, in colours of accidental, though transitory, brightness. The green, when it is persistent, is more grave and sober in its hue; it becomes brown in the Walnut, it takes a whitish tone in the Honeysuckle. The leaves of other plants, as the Ivy, the Sumach (*Rhus*), the Dog-wood tree (*Cornus*), become clothed with a reddish tint; they become yellow in the Maple (*Acer campestris*), and many others of our forest trees. But whatever may be the variety of shades which leaves take in their decay, a certain air of sadness pervades these ornaments of our fields, which proclaims their approaching dissolution, and betrays the imminence of the cold season. Cold and humidity will soon arrest the sap and disorganise the petiole; the leaves, withered and deformed, will soon cumber the ground, to be blown hither and thither by the wind. It is the season of the fall of the leaf with all its melancholy associations, which the poets have so often depicted.

Nevertheless, leaves, when separated from the vegetable which has given birth to and matured them, are not lost to the earth which receives them. Everything in Nature has its use, and leaves have their uses also in the continuous circle of vegetable growth. The leaves which strewed the ground at the foot of the trees, or which have been disseminated by the autumn winds over the naked country, perish slowly upon the soil, where they are transformed into the *humus*, or vegetable mould, indispensable to the life of plants. Thus, the *débris* of vegetables prepares for the coming and formation of a new vegetation. Death prepares for new life; the beginning and the end join hands, so to speak, in vegetable Nature, and form the mysterious circle of organic life which has neither commencement nor limit.

But let us return to the general study of leaves. We have still to note a last and most important phenomenon in the variety of their functions. We speak of the spontaneous movements executed by leaves under many circumstances.

Leaves almost always assume the horizontal position. They have an upper surface turned towards the heavens, and a lower surface looking to the earth. This position is so natural, and hence so necessary, that leaves assume it of themselves during day and night when from any accidental cause that position has been lost. If we place a plant in an apartment lighted by a single window, it is soon observed that all the leaves direct their upper surface towards the light. This is an experiment which our readers can easily repeat for themselves. But leaves perform other movements, equally remarkable, on which we must pause an instant. The study of these movements has been, as we shall see, the subject of some curious and interesting experiments.

Dutrochet, having placed a young pea in a chamber lighted on one side only, observed that the leaf sometimes inclined itself towards the light, and directed its petiole towards the heavens, and sometimes inclined it towards the dark part of the chamber. The tendril was now nearly straight, at another time curved and arched, presenting very irregular motions. Dutrochet placed fixed indicators both over the tendril and over the petiole, at the insertion of the two leaflets. He was thus able to state what directions they took by standing at a little distance from the fixed indicators. He soon observed that the summit of the petiole described an elliptical curve in the air, while the tendril which terminated it had various motions. He soon observed also that the internode belonging to this leaf participated in this movement of revolution, and that it was even the principal agent in it. The *merithal* or internode and the leaf then produced by their general movements a sort of cone, whose summit occupied the lower part of the merithal, and whose base was the curve described in the air by the summit of the petiole. The tendril, during the movement of revolution, constantly directed its point towards the bottom of the chamber, thus shunning the light, turning itself when the movement of its revolution in bringing its point near the window carried it in the direction of the light.

This revolution was effected in a period varying with the temperature and the age of the leaf. It lasted from twenty minutes to an hour at a temperature of 75° F.; from seven to eleven hours when the temperature was lowered to 41° or 42°. The extent of the revolutions diminished in proportion as the temperature decreased. "What is the cause of this revolving movement?" asks Dutrochet. "It is not revealed to our eyes. It is some vital and internal exciting cause. Not only does the light contribute nothing towards the production of the movement, but it operates against it, and when unusually vivid it seems to stop it."

Dutrochet observed the movement of revolution in the tendril of *Bryonia*, and also in the cultivated Cucumber. In *Bryonia* the tendril moved in very varied directions, sometimes horizontally, sometimes upwards, sometimes downwards, and sometimes directing its point towards the heavens, then taking any curve whatever, to take immediately afterwards a curve in the contrary direction. The tendrils of the Cucumber moved like the hands of a watch placed face upwards, directing its points successively towards every point of the compass, from right to left, and from left to right. But it needed all the sagacity, all the quick and deliberate observation of Dutrochet, to discover the slow and obscure movements which have been described, or their origin.

The spontaneous movements which we have now to note in certain vegetables are much more apparent. Let us speak first of the movement of the plant known as the *Desmodium gyrans*, Fig. 120. This plant belongs to the family of *Leguminosa*. It was discovered in Bengal, in the neighbourhood of Dacca, by an Englishwoman, Lady Monson, whose devotion to natural history had led her to undertake the voyage to India, and who died on one of her botanical excursions.

Fig. 120.—*Desmodium gyrans.*

The leaves of the *Desmodium* are composed of three leaflets. The terminal leaflet is very large, and the lateral ones very small, but these last are almost always in motion. They execute little jerks somewhat analogous to the movements of the second hand of a watch. One of the leaflets rises and the other descends at the same time, and with a corresponding force. When the first begins to descend, the other begins to rise. The large leaflet moves also, inclining itself now to the right, now to the left, but by a continuous and very slow movement when compared with that of the lateral leaflets. This singular mechanism endures throughout the life of the plant. It continues day and night, through drought and humidity. The warmer and more humid the day, the more lively are its movements. In India the plant has been known to make sixty jerks in the minute.

This curious plant, which was introduced into Europe for the first time in 1777, is cultivated in our principal botanic gardens. The auditors of every botanical course of lectures have frequently their attention directed to the strange phenomenon of which it is the subject. Its movements occur spontaneously and without any apparent cause.

But there are movements in other plants which are produced in response to external irritation. Such are those of the Venus's Catchfly and the Sensitive Plant.

The *Dionæa muscipula*, Fig. 121, is a native of the swamps of New Carolina. Its leaves, which are spread out on the soil near the roots, are composed of two parts—the one elongated, which may be considered as a sort of petiole, the other larger and broader and nearly circular, formed of two plates, which are united by a rib, fashioned like a hinge, and are furnished round the edge with rough hairy cilia. In the upper surface these plates are furnished with certain small glands, whence exudes a viscous liquid which attracts the insects. If a fly lights on this singular apparatus, the plates turn rapidly upon their hinge. They approach, their long cilia interlace, and the insect is a prisoner. The efforts of the fly to escape increase the irritability of the plant, whose fangs only open when the movements of the insect have ceased with its life.

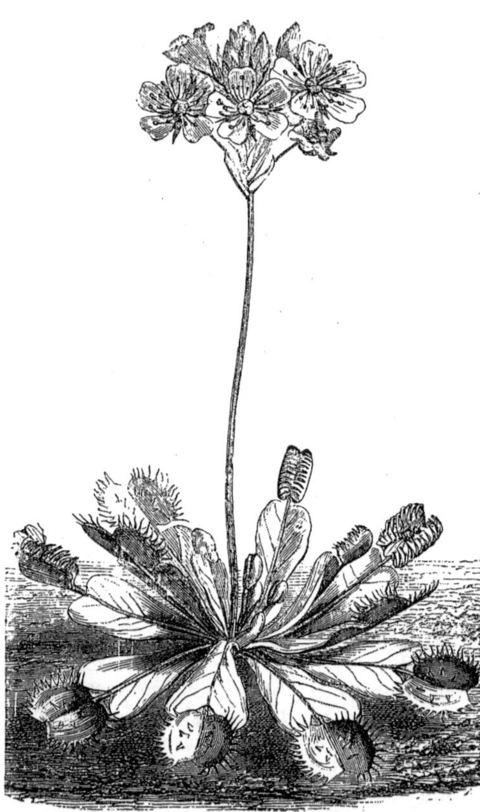

Fig. 121.—*Dionæa muscipula.*

Who that knows, who that has seen the Sensitive plant, *Mimosa pudica*, Fig. 122, has not also remarked on the sensibility of its leaves? The lightest touch suffices to make its leaflets close upon their supports, the petiolules, and the petiole upon the stem. If we cut with scissors the extreme end of one leaflet, the others will immediately close in succession. De Candolle was in the habit of placing a drop of water upon one of the leaflets of the Sensitive Plant, applying it with so much delicacy as to excite no movement whatever. But when he substituted for the water a drop of sulphuric acid, he observed that the leaflets closed, the petiolules

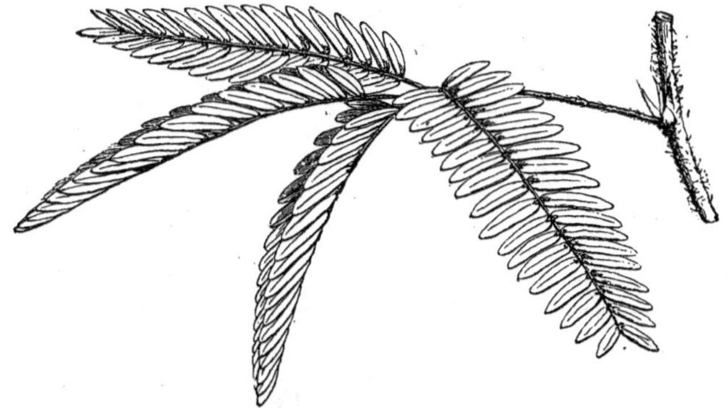

Fig. 122.—Leaf of the Sensitive plant.

as well as the petiole were lowered, and submitted to its influence, and gradually all the leaves situated above the one touched have undergone the same fate, without the leaves situated below participating at all in the movement. This experiment shows very clearly that the irritation is not local, but communicates itself gradually in the various elements of a leaf, and propagates itself similarly from one leaf to another.

During the time that these movements are in operation, it will be observed that the limb of the leaflet neither curves itself nor shrivels. In short, the contractile power resides at the point of insertion of the leaflets upon the petiolules, these upon the common petiole, and these again upon the stem. These various points of insertion correspond with the very perceptible cylindrical cushions which during the season of rest swell below, while in the state of irritation they are distended above. The movement which we provoke in the case of the Sensitive Plant manifests itself with much greater rapidity when irritated upon this cushion than in any other part of the plant.

We have remarked that the more vigorous the Sensitive Plant is in its habit, the more susceptible is it; the higher the temperature, the more promptly does it respond to the touch. We may observe besides that the Sensitive Plant can, up to a certain point, get accustomed to the movement. Desfontaines, carrying a Sensitive Plant in a carriage, observed that the plant closed its leaflets, and all its leaves drooped as soon as the carriage began to roll over the pavement, but by degrees it seemed to recover from its fright, became habituated, so to speak, to the movements, its leaves resumed their erect position, and its leaflets their full expansion. Desfontaines now caused the carriage to stop for a time. When it resumed its motion the plant responded by dropping its leaves as before, but after a time they expanded again, and so continued during the remainder of the journey. These phenomena of irritability under the influence of direct chemical or mechanical action the plant repeats of itself during the night. The Sensitive Plant closes its leaflets when the obscurity of night sets in.

This habit of folding up its leaves during the night is not confined to the Sensitive Plant exclusively; it appertains to other plants whose leaves occupy different positions during day and night. These are the plants to which Linnæus alludes when he writes of the *sleep of plants*. "But we must remark," says De Candolle, "that this term, borrowed from the animal kingdom, does not represent the same idea in both. In animals sleep indicates a flaccid, drooping state of the members, of limpness in the articulations; in vegetables it indicates a changed state; but the nocturnal state maintains the same degree of rigidity and the same constancy as the diurnal position. We may break the sleeping leaf rather than maintain it in the position which belongs to it during the day."

It was in the Bird's-foot Trefoil, the pretty *Trigonella ornithopodioides*, that Linnæus remarked for the first time the difference between the attitude of the leaves during the day and night. Scarcely had he made this discovery when he came to the conclusion that this phenomenon would prove not to be confined to this single plant, but would be found general in vegetable life. From that time, every night Linnæus tore himself from sleep, and in the silence of Nature studied the plants in his garden. At each step he discovered a new fact. Each natural fact, when put in evidence by a first observation, was rapidly confirmed by crowds of facts quite analogous to the first, and Linnæus very soon satisfied himself that the change in the position of leaves during the night was observable in a considerable number of vegetables, and that in the absence of light, plants quite changed their physiognomy, so that it became very difficult to recognise them from their bearing. He further states that it was the absence of light, and not the nocturnal cold which was the principal cause of the phenomena, for plants in hot-houses closed themselves during the night just like those which were exposed in the open air. He recognised also that this difference is much less apparent in young plants than in more matured ones.

The illustrious Swedish botanist made many observations on the diversity of position that leaves affect during the night, and he has even attempted a classification of these differences. The most general idea which he sought to establish was, that the positions differed according as the leaves were simple or compound. Linnæus thought that the object in these circumstances was to place the young shoots under shelter from nocturnal cold and from the effects of the air. It is among the compound leaves, in short, that the difference between the waking and sleeping is most clearly indicated.

The leaflets of the Trefoil stand erect, curving in a longitudinal direction in such a manner as to form a sort of cavity or cradle. The leaflets of the *Melilotus* are half erect, but divergent at their summits. In the *Oxalis*, Fig. 123, the leaflets usually rest upon a common petiole in such a manner as to turn their lower surfaces inwards, and show only their upper surfaces. In the Bladder Senna (*Colutea arborescens*) the leaflets rise vertically in such a manner as to rest perpendicularly upon the common petiole, the upper surfaces applied upon each other. The Cassias have, on the contrary, the leaflets depressed and folding back on the lower surfaces. The leaflets of the *Mimosa* lie along their petiole, directed towards its apex in such a manner that the two extreme leaflets are directed forward, applying together their upper surfaces, and the others overlap the backs of the leaflets nearer, in the series of them, to the apex of the petiole.

The leaves of the Orach (*Atriplex hortensis*) fold themselves upon the young shoots, and enclose them, as if to protect them from the effects of the atmosphere. The Chickweed closes its leaves during the night, and only opens them in the morning. The Evening Primrose (*Œnothera*) has similar properties, and, like the Trefoil, forms during the night a sort of cradle by the approximation of the leaves. On the contrary, the genera *Sida* and *Lupinus* reverse their leaves. Many of the Mallows roll their leaves into a horn. The

Vetch, the Sweet Pea, the Broad Bean, rest their leaves during the night one against the other, and seem to sleep.

This strange repose of plants vaguely reminds us of the sleep of animals. In its sleep the leaf seems, by its position, to recall the

Fig. 123.—Closed leaves of *Oxalis*.

age of its infancy. It folds itself up, nearly as it lay folded in the bud before it opened, when it slept the lethargic sleep of winter, sheltered under the robust and hardy scales, or shut up in its warm down. We may say that the plant seeks every night to resume the position which it occupied in its early days, just as the animal rolls itself up, lying as if it lay in its mother's bosom.

What is the cause of this phenomenon which we designate the *sleep of plants?* It occurs in all hygrometrical conditions of the atmosphere, and the hours during which it affects them are not influenced by any change of temperature. De Candolle supposed that the absence of light was the direct cause of the phenomenon. To assure himself of this, he subjected plants whose leaves are disposed to sleep, to the action of artificial light, furnished by two lamps. The results were very varied. "When I exposed the Sensitive Plants to the light during the night, and to the shade during the day," says De Candolle, "I observed that at first the plants opened and closed their leaves without any fixed rule, but after a few days they seemed to submit to their new position, and opened their leaves in the night, which was day to them, and closed them in the morning, which was their night. When exposed to a continuous light they had, as in their ordinary state, alternations of sleeping and waking; but each of the periods was shorter than ordinary. On the other hand, when exposed to continued obscurity, they still presented the alternations of sleeping and waking, but very irregularly."

De Candolle adds that he was unable to modify the sleep of two species of *Oxalis* either by light or darkness, or by light at other than the natural periods. We may conclude with him, from these facts, that the movements of sleeping and waking are connected with some disposition inherent in the vegetable, but which is thrown into special activity by the stimulating action of light, which acts with different intensity on different vegetables, so that the same amount of light produces different results in different species.

Having thus minutely studied the exterior character of leaves, we shall endeavour to penetrate a little into their structure, and unveil the delicacy of their arrangements.

A cellular tissue, to which we give in this case the name of *parenchyma*, from the soft, pulpy, and closely approximating cells of which it is composed, and which have been aptly compared to a mass of soap bubbles pressing against each other, fills all the interstices of the leaves left by the spreading reticulations of the veins or nerves. It is covered, consolidated as it were, and protected against all external influence by the *epidermis*, a covering which spreads itself like a mantle over the whole surface of the plant. Let us submit this *parenchyma*, *venation*, and *epidermis* to microscopical examination, first considering the leaves of those vegetables which live in the air.

If we pull to pieces with some care any leaf whatever, a fragment of transparent membrane without colour will be observed to detach itself from the leaf. If we place this moist shred upon a glass plate, and subject it to a magnifying glass, it will be found composed of large flattened cells, having a contour, sometimes rectilinear and square, sometimes irregular and sinuous, as in Fig. 124. The granular contents of these cells are neither very apparent nor very important, but we find in them sometimes an aqueous liquid variously coloured. The cellular elements of this epidermal membrane are intimately united and pressed one against the other, in such a manner as to give it a certain solidity and power of resistance. The cells present an exterior wall in connection with the air, which is much thicker than the lateral or interior walls. Some of these cells are

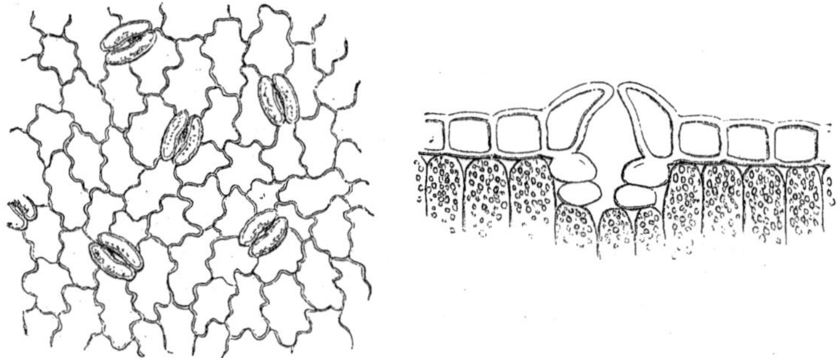

Fig. 124.—Structure of the Epidermis of a Leaf.

Fig. 125.—Vertical section through one of the stomata of a *Cycas* magnified.

sometimes observed to be elongated, ramified, partitioned off, as it were, so as to constitute hairs or down of various forms.

The epidermal membrane is not continuous or perfectly close. It presents, on the contrary, from space to space, small openings formed by the separation of two cells. These openings being elastic, and capable of expansion or compression, according to exterior circumstances, are intended to exhale the gaseous and vapoury products of perspiration in the plant, and also permit the entrance into the interior of the leaf of the gaseous constituents and moisture of the atmosphere. They bear the name of *stomata*, from the Greek word στόμα, "mouth." A vertical section through one of the stomata of *Cycas* as they appear under the microscope is seen in Fig. 125.

The stomata are most abundant on the lower surface of leaves. Their number varies much according to the plant; and the smaller they are, the more numerous. In the Pink they present 40,000 in the space of a square inch; the Iris 12,000, and the Lilac 200,000. The epidermis which covers and protects the parenchyma of the leaf is itself covered with an extremely delicate pellucid membrane, whose structure is almost inappreciable, the discovery of which we owe to M. Adrien Brongniart. It is termed the *cuticle.* It adheres closely to the epidermis, moulds itself exactly on this membrane, and even upon its hairs, to which it forms a sheath, much as the glove does to the finger, covering the epidermis, but presenting minute openings of great delicacy, corresponding with the *stomata.* It is probably, however, not a distinct structure but only the confluent portions of the external walls of the epidermal cells.

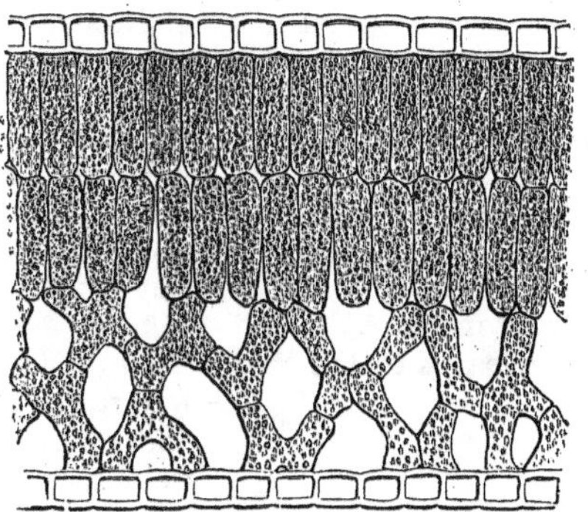

Fig. 126.—Transverse Section of a Leaf, showing the structure of the Parenchyma.

In the parenchyma of the leaves of the greater number of vegetables, two distinct regions, an upper and a lower, may be observed (Fig. 126). In the upper regions, one, two, or three rows of oblong cells may be traced perpendicular to the surface of the leaf, and closely pressing one against the other, spread out sometimes in such a manner as to leave between many of them air passages, which are generally found to correspond with the stomata. The lower layer is composed of irregular cells, often branching out and touching each other only at the extreme points of their branches, and leaving numerous air passages, which communicate one with another, forming a sort of spongy tissue. Among these cells, many are situated immediately upon the epidermis of the lower surface, which is furnished with a much greater number of stomata than the epidermis of the upper surface. It is with the stomata that the air passages correspond.

These parenchymatous cells, whose walls are always very thin, contain globules of the green substance to which young plants and leaves owe their colour. This *chlorophyll* is found in great quantities, and is much denser in the upper zone than in the lower and more spongy parenchyma. The aggregate of cells thus coloured by the chloropyhyll gives to the vegetable leaf that green and uniform tint which is its characteristic.

We have noted these peculiarities of organisation, and above all the existence of these intercellular passages—cavities which communicate freely between one another, and which are put in connection with the ambient medium by means of the stomatal openings, so admirably organised for carrying out the vital phenomena of which the passages are the seat, phenomena which we shall study more minutely by-and-by.

We find in the veins of leaves, vessels of different kinds bound together by cells of various forms. The structure of the vein becomes more simplified in proportion as it is divided and diminished in its dimensions.

If from plants with aërial leaves we pass to plants with leaves which float upon the water, or whose natural position it is to be submerged, we shall see the structure of these organs modifying themselves according to the medium in which Nature has placed them. These curious modifications have been carefully studied in our own day by Brongniart.

The leaves of the Water Lily (*Nymphœa*), which float upon the water, present, it is true, both epidermis and parenchyma, differing very slightly from those with aërial leaves, but the lower epidermis in contact with the water has no stomates. The submerged leaves of the *Pond Weed* are generally very thin, and quite destitute of any epidermis, and consequently of stomates also. They are channelled with air-cells, which have no communication with each other, but are formed of polyhedric or many-sided cells, compressed and gorged with green matter. These air-cells have no analogy with those of aërial leaves; they can only be considered as reservoirs of air furnished by the plant itself, and calculated, no doubt, to reduce its weight. They are floating apparatus, which seem to play a part analogous to the swimming-bladder of fish.

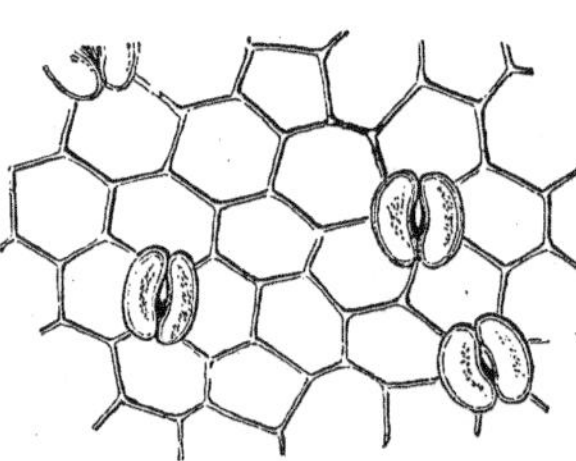

Fig. 127.—Epidermis of upper surface of Leaf of *Ranunculus aquatilis.*

The *Ranunculus aquatilis* presents at once, as we have seen, the aërial leaves which float on the surface of the water, and the much-divided leaves which are submerged. The aërial leaves, furnished with an epidermis provided with stomata, present a parenchyma whose structure scarcely differs from that of the aërial leaves already described. Aquatic leaves have no epidermis properly so called, but only greenish parenchymatous cells pressing one against the other (Fig. 128), constituting thus a parenchyma uniformly dense, hollowed here and there with isolated air-containing cavities.

Fig. 128.—Young leaves and stipules of Tulip-tree.

We cannot conclude our study of leaves without saying a few words upon the *stipules*. These stipules are organs of only secondary importance, which accompany the leaves in many plants. They are organised like leaves, but they are not true leaves. At an early period in the growth of certain plants, two small tumours appear, one at either side of the base of the leaf-stalk. They appear earlier than the leaf, at whose base they are developed, and at some period of their existence they almost invariably enclose the leaf, being of quicker growth than the leaves. But their form and functions are altogether different, being small foliaceous organs with membranous appendages, whose points of insertion are very various. In the Tulip-tree (Fig. 129) we observe two *stipules*, placed the one on the right and the other on the left of the point of insertion

Fig. 129.—Leaf of the Rose.

of the leaf-stalk. In the Rose (Fig. 129) the two stipules are

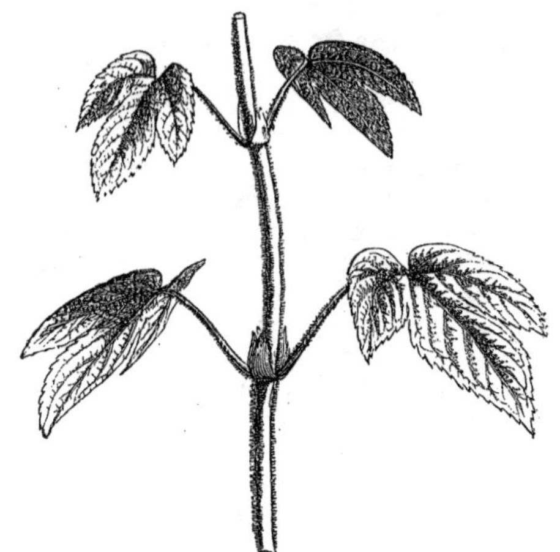

Fig. 130.—The Hop-plant.

Fig. 131.—Buckwheat (*Polygonum Fagopyrum*).

attached to the petiole of the leaf. In the Hop-plant (Fig. 130) the two stipules on the same side of the stem, and belonging to two different leaves, are so combined together to appear to form single stipules. In the Buckwheat (Fig. 131) we see a stipule placed between the leaf and the stem.

The stipule appears, as we have said, usually before the leaf which accompanies it. But the stipules often grow much more rapidly than the leaf, and in the bud they often completely cover these organs. They are probably intended in this case to shelter the young leaves. They are also very fugitive. Sometimes the stipules have neither this rapid development nor this fugitive character. They are then called *persistent.* It is probable that in their persistent character they are useful to the plant, either in sheltering its more delicate organs, or in taking the place of the leaves when they are prematurely shed.

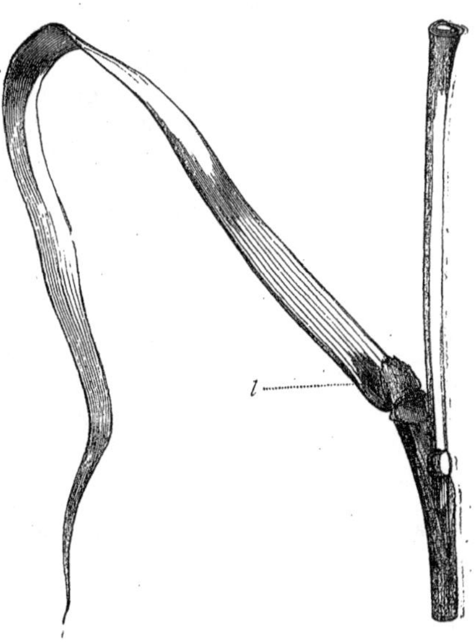

Fig. 132.—Ligula of Millet.

In the Grasses (*Gramineæ*) there is a little membrane which appears to be a continuation of the inner lining of the petiole or sheath beyond the origin of the lamina of the leaf. It is called the *ligule.* As an example of the ligule we give a figure from a plant furnished with a ligulate leaf, (*Milium multiflorum*), Fig. 132. Here the leaf-stalk or petiole, which is sheathed as in all the Grasses, has its origin in one of the nodes of the stem; while at *l*, the apex of the sheathing petiole, there is a small membrane; this is the ligule.

PHENOMENA OF THE LIFE OF PLANTS.

EXHALATION, RESPIRATION, CIRCULATION.

HAVING acquired some knowledge of the external arrangements and of the internal structure of the roots, stem, and leaves, we may now consider the essential phenomena of the life of plants (constituting vegetable physiology), in some of its most important points.

Vegetation presents, in exhalation, respiration, and circulation of the sap in the interior of the tissues, three functions only secondary in their importance to reproduction.

EXHALATION.

Exhalation in plants is performed by the leaves and branches. Plants exhale water or vapour by their leaves; it is retarded by the presence on the surface of the leaves of a coating of wax, which gives them their greyish-blue glossy appearance, and it varies according as the epidermis is thick or thin.

The medium in which the plant is placed also greatly influences the function of exhalation. If the air is very dry, the exhalation is abundant and rapid. It is less active where the air is charged with humidity; it increases as the temperature rises; it is diminished during the night.

It is not alone by the stomata of the leaves, but also through means of the epidermal membrane itself, that the exhalation takes place. The result of the perfect equilibrium which exists between the absorption of the roots and the foliaceous exhalation is proof of a normal state of healthy vigour in the plant. If the exhalation exceeds the absorbing powers of the roots the plant must fade.

It has long been believed that, at the same time that leaves transpire, they also reciprocate and absorb water by their own surface. Numerous experiments, however, which have been made of late appear to prove that leaves do not absorb water either in its vaporous condition or in the form of dew or rain to any appreciable extent.

Respiration.

If we place an entire plant or a leafy branch in a closed glass vessel filled with air which remains unrenewed, and leave the whole in darkness for some ten or fifteen hours, we may assure ourselves at the expiration of this time that the air contained in the vessel is no longer of the same composition as before the experiment. Carbonic acid will be there in greater quantity, and the proportion of oxygen will be less. But if in place of leaving the plant in darkness we expose the apparatus to the influence of the sun's rays, the phenomena will be reversed; after a few hours the air in the receiver will have lost a noticeable quantity of its carbonic acid, and will be enriched by a greater proportion of oxygen.

In order to test these phenomena more closely, let us fill a bell glass with water, through which has been previously passed, so as to ensure its dissolving in it, some carbonic acid gas, and place in it a branch or an entire plant covered with leaves; expose the whole to the rays of the sun for some hours, as is represented in Fig. 133. The air which collects (Fig. 134), if analysed after the experiment, will be found to contain scarcely any carbonic acid, but it will contain a large proportion of oxygen. If a branch of a plant, with the roots fixed in the soil, and consequently in its normal state of vegetation, is placed in a glass vessel, and by means of an aspirator a limited quantity of air is caused to circulate round it, this air, which, before the experiment, contained from four to five ten-thousandth parts of carbonic acid, after the apparatus has been exposed to the influence of the sun's rays for a certain time may not be found to contain more than from one to two. If, on the contrary, the experiment is made during the night, it will be found that the quantity of carbonic acid would be increased, and at the expiration of a certain time may have risen to eight ten-thousandth parts. These experiments, in which there is an interchange of gas between the plant and the atmosphere, exhibit the double phenomena of absorption and exhalation in plants; in fact, there is *respiration.* But the respiration of plants is not always the same, like that of animals, in which carbonic acid gas, water, and vapour are exhaled without cessation either by day or night. Plants possess two modes of respiration: one diurnal, in which the leaves absorb the carbonic acid of the air, decompose this gas, and return to the air the oxygen whilst the carbon remains in their tissues; the other nocturnal, and of a reverse kind, in which the plant absorbs the oxygen and exhales the carbonic acid; that is to say, they breathe much in the same manner as animals do. The carbon which is gained from the air by plants during the day is indispensable to the perfect development of their organs and the consolidation of their tissues. By respiration plants live and grow.

It is necessary to remark here that it is only the green parts of vegetables which respire in the manner described; that is to say, by

Fig. 133.—Respiration of Plants exposed to Light. Commencement of the Experiment. Fig. 134.—Respiration of Plants exposed to Light. Result of the Experiment.

absorbing carbonic acid and disengaging oxygen under the influence of light. The parts not coloured green, such as flowers, fruit, red and yellow leaves, &c., always respire in one and the same manner: whether exposed to light or left in darkness, they always absorb the oxygen and disengage carbonic acid. They respire in the same manner as animals. If we consider that the green parts of the plant are far more numerous than those which are otherwise coloured—that the clear light nights of hot countries may rather be said to diminish than to interrupt their respiration—that the season of long days in northern countries is that of the greatest vegetative activity—we shall be led to the conclusion that the great mass of plants are far more active in light than in darkness, and consequently that their diurnal respiration greatly preponderates over their nocturnal. The diurnal respiration of plants, which pours into the air considerable quantities of oxygen gas, happily compensates for the effects of animal respiration, which produces carbonic acid gas, injurious to the life of man. Plants purify the air injured by the respiration of men and animals. If animals transform the oxygen of the air into carbonic acid, plants convert this carbonic acid back again into oxygen by their diurnal respiration. They fix the carbon in their tissues, and return oxygen to the air in reparation.

Such is the admirable equilibrium which exists between animals and plants; such the beneficial communication which assures to the air its constant soundness, and maintains it in that state of purity which is indispensable to support the life of the living creatures which cover the globe.

We have now been speaking of the respiration of aërial plants. Water plants cannot respire by the same organic mechanism. In the former the air circulating across the intercellular channels of the leaves acts directly upon the contents of the cells. The leaves of aquatic plants, which are destitute of epidermis, and are in general very slight, borrow air which the water holds in solution, in such a manner that, according to the ingenious remark of M. Brongniart, they respire in a manner analogous to that presented by fishes and other animals which breathe by gills.

Plants may live for a time at the expense of their stored-up supplies of food, in perpetual darkness, and consequently are, in that case, always subject to nocturnal respiration. In this anomalous condition they lose some of their carbon, which passes into the state of carbonic acid, and they exhale large quantities of water. The result is a decided elongation of the plant, a greater softness in its tissues, and the absence of green colouring.

The juices of plants kept constantly in the dark modify themselves in a sensible degree. Often acrid and bitter in their normal condition, they are rendered sweet and succulent under such modifications, and market gardeners turn these facts largely to their profit. They treat Lettuce plants in order to make the hearts white, by binding them closely together and tying the leaves one against the other, to the exclusion of light. Sea Kale, in its wild state a useless weed on the sea-shore, becomes a delicate vegetable when blanched under the gardener's care.

CIRCULATION.

The manner in which nourishing juices circulate in the interior of plants has long been the subject of discussion amongst botanists, and science is still far from being agreed on this important point of vegetable physiology. Limiting ourselves, however, to the consideration of dicotyledonous vegetation, such as our indigenous forest trees, we may enunciate the following simple facts on which all botanists are agreed.

From the moment when the water which impregnates the earth has penetrated into the roots of a plant, and mingled with the juices which are contained in the cells of the vegetable, it constitutes what botanists call the *sap*—a complex fluid, which, at certain periods in the life of a plant, circulates constantly through its tissues. Inasmuch as it has now been ascertained with tolerable certainty that plants gain their *liquid* supplies of food from their roots alone, and since also there is a constant loss of water by exhalation from the foliage, it follows that there must be a continual upward stream of fluid through the plant. This is what is meant by the *ascending* sap. Besides water, its main ingredient, it contains various matters taken up in solution from the soil, and also other matters, which are washed, as it were, out of the parts of the plant through which it passes. What is the force which causes the water to penetrate into the roots, impels the sap into the stem, and finally to its last ramification—namely, the leaves? What is the route which the sap follows in ascending? Does it traverse the pith, the bark, or the wood? or, finally, does it permeate through all three at the same time?

When a tree is cut down in spring it is easily seen that the sap flowing in it is then in the wood. If a plant is made to absorb coloured liquid, or if the branches are plunged into the same liquid, it is easily seen that it does not rise first either in the bark or pith. It is the wood or ligneous body through which it manifestly takes its passage. This passage is effected through all the ligneous elements, cells, fibres, and vessels. The anatomical construction of these vessels, their large number, their strength in the prostrate filiform and slender stems, which often attain a very considerable length, and which require to be traversed by a large quantity of sap in order to supply what is necessary for evaporation by the leaves—all these general facts leave no doubt as to the part which the vessels play in the circulation of the sap.

Dr. Hales, an English clergyman, to whom science is indebted for numerous experiments throwing light upon the movement of the nourishing juices in plants, was anxious to discover the force with which the sap rose to the stem. In order to ascertain this, he fastened a bent tube on the top of an ascending branch of a vine-stem in the spring; having carefully fixed the tube upon the transverse section of the vine-stock, the lower bend being filled with mercury. The flowing sap accumulated in the interior branches began little by little to move the mercury, depressing it in one tube and consequently raising it in the other, till the difference which indicated the pressure exerted by the sap rose to thirty-eight inches (Fig. 135). The flow of sap then was sufficiently powerful to move a column of mercury, in addition to the weight of the atmosphere, this height. Hales calculated from this that the force which impels the sap in the vine is five times as great as that which impels the blood through the large arteries of the horse. Having reached the leaves, the sap becomes exposed to the air by the innumerable openings or stomata, which communicates with the hollow channels in the substance of the parenchyma. The respiration of the plant—that is to say, the chemical action which the air exercises upon the liquids which supply the leaves—together with the exhalation of vapour which proceeds from the same organs, modifies the ascending sap by methods of precisely the same kind, only with exactly opposite results, as those by which air modifies venous blood in the lungs of animals, changing it into arterial blood. Thus it is with the leaves in the phenomena of exhalation and respiration, of which they are the seat. In ascending, the sap changes its nature, and becomes transformed into a nourishing fluid.

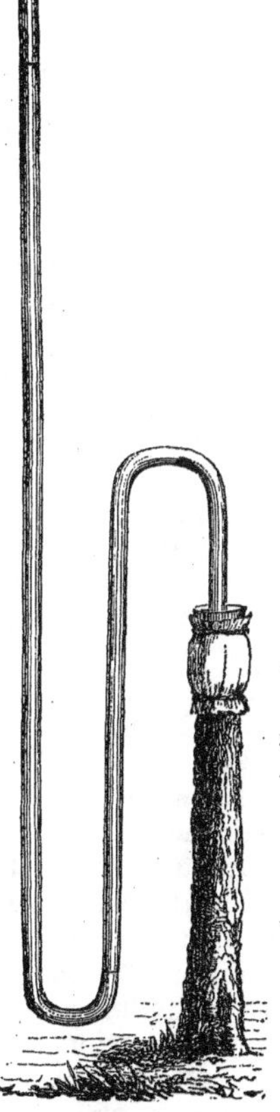

Fig. 135.
Hales' apparatus for measuring the force of ascending sap.

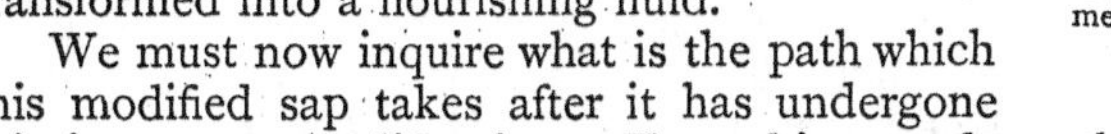

We must now inquire what is the path which this modified sap takes after it has undergone this important modification. Everything tends to the belief that it circulates in the general tissues of the plant, its movements being directed by the demand which is created for it at any point where active growth is going on, or where food is being stored up for future purposes. If a stem or branch is strongly bound in such a manner as to compress the bark, a cushion or excrescence is formed above the ligature, which continues to increase, and appears to proceed from the partial stoppage of the downward distribution, showing that the nutritious matters come from the leaves, for the parts beneath the ligature show much less increase. The same phenomenon is produced upon the trunk of the tree when annular or spiral incisions are made all around it. Again, the trunks of trees round which climbing or twining plants wind themselves, demonstrate the same physiological fact in a very peculiar manner. But the downward movement of elaborated matters does not, as was once generally supposed, take place in the bark exclusively, but goes on also through the wood. This is proved by the fact that cutting through the bark right round the stem does not by any means entirely arrest the growth of new wood *below* the incision, nor is the distribution of elaborated matters exclusively downwards; it takes place in any direction where a demand for them has been set up.

In speaking of the structure of roots, we have enumerated the causes under the influences of which the sap ascends; the causes which determine its descent are, we must acknowledge, imperfectly known to us. It appears most probable that it is through the sieve-cells in the deep layers of the bark, the admirable structure of which we have before mentioned, that the mucilaginous and albuminous sap takes its way. Cutting down through the bark arrests the course of these matters, but the starch and sugar appear to take their way through the wood. The movement of sap is in its most active state in spring-time, when the plant is gorged with nutritive matter preserved in deposit during the winter. It is then full of liquid, and in some plants the juices flow from the slightest incision. In spring, according to the poetical expression consecrated by use, the vine and other plants bleed; but when the leaves are fully developed, the active evaporation which takes place on their surface impels the liquid to the extremities of the vegetable, whence it exhales in vapour; they will no longer bleed when wounded.

When the branches develop themselves and consolidate, the movement of the sap becomes slower; it is sometimes roused towards the end of summer, when, the spring having been premature, the materials which the plant has elaborated for the vegetation of the following year have been set to work before their time. After the fall of the leaf, and when the approach of winter lowers the temperature, the movement of the sap is stopped entirely; the tree arrives by little and little at a state of almost absolute repose; this is not death, but life, which awaits its re-awakening.

It thus appears that all plants in a healthy state must be so situated as to be able to absorb from the air and soil surrounding them the elementary bodies which constitute their tissues, namely, carbon, oxygen, hydrogen, and nitrogen. The principal form in which these bodies are absorbed is as water, carbonic acid, and ammonia.

Water is supplied to plants in the form of rain and dew; these moisten the soil, and from the soil the plants take up water by their roots.

Carbonic acid, which supplies plants with the carbon of their tissues, is supplied to the air whence it is absorbed by the leaves, by the respiration of man and animals, by combustion, and by the decomposition of saccharine matter, &c., volcanic action, and hot springs.

The sources of ammonia are in the main the decomposition of animal matters, which are far richer in nitrogen than vegetable.

The special constituents of sap in plants are the salts of the metals, such as those of potassium, sodium, calcium, and magnesium. These substances permeate through the tissues of plants dissolved in the water taken up by the roots from the soil.

ON THE GROWTH OF VEGETABLES.

Vegetables increase by means of the nourishing materials elaborated by the leaves and distributed in the sap. It would be entering upon a long and difficult task to take into consideration all the points which botanists have made out concerning the growth of vegetables. We must, therefore, limit ourselves here to the consideration of what actually comes under our own knowledge in the growth of trees. Trees grow in length by the development of their terminal shoots. We observe the intervals between the points of the insertion of the leaf, which at first are very short, gradually become longer as the shoot upon which they are attached expands after emerging from the bud.

But how do trees increase in circumference? This demands a closer examination.

If we study the internal structure of a branch of one year old in one of our forest trees, we shall find that towards the end of the year the branch is arranged in the following manner:—There is a pith, a fibro-vascular circle, furnished with spiral vessels on its internal face, with straight medullary rays traversing the mass of wood, losing themselves in the bark, which is composed of an epidermis, of the herbaceous or green layer envelope, and of bast. But between the ligneous and cortical system we can distinguish the presence of a special zone formed of very delicate cells, with walls soft and transparent, which in spring-time are bathed in liquid supplied abundantly by the sap, to which botanists give the name of *cambium*. This last layer is of the utmost importance, for we can perceive that in the course of the second year, by the process of vegetation, this intermediate zone, which is situated between the cortical and ligneous system, becomes the seat of a double formation, the one cortical, the other ligneous. Fig. 136, which we have given to show the different elements of the stem of a tree, serves also to show its mode of development. That part included in the lines marked 1 and 3, represents the wood and bark of the first year, that marked 2 the wood and bark of the second year. The *cambium* layer separates these elements, and is indicated by the letter *c*. Cells, fibres, and vessels result from the transformation of the delicate elements of this generating tissue. In other words, at this point the transformation of *cambium*, into wood from without inwards, and into bark from within outwards, takes place. The medullary rays are continued without interruption or modification across the new layers, new ones also forming themselves without being in connection with the pith. What passes in the second year is repeated during the third and fourth. A very useful consequence results from

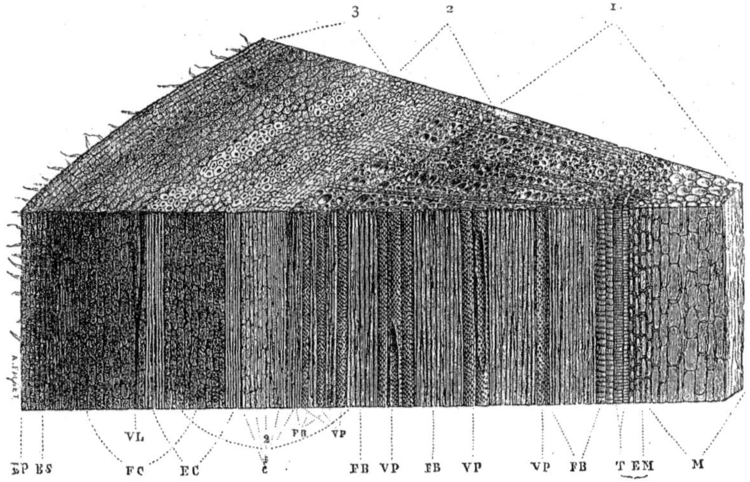

Fig 136.—Horizontal and transverse section of the trunk of an Elm.

this. We are enabled by observing certain characters to distinguish the successive layers which form themselves year by year. The age of a tree is thus, so to speak, inscribed upon its cut surface. In the oak, for example, it is very easy to distinguish the annual layers. The transformation of the generating zone in the wood continues from spring to autumn, and consequently under climatic influences very different. The great vessels are formed at the season when the sap is most active in its circulation; then come vessels of smaller calibre and much more numerous. Towards the end of the year, when vegetation is less active, principally ligneous fibres are formed. It follows from these differences between the wood of spring and of the autumn, that it is very easy to distinguish the passage from the former to the latter, and that the different annual formations appear consequently as so many concentric zones on the horizontal section of the tree. If the annual growth of a tree were equal during all the time of its life, there would be no line of demarcation between each annual period, and consequently no concentric zones. It is thus with trees in some hot countries, where vegetation is neither checked nor accelerated by change of temperature. The age of a tree is not there, as in temperate countries, to be found inscribed in the interior of its stem.

We must, however, offer two important remarks. The new wood alone participates in the nature of the exterior parts of the ligneous circle; the medullary sheath is not renewed. On the other hand, new cells are constantly formed in the middle portion of the cambium layer, in default of which the growth would cease as soon as all the cells of this zone were transformed into the elements of new wood and bark.

Fig. 137.—Flower of *Rafflesia Arnoldi.*

THE FLOWER.

We have admired the roots, with their innumerable fibrous tufts, which by a marvellous faculty, scarcely explicable to us, imbibe the liquids contained in the earth, and convey the nourishing fluid into the cells of plants; the stems and branches which support the plant in mid air; the leaves, organs at once of respiration and evaporation; the vessels, so variable in form; the breathing pores (*stomata*); in short, all the appliances, all the living mechanism, by means of which the vegetative functions are carried on; all tending to the production of flowers, which, in their turn, live only for the production of fruit; while the fruit itself only exist for the purpose of developing the seed, that ultimate end and essential design of vegetation; for Nature seems to concentrate all her efforts with a view to the reproduction of the individual, and consequent preservation of the species.

But what is a flower? What definition shall we give of a flower, that can pretend to exactness, and at the same time be framed in scientific terms? A rigorous definition of the flower is more difficult than one would think. It was a saying of Linnæus that minerals grew, plants grew and lived, animals grew, lived, and felt; but this is now known to be altogether incorrect, for certain plants not only grow and live, but give every indication of feeling: witness the *Mimosa*, or Sensitive plant, which closes its leaves on being touched, and some of the most obscure plants among the *Confervæ*, which move about by the action of their own *cilia*, or mobile hairs, until they have found a resting-place for themselves. He would, in short, be a rash man who should now attempt such a definition of plant or mineral. This, however, was not the opinion of Jean Jacques Rousseau, the celebrated philosopher of Geneva, who was indebted to the study and cultivation of botany for some of the happiest hours of his life. He has left, in his "Letters on Botany," a book full of interest and sound science, a passage in which he thus expresses himself on the definition which can be given of a flower:—

"If I resigned my imagination to the pleasing sensations which this word seems to call forth, I should write a paper agreeable, perhaps, to sentimental shepherds, but very unsatisfactory to botanists. Let us put aside for a moment the vivid colours, sweet odours, and graceful forms of flowers, and try, in the first place, to understand the organised being which unites these attributes. Nothing at first sight appears easier. Who thinks he requires to be taught what a flower is? 'When no one asks me what is time,' said St. Augustine, 'I know it very well; but I do not know it when I am asked.' One might say as much of a flower, perhaps of its beauty even, which is the prey of time. I am presented with a flower, and I am told, 'Here is a flower.' This is showing it to me, I confess, but not defining it; nor will this inspection enable me to decide, in the case of any other plant, whether what I see is, or is not, a flower; for there are multitudes of vegetables which have in none of their parts the colour which Ray and Tournefort have introduced into their definitions of a flower; and yet these vegetables bear flowers not less real than those of the rose-tree, although much less apparent."

Nevertheless, although the definition of the flower appeared to Rousseau so surrounded with difficulties, he does not hesitate to propound the following:—

"The flower," he says, "is a local and temporary part of the plant, which precedes the fecundation of the germ, in or by means of which it is effected." This was an unexceptionable definition, which was scarcely modified a century later by Moquin-Tandon, when he said, "A flower is that temporary apparatus, more or less complicated, by means of which fecundation is effected."

The flower, then, is an apparatus composed of two envelopes, the *calyx* and the *corolla*, and the *essential organs* proper to ensure the reproduction of the plant, namely, the *pistil*, in which the seeds are afterwards enclosed, and the *stamens*, destined to fecundate the pistil. The position of the organs will be better understood by examining the following diagrams. Fig. 138 is a portion of the

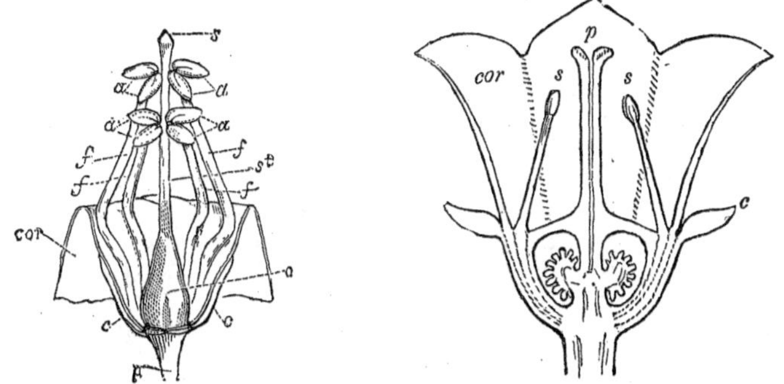

Fig. 138.—Section of a Flower of the Foxglove. Fig. 139.—Section of a Flower of *Campanula*.

flower of the Foxglove (*Digitalis purpurea*), showing the stamens *f a* united with the corolla, *cor*; *p*, the peduncle, or floral axis; *c*, the calyx; *o*, *st*, *s*, the pistil. Fig. 139 is a *Campanula* showing how the calyx *c*, the corolla *cor*, the stamens *s*, and the pistil *p*, are consolidated throughout their basal portions.

The calyx, corolla, stamens, and pistil are inserted on an axis called the *receptacle*, the form of which varies according to the plant. In the Strawberry, for instance, the receptacle is the succulent fleshy part which is eaten under the name of fruit; the real fruit, in fact, being the little yellow specks which cover it. It is also the core of the Raspberry.

The flowers of every plant present only these five sets of organs. There are some which have no stamens, and others without pistil. In these two cases the flowers are said to be *unisexual*. In the first they are called *female* flowers, in the second, *male*; but both are always present, either on the same plant or on distinct plants. The Box-tree has unisexual flowers, some furnished with stamens without a pistil, others with a pistil without stamens. Other flowers are without a corolla, and some even without both calyx and corolla. The former are *incomplete* (*monochlamydeous*), the latter are called *naked* (*achlamydeous*). The Marsh Marigold (*Caltha palustris*), which expands its magnificent golden flowers in spring along the marshy borders of rivers, ponds, or lakes, is without a corolla; the flower of the common Ash has neither corolla nor calyx, but reproductive organs only. Another species from Southern Europe (*Fraxinus Ornus*) gains its name of Flowering Ash from the presence of the calyx and corolla in the flowers.

In short, some flowers have neither calyx, corolla, nor stamens; others, neither calyx, corolla, nor pistil. They are at once *incomplete* and *naked.* The flowers of the Willow are of this sort; some are possessed of two stamens, and others of a pistil only.

A flower provided with stamens and pistil is said to be *hermaphrodite*, whether it has floral envelopes or not. There are a great many plants which bear hermaphrodite flowers only; there are others bearing, on the same individual, male, female, and hermaphrodite flowers; these are *polygamous* plants; the Ash is an example. Some plants present only male, others, again, only female flowers, and these sometimes occur on the same, and in other cases on different plants. In the former case, as in the Chestnut, Hazel-nut, and *Ricinus*, to which the Castor-oil plant belongs, the plants are said to be *monœcious.* In the latter, as in the Hemp-plant, Date-tree, and *Mercurialis* (one of the *Euphorbiaceæ*), the plants are *diœcious.*

Plants present themselves with flowers varying as much in their dimensions as in their structure. There are flowers only the thousandth part of a foot in diameter, and some which are celebrated for their immense bulk. We find in Sumatra and the Sunda Islands a parasitical plant consisting of little more than a flower; but is nearly nine feet in circumference: it is the *Rafflesia Arnoldi* (Fig. 131). The calyx of some of the *Aristolochiaceæ*, on the banks of the Rio Magdalena, is so voluminous, that the children of the inhabitants sportively use it for a cap. The flower of *Victoria regia*, represented in PLATE III., is about forty inches in circumference. The effect produced upon Sir Robert Schomburgk, when he first saw this magnificent plant on the River Berbice, is thus described:—

"It was on the 1st of January, while contending with the difficulties Nature opposed in different forms to our progress up the River Berbice, that we arrived at a point where the river expanded,

III.—*Victoria regia*, on a river of Guiana.

and formed a currentless basin. Some object on the southern extremity of this basin attracted my attention; it was impossible to form any idea what it could be; and, animating the crew to increase the rate of their paddling, we were shortly afterwards opposite the object that had raised my curiosity—a vegetable wonder. All calamities were forgotten; I felt as a botanist, and felt myself rewarded: a gigantic leaf, from five to six feet in diameter, salver-shaped, with a broad brim, of a light green above and a vivid crimson below, rested on the water. Quite in character with the wonderful leaf was the luxuriant flower, consisting of many hundred petals, passing in alternate tints from pure white to rose and pink. The smooth water was covered with the blossoms, and, as I rowed from one to the other, I always observed something new to admire."

The leaves are of an orbicular form, the upper surface is bright green, and they are furnished with a rim round the margin from three to five inches in height; on the inside the rim has a green colour, and on the outside, like the under surface of the leaf, it is of a bright crimson; they have prominent ribs, which project an inch high, radiating from a common centre; these are crossed by a membrane, giving the whole the appearance of a spider's web; the whole leaf is beset with prickles. The stem of the flower is an inch thick, and studded with prickles; the calyx is four-leaved; each sepal is seven inches in length and four inches broad; the corolla covers the calyx with hundreds of petals; those first expanded are of a white colour, but the innermost are of a beautiful rose-pink, which eventually suffuses the whole flower; the flower is very fragrant. Like all other water lilies, its petals and stamens pass into each other, a petal often being found surmounted with an anther. The seeds are numerous, and embedded in a spongy substance. This plant has by some botanists been placed in the genus *Euryale*, whilst Lindley thinks it is nearer *Nymphœa*, from which it differs in the sepals and petals being distinct, and the stigmas being prolonged into horn-like bodies. This splendid plant has now been successfully cultivated in many of the hot-houses of this country. Beautiful specimens are to be seen in the Royal Gardens at Kew, and at the Crystal Palace, Sydenham, at Chatsworth, Sion House, and elsewhere.

The dimensions of flowers are by no means in proportion to the plant which produces them. The flowers of the greater part of our forest trees are very inconspicuous, and little valued except by the botanist, being generally so small as to escape the ordinary observer; to study some of them, indeed, a lens must be employed. On the other hand, smaller plants often produce magnificent flowers; witness the Daisy and the floral ornaments which decorate our meadows, woods, and gardens, often dazzling us by the elegance of their forms, and brilliancy of their colours.

It is on the corolla especially that Nature has expended all the riches of her inexhaustible palette. The corolla is also peculiarly the seat of the sweetest perfumes of the vegetable world.

Plants with sweet-smelling flowers are believed to be more common in dry than in moist countries. On the burnt-up and naked hills of Southern France, the Thyme, Sage, and Lavender perfume the air with their aromatic scents; whilst the moist plains of Normandy exhale no vegetable aroma.

Before a flower blows, the different parts constituting it are packed closely together and compressed one against the other; they then form a *flower-bud.* The buds of all *annual* plants—that is, plants germinating, growing, flowering, and dying, all in the same year—continue to develop themselves up to the time of their full bloom. The flower-buds of certain ligneous plants, as the Lime-tree, also act in the same way. But there are other plants, as the Almond-tree, the Plum-tree, the Pear-tree, &c., in which the flower-buds appear during the summer, and increase in size up to the time of autumn. They remain stationary during winter, and come into bloom the following spring with the first rays of the warm sun. These flower-buds are *scaly* (*squamose*), that is, shut up in scale-like coverings, and bear the name of *hibernacula.* The buds which spring and are developed during the warm season are *naked*, because destitute of these protections.

The flower-bud at last opens, blows, and passes into the state of a flower. This blooming does not take place at all times of the day indifferently. Linnæus has drawn up a list of plants arranged according to the hour at which their flowers expand. He called this list the *Floral Clock.* De Candolle has also noted the times at which the following flowers blow at Paris :—

Between 3 and 4 A.M.	Bindweed.
At 5 A.M.	Naked-stalked Poppy and most of the Chichoraceæ.
Between 5 and 6 A.M.	Nipple-wort and *Convolvulus tricolor*.
At 6 A.M.	Many of the Solanaceæ (Nightshade family).
Between 6 and 7 A.M.	Sow Thistle and Hawkweed.
At 7 A.M.	Water Lilies and Lettuces.
At 7 to 8 A.M. . . .	Venus's Looking-glass.
At 8 A.M.	Wild Pimpernel
At 9 A.M.	Wild Marigold.
At 9 to 10 A.M. . .	Ice Plant.
At 11 A.M.	Purslain, Star of Bethlehem.

At 12	Most of the Ficoid, or *Mesembryanthemum* family.
At 2 P.M.	*Scilla pomeridiana.*
Between 5 and 6 P.M.	*Silene noctiflora.*
Between 6 and 7 P.M.	Marvel of Peru.
Between 7 and 8 P.M.	Night-blooming Cereus (*Cereus grandiflorus*).
At 10 P.M.	Purple Convolvulus.

Some plants remain in bloom many days in succession. There are others which are ephemeral, and, opening at a fixed time, finally close up and fall off the same day at a nearly settled hour. The Cistus and Flax plant expand their flowers about five or six o'clock in the morning, and are withered before mid-day. The *Cereus grandiflorus* blows at seven in the evening and closes about midnight.

Equinoctial flowers open and close at a fixed time in the same day; on the morrow and for several following days, they again open and shut at the same regular hours. The Star of Bethlehem opens several days in succession at eleven in the morning, and closes at three. The night-flowering *Mesembryanthemum* blows several days in succession at seven in the evening, and closes about six or seven in the morning.

"The regularity of these phenomena," says De Candolle, "has struck all observers, but although the cause evidently is referable to the action of light, it is still difficult to verify it with precision. . . . . I have subjected the Marvel of Peru to continuous lamplight, and by that means have obtained an inflorescence altogether irregular; but having also placed them where they were lighted during the night, and kept in darkness during the day, I noticed that at first their flowering was very irregular. They soon became accustomed, however, to their new circumstances, and ended by blowing in the morning, that is, at the end of the day artificially made for them, and closing in the evening as their period of obscurity approached."

Heat, however, appears to have a certain influence on the time and the duration of the blowing of flowers; and it is observed that these phenomena vary in different countries, according to their latitudes and in the same country according to the seasons. The *Floral Clock* drawn up by Linnæus at Upsal goes slower than the clock arranged by De Candolle at Paris.

In a few flowers the time of blowing is modified by the state of the atmosphere. These may be called *meteoric* plants. The Siberian Sow Thistle never closes in the evening, it is said, when it is going to rain next day. Several of the *Chichoraceæ* do not open in the morning when it is going to rain. The Cape Marigold (*Dimorphotheca annua*) is said to close its petals when the weather indicates rain, but its flowers remain open in sudden storms, which seem to take it by surprise. Facts of this kind, which are, however, neither numerous nor very trustworthy, have been used in the arrangement of a floral weather-glass.

The duration of inflorescence varies much in different species. In the Peach, Almond, and Apricot, among trees, and in the Hyacinth and Tulip among herbs, the blossom remains for a few days only. But the Winter Hellebore remains covered with flowers the whole winter, and the Shepherd's Purse (*Capsella Bursa-pastoris*) flowers from March till November.

The period at which flowering commences varies in the same way, according to species. Linnæus arranged a table of the flowering of different vegetables in the climate of Upsal, in Sweden, for the year 1755, and gave to this list the name of the *Floral Calendar*. But this calendar necessarily varies with each climate, for the date of the flowering of a plant is sooner or later, according to the latitude of the country. At Smyrna the Almond-tree blooms in the first half of February; in central France it blooms at the beginning of April; in Germany in the second half of April; at Christiania in the early days of June.

It is hardly necessary to observe here how very indispensable an exact acquaintance with the times of flowering is to those who wish to see flowers succeeding each other harmoniously and without intermission in their gardens.

Inflorescence.

The arrangement of the flowers on the plant is called its *inflorescence*, from *infloresco*, "I begin to blossom."

The various kinds of inflorescence noticed by botanists may be enumerated as follows:—

I. Centri-Petal, when the lowest flowers open first, and the main stem continues to elongate, developing fresh flowers (hence also called *Indefinite*).
  - i. *Spike*, when the flowers are sessile along an unbranched axis.
    - *a. Catkin*, a variety of the spike falling off (deciduous) after the time of flowering.
    - *b. Spadix*, a fleshy spike enclosed in a large floral leaf or *spathe*.
  - ii. *Raceme* differing from the spike in the flowers being stalked (*pedicelled*).
    - *a. Corymb*, a variety in which, the lower pedicels being longer the flowers are all brought to the same level.
    - *b. Panicle*, a term applied, amongst other cases, to racemes in which the lateral stalks are branched.

iii. *Head* or *Capitulum*, in which a number of sessile flowers are collected into a compact cluster; it may be regarded as an extremely reduced form of the spike.

iv. *Umbel* differs from the head in the flowers being stalked, and may be regarded as a reduced raceme.

II. CENTRIFUGAL, when the terminal flower opens first (hence called also *Definite*) and the lateral ones successively afterwards.

i. *Cyme* may be corymbose, paniculate, umbellate.

a. *Glomerule*, a contracted form.

Flowers may be *sessile*, in the language of botanists, when they are placed immediately on the main stem or *axis*, or they may be attached to it by a small support, called a *pedicel*. The *peduncle* is the *stalk* of a solitary flower, or of an inflorescence. The peduncle therefore is to flowers what the petiole is to leaves, that is, a means of attachment to the stem or branch. Nevertheless, the analogy between the peduncle and the petiole is limited to the external shape, for the two organs differ essentially in their special organisation.

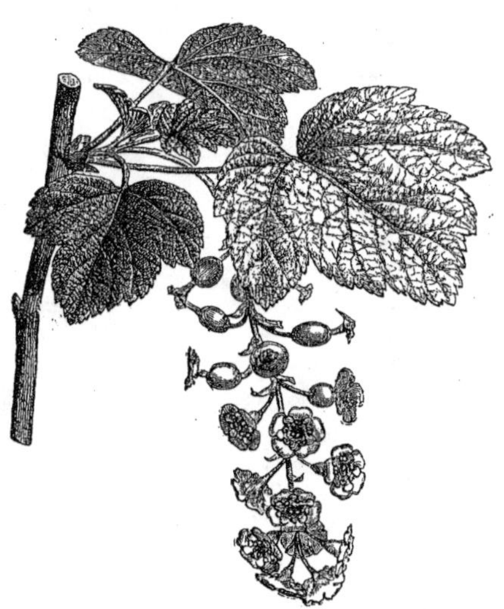

Fig. 140.—Raceme of the Red Currant (*Ribes*).

Do the flowers, whose constituent parts we shall soon consider in detail, spring at random from the stem, the branches, or the boughs supporting them? When we consider the admirable regularity and the rigorous laws by which leaves are arranged on the boughs, we are led *à priori* to confess that the distribution of flowers on their vegetable axes must obey very determinate laws also. This is, in fact, what is shown us by Nature. Her laws are generally easily recognised, and though sometimes hidden, are never violated. Flowers are always the termination of an axis, or branch, and the order governing their arrangement is only a repetition of that which regulates the ramification of the plant.

In order to study inflorescence more in detail, let us take some of the more common plants and examine the arrangement of the

Fig. 141.—Inflorescence of *Verbena officinalis*

Fig. 142.—Branching raceme of the Oat.

flowers on each of them. In the Red Currant (Fig. 140), the floral axis carries at intervals modified leaves, called *bracts*; at the axil of

Fig. 143.—Male catkin of the Willow.

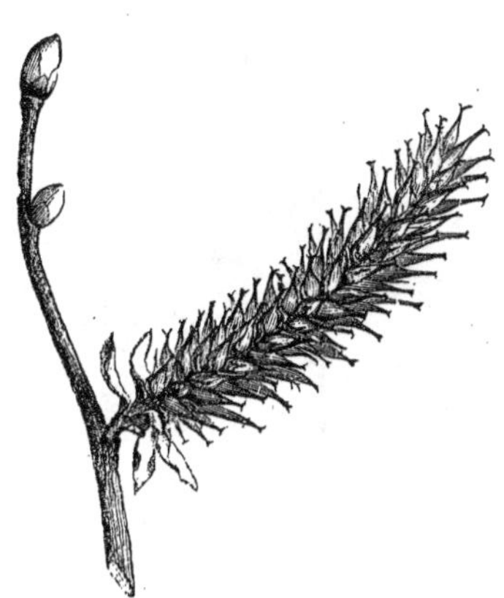

Fig. 144.—Female catkin of the Willow.

Fig. 145.—Corymb of the Cherry

Fig. 146.—Simple umbel of *Astrantia*

each of these bracts the *pedicel* takes its rise, and it is terminated by a flower. This is the type of the *raceme.*

The inflorescence of the Vervain (*Verbena officinalis*), Fig. 141

Figs. 147 and 148.—Capitulum of the Ox-eye Daisy (*Chrysanthemum leucanthemum*).

Fig. 149.—Compound corymb of the Wild Service (*Pyrus torminalis*).

is only distinguished from the above by the extreme shortness, or absence, of the pedicels ; this constitutes a *spike.*

The Oat (Fig. 142) furnishes us with an example of the modifications which the raceme experiences. Its inflorescence is called a compound raceme, or *panicle.*

The *spike* takes the name of *catkin* or *ament* (from *amentum*, "a strap,") when it is formed of unisexual flowers, as represented in the *male* and *female* catkins of the Willow (Figs. 143 and 144), and falls off after flowering.

In the cherry of Ste. Lucie (Fig. 145) the pedicels originating from the principal axis are longer at the lower part than they are towards the summit, so that the whole of the flowers form together a sort of flat-topped cluster; this cluster is called a *corymb*.

Fig 150.—Compound umbel of the Chervil.

In the *Astrantia* (Fig. 146), so called from its star-like flowers, the axis of inflorescence is very short, and bears at the extremity, which is enlarged, a certain number of secondary axes or stems, rather elongated, and of equal length, so that the flowers seem to spring from the same point in order to attain the same height. Here the group of flowers is said to be *umbelliferous*.

In the Ox-eye Daisy (Figs. 147, 148) the numerous sessile flowers are inserted on the surface of an enlarged and flattened axis, and form a *capitulum*, or *head*. The capitulum is influenced by the form of the receptacle: when the latter is spherical, the capitulum is globular; when flat, concave, or convex, the capitulum varies accordingly.

All these modes of inflorescence are only modifications of one of them, which may be taken as a type, namely, the *cluster*. They sometimes seem to attain a certain degree of complication, without having the simplicity of their plan thereby diminished.

Thus, in the Service (Fig. 149), the corymbs of flowers are themselves arranged corymbosely. In the Carrot and Chervil (Fig. 150), the umbels rising from the summit of the stem, each peduncle has an umbel on its summit, and they form a compound umbel. In the Privet and the Vine (Fig. 151), the inflorescence is formed of small racemes, arranged in larger racemes, or *panicles*. In the Wheat-plant (Fig. 152) the spikelets are grouped in spikes. There are, then, *compound corymbs*, *compound umbels*, *compound racemes*, and *compound spikes*.

Fig. 151.—Compound raceme of the Vine.

In every case hitherto mentioned, the number of flowers of the same generation is indefinite in each group. This will not be the case in the following instances.

Let us examine, for example, the method of arrangement in the flowers of the Centaury (*Erythræa centaurium*), Fig. 153. The main stem terminates with a flower, and a little below this the stem bears two opposite leaves, or *bracts*. From the axil of each of these arises a secondary branch, terminating also in two leaves and a flower. Each of these branches comports itself in the same way as the stem, that is, it gives origin to two tertiary branches, each terminating in a flower, and thus it goes on. We thus see that at each branching out the number of flowers is doubled. We see besides that the flowering in this plant goes from the centre to the circumference.

Fig. 152.—Compound spike of Wheat

Fig. 153.—Dichotomous cyme of Centaury.

This sort of inflorescence bears the name of a *cyme*, and in this particular case is said to be *dichotomous*, from δίχα and τέμνω, "by pairs I divide." In some cases three whorled bracts develop axes from their axils, when the cyme is said to be *trichotomous*, divided by threes.

The inflorescence of the Forget-me-not (*Myosotis*), and the Heliotrope, also form *cymes*. As in these plants the multiplying

Fig. 154.—Scorpioid cyme of *Myosotis*.

Fig. 155.—Spadix and spathe of *Arum maculatum*.

axis of inflorescence often tends to form a curling inflorescence, rather unfitly compared to the tail of a scorpion, the cymes of the Forget-me-not and Heliotrope are said in consequence to be *scorpioidal* (Fig. 154).

In the Horse Chestnut the principal axis bears an indefinite number of small scorpioidal cymes. This is a mixed inflorescence of the two principal forms we have already pointed out, and is a *raceme of scorpioidal cymes.*

When a succulent spike is enclosed within a large floral leaf or *spathe*, it is called a *spadix*, from σπάδιξ, "a palm branch." In this case the axis is sometimes elongated beyond the flowers in the form of a club-shaped body, as in the Cuckoo-pint (*Arum maculatum*), Fig. 155; or the axis only bears flowers at its base, which is here enclosed by the spathe.

The general name of *involucre* is given to a more or less considerable collection of bracts, disposed in *whorls* (*verticils*) in either one or several ranks, which surround and seem to protect the flowers (the *capitulum* of composite flowers); but in the natural group *Aroideæ* the involucre is monophyllous, and is called a *spathe*. Fig. 155 shows the spathe of *Arum maculatum*.

We should exhaust the attention of the reader were we to enlarge more on inflorescence—a subject which we have only glanced at, but which has been the object of considerable study on the part of botanists. We now come to the consideration of the constituent parts of the flower.

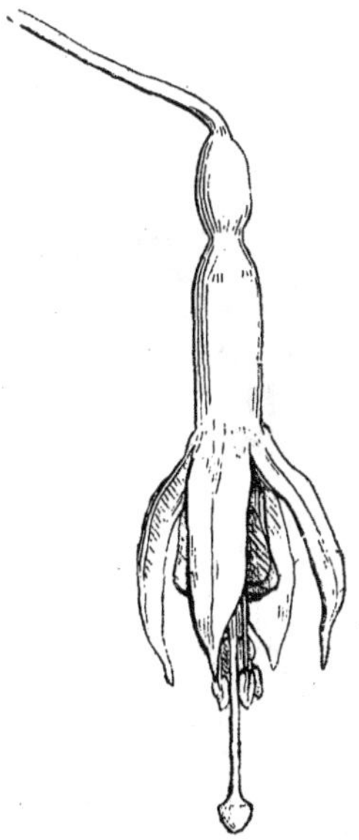

Fig. 156.—The Fuchsia.

THE CALYX.

The *calyx*, κάλυξ, "the cup which holds the flower," is the external envelope. It has neither the elegant shape nor the varied colours of the corolla, usually appearing as a little green receptacle, on which the corolla rests. On account of its appearance and colour, it is almost always confounded with the peduncle, of which it appears to be a mere prolongation—an expansion which subdivides into several lobes.

The simple form of the calyx is suited, however, to its functions. It is, in general, not elegance or delicacy, but firmness and solidity, which are able to protect or defend. As the calyx forms the external envelope of the flower, it must be constituted so as to resist action from without. It is true that the calyx of some plants, such as the Fuchsia (Fig. 156), rivals the corolla in colour and beauty, but these are exceptions to the general rule.

The calyx is, then, the external envelope of the flower, and its different parts bear the name of *sepals*.

These sepals, however, are only modified leaves. If we glance at the Camellia-bud (Fig. 157), the same structure, the same nervation, almost the same form, will be observed to belong both to the five sepals of the flower, and to the bracts accompanying them. In the Peony and Foxglove (Fig. 158) there is still a resemblance between the bracts and the sepals. When we notice the various

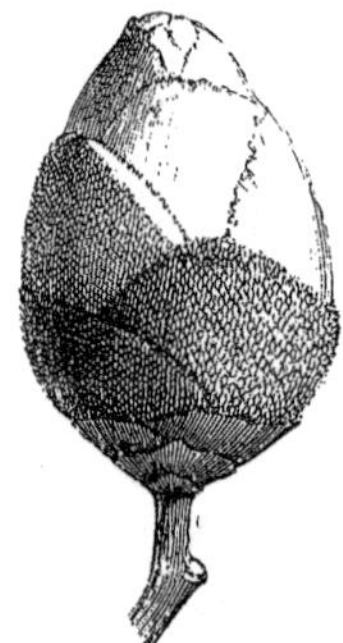

Fig. 157.—Calyx of the Camellia.

Fig. 158.—Calyx of Foxglove.

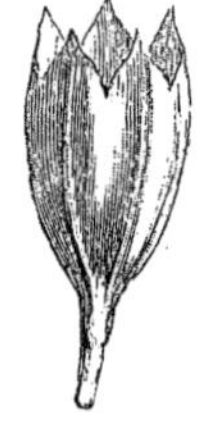

Fig. 159.—Monosepalous calyx of the Primrose.

intermediate conditions of texture, form, and size existing between bracts and leaves, we are led of necessity to consider the calyx of flowers as proceeding from a modification of the leaves.

The calyx seems sometimes to be formed all in one piece, although more or less deeply divided. In this case it is *monosepalous*; if composed of perfectly distinct parts or sepals it is *polysepalous*.

The flower of the Primrose (Fig. 159) has a monosepalous calyx; that of the Flax-plant (Fig. 160) has one which is polysepalous.

Ancient authors considered the calyx as an entire organ, which was sometimes cut more or less deeply. To this false idea we owe the defective expressions, *segments*, *lobes*, *teeth*, &c., by which the free parts of the combined leaves of the calyx have been described.

The incisions, in fact, are not made from above downwards.

When a calyx begins to show itself, its elements, the *sepals*, are always free. In a polysepalous calyx they remain isolated until fully developed, but if the calyx is monosepalous, they are at a particular period of their growth elevated upon a kind of ring.

Without dwelling on the different shapes the sepals assume, we will content ourselves with stating that these organs become irrecognisable in the Valerians (Fig. 161) and Groundsel (Fig. 162), and a host of other similar plants. They appear in these plants like a tuft of silk or hair called a *pappus*, and if we did not arrive gradually at this curious modification by a series of illustrative examples, it would be very difficult to ascribe to the sepals their real origin.

Fig. 160.—Polysepalous calyx of *Linum*.

Fig. 161.—Pappus of Valerian.

Fig. 162.—Pappus of Groundsel.

The number of sepals in a calyx is extremely variable. There are two in the Celandine, three in the Virginian Spider-wort, four in the Willow-herb, five in the Hellebore, six in the Barberry, and a considerably larger number in the Cacti.

With regard to their arrangement on the receptacle, the sepals are sometimes in a whorl (or verticil), that is, several placed at the same level, if the receptacle is conical, or at an equal distance from the centre, if the receptacle is flat, sometimes disposed *spirally*, that is, at different heights, so that the line uniting their bases is a spiral.

Sepals, are, in short, either alike in size, inserted at the same height or distance, free or united, or else they do not show a perfect agreement in these points, and thus determine whether the calyx be *regular* or *irregular*. That of the Loosestrife (Fig. 163) is regular, and that of the Aconite (Fig. 164) is irregular.

In the Poppy the calyx falls before the flower expands; in the Ranunculus it is not detached until after the fecundation of the flower; in *Physalis* it remains round the fruit, enlarges very much, and becomes yellow or reddish.

This last phenomenon of coloration brings us to a reflection which

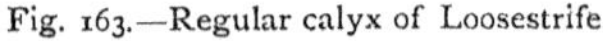

Fig. 163.—Regular calyx of Loosestrife.

Fig. 164.—Irregular calyx of Aconite.

has its interest. In some cases where the calyx and the corolla exist simultaneously in the flower, the calyx becomes coloured, and thus puts on the appearance of a corolla. The calyx of the Pomegranate and that of the Fuchsia are red, that of the Larkspur and of the Aconite are blue. The calyx may thus be coloured even when the corolla is wanting.

THE COROLLA.

All that we have said about the charm of flowers applies particularly to the *corolla*, for on that organ Nature lavishes her brightest colours. In spite, however, of the beauty and elegance of shape which we admire in it, the corolla is only the immediate envelope of more important organs, which, with the help of the calyx, it protects against the action of external causes. When the fundamental phenomenon of fecundation is effected, when the fertilised ovary begins to enlarge, and can of itself oppose a sufficient resistance, then Nature, which suffers nothing useless to exist, throws away this graceful decoration: the corolla fades, withers, and falls. If it remains occasionally a short time after fecundation, it is probably only to reflect the rays of external heat, and concentrate them on the fertilised ovary, thus accelerating its development.

The corolla is the inner envelope of the essential organs of the flower. It differs generally from the calyx in being of a more delicate tissue. The corolla alone constitutes the flower in the eyes of the world generally, but to a botanist the stamens and pistil are its essential parts, for under the influence of the stamens the pistil produces fruit, the seeds in which will perpetuate the species.

The *petals* are organs which, when taken together, constitute the corolla. They take their rise, like sepals, from modified leaves, a fact which is easily established. In some flowers—the Calycanthus, for instance—the petals are so completely shaded off into the sepals, that it is impossible to say where the calyx ends or where the corolla

Fig. 165.—Petal of *Dicentra.* Fig. 166.—Petal of *Nigella.* Fig. 167.—Petal of the Columbine. Fig. 168.—Petals of Aconite.

begins. In fact, the external divisions of these flowers are of a greenish hue, the internal parts being of a purple tint; but it is impossible to allot the intermediate divisions to one of the floral envelopes more than to the other. As the petals are shaded off into the sepals, the latter into the bracts, and the bracts into the leaves, we must conclude from this fact also that the petals are really modified leaves.

Like leaves, petals offer to us very different forms and most varied dimensions. They are generally either linear, oblong, elliptical, oval, or rounded. Sometimes they are boat-shaped, as in the *Blumenbachia insignis.* Sometimes they take the form of a spoon, as in the *Dicentra spectabilis* (Fig. 165); sometimes they show two lips, as in *Nigella* (Fig. 166); sometimes they are elongated like a horn, as in the Columbine (Fig. 167); sometimes they are helmet-shaped, as in the Aconite (Fig. 168).

Petals, like leaves, are either entire or indented. They present, like them, a sort of skeleton, if one may give that name to vascular and slender ramifications, which can only be perceived clearly in some cases by placing the petal between the eye and the light, so as to look at them as in a transparency.

The veins determine the shape assumed by the petal. Figs. 170, 171, and 172 give an idea of the three principal forms for the distri-

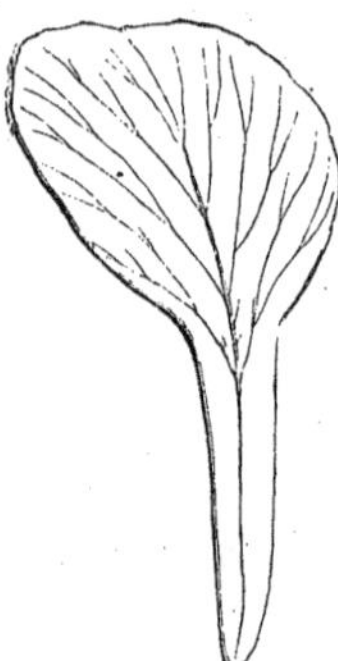

Fig. 169.—Petal of Wallflower.

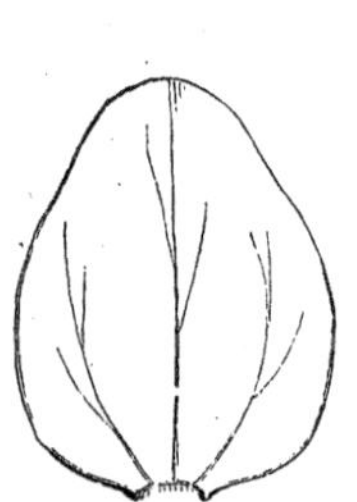

Fig. 170.—Petaloid sepal of Winter Hellebore.

Fig. 171.—Petal of *Cerastium præcox*.

bution of the veins. We know that in the Wallflower (Fig. 170) the petal is elongated at its lower part by a slender portion called the claw (*unguis*), the larger part of the petal taking, in this latter case, the name of the *limb*. The petal of *Cerastium præcox* (Fig. 171) and the petaloid sepal of the Winter Hellebore (Fig. 170) have no claw, and, as we see, a limb only.

The number of petals in a corolla vary much; they are sometimes very numerous, and are then arranged in a spiral; but they are oftener few in number, and are then arranged in one or more whorls. In the Cactus the petals are extremely numerous, and arranged in a spiral, continuous with that of the sepals. In the Geranium (Fig. 172), Violet, and Wallflower (Fig. 173), there are only five petals arranged in a whorl.

Just as the calyx may be monosepalous or polysepalous, so the corolla is monopetalous or polypetalous. The flower of the Geranium (Fig. 172), and of the Rose and Pink, have their petals perfectly distinct, so that one can be detached without interfering with the others. On the contrary, the Lilac (Fig. 174), the Primrose, and the Belladonna have their petals united together at their edges, so that one cannot pull off a petal without severing it from an adjacent one.

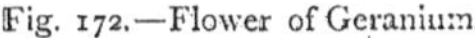

Fig. 172.—Flower of Geranium.

Fig. 173.—Flower of Wallflower.

When a flower begins to develope, the petals are always free. The transformation of a corolla, at first polypetalous, into a monopetalous corolla takes place in the bud, just as we have already shown to be the case with the calyx; that is, the free extremities of the petals are supported and united below in one whole by a common and continuous membrane.

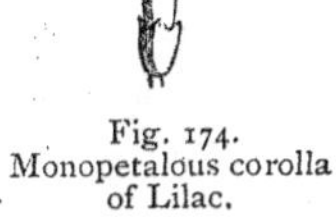

Fig. 174. Monopetalous corolla of Lilac.

We have remarked that the sepals are developed on the receptacle in succession, whilst the petals appear, on the contrary, simultaneously. This fact may help us to solve a problem which much occupied the attention of the older botanists.

In the Lily (Fig. 175), for instance, the floral envelopes are composed of six divisions, which are white, and of a delicate tissue, analogous to petals. Do the whole of these divisions constitute a corolla? By no means. Without mentioning the differences of shape, size, structure, and position, which could not escape the eyes of an attentive observer, we can show that the pieces of the external whorl of the Lily are developed in succession, like sepals, and that the pieces of the internal whorl are developed simultaneously. It has been decided from this fact that, in spite of appearances, there are in the Lily a calyx and a corolla; in other words, the Lily has a *petaloid calyx.*

In Rushes, on the other hand, contrary to what takes place in the

Lily, we find a sort of double calyx. Considerations analogous to

Fig. 175.—Petaloid corolla of *Lilium*.

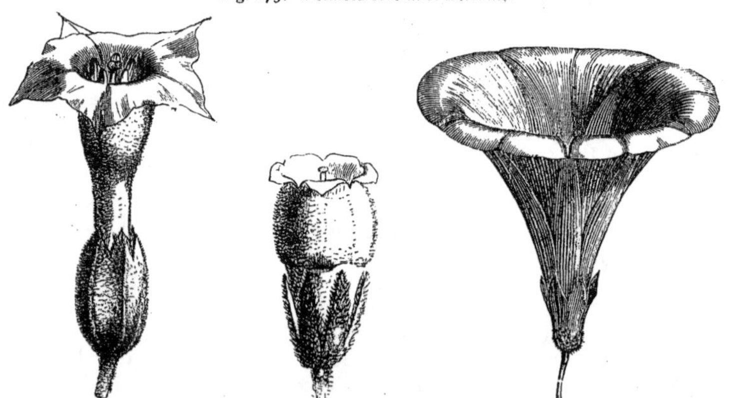

Fig. 176.—Infundibuliform corolla of Tobacco.

Fig. 177.—Tubular corolla of *Symphytum officinale*.

Fig. 178.—Corolla of Bindweed.

those mentioned have led botanists to concede to these plants a true corolla. We must admit, then, that in the Lily the calyx is white and petaloid, and in Rushes that the corolla is green and sepaloid.

Let us glance at the principal forms of the corolla when it is

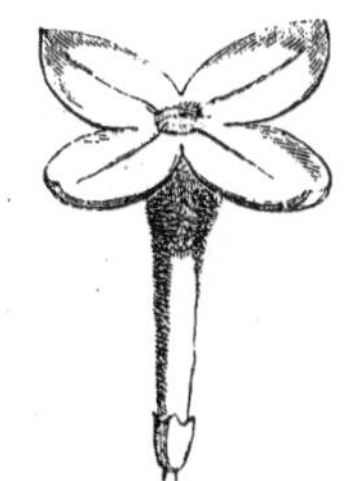

Fig. 179.—Hypocrateriform corolla of Lilac.

Fig. 180.—Rotate corolla of Borage.

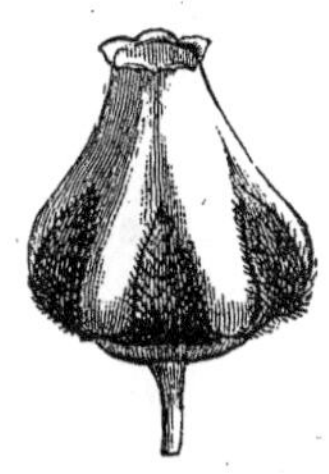

Fig. 181.—Urceolate corolla of Arbutus.

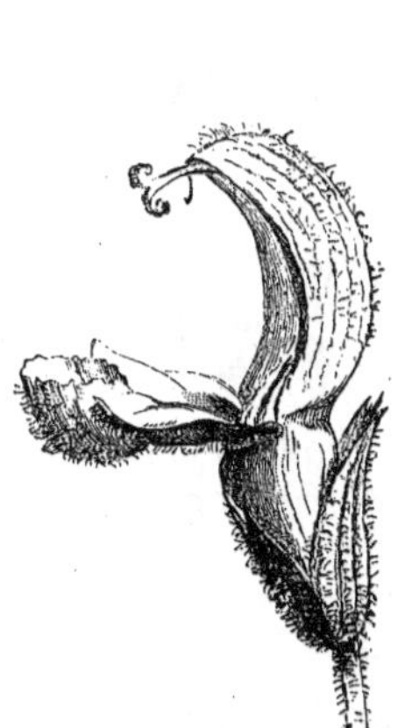

Fig. 182.—Bilabiate corolla of the Sage (*Salvia*).

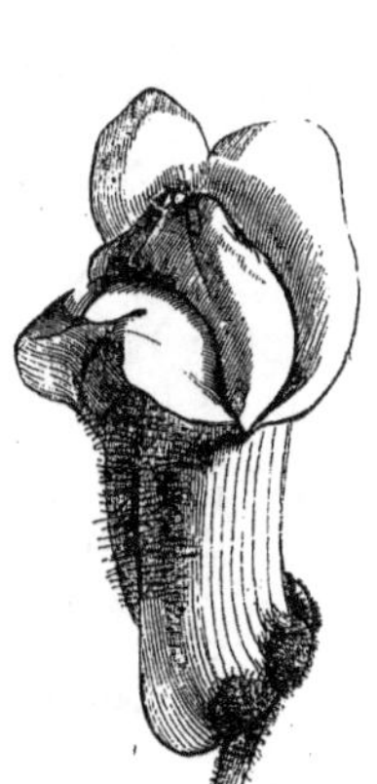

Fig. 183.—Corolla of Snapdragon.

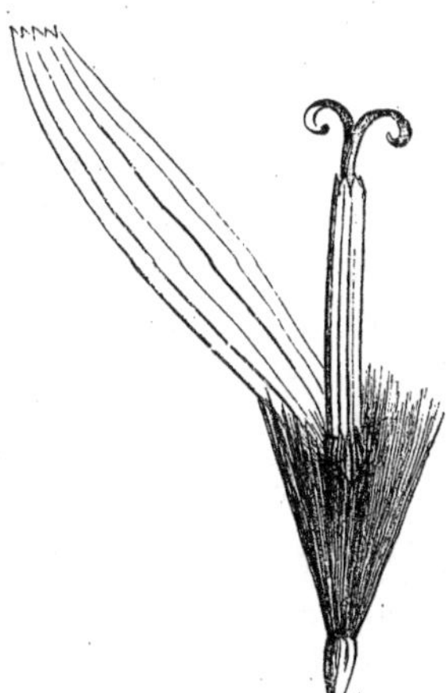

Fig. 184.—Ligulate corolla of the Dandelion.

monopetalous and regular. The six principal forms which the corolla assumes need not be otherwise described than by the annexed figures and the denominations which indicate them. The corolla is *infundibuliform*, that is, like a funnel, in the Tobacco-plant (Fig. 176), *tubular* in the Comfrey (Fig. 177), *campanulate* or bell-shaped in

Bindweed (Fig. 178) and the Campanula, *hypocrateriform* or salver-shaped in the Lilac (Fig. 179) or in the Jasmine, *rotate* in Borage (Fig. 180), or in the Willow-herb, *urceolate* or urn-shaped in the Arbutus (Fig. 181).

When the corolla is monopetalous and irregular, its principal forms are reduced to three. In the Sage (Fig. 182), or the Dead Nettle, &c., the limb of the corolla, placed at the summit of a more or less elongated tube, is divided transversely into two parts called lips (*labia*); the upper lip, presenting two divisions, is formed by two petals united almost to the summit; the lower lip, presenting three divisions, is formed by three petals joined more or less high up. This form of corolla, called *labiate*, characterises a very important group of the vegetable kingdom.

In the Snapdragon (Fig. 183) the mouth of the labiate corolla, instead of being wide open, is closed by a swelling of the upper lip.

"Among the regular monopetalous plants," says Jean Jacques Rousseau, "there is a family whose appearance is so marked, that we can easily distinguish its members by their look. It is that to which we give the name of gaping plants, because their flowers are divided by two lips, the opening of which, whether natural or produced by a slight pressure of the fingers, gives them the look of a gaping mouth. This family is subdivided into two sections or races, one with flowers having lips, or *labiate*, the other with flowers resembling a mask, or *personate*, for the Latin word *persona* signifies a mask—a very fit name, assuredly, for many of those whom we call persons."

In the Endive, or the Dandelion (Fig. 184), the corolla, cylindrical at its lower part, is split on one side, and forms a flat tongue, terminating in several small teeth. This form of corolla is said to be *ligulate*, and belongs to a considerable number of plants composing the largest natural group in all the vegetable kingdom, that of the *Compositæ*.

In regular polypetalous corollas, Tournefort distinguished three principal forms, which we meet with in a great number of flowers, in general those of the same family.

Cruciform corollas have four petals arranged in a cross, and generally provided with a claw—the characteristic peculiar to plants of the group of *Cruciferæ*. Fig. 185 represents the cruciform corolla of the Mustard Plant. Caryophyllaceous corollas have five petals, with a very long claw, hidden by the calyx. Fig. 186 represents a corolla of this kind, that of the Pink. The *rosaceous* corollas have five petals without claws, and open, arranged as in the Common Rose (Fig. 187).

Tournefort also classed all the modifications of the irregular

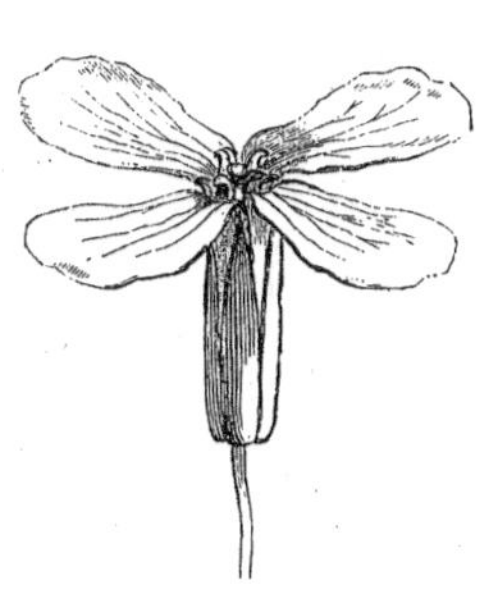

Fig. 185.
Cruciform corolla of Mustard.

Fig. 186.
Caryophyllaceous corolla of the Pink.

Fig. 187.
Rosaceous corolla of the Red Rose.

Fig. 188.
Papilionaceous corolla of the Pea.

polypetalous corolla under two names: these were *papilionaceous* and *anomalous* corollas. The Pea (Figs. 188 and 189) has a papilionaceous, the Aconite an anomalous, corolla. Let us dwell a little on the first of these shapes.

"The first portion of the corolla," says Rousseau, in his third "Letter on Botany," "is a large and broad petal covering the others, and occupying the upper part of the corolla, for which reason this large petal has taken the name of the *standard*. We must close both our eyes and our minds not to perceive that this petal is placed there as an umbrella to guard that which it covers from atmospheric injuries. By lifting up this canopy, you will remark that it is jointed into the side pieces on each side by a little ear, in such a manner that its position cannot be disturbed by the wind.

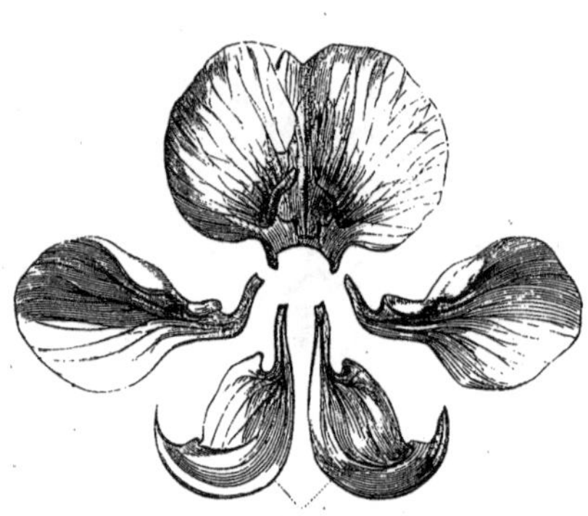
Fig. 189.—Separate parts of the corolla of the Pea.

"The canopy being removed, discovers the two lateral limbs, to which it was attached by its little ear-pieces. You will find, on detaching them, that they are fastened still more strongly into the remaining part, and cannot be separated from it without some effort. So the wings are not at all less useful in strengthening the sides of the flower, than the canopy is in covering it. When the wings are taken away, you see the last piece of the corolla, the piece which covers and defends the centre of the flower, and envelops it, especially below, as carefully as the three other petals envelop it above and at the sides. This last piece, which, on account of its shape, is called the *keel*, is like a strong box in which Nature places her treasure beyond the reach of injuries arising from the air and water."

Rousseau describes here the flower of the Pea, as an application of the principles which he had first laid down.

The Stamens.

Immediately within the petals, spring the *stamens* and pistil—the organs of reproduction and essential parts of the plant. The stamens, forming collectively the *andrœcium*, vary greatly in number, from one to fifty, and even more, and constitute the third verticil of the floral organs.

The stamen generally consists of two parts, an upper and thick portion, and a lower portion generally elongated and slender. The former is called the *anther;* the latter the *filament*. The filament is much less important than the anther, and is often wanting. The rudiment of the anther is at first a uniform cellular mass, but when it has attained a given size larger cells appear in its interior, usually at four distant points. By their division and augmentation four separate groups are produced, around which smaller cells are arranged in a given order, forming a special covering. In due time these groups of cells absorb the surrounding cellular matter. Usually they unite into two masses (two becoming one) by the absorption of the cellular matter between them. In the cells which compose them *pollen* is now formed by the division of each cell, first into two and then into four smaller cells, the contents of which gradually change into pollen grains, the mother cells being either absorbed or remaining in the form of filamentary, gelatinous, elastic matter among the ripe pollen. The lobes of the anther are separated by the *connective*, which is the direct continuation of the filament in the anther. The lobes of the anther are its halves, each half being a hollow cavity, or anther-cell. It can easily be understood that if the filament and the connective were prolonged into one another, and of nearly the same thickness, as in the Iris (Fig. 190), the anther would be fixed; but that this would not be the case if the connective were inserted on the attenuated extremity of the filament at a point only, as happens in the Amaryllis (Fig. 191). The powdery matter or fine dust contained in the cells of the anther is pollen; the membranous sides or walls of the cells of the anther are its valves; and the lines which pass down the sides of the anther are the sutures or seams.

We have stated above that the anther generally possesses two lobes, yet in some plants the anthers are *unilocular;* either the two lobes existing at first have been blended in one, or there was actually but one lobe, as in the *Epacridaceæ*, a family of elegant Heath-like plants, from New Holland. In other plants, as the Laurels and the *Ephedra*, the anthers are *quadrilocular*.

The lobes of the anther open at the sutures in order to discharge the fertilising pollen. Generally each lobe presents a longitudinal cleft, along which the opening of the anther takes place, this being called its *dehiscence* in botanical language. Sometimes the cleft only extends a short distance towards the summit of the lobe, and constitutes a sort of pore, as we see in the Heaths and Solanums (Fig. 192).

In the Barberry (Fig. 193) and the Laurel (Fig. 194) a very remarkable and elegant mode of dehiscence is observable: a certain restricted portion of the walls of the anther opens upwards, so as to form little trap-doors or valves. There is one valve for each lobe in the Barberry, and two for each lobe in the Laurel, as represented in the engravings.

A familiar acquaintance with the microscopic structure of the pollen of vegetables reveals some very curious structures. When a microscope of considerable power is employed, we find that the forms of these grains vary considerably in different species, some of their forms being of a very elegant description.

The pollen-grain is generally composed of a double covering, the innermost containing a mucilaginous liquid, named *fovilla*.

DIFFERENT FORMS OF STAMENS (MAGNIFIED).

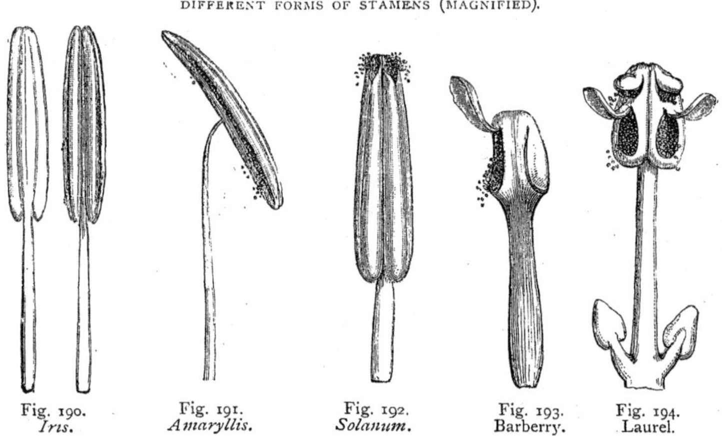

Fig. 190. *Iris.* Fig. 191. *Amaryllis.* Fig. 192. *Solanum.* Fig. 193. Barberry. Fig. 194. Laurel.

Figs. 195 and 196 are the pollen-grains of the Hollyhock, having a double *sac* or covering to each grain. The external membrane of the globule of pollen is smooth, dotted, or granulated. It is covered with small prickles, or finely reticulated, according to the species. It also exhibits folds and pores. In the pollen of Wheat (Fig. 197) there is only one pore; in the Evening Primrose (Fig. 198) there are three. Figs. 199, 200, and 201 show the pollen-grains of Garlic, the Phlox, and the Melon. The number of pores in a pollen-grain may reach five, or even eight. These pores perform important functions, as we shall soon see.

When a grain of pollen is placed in water it swells, because it absorbs a certain quantity of the liquid. Its membranes expand, and the internal one protrudes through the pores of the external membrane. The vessel bursts, and the fovilla escapes in a sort of mucous and granular jet. This is an anomalous but very curious phenomenon to observe. It is anomalous, because it is not thus that the matter takes place in Nature. When a pollen-grain falls on the moist and viscous surface of the uppermost part of the pistil, which we shall soon describe, and which bears the name of *stigma*, it expands slowly, appearing to absorb its humidity; the interior membrane

DIFFERENT FORMS OF POLLEN-GRAINS (MAGNIFIED) DISCHARGING FOVILLA.

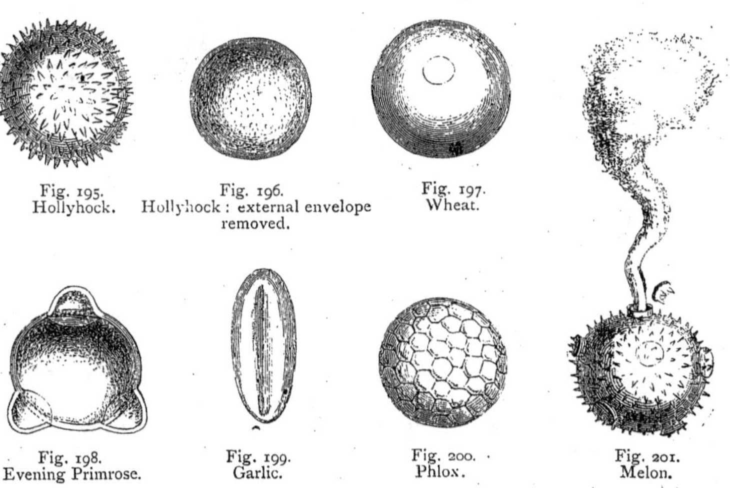

Fig. 195. Hollyhock.
Fig. 196. Hollyhock: external envelope removed.
Fig. 197. Wheat.
Fig. 198. Evening Primrose.
Fig. 199. Garlic.
Fig. 200. Phlox.
Fig. 201. Melon.

becomes gradually extended through one or two pores in the exterior coating, in the form of delicate tubular protrusions, which lengthen by degrees, and end by forming real tubes, called *pollen-tubes*.

The length of these tubes varies considerably; they attain in certain cases many hundred times that of the pollen-grain which gave it birth. This prodigious lengthening evidently cannot proceed from a mere elongation of the internal membrane of the pollen-grain, but is the result of an actual growth in this membrane. The pollen-tube is nourished and grows, that is to say, it *vegetates;* so that, leaving the stigma, it penetrates into the tissues which it is intended to traverse. We shall have occasion to return to the pollen-tube when we speak of *fecundation* in the next section.

Although pollen-grains are almost always free and distinct, there are some plants in which these grains are joined together, and often very closely. In the *Orchidaceæ* (Fig. 202), the pollen is gathered

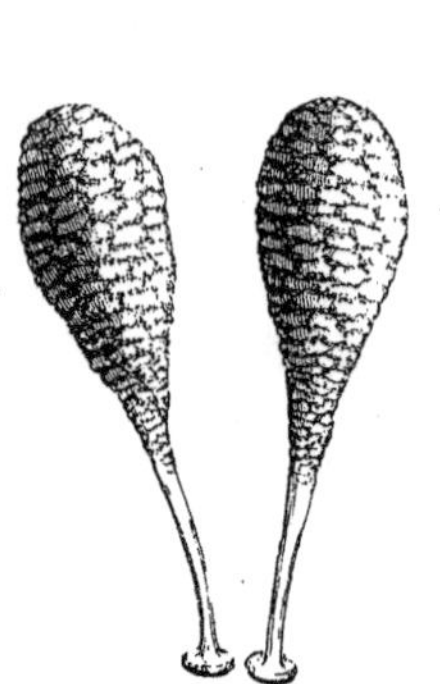

Fig. 202.
Pollen masses in *Orchis maculata* (magnified).

Fig. 203.
Pollen mass of *Habenaria chlorantha* (magnified).

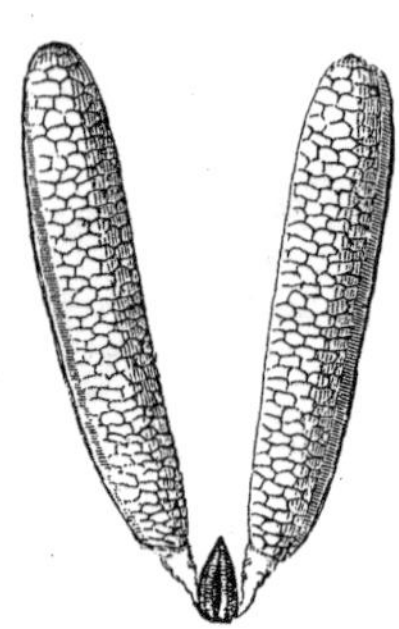

Fig. 204.
Pollen mass of *Asclepias floribunda* (magnified).

together into masses, sometimes almost pulverulent, with loosely cohering granules; sometimes it is formed of numerous small angular masses, joined together by means of glutinous matter. In *Habenaria chlorantha* and *Asclepias floribunda* the masses of pollen show the arrangement as represented by Figs. 203 and 204.

The number of stamens in each flower varies according to the species. When they are arranged in *whorls*, they are generally definite in number, as in the Vine (Fig. 205) and Primrose, which have five. When they are in a spiral, they are usually very numerous, as in the Magnolia and the Ranunculus.

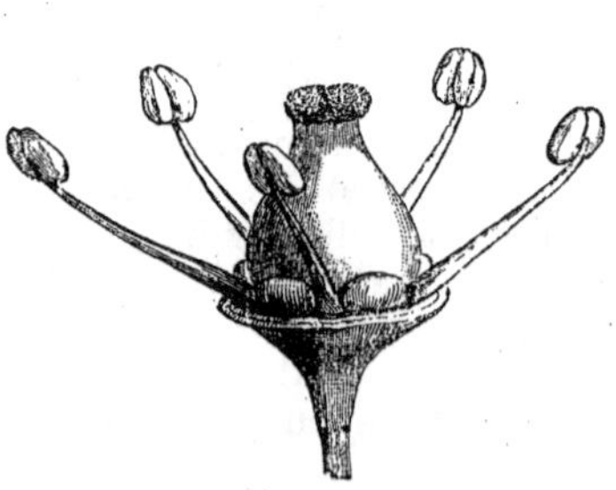

Fig. 205.
Andrœcium of the Vine (magnified).

Stamens may be all of the same height, as we see in the Lily, Tulip, and Borage, or else very unequal. In the Geranium there are five stamens bigger than the others, which are also five in number. In the Wallflower (Fig. 206), which has six stamens, four are bigger than the others; Linnæus called these *tetradynamous.* In the Snapdragon (Fig. 207), there are four stamens, two of which are larger than the others; Linnæus called them *didynamous* stamens.

The stamens of the same flower may be completely independent one of the other, or more or less united either by their filaments or by their anthers. In the Mallow (Fig. 208), all the stamens are united to each other by their filaments in a single bundle. In the French

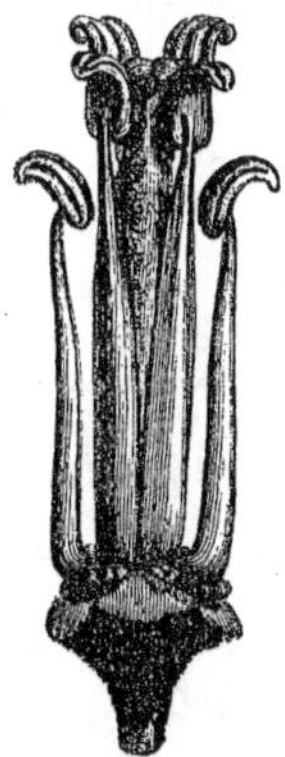

Fig. 206.
Andrœcium of the Wallflower (magnified).

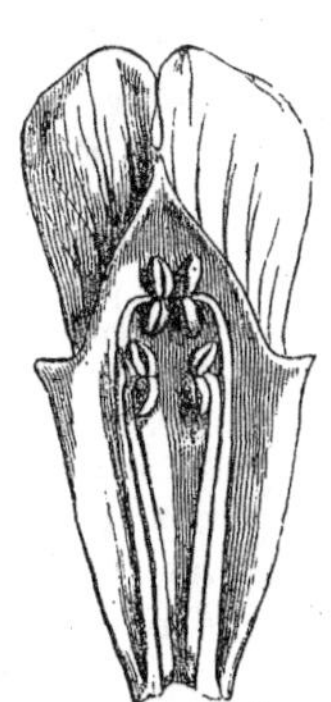

Fig. 207.
Andrœcium of the Snapdragon.

Fig. 208.
Andrœcium of the Mallow.

Bean and the Fumitories they are united in two bundles; in the St. John's Wort (Fig. 209), in three or five bundles; in the *Ricinus* (Fig. 210), in several bundles. After Linnæus, the stamens are called *monadelphous*, *diadelphous*, *triadelphous*, *polyadelphous*, as they form one, two, three, or more bundles. In the Dandelion, Artichoke, and Thistle, the stamens *cohere* together by their anthers, so as to form a kind of tube, supported by free filaments: these are said to be *synantherous* (Fig. 211).

Finally, stamens may form adhesions with the floral envelopes, In the Squill, for instance, six stamens *adhere* by their base to the six divisions of the flower. In the Primrose, five stamens are attached to the tube of the corolla, which is monopetalous.

We shall conclude our remarks on the stamens by some inquiry into the morphological nature of this part of the flower.

Bracts, sepals, and petals are modified leaves, as we have seen. It appears difficult to believe, at first sight, that the same should be the case with stamens. Yet, pick the petals from a flower of the White

Fig. 209.
Andrœcium of St. John's Wort.

Fig. 210.
Stamens of *Ricinus* (Castor-oil tree).

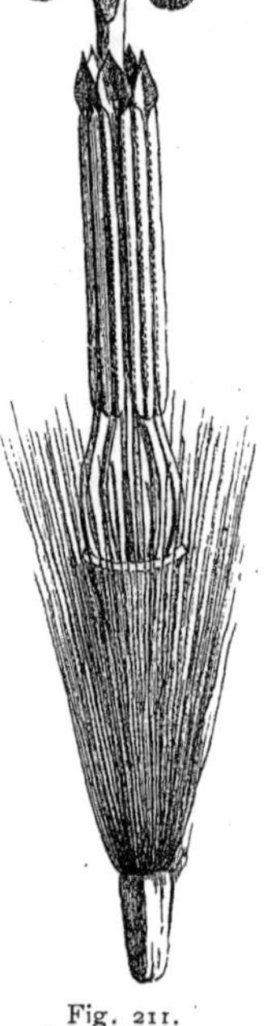

Fig. 211.
Synantherous Stamens.

Water-Lily; you will observe, as you approach the centre of the flower, that the petals diminish in length and breadth, and present towards the summit an anther, at first rudimentary; this becomes more and more complete as the supporters pass insensibly from the form of a petal to that of a filament. In the Columbine, under the influence of cultivation, we see the stamens changing by degrees into forms analogous to those constituting its elegant corolla. In the Rose we often find organs which are half petal and half stamen.

There is a very curious monstrosity in one species of the Rose, in which all the organs of the flower are transformed into leaves, so as to constitute what horticulturists call a *monster rose*. In this production we can follow, step by step, all the transformations between an almost perfect stamen and a petal which has been transformed into a green leaf.

All these facts demonstrate that the stamen is only a metamorphosed petal. But we have already shown the analogy of petals with sepals, of sepals with bracts, and of bracts with leaves. Stamens, therefore, like these organs, are only metamorphosed leaves. In short, the *filament* of the stamen has been compared to the claw of the petal, or the *petiole* of the leaf, the *limb* to the blade of the leaf, the pollen to a special modification of the *parenchyma* of the leaf, lastly the *connective* to the central part, that is, the *midrib* of the leaf.

The Pistil.

As we advance in the study of the organic parts of plants, it will be observed that Nature is constantly approaching her essential object, the propagation and preservation of the species. The pistil is the most essential organ in the reproduction of plants; Nature, therefore, has taken care to collect round the pistil all possible means for its protection and defence. It is placed in the centre of the flower, sheltered under several concentric coverings, and defended, besides, externally by the filaments of the stamens, which form a rampart round it. These various floral envelopes last as long as the pistil needs their protection and shelter. They disappear after fecundation, when the ovary is strengthened by its own proper development.

The pistil is the female structure in vegetables, the *gynœcium*, as it is sometimes called, as opposed to the *andrœcium*, the name which designates the whole of the stamens.

The gynœcium presents one of the most remarkable applications of the doctrine of *vegetable metamorphosis*, made popular by Goethe, the celebrated German poet, and also a profound naturalist. We can easily understand the structure, origin, and arrangement of the gynœcium, if we consider it as constituted by the transformation of a single leaf, or from the union, blending, and combination of several leaves in one single organ.

The elementary organs, the junction of which form the pistil, are called by De Candolle *carpels*. The carpel is to the pistil what the sepal is to the calyx, the petal to the corolla, and the stamens to the andrœcium. The aggregate of the carpels forms the pistil, as that of the petals forms the corolla, and that of the stamens the andrœcium. Sepals, petals, and stamens are only modified leaves; it is just the same with the carpels, which take their rise by the metamorphosis of the leaves.

Three parts are observable in the pistil—the *ovary*, the *style*, and the *stigma*. These three parts are very apparent in Fig. 212, representing the pistil of the Chinese Primrose, where the letters *stig.* indicate the stigma, *sty.*, the style, *o*, the ovary, *r*, the receptacle, *p*, the peduncle.

The *ovary* is the part of the plant destined to contain the seed, that is, the *ovules*, which, when fertilised and developed, become the *seeds*. The part which supports the ovules, which is generally rather thick, is called the *placenta*.

The top of the ovary is prolonged by a filament, either long or short, which is called the *style*, and is analogous to a prolonged apex in leaves. The style carries a glandular appurtenance on its summit, destined to receive the pollen-grains, and to help fecundation; this is the *stigma*.

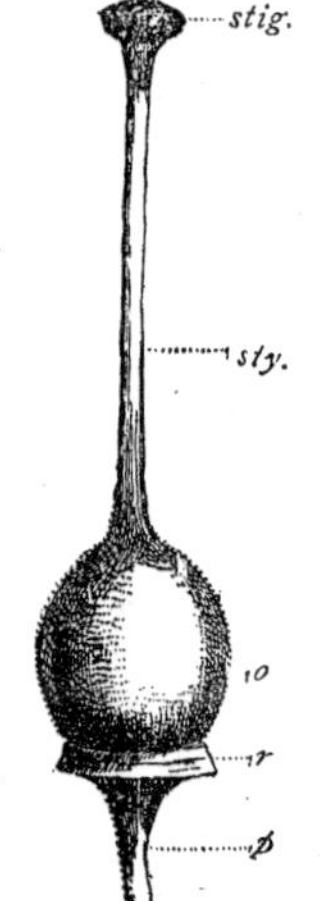

Fig. 212.—Pistil of Chinese Primrose (magnified).

The style is not a solid cylinder, as we might think at first sight; on the contrary, its axis forms a sort of canal leading into the ovary to the vicinity of the ovules. The stigma, which is the upper part of the pistil, is very variable in form; it is essentially formed by a mass of thin, transparent cells, loosely united, and coated over with a gummy, mucilaginous matter. It is thus well fitted to receive and retain the pollen-grains.

The carpels have a greater tendency than the more external organs to unite with each other: this is, no doubt, either owing to their proximity, or to the pressure upon them of the external organs, assisted by their position in regard to each other. This junction takes place either by the ovaries alone, or by the ovaries, styles, and stigmas, or by the stigmas alone.

When two or more carpels are united by means of the ovaries, an ovary results composed of several partial ones, there being as many cells as there were carpels at first. In the Hellebore (Fig. 213) the junction of the ovaries takes place at the base only; in the Fennel-flower (Fig. 214), about half way up. But most frequently the junction takes place at the summit.

When the styles are joined together, at least in some observable portions of their length, there results from this cohesion a style single in appearance, but in reality constituted by as many partial styles as there were carpels. In this case the number of free stigmas will, if they are simple, indicate the number of cells in the ovary. The partial stigmas may also unite, and constitute a stigma, in appearance single, but often so divided as to indicate by the number of its divisions the number of carpels constituting the pistil. The absolute number of cells in a pluri-locular ovary is subject to variation, but is generally three; next come the numbers two and five, but four very rarely. The number, besides, is not always the same in the different stages of the flower; it sometimes happens that they are multiplied by the formation of partitions, afterwards further developed, as in the Vervains and the *Labiatæ*, which at first have only two cells, but later show four, by the partition of each of the primitive cells into two compartments. This is also seen in the Flax-plants, the five primitive cells of which divide at a given time into two, by a partition of new formation. These supplementary partitions, which thus mask the first structure of the ovary, are called *spurious dissepiments.*

The ovary is usually free, and easily seen by looking into the

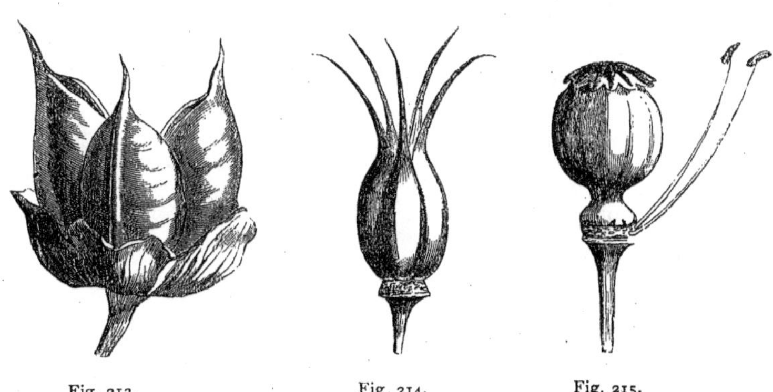

Fig. 213. Pistil of Fetid Hellebore.

Fig. 214. Pistil of Fennel flower (*Nigella sativa*).

Fig. 215. Ovary (superior) of the Poppy.

bottom of the flower; it is then called a *superior ovary*, as in the Poppy (Fig. 215) and Lily. At other times only the summit of the ovary shows itself at the bottom of the flower, and it is then united with the receptacle, and must be looked for underneath the flower; the ovary is, in this case, called *inferior* or *adherent*, as in the Coffee-plant, Madder (Fig. 216), and Melon.

We have stated above that the small bodies attached to the placenta are called *ovules*, and that afterwards they become seeds. These ovules are composed of a small central body or *nucleus*, adhering by its base to a double sac, showing only a very small aperture corresponding to the free summit of the *nucleus*. The external sac is called the *primine*, and the internal sac the *secundine*. The aperture of this double envelope is the *micropyle*. The point of adhesion of the nucleus with its integuments is called the *chalaza*.

Some ovules have no *primine*, and some have neither *primine* nor *secundine.* But these cases are rare. The point at which the ovules

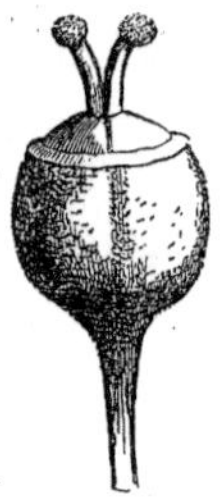

Fig. 216.
Ovary (inferior) of the Madder (enlarged).

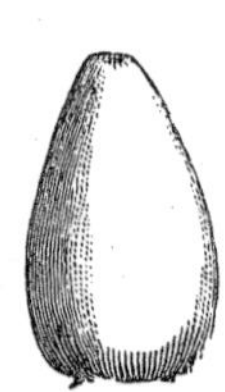

Fig. 217.
Ovule of Rhubarb.

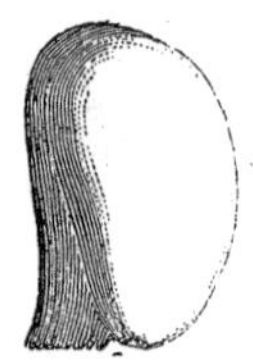

Fig. 218.
Ovule of Hellebore.

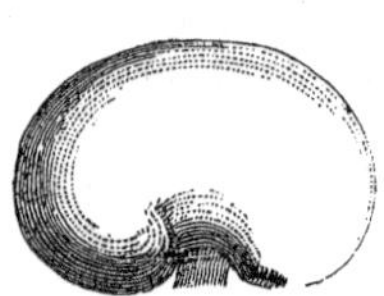

Fig. 219.
Ovule of the Haricot.

are attached to the placenta, either directly or indirectly by means of a small thread or *funiculus*, bears the name of *hilum*.

Ovules are not all of the same form. The ovule of Rhubarb

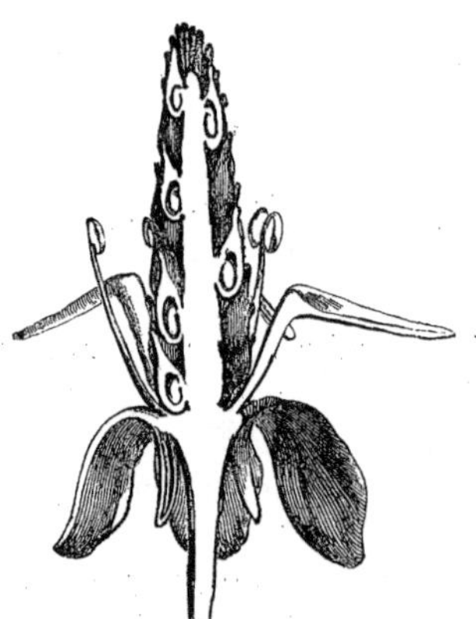

Fig. 220.
Receptacle of *Myosurus* (magnified).

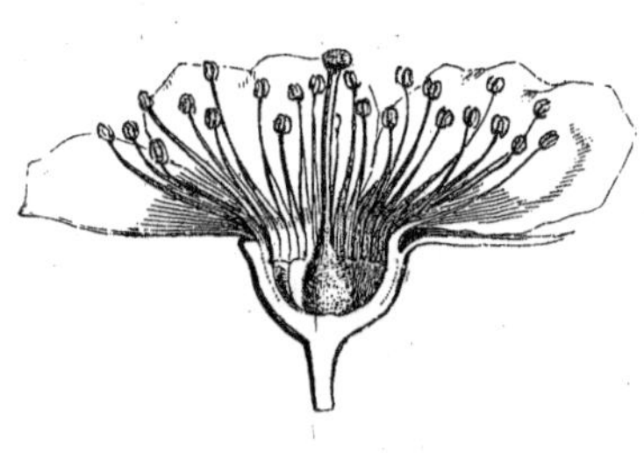

Fig. 221.
Perigynous stamens in the Peach-flower.

(Fig. 217) is shaped like an egg. Its *hilum* is diametrically opposite to the *micropyle.* This kind of ovule is called *orthotropal.*

In the Hellebore, on the contrary, the ovule has its point of attachment placed near the micropyle, and we notice a cord-like swelling on one of its sides, which reaches all along it, and is called the *raphe* (Fig. 218). This kind of ovule is called *anatropal.*

When the ovule has in the same way its point of attachment placed near the micropyle, but has no *raphe* and is bent, it is said to be *campylotropal.*

In the Haricot (Fig. 219) there is a form—the *amphitropal*—

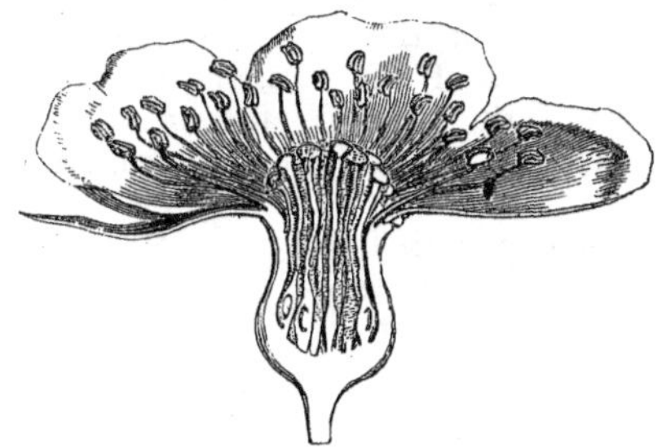

Fig. 222—Receptacle in the Burnet Rose.

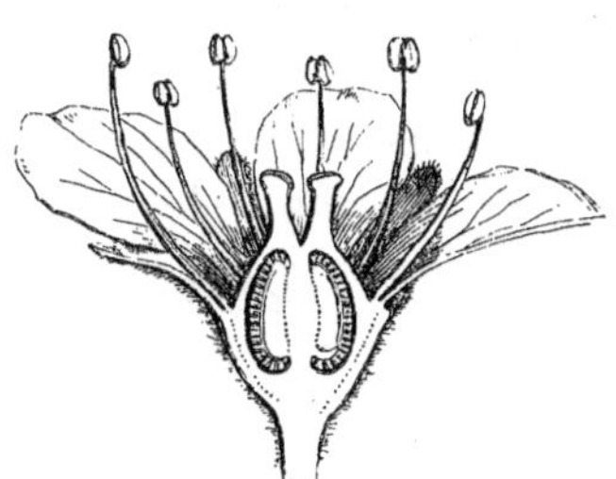

Fig. 223.
Semi-adherent Ovary of Saxifrage.

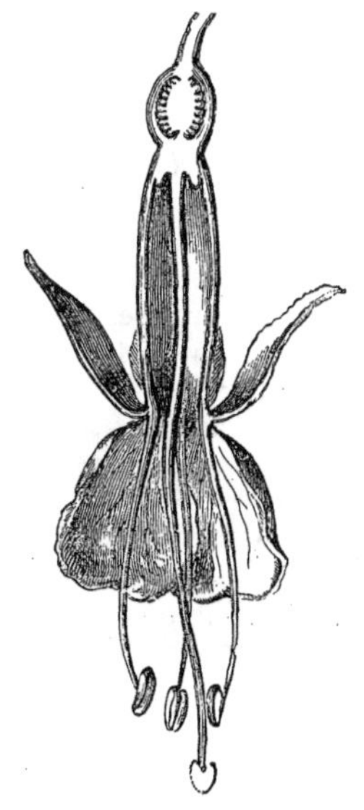

Fig. 224.
Adherent Ovary of the Fuchsia.

intermediate between the two first; the ovule is parallel to the placenta instead of at right angles to it.

Such are the principal forms of the ovule. The second is the most common, and the first is the most rare.

The Receptacle.

The calyx, corolla, stamens, and pistils are inserted on the extremity of the peduncle, which is called the *receptacle.* Its form is very variable. In the Ranunculus it is conical; the calyx, corolla, stamens, and pistil are inserted and ranked in succession on its sides, the last organs nearly at its summit. In the *Myosurus* (Fig. 220) it is so much lengthened, that it resembles a small *spike*, of which the flowers would be the carpels. As in these circumstances the stamens are inserted underneath the pistils, these stamens are said to be *hypogynous*. In the Peach (Fig. 221) and the Apricot, the calyx, corolla, and stamens are blended together at the base, and, being continuous with the receptacle, form a kind of cup at the bottom of which is the pistil, whilst the calyx and the stamens appear to be inserted in the side. These last surround the pistil, and are called *perigynous*.

In the Rose (Fig. 222) the receptacle is hollowed out, so as to take the form of a bottle, the bottom of which is occupied by the carpels, and on the upper edges are inserted the sepals, the petals, and the stamens. These are also perigynous.

In all the examples we have hitherto cited the pistil does not form any adhesion with the other floral organs. Also, in every case, even in that where the receptacle is hollowed out like a bottle, the ovary is *free* or *superior*. But it is not always thus. The tube of the calyx is joined rather frequently with the ovarian part of the carpels, which it encloses; and this junction is made more or less high up, so as to show every possible degree of adhesion. We see this in the flower of the Saxifrage (Fig. 223), and Myrtles, also in the Fuchsia (Fig. 224). The ovary is then called inferior.

Fig. 225.

THE FRUIT.

Flowers have but a short-lived existence; after fecundation they disappear; the ovary, rendered fruitful and enlarged in size, alone remains. The withered and dried-up fragments of the corolla strew the ground, or are carried about by the wind. But though the plant has lost much that embellished it, though it no longer possesses the brilliant ornaments which attracted observation and charmed all eyes, it still retains an interest of its own. A new decoration replaces the former one, leaving nothing to regret in the change. To the white flowers of the Wild Rose succeeds the young fruit, tinted with a pleasing green. The Mountain Ash and the Hawthorn, in casting off a delicately-tinted corolla, display their fruit, which soon changes to a bright red colour. The perfumed flowers of the Orange-tree are succeeded by the golden apples of the Hesperides; the delicate corolla of the Cherry-tree is followed by the purpled globes of its fruit. The verdure of our corn-fields, ripened in the summer's sun, now bends under the weight of the golden grain. We can now admire the soft down of the peach, the enormous globes of the Melon tribe, the firm and juicy pulp of the sweet-tasting plum, the nutritious substance of the legumes, the purple bloom of the grape, gilded by the autumnal sun! If flowers awake in us a feeling of happiness and joy, fruits bring with them the promise of abundance and wealth.

When fecundation is effected, or the fruit is *set*, as the gardeners say, life is concentrated in the ovules, and in the ovary enclosing and protecting them. These organs continue to grow, and soon present new characteristics. The ovule becomes the *seed*, the ovary becomes the *pericarp*, and the two together constitute the *fruit*. The fruit is, then, the ovary which has ripened.

The appearance of the fruit differs according as the ovary is *free* or *adherent*. In the former case, the fruit only shows on its surface the scar of the style, and sometimes at its base the remains of the calyx, the corolla, and the andrœcium. In the second, the fruit presents at its surface, and near the summit, the scars left at the insertion of the sepals, petals, and stamens. Thus it is, that an apple, a quince, or a gooseberry, all resulting from the ripening of an adherent ovary, are provided with an *eye*, which is completely wanting in the plum, the cherry, and the peach, these latter fruits resulting from the ripening of the free ovary.

"The analogy between fruits and leaves," says A. de Jussieu, "is as much shown in their nutrition as in their outward characters. Green fruits, like leaves, though in a less degree, when acted upon by light, take up carbonic acid from the surrounding air, and throw off oxygen; during the night, on the contrary, they take up oxygen and throw off carbonic acid. Their life passes through the same phases; their tissues, at first soft and rich in juices, gradually solidify, and at a certain period begin to dry up, changing their green hue for some other, either that of the dead leaf, or one of the various tints, analogous to those assumed in autumn by certain leaves; the withered pericarp remains attached to the tree, or falling to pieces, drops to the ground."

Fruits are divided into two great sections—Dry Fruits and Fleshy Fruits.

Dry Fruits.

There are some among the dry fruits which open their shell at maturity to allow the seed to escape; others, on the other hand, remain always closed up. Thence arises the division of dry fruits into *dehiscent* and *indehiscent*. The fruits of the Dandelion, Chicory, Buckwheat (Fig. 226), Corn-flower (Fig. 227), and Ranunculus (Fig. 228),

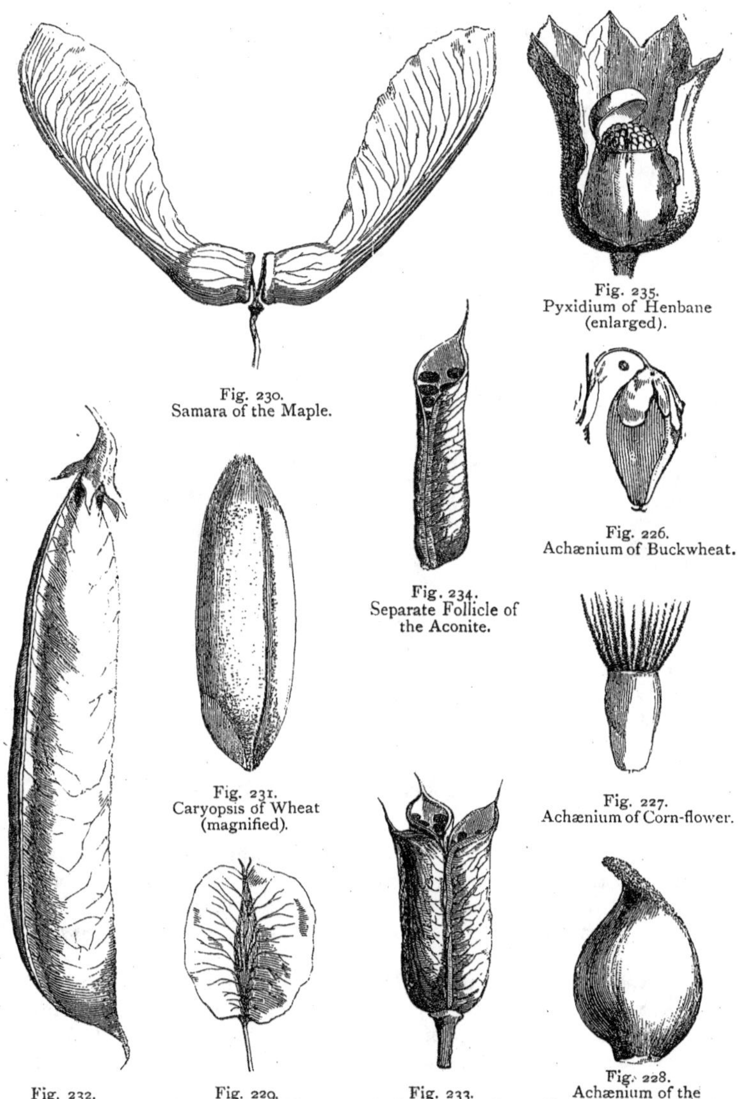

Fig. 230. Samara of the Maple.

Fig. 235. Pyxidium of Henbane (enlarged).

Fig. 226. Achænium of Buckwheat.

Fig. 234. Separate Follicle of the Aconite.

Fig. 231. Caryopsis of Wheat (magnified).

Fig. 227. Achænium of Corn-flower.

Fig. 232. Legume of the Pea.

Fig. 229. Samara of the Elm.

Fig. 233. Fruit of the Aconite.

Fig. 228. Achænium of the Ranunculus (magnified).

are *dry* and do not open. The single seed that they contain does not adhere to the pericarp; this kind of fruit is called an *achænium*, from α, "not," and χαίνω, "to open." The Elm has for its fruit an achænium; but being surrounded with a dilated membrane, somewhat like wings, it is called a *samara*. Fig. 229 represents the samara of the Elm; Fig. 230 the samara of the Maple. The fruit of Wheat,

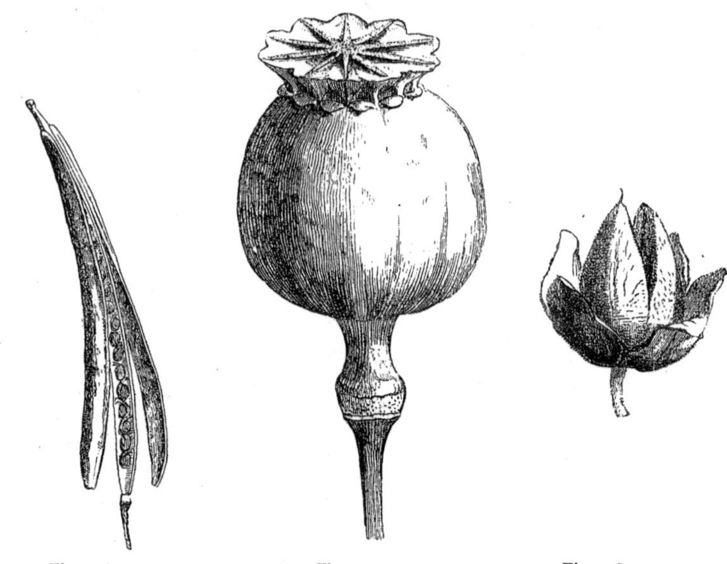

Fig. 236. Siliqua of the Gillyflower.

Fig. 237. Capsule of the Poppy.

Fig. 238. Capsule of *Digitalis*.

Barley, Oats, &c., is, like the achænium, dry and indehiscent; but the single seed that it encloses adheres to the pericarp, so as to form one body with it. This fruit is called a *caryopsis*, from κάρυον, "a nut," ὄψις, "appearance." Fig. 231 represents the fruit, or *caryopsis* of Wheat.

What a variety there is among dry fruits in their mode of opening! Some open with two valves, each carrying, on one of its edges, a row of seeds. Such are the *pods* of the Pea (Fig. 232) Bean, and other *legumes*. Others split up longitudinally on one side, and, in opening out, take the form of a leaf, carrying seeds on their two edges; this is called a *follicle;* of this sort is the Aconite (Figs. 233, 234). Some dry fruits open in two parts by a circular horizontal fissure, so that the upper part of the fruit is detached like a lid. This kind of fruit is called *pyxidium*, from πυξίς, "a box." We see it in the Red Pimpernel and Henbane (Fig. 235). In others, the pericarp comes away in two valves, which, by their fall, uncover a frame formed by the placenta furnished with their seed; this kind of fruit is called a *siliqua;* the fruit of the Gillyflower is of this sort (Fig. 236).

Can anything be more ingenious than the plan of opening in the

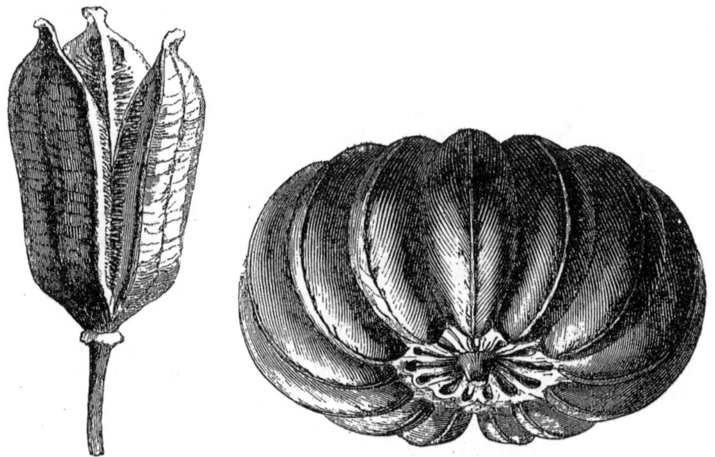

Fig. 239.—Fruit of the Tulip. Fig. 240.—Fruit of the Sandbox-tree.

capsule of the Poppy (Fig. 237)? In this instance dehiscence takes place by means of a certain number of small reflex valves disposed in a circle on the flattened top of the fruit. The seeds in it are very numerous; but, in consequence of the beautiful arrangement just mentioned, they only fall, one by one, when the capsule is bent over by the wind, thus forming a kind of natural seed-drill.

The fruit of the Foxglove (*Digitalis*), Fig. 238, which is also a capsule, opens with two valves, by the severance of the partitions in it, and each valve corresponds with a carpel; this is called *septicidal dehiscence.* The capsule of the Tulip (Fig. 239) opens with three valves, each valve corresponding with the two halves of two of the carpels, and having a partition in the middle. This is called *loculicidal dehiscence.*

In some plants the scattering of the seeds is assured by means rather difficult of explanation. Every one knows that by merely touching the fruit of the Balsams their valves are suddenly thrown back, and the seeds are dispersed with great force. This peculiarity has given to one species of this order of plants the common name of Touch-me-not, and the generic name of *Impatiens.*

The capsular woody fruit of the Sandbox-tree (Fig. 240), an American tree of the order *Euphorbiaceæ*, is composed of from twelve to eighteen *cocci*, which having become desiccated, open suddenly at the back, with two valves, and are detached from their axis with a kind of detonation. These fruits have actually been surrounded with iron wires, yet the force with which they expand has been such that the valves have been separated from each other notwithstanding. We will take our last example from a nearer source; the seeds of the Geranium (Fig. 241) are enclosed in little membranous cells, which are inserted in the lower part of an axis, which is elongated and supported by a filament coming from the summit. At maturity, this filament bends in a spiral, and lifts with it the case with the seed contained inside. Thus the fruit of the Geranium, or Crane's-bill, met with in the woods and green lanes, resembles a sort of candelabrum with five branches, hung from the summit of a central column.

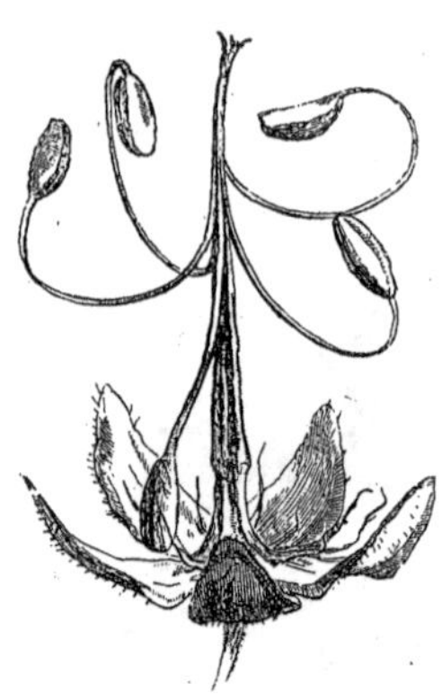
Fig. 241.—Fruit of the Geranium.

Fleshy Fruits.

When the parenchyma of the fruit is largely developed, and it swells with juice, the fruit is said to be *fleshy.* Man derives from this kind of fruit so great a part of his nourishment, that he has exclusively styled the trees furnishing it *fruit trees.* This singular illogical use of the term might lead to the inference that the Apricot, the Peach, the Apple, and such-like trees alone produced *fruit.* There is an obvious disagreement in this case between science and sentiment; all flowering plants bear fruit.

Fleshy fruit is green in the first phase of its development. It then, like all the green parts of vegetables, gives out oxygen during the day, and carbonic acid in the night. But its bulk soon increases, and it receives through its peduncle the moisture and other substances indispensable to its growth. During this first period, the principles immediately soluble take their rise, and their proportions increase as the fruit is developed. These soluble bodies are :—Tannin, the organic acids, which vary with the fruits (malic, citric, or tartaric acid preponderating, as the case may be), sugar, dextrine, and pectine, &c. The formation of *pectine*, the substance from which the *jelly* of our household conserves is prepared, is the result of a sort of reaction of the acids on *pectose*, a substance insoluble in water, alcohol, and ether, and which almost always accompanies the cellulose in the tissue of vegetables.

Sugar proceeds from the modification of certain neutral matters, such as gum and dextrine. In fact, starch exists in large quantities in some green fruits, but it completely disappears at the time of ripening. It is extremely probable, therefore, that it is the starch which is transformed into sugar (*glucose*) under the influence of acids. Tannin itself, existing in almost all green fruit, is not found in the mature state, but seems also to be changed into glucose under the influence of acids.

The absence of acidity in fruit is the most curious fact attending its maturity. It has been stated that this disappearance is not owing to the saturation of acids with mineral bases, that the acids are not hidden by the sugar or mucilaginous matter existing in the ripe fruit, but that they are really destroyed by oxidation during the ripening process. Tannin disappears first, and then the acids.

The moment when the tannin and acids have disappeared is that in which the fruit is most delicious; in a short time the sugar itself disappears, and the fruit becomes insipid. About the period of maturity, fruits exhale carbonic acid; they no longer disengage oxygen during the day.

Fruit at last undergoes a third modification, that of *bletting;* it becomes woolly. This new change has the effect of expelling from the fruit certain principles which belong to it. A Medlar for example, at first very acid and astringent, loses its acid and tannin, and becomes eatable when it has undergone the process of bletting. But the great difference established between the ripening and bletting of fruit is, that the latter state is only manifested when, the skin of the fruit being somewhat decayed, the air has been able to penetrate the cells of the pericarp, colouring them yellow, and partly destroying them.

We need not mention here the important part played by fleshy fruit in the production of alimentary beverages. The juice of the grape, having undergone fermentation, becomes wine; the fermented juice of numerous varieties of apples and pears yields cider and perry, and almost every known fruit may be formed into a fermented liquor.

It is in the fleshy fruits that we can most readily distinguish the three parts constituting the pericarp, that is to say, that portion of the fruit which forms the walls of the ovary. These three parts are, tracing them from outside, the *epicarp* (ἐπί, over, καρπός, fruit), an epidermal membrane varying in thickness; the *mesocarp* (μέσος, middle, καρπός, fruit), constituting ordinarily the flesh and pulp of the fruit; and the *endocarp* (ἔνδον, inside, καρπός, fruit), often forming the kernel, but the consistency of which varies; as we shall soon see.

As the ovary results from the transformation of a leaf, and as the fruit is nothing but a ripened ovary, we may consider the epicarp and endocarp as representing the two epidermal surfaces of the leaf, and the mesocarp as the intermediate parenchyma. Most practical botanists only admit two classes of fleshy fruits, the *drupe* (stone fruit) and the *berry*. The Peach, Cherry, Plum, Medlar, and Cornel are drupes; the Grape, Gooseberry, Apple, Orange, and Pomegranate are berries. All these fruits are more or less fleshy or pulpy; they are besides *indehiscent*.

Let us first take a glance at the drupes. In the Peach, the Cherry, and the Plum, resulting from the ripening of a simple and superior ovary, it is easy to distinguish three parts—first, an exterior skin, more or less thick, smooth, or hairy, or covered with a waxy secretion, known as the bloom, this is the epicarp; second, a thick, pulpy, succulent flesh, this is the mesocarp; third, a woody kernel, either smooth or furrowed with deep-winding dents, constituting the solid cavity and protection of the seed, this is the endocarp. Fig. 242, showing the fruit of the Cherry, and Fig. 243, giving a vertical section of the same fruit, enable us to see the interior and exterior arrangements of this drupe.

The fruit of the Medlar proceeding from the ripening of an inferior ovary, is composed of five compartments, and joined together by an external covering considered as an expansion of the floral receptacle. This fruit is also crowned with the sepals of the calyx. The Medlar presents five bony kernels, embedded in a pulpy mass, resulting from the transformation and fusion of all the ovarian walls (except the woody endocarp), added to the expansion of the floral receptacle.

The small oblong and red fruit of the Cornel is also a drupe resulting from the ripening of an inferior and compound ovary. But the kernels are joined together in such a way that we find in the centre one plurilocular stone, presenting two or three cells.

It results from what we have said, that in the Peach, Cherry, and Plum, the eatable part proceeds exclusively from the ripening of the pericarp or the ovarian walls, whilst in the Medlar or the fruit of the Cornel, the eatable part results not only from the ripening of the

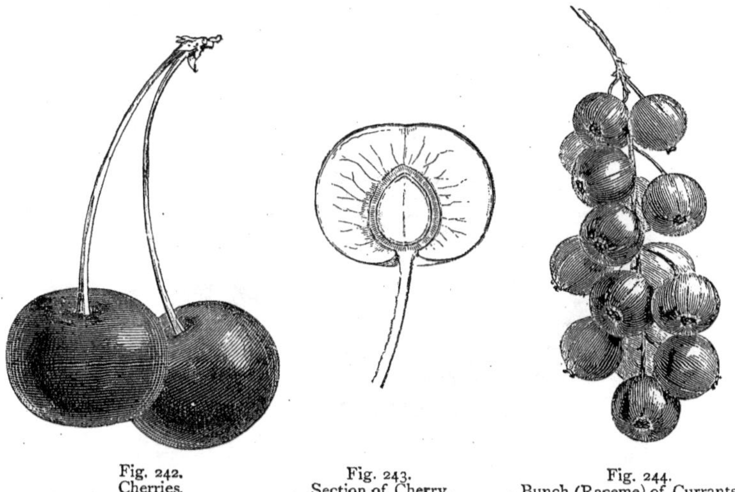

Fig. 242. Cherries. Fig. 243. Section of Cherry. Fig. 244. Bunch (Raceme) of Currants.

pericarp, but also from the transformation of the tube of the calyx of the flower, which increases and becomes succulent.

Berries, like drupes, are fleshy and indehiscent, but without stones. Such are the berries of the Vine or the Currant (Fig. 244) and the Gooseberry; only we must remark, with regard to this latter fruit, that its eatable and pulpous part does not belong only to the pericarp, but also to the seeds, which afford a gelatinous *testa*, when sufficiently developed. The seeds of the Pomegranate also present a testa full of pulp.

There are other berries, the structure of which is so peculiar, that they have received special names. We will content ourselves with mentioning here the fruits of the Apple and Orange.

The Apple results from the ripening of an inferior and compound ovary with five carpels originally free. It is wrapped, like the fruit of the Medlar and Rose, by an expansion of the floral receptacle. This covering becomes fleshy and succulent, like the ovary with which it is joined, of which the endocarp alone, lining the hollow of the five cells, is thin and cartilaginous. The endocarp forms that sort of scale which often sticks between the teeth when we eat an apple.

The fruit of the Orange (Fig. 245) results from the ripening of a superior and compound ovary with several cells. The external skin, yellow-coloured, dimpled, and strewn over with glands secreting an

Fig. 245.—Section of an Orange.

odoriferous liquid, is the epicarp. The white, spongy, and dry layer immediately under the external skin is the mesocarp. The thin membrane lining the "*quarters*" is the endocarp. These quarters or compartments, towards their inner angle, contain seeds, and are filled with a novel and peculiar tissue, which is developed on the opposite wall of each compartment. It appears at first like slender hairs, which increase by degrees, filling up the entire cavity, and, swollen with juice, constitute ultimately a succulent parenchyma, which forms the delicious pulp of the orange.

Thus, in this much-admired fruit, the eatable part does not belong to the mesocarp, as in the Cherry or Grape; we can only say that it belongs to the pericarp as an accessory, since we reject the three principal parts constituting this integument. The eatable part is an additional tissue, so to speak, which does not exist in other fruits.

We see, by this example, how various is the structure of fruits, and what difficulties their study presents even on a limited scale. Here we must confine ourselves to a rapid sketch of some of the common fruits, whose diverse and peculiar appearance requires a few words of explanation.

What constitutes the Strawberry? Is it that fleshy, succulent part essentially forming it, which is the fruit? Certainly not: the true fruits of the Strawberry (Fig. 246)—and they are very numerous —are those little brownish, dry, insipid grains, crunching between the teeth, which remain at the bottom of the vessel, mixed with small dark threads, when you beat up strawberries with wine. The little

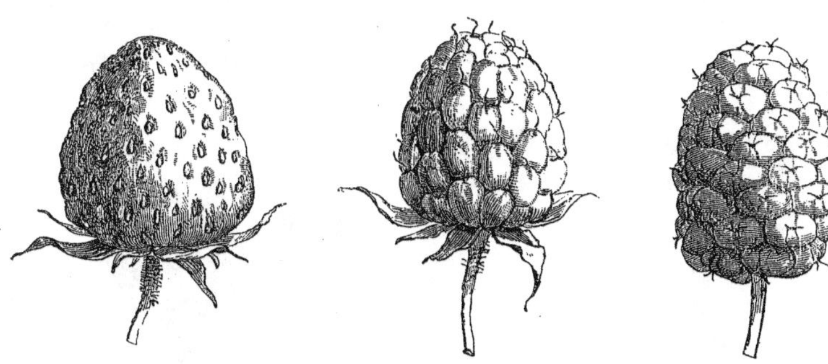

Fig. 246.—Strawberry. Fig. 247.—Raspberry. Fig. 248.—Mulberry.

brownish grains are *achænia*, the small dark threads are the styles of the withered flower. What we eat, then, in the strawberry, is the receptacle, which is gradually filled with juices; it increases in size, pushes out the little achænia, setting them into its parenchyma; it then assumes a rich colour as well as a most pleasant odour, and a sweet, aromatic, and slightly acid flavour.

In the Raspberry, on the contrary (Fig. 247), the receptacle is dry and bears several fruits, which, far from being achænia, as in the Strawberry, are little *drupes*. The seat of the fleshy and eatable part here occupies an entirely different position.

In the Fig (Fig. 249) the eatable part is formed, as in the Strawberry, by a thick, fleshy, and succulent receptacle of gourd-like shape. The real fruits, which the reader will have no doubt taken for the mere seeds, are achænia, and are inserted in the inside surface of the receptacle. But there is this difference between the Fig and the Strawberry, that all the fruits of the Strawberry appertain to one flower, while the fruits of the same Fig belong to different flowers.

The Mulberry (Fig. 248) is not a fleshy fruit, properly so called ;

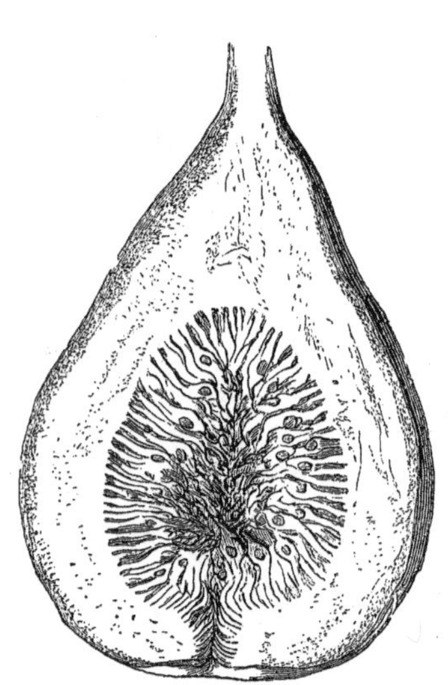

Fig. 249.—Section of Fig.

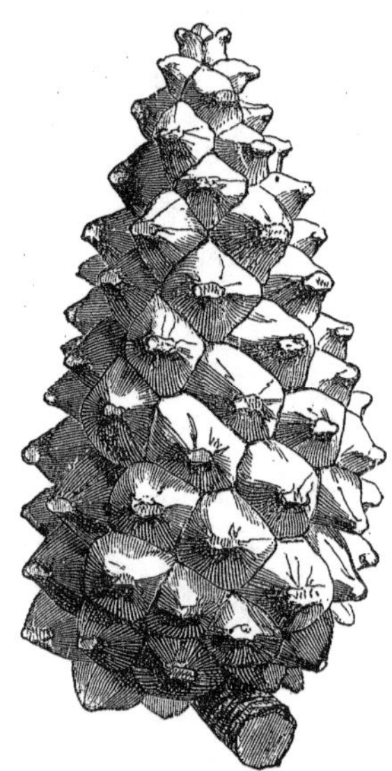

Fig. 250.—Pine-cone.

it is an achænium, enclosed in a persistent calyx, which has become fleshy.

The name of *cone* has been given to the fruit peculiar to a natural group of plants, the Pines, for this reason designated *Coniferæ.* The cone is a dry fruit, composed of a great number of winged seeds borne upon the upper surface of the hard, highly-developed, woody scales. Fig. 250 represents the Pine-cone.

THE SEED.

The seed is the essential part of the fruit. It represents outwardly a system of protecting coverings, which are generally double. The various appearances presented by different seeds are owing to this covering.

Seeds are sometimes smooth, like those of Tobacco (Fig. 251) or the Pear (Fig. 252), sometimes wrinkled and rough-skinned, as in the Nigella (Fig. 253); or papillous or warty, as in the Chickweed (*Stellaria*), Fig. 254; honey-combed with alveolar depressions, as in the Field Poppy (Fig. 255); winged, as in the Pine (Fig. 256), or

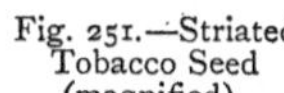

Fig. 251.—Striated Tobacco Seed (magnified).

Fig. 252.—Seed of the Pear (magnified).

Fig. 253.—Seed of Nigella (magnified).

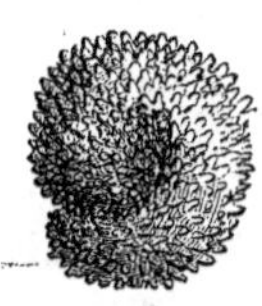

Fig. 254.—Seed of Chickweed (magnified).

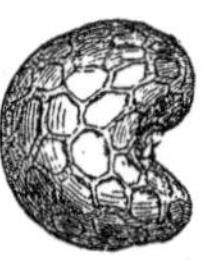

Fig. 255.—Seed of the Field Poppy (magnified).

hairy, as in the Cotton-plant (Figs. 257 and 258), with hairs rising from the apex, as in *Asclepias* from the base in the Willow.

That part of the seed which is enveloped and protected by these integuments is called the *kernel*, or *nucleus*.

The essential character of the kernel is, that it contains the *embryo*, that is, the germ of a new individual, a rudimentary plant in miniature, which will soon present all the characteristics of the parent plant, whose species it is to perpetuate. The embryo is composed of cells and spiral vessels; a small stem, or *caulicle;* a rudimentary descending portion, which becomes the root, or *radicle;* and a rudimentary bud, or *plumule.* Between the radicle and the plumule, the first leaves are developed, termed *cotyledons.* Fig. 259 shows these different parts in an embryo of the Almond-tree. When the plant has only one germinating leaf, or cotyledon, as the Wheat-plant, Tulip, Palm, and Pond-weed (Fig. 261), we say that the embryo and the plant are *monocotyledonous*: when there are two, as in the Castor-oil-plant (*Ricinus*), Fig. 260, Rose, Almond-tree (Fig. 262), and the Bean, we say that they are *dicotyledonous*.

The cotyledons of *Ricinus* (Fig. 260) are very thin, and offer, on their surface, very distinct traces of veins; they resemble small leaves, while those of the Almond-tree and the Bean are thick and fleshy, and present nothing like a leaf at first sight. They have undergone deep and essential modifications, appropriate to the functions they are called on to perform in the act of germination.

In a great number of cases the *kernel* is formed exclusively by the

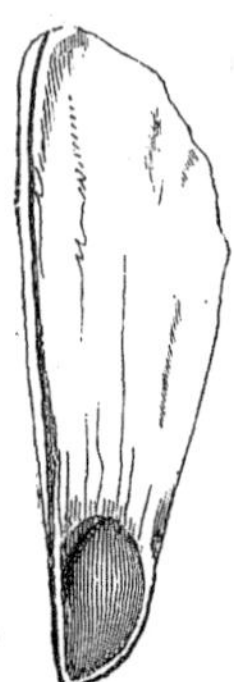

Fig. 256.—Seed of the Pine.

Fig. 257.—Section of Seed of Cotton.

Fig. 258.—Seed of Cotton.

embryo, that is, the entire seed is made up of the embryo and the integumentary covering only. But there is often developed, either around or by the side of the embryo, an accessory and completely independent body, which is a sort of reservoir of nutritious matter, from which the embryo draws the substances necessary for its first growth. This body is the *albumen*. When this is wanting, the cotyledons perform the functions of the nurse, nourishing the young plant, and it is to this end that they undergo the modification of which we have just spoken. Thus, in the seed of the Bean, which has no albumen, the cotyledons are much developed and full of a nutritive substance, of which the embryo takes a considerable portion. In the seed of *Ricinus*, which encloses a considerable portion of albumen, the cotyledons preserve the characteristics peculiar to the organs they represent; they are thin and foliaceous. The albumen varies very much in its bulk, nature, and position, in regard to the embryo, consisting of amylaceous, ligneous, gummy, and saccharine

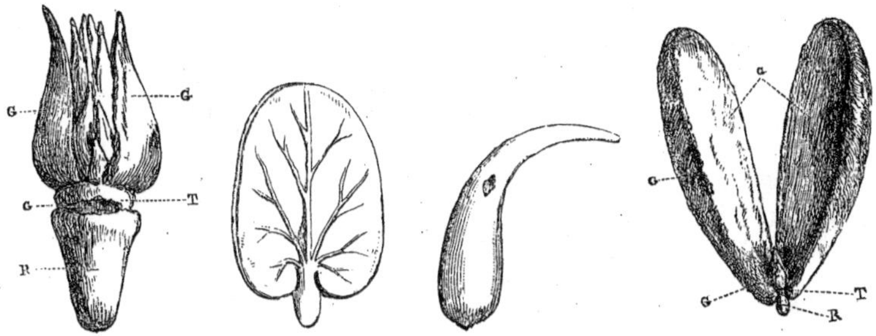

Fig. 259.—R, Radicle; C, Cotyledons; G, Plumule.

Fig. 260.—Embryo of Castor-oil plant.

Fig. 261.—Embryo of *Potamogeton*.

Fig. 262.—Embryo of the Almond.

matters, with oils and other heterogeneous substances. It is very considerable in Wheat (Fig. 263) and in Ivy (Fig. 264); it is reduced to a thin layer in *Hibiscus*. In Wheat the embryo is placed laterally at the base of the albumen, completely enclosing it in the Corn Cockle (*Githago segetum*), Fig. 265; it is, on the contrary, surrounded on all sides in the seed of the Wood-Sorrel (*Oxalis*), Fig. 266.

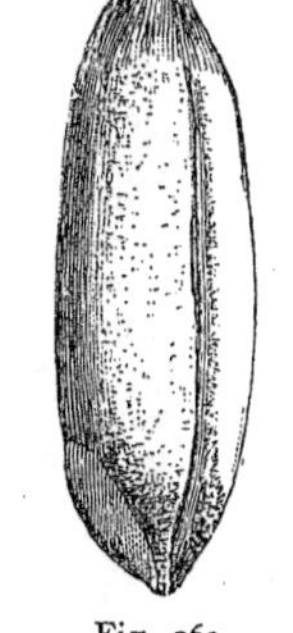

Fig. 263. Caryopsis of Wheat (magnified).

Albumen is almost exclusively formed of cellular tissue. We observe in it neither fibrous cells nor vessels. These cells have sometimes thin walls, as in the Ricinus, Wheat, and other cereals; sometimes their walls are very strongly thickened, as may be seen in the horny and firm tissues of the Date-stone (Fig. 267), which is only the albumen of the seed. In the albumen of Wheat and other cereals, starch predominates in the cells. The form of the starchy grains, varying with the species, is not unimportant. If we add to this characteristic some other considerations taken from their size, and the structure of their grains, we might detect the adulteration of flour by a simple microscopic observation, and at a mere glance.

The grains in Wheat (Fig. 268) are lenticular, elliptical, and egg-shaped, and about $\frac{1}{1000}$th of an inch in diameter. It is easy to distinguish these from the grains of the Potato (Fig. 269), which are generally larger, about $\frac{1}{300}$th of an inch in diameter, egg-shaped also, but with punctations, surrounded by certain zones, more or less

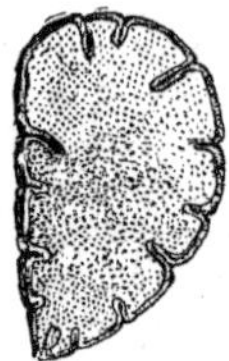

Fig. 264.—Section of Seed of the Ivy (magnified).

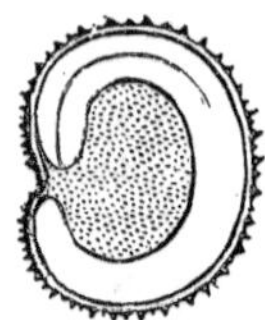

Fig. 265.—Section of the Seed of Corn Cockle (magnified).

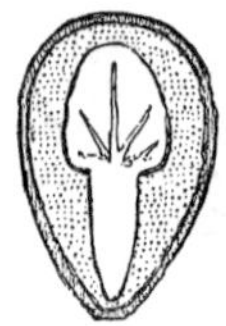

Fig. 266.—Section of the Seed of *Oxalis* (magnified).

regular and defined. In Maize (Fig. 270) the starchy granules of the horny part of the albumen are polyhedral, and nearly always show a punctation placed in their centre.

In the Oat-plant the starch grains are of several sorts. Some are

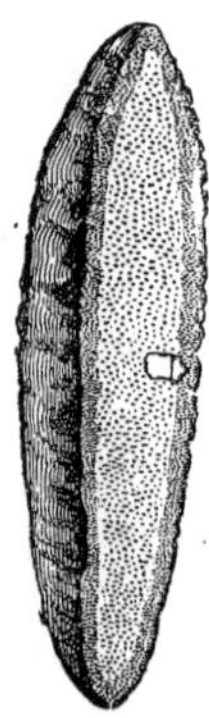

Fig. 267.—Section of the seed of the Date.

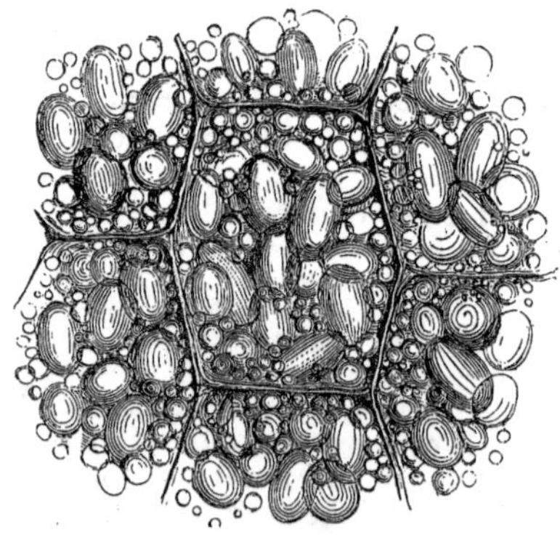

Fig. 268.—Starch Grains of Wheat (magnified).

simple, and their form is rounded, egg-shaped, and fusiform. Some are formed of two, three, four, or a higher, but still limited number of elements. There are some, also, compound, either spherical or egg-shaped; their surface, under the microscope, resembling a mosaic of polyhedral segments. We find other substances besides starch in the thin-walled cells of the albumen of the Castor-oil plant and in the thick-walled cells of the albumen of the Date. Oily matter abounds there. They are filled with corpuscles of a complex structure, whose chemical nature is not yet determined. These corpuscles, which in

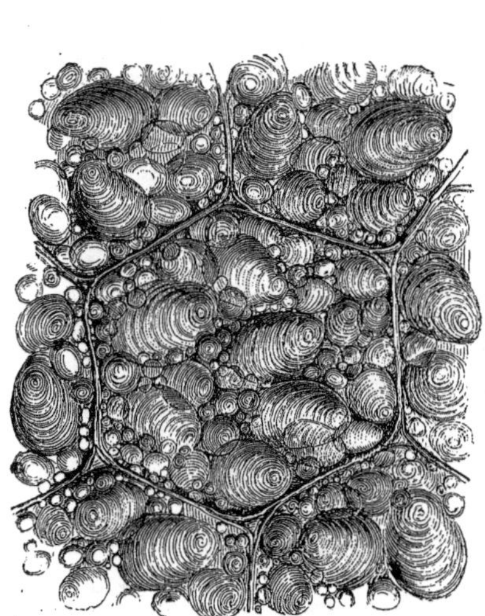

Fig. 269.—Starch Grains of the Potato (magnified).

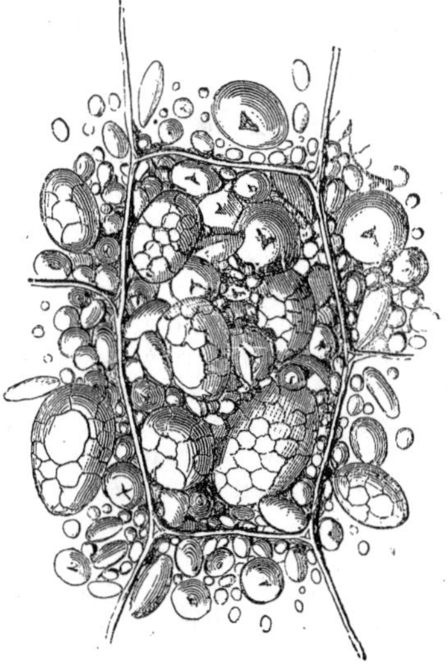

Fig. 270.—Starch Grains of Maize (magnified).

certain plants somewhat resemble grains of starch, are termed *grains of aleurone.* They are more or less soluble in water, and are coloured yellow by iodine. Grains of starch are, on the contrary, insoluble in water, and are coloured blue by iodine.

A very natural question here suggests itself as to the mode of transport and vitality of seeds; in short, as to the physiological phenomenon of *germination.*

Wind, running water, blocks of ice drifting in the Polar seas, the action of animals and men—that is, cultivation, merchandise, and voyages; such are the causes, more or less powerful, which effect the conveyance of seeds from one place to another. If we consider how many seeds are light, hairy, and provided with a sort of wings in their downy tufts, we can understand that the wind may be the most general and ordinary means for disseminating vegetable germs over a country. Rivers also carry away the seeds of plants to great distances. If their course runs from north to south, or the contrary way, the wandering seeds would be carried to a climate where they, in many cases, could not live; but if the river flows from east to west, or from west to east, the seeds thus transported would much extend the limits of the species. The currents of the sea, which skirt the coast, or extend from one country to a neighbouring one, carry the seed, so to speak, from one place to another step by step. In the latter case the seeds remain but a short time in the water, and are little altered in consequence; besides, the graduated temperature of the successive localities which they reach is favourable to their acclimatisation and to their further development. The operation of blocks of ice in the transport of seeds is not without a certain importance. Navigators of the Polar seas often meet with icebergs loaded with an enormous mass of *débris*, mixed with earth and seeds; if the iceberg runs aground on some distant coast, where it melts, the seeds are deposited; they soon produce plants, which are then spread over the country by the other influences already hinted at.

The dissemination of seeds is helped, it is said, by the distant migration of granivorous birds. Yet the influence of birds appears to us of very little importance in the matter we are now considering. Most birds completely destroy seeds in the act of digestion, and it is only exceptionally that seeds can traverse their intestinal canal without being destroyed.

"Omnivorous birds," says De Candolle, "often search for berries containing little hard seeds, as Grapes, Figs, Raspberries, Strawberries, Asparagus, Mistletoe, &c. Their stomach is not so destructive as that of the gallinaceous birds, and it appears that small seeds can traverse their alimentary canal without alteration. When these birds are migratory, which is often the case in temperate and northern regions, they carry the seeds to a great distance, particularly when, in the autumn, they leave northern climates to seek the sunny south, for at this season ripe fruits abound in the country. Thrushes, many of which change their country either in Europe or in America, can thus transport some sorts. When they eat too large a quantity of stone fruit, they digest it badly, and disperse the kernels." The great Linnæus assures us that the skylark scatters a great many seeds in the field during its flight.

By the same process, namely, by the imperfect digestion of seeds which have served them for nourishment, certain quadrupeds, particularly the herbivorous, are able to transport seeds from one part of the country to another. This happens in the case of the reindeer, an animal living in herds in the plains of Siberia, which, at a certain period, migrate in large numbers. Such also is the part played by the herds of cattle driven often to great distances in our European climates, and in general in all civilised countries.

The action of men in the dissemination of vegetable seeds is shown in a thousand ways. We will borrow from M. Alphonse de Candolle some interesting remarks on this subject.

"The first colonies which were sprinkled over each continent," says this learned botanist, "probably carried with them several species of useful plants, and especially some of those seeds which get attached to clothes and to domestic animals, and which will grow easily in the neighbourhood of dwellings, near dunghills, burnt ground, and rubbish heaps. The scantier the population, and the more foreign it is to the arts of civilisation, the more insignificant become these primary seed-carryings. When the population becomes denser and more civilised, when agriculture begins and extends its rule, then the occasions are multiplied for the transport of seeds. A hunting or pastoral people, no doubt, traverse a vast extent of country, but an agricultural people prepare ground fit for the reception of new species, and bring the seed for their fields from more or less distant countries, introducing with it different plants, many of which naturally grow wild. In short, war has created vast empires, and compelled men to make numerous journeys; navigation has extended itself, new countries have been brought in communication with the old, agriculture has exported its products, and horticulture has stocked our gardens with thousands of foreign species. By all these means the transport of seeds has become increasingly great, and an influence has been exercised quite preponderating over natural causes."

Commerce, which carries in its ships the products of the trading of nations, furnishing Europe with the produce of the New World and returning to it European productions in exchange, is sometimes an indirect agent in the transport of vegetable seeds. The wool from the sheep of Buenos Ayres, Mexico, or La Plata, when it is brought into Europe, carries entangled in the fleeces the seed and remnants of plants in those countries. When the fleeces arrive in Europe they are cleaned, beaten, and washed, and the seeds fall off; they may then shoot in this new soil, and transplant into our climate the vegetable species from regions across the Atlantic. At the edge of the river Lez, near Montpellier, at a place called Port Juvenal, the American wools are received to be cleaned and purified, and then sold to the cloth-makers of Lodève. Seeds of American plants which have been brought in these fleeces have actually sprung up in the environs of Montpellier; so much so, that all the celebrated botanists of Montpellier, such as De Candolle, Dunal, Delille, Gordon, and Ch. Martins, have examined and studied, in this small place in southern France, many vegetable species belonging to the Flora of Buenos Ayres and Mexico.

How long does the power of germination last in a seed? Some seeds rapidly lose their latent life, or, which comes to the same thing, the faculty of germination. There are others which, placed in the same circumstances, preserve their vitality for ages.

The seeds of most plants of the Leguminous tribe will germinate many years after they are gathered. Every one has heard that the seeds of Beans, taken lately from the herbal collection of Tournefort, a celebrated botanist in the seventeenth century, germinated perfectly. In 1824 there were sown, at the Jardin des Plantes of Paris, some seeds of the *Mimosa pudica* which had been gathered at St. Domingo in 1738.

If seeds are placed under special conditions, sheltered from atmospheric agency, in ground more or less dry and heaped up—for example, in tombs or catacombs—their vitality may be preserved for a prodigious time. It is a recognised fact that, after the destruction of a forest, we see a new sort of growth appear on the ground the forest used to occupy. It has been admitted, to explain this fact, that seeds of trees, buried in the ground while the forest existed, were preserved in the soil with their life suspended during a considerable number of years; that, then awaking from their lethargic sleep, they have been developed under the influence of new conditions favourable to their germination. This hypothesis is plausible enough in some cases, yet, as no scientific experiment has been made on the subject, it might happen, and is more probable, that in this case the second germination of trees was owing to the transport of foreign seeds, which germinated as soon as the soil had become free and restored to light.

Almost marvellous examples are quoted of the longevity of seeds. Dr. Lindley, the learned botanist, asserts that some Raspberry seeds which were found in a Celtic tomb, and which numbered about seventeen hundred years of existence, had germinated perfectly, and produced Raspberry plants, which still exist in the Royal Horticultural Society's Gardens at Chiswick. M. Ch. Desmoulins asserts that the seeds of the Black Medick, Cornflower, and Heliotrope found in some Roman tombs, dating from the second or third century of the

Christian era, have not only germinated, but produced individual plants, which have in due course flowered and borne fruit.

We must be on our guard against such prodigies as these. Although Nature, as has been said, performs wonders, we must only believe in facts rigorously established, in order to avoid the mystifications and tricks which the malignity of the vulgar loves to play upon the credulity of the *savants*.

We must not forget to speak of those wonderful seeds of Wheat found in the tombs of ancient Egypt. It is now acknowledged that in this affair some one must have abused the confidence and credulity of travellers. A variety of Wheat called Mummy Wheat is common, it is true, among farmers; but no authentic fact justifies its name. Though there is, as we have just seen, so great a difference in seeds as to the duration of their vitality, the difference is not less as regards the time necessary for their germination. Some seeds, as Garden Cress, Poppy, and the cereals, germinate in a few days. Others, as those of the Peach, Almond-tree, Nut, and Rose, require one or even two years before they "come up." This difference is owing partly to the size of the seeds, their hardness, and the woody nature of their integuments; it is due also to the presence of a shell round the seed.

There are some seeds which seem in such a hurry, so to speak, to develop themselves, that they even germinate in the fruit enclosing them. This very often happens among some sorts of Lemon-trees, and certain of the Cucumber family. The embryo of the Mangrove, a tree inhabiting swamps, mouths of rivers, and sea-beaches in the equinoctial regions of America, is developed inside the fruit still hanging on the branches, and there can often be seen hanging from it a root more than a foot in length.

PHENOMENA OF REPRODUCTION IN PLANTS.

FERTILISATION—GERMINATION.

THE consideration we have given to the subject of the flower and fruit enables us now to enter on two great questions in vegetable physiology: firstly, the influence of the stamens on the pistil, or *fertilisation* in plants; secondly, *germination*.

FERTILISATION.

Of all the phenomena in the life of plants, there is none more interesting or more remarkable in itself than fecundation. When the existence of sexual differences in vegetables was first propounded, the discovery produced general astonishment. If the most convincing proofs had not established it, if the commonest observation had not allowed every one to verify its reality, it would certainly have been classed among the most singular inventions that ever issued from a poet's imagination. But the proofs were convincing. The demonstration of the existence of sexual organs in vegetables became a brilliant and unexpected fact, exhibiting a wonderful analogy between animals and plants; filling up in part the gulf which had hitherto existed between the two great classes of organic beings, yielding an inexhaustible fund of reflection and comparison to naturalists and thinking men.

The ancients had very vague ideas on this subject. Yet we learn from Herodotus that, in his time, the Babylonians already distinguished two sorts of Date Palms; they sprinkled the pollen of one on the flower of the other, in order to perfect the production of the fruit of that valuable tree.

Cæsalpinus, an Italian philosopher, physician, and naturalist, who, in the sixteenth century, was professor of medicine and botany at Pisa, remarked that certain plants of *Mercurialis* and Hemp remained sterile, while others were productive. He considered the first as the male plants and the second as the female. In the seventeenth century, a learned Englishman, Nehemiah Grew, Fellow of the Royal Society of London, who published in 1682 an "Anatomy of Plants," and above all, Jacobus Camerarius, a German botanist, born at Tübingen, showed the precise use of the two essential parts of the flower, and the part that each plays in producing the fecundation of germs. In a letter, now become celebrated, addressed to Valentini, "De Sexu Plantarum," published in 1694, Camerarius completely proved the great fact of the existence of the sexes in plants, just as in animals. This discovery made an impression on the minds of naturalists: it was, in fact, one of the most striking victories which natural science ever obtained.

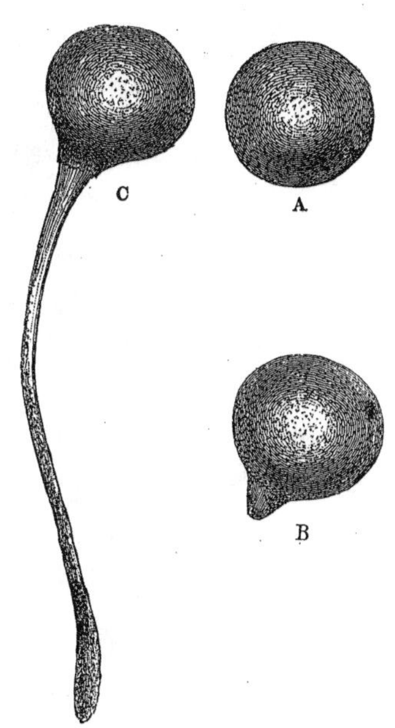

Fig. 271.—Pollen-grains emitting the Pollen Tube (magnified).

After the labours of Camerarius, the existence of sexes in vegetables was generally admitted. Tournefort was incredulous; but Sebastien Vaillant, one of his most brilliant pupils, in 1717 publicly professed in the Jardin des Plantes at Paris the theory of separate sexes in plants. In 1735 the celebrated Linnæus rendered it popular by basing on the sexual characteristics of vegetables his vast and admirable system of classification, the importance of which we shall appreciate further on.

The pollen having been recognised as the matter which fecundates the ovary, the next question was to discover in what manner the grains of pollen produced the fecundation of the vegetable germ. It was at first thought that the grains of pollen simply opened on the stigma, and that the granules which they contained, being absorbed by the stigma, went to form the embryo, or concurred in its formation. It was the most natural opinion to form *à priori*, yet observation has since proved that a much more complicated process takes place.

In 1822, Amici, an Italian natural philosopher, while observing the Purslane (*Portulaca oleracea*), perceived that the grains of pollen, far from opening, as was thought, on the stigma, in order to pour out the fecundating matter, changed by degrees into a sort of membranous tube, which he called the *pollen-tubes*, as represented in

A, B, C, Fig. 271, which shows the successive stages through which the pollen passes in emitting the *pollen-tube.*

In 1826 the celebrated botanist, M. A. Brongniart, in his researches on this subject, perceived that the same fact recorded by Amici occurred in numerous plants; he observed also that the

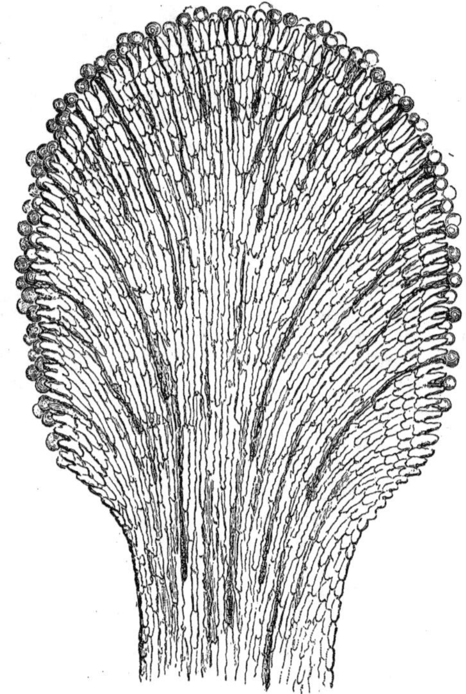

Fig. 272.—Vertical Section of the Pollen Tube of *Datura* (magnified).

pollen-tubes generally penetrated more or less into the style. He instanced the *Datura* as one of those plants in which the action of the pollen on the stigma is very observable. "These tubular sacs," he says, "are for the most part already filled with granules, and easily distinguished from the tissue of the stigma by their brownish colour and opacity. I could not find a better comparison for one of these stigmas," he adds, "than a pincushion entirely filled with pins stuck into it up to the head."

Fig. 272 represents, according to the account of M. Brongniart, a vertical section of a stigma of *Datura* penetrated by pollen-tubes

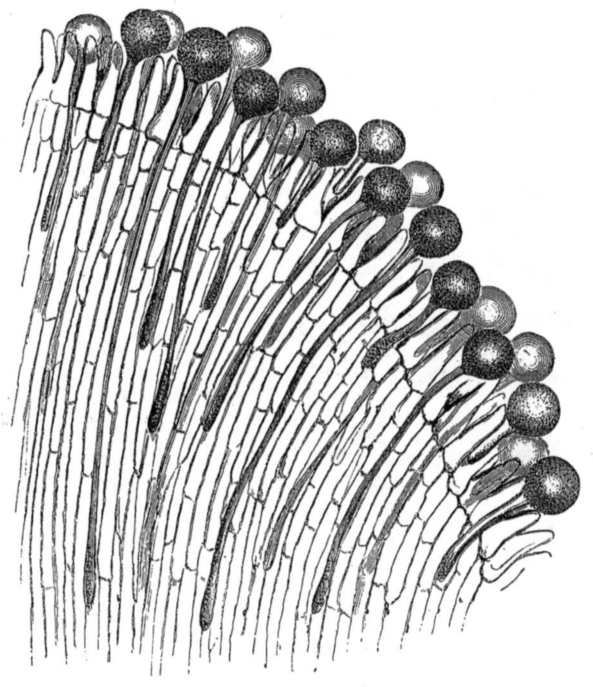

Fig. 273.—Conducting Tissue in *Datura* (magnified).

through all its thickness. Such is the appearance which the stigma of the *Datura* presents when strongly magnified.

Fig. 273 is intended to show the same arrangement in the same plant, but still more strongly magnified. The grains of pollen and the pollen-tube are here still more enlarged, the better to show the structure and passage of the tube through the substance of stigma.

In order to understand this curious organic peculiarity, Fig. 274 represents a similar stigma of the *Datura* seen externally, and resembling, as M. Brongniart says, a pincushion full of pins. But such is the incessant progress of science, that in our days these early observations of M. Brongniart have been carried much further, and

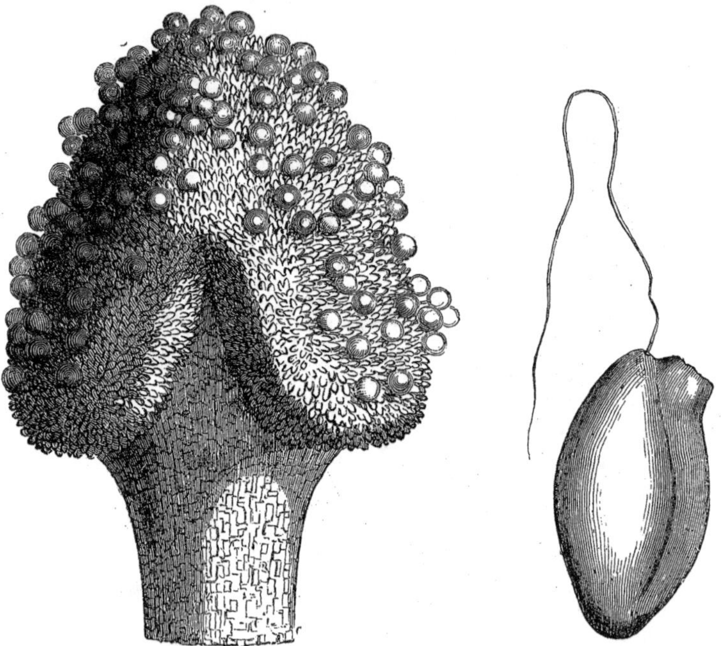

Fig. 274.—Stigma of Datura covered with Pollen (magnified).

Fig. 275.—The Ovule (magnified).

recent investigations show still more clearly the system of progression in the pollen-tube.

This tube, as M. Brongniart has shown, elongates itself by a most remarkable vegetative process, insinuating itself into the interstices of the cellular tissue, which has been designated from this cause the conducting tissue, and by which doubtless it is nourished. Occupying the centre of the style, this tube traverses its whole length,

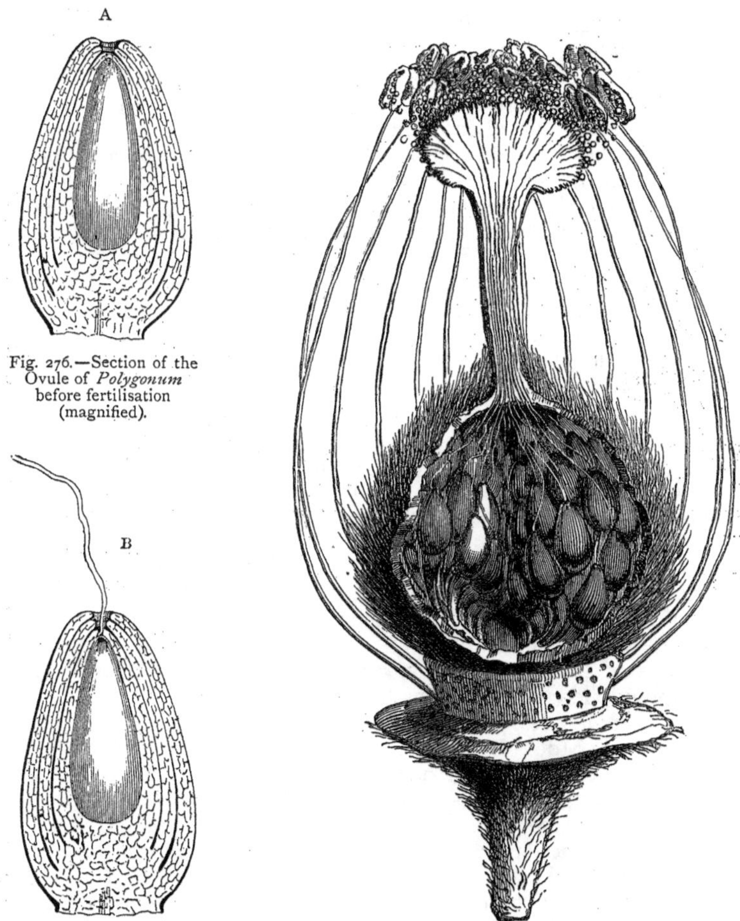

Fig. 276.—Section of the Ovule of *Polygonum* before fertilisation (magnified).

Fig. 277.—The same after fecundation (magnified).

Fig. 278.—Fertilisation of the Ovule.

entering into the ovary, and is there brought in contact with the ovules, penetrating by their micropylar openings.

Fig. 278 is a section of the stigma, style, and ovary, and is intended to point out the long course followed by the pollen-tubes in penetrating from the stigma to the interior of the ovary, where each of them comes in contact with the ovules.

One of these ovules is represented in Fig. 275, taken singly and magnified, to show this phenomenon more clearly. The ovule here represented is that of *Viola tricolor*. The extremity of the pollen-tube, in contact with the summit of the ovule, proceeds to place itself in still nearer connection with one of the component cells of its nucleus, now excessively developed, in which state it bears the name of the *embryo-sac*, because there the embryo undergoes in it the process of development. The same organ is represented, at the moment of fecundation, in Fig. 276. Here an internal section of the ovule of the *Polygonum* is given, both before and after fecundation; A is the ovule before fecundation, B, Fig. 277, the same organ after it. We see on the fecundated ovule, B, the commencement of the formation of the *embryo sac*, near the terminating point of the pollen-tube.

About the year 1837, two German botanists, MM. Schleiden and Horkel, announced that the vegetable embryo pre-exists as a germ within the grain of pollen, and that it is formed of the end of the pollen-tube itself, when this extremity is lodged in the embryonic sac, which is driven back before it.

This theory, which reproduced, and seemed to take for granted, in the vegetable kingdom the celebrated hypothesis on the *enclosure of germs* put forth by Buffon for the animal kingdom, made much noise among the learned in Europe. It was supported by the personal observations of many of our best botanists; but it could not long resist the multiplied investigations that the importance of the subject called forth on all sides.

MM. Amici, Mohl, Unger, and Hoffmeister soon demonstrated that, in fact, when the pollen had once reached the embryo-sac, it remained there, attached by its external wall, and that there its functions ended with its life; whilst a little vesicle plunged in the mucilaginous juice with which the embryo-sac is filled, absorbs by endosmose the fertilising elements which the pollen-tube has doubtless transmitted through its constituting membrane, and that this element is then developed so as to form the embryo.

Schleiden's theory of the pre-existence of vegetable germs received its final blow when, in 1849, M. Tulasne, one of the ablest of French anatomists, published his magnificent studies on vegetable embryogeny. M. Tulasne had always observed that the obtuse extremity of the pollen-tube was brought close to the membrane of the sac, strongly adhering there without causing any perceptible depression. At some distance from the point of contact, there was developed, on the membrane of the sac, a vesicle with a circular base, at first like a blister, which by cellular growth was soon transformed into the embryo. Fig. 279 represents the result of M. Tulasne's observations, and the manner in which the extremity of the pollen-tube is introduced into the *nucleus.* Fig. 281 is an internal section of the same organ, showing the formation of the vesicle about to become the embryo, and Fig. 280 shows this vesicle about to become a small globe of parenchymatous tissue, a sort of rough sketch of the embryo. The embryo thus formed may acquire considerable development, and absorb

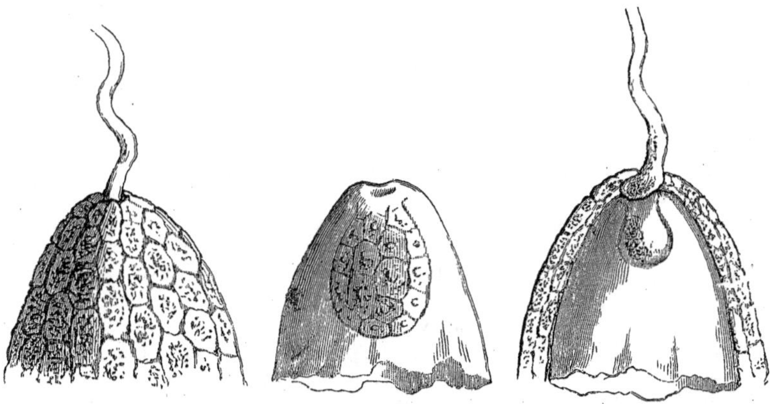

Fig. 279.—Pollen Tube penetrating the Nucleus (magnified).

Fig. 280.—Formation of the Embryo (magnified).

Fig. 281.—Pollen Tube after penetrating the nucleus (magnified).

for its use all the soft matter contained in the embryo-sac; or it may be limited in size, and this soft matter, becoming a permanent and cellular tissue, soon constitutes itself an accessory but important part of the seed, which is known by the name of *albumen.*

We have now rapidly set forth the functions of the pollen and the ovule in the great phenomenon which secures the perpetuity of the species; but in this rapid glance at some of the most secret mysteries of vegetable fertilisation, we have stated the facts without occupying ourselves with any of the external circumstances, that is to say, the influences acting from without, which prepare for it, and which determine and favour it. We now enter into some details on this subject, and of some of the phenomena accompanying fecundation.

In a great number of hermaphrodite flowers, the stamens at the period of fertilisation elevate their anthers higher than the stigma; so that the pollen falls naturally upon it at the moment of the opening of the anthers. In other flowers, the stamens carry their anthers lower than the stigma, but the flower is habitually inclined or suspended, as in the Fuchsia; the deposition of the pollen on the stigma is then made without any obstacle.

When the stamens and pistils are not close to one another, Nature sets the necessary means to work to promote their near approach. Thus, we observe in different plants some very curious and varied movements in the stamens at the period of fecundation. In the Nettle, the Mulberry, and the Pellitory, for instance, the filaments of the stamens are bent backwards on themselves, under the pressure of the floral envelope; but as soon as full bloom takes place the filaments unroll, and the pollen is projected to a distance of thirty or forty inches or more. This movement is simply the result of the elasticity of the organs. In the Rue, at the moment of fecundation, each of the numerous stamens constituting the andrœcium bends itself over the stigma, deposits the pollen there, and resumes its former position. Here is an individual and really spontaneous movement.

In the Passion-flower the styles are at first erect, but at the moment of the opening of the anthers, they are observed to curve downwards, and lower themselves towards the stamens, and then to rise up and resume their former position.

In the flower of the Barberry, if a stamen is touched with the point of a pin, it is brought close to the pistil by a sudden movement, and then, in a little time, resumes its former position; and this it will do again if fresh irritation is produced. A phenomenon of irritability is shown here, which does not exist in the other cases just specified.

The hairs which cover the styles of the Campanula show a very singular property. They fold back on themselves, like the finger of a glove, the end of which is pushed inwards, and they occasionally draw with them into this retreat the grains of pollen; Hartig has suggested that they may then penetrate the tissue of the style, but this is very improbable.

In a pretty little plant of New Holland, known under the name *Leschenaultia*, the stigma is in the form of a cup, and it is edged with rather long hairs. At the moment of the anthers opening, part of the pollen grains falls into the cup of the stigma, which contracts in order to grasp them, whilst the hairs approach each other so as to prevent the exit of the fertilising dust.

In the facts we have just pointed out, the organs themselves appear to act to produce the fecundation of the flower. But this physiological action is often facilitated by the concurrence of exterior agents. The wind has power to transport the pollen to a certain distance, and thus favours the fecundation of the flowers in *monœcious*, *diœcious*, or *polygamous* plants. Insects, while flitting from flower to flower, often become the active instruments of vegetable fecundation.

In the *Orchidaceæ*, in which the structure of the pollen is very peculiar, the intervention of insects appears not only favourable, but in most cases indispensable, to fecundation.

When the doctrine of sexual organs in vegetables was first made public by Linnæus, it was disputed by many. Conrad Sprengel, a patient observer, watched during many long hours for the instant when an insect, settling on a flower, should suck out its sweet-smelling juices and deposit the pollen-grains on the stigma of the flower. Sprengel succeeded in this way in showing that even in hermaphrodite flowers—inasmuch as the style and anthers are rarely matured at precisely the same time—the intervention of insects is often taken advantage of in fertilising one flower by the pollen of another.

In certain climates the Humming-birds are useful auxiliaries also in the fecundation of flowers. The hand of man frequently intervenes in practising artificial fecundation—bringing in this way the most convincing of all arguments in favour of the doctrine. We may instance as an example the fecundation of the Date-tree, which is practised in Algeria and all over the East, as related by a botanist who has studied the subject on the spot:—"Towards the month of April," says M. Cosson, "the Date-tree begins to flower, and then artificial fecundation is practised extensively. The male spathes are opened at the time when a sort of crackling is produced under the finger, which indicates that the pollen of the flowers in the cluster is sufficiently developed, yet has not escaped from the anthers; the cluster is then divided into fragments, each containing seven or eight blooms. Having placed the fragments in the hood of his *burnous*, the workman climbs with marvellous agility to the summit of the female tree, supporting himself by a loop of cord passed round his loins, and at the same time round the trunk of the tree. He glides with great address between the stalks of the leaves, the strong and sharp thorns of which render the operation rather dangerous; and having split open the spathe with a knife, he slips in one of the fragments, which he interlaces with the branches of the female cluster, the fecundation of which is now made certain."

Another phenomenon sometimes exhibits itself at the time of

flowering, which bears an intimate relation to fecundation; this is the production of heat. M. Ad. Brongniart has made some experiments on this subject which have become famous. At the time of opening, the inflorescence of the sweet-smelling Colocasia presented to this observer an increase of temperature that might almost be compared to an attack of daily fever. This increase was repeated for six following days with a considerable intensity, and almost at the same hour; for it was between three and six in the afternoon that it reached its maximum. Analogous phenomena have been noticed at the time of fecundation in the spadix of our common Arum (*Arum maculatum*), the flowers of the splendid *Victoria regia*, the Magnolia, and some other plants.

It is impossible to conclude our remarks on the fecundation of plants without instancing the aquatic plant known as *Vallisneria spiralis*, which has long been the admiration of naturalists, while poets have sung its praises. The *Vallisneria* is a diœcious plant; that is, it has male and female individuals existing separately, in the tranquil waters of some countries in central Europe, principally France and Italy (Fig. 282). In the female plant the peduncle of the flower is very long, having the form of a spiral twisted thread-like filament. A few days before fecundation the spiral turns untwist themselves, and the peduncle lengthens, until the female flower terminating it reaches the surface of the water, on which it floats. The male plant presents, on the contrary, a very short peduncle, which is not capable of any extension: it bears a multitude of little flowers, provided with stamens only, and enveloped by a closed transparent spathe. At the time of full bloom the spathe is torn, the peduncle of the male flower severs itself towards its upper part, and the flowers separated from the stalk rise; all shut up, like very small white pearls; they float on the surface of the water, and proceed to open near the female flower, which seems to wait for them. When fecundation has been effected, the peduncle of the female flower contracts; it brings together its spiral turns, and carries its ovary to the bottom of the water, in order to ripen its seed.

This is a phenomenon which has always excited the just admiration of naturalists and observers of every class. I was initiated into the first elements of natural science, at the Lyceum of my native town, by M. Joly, now Professor of the Faculty of Science at Toulouse, a young professor then, who excelled in inspiring his pupils with a taste for this sort of study. The wonderful incidents attending the process of flowering in the *Vallisneria spiralis*, were a favourite text for the discourses of M. Joly during our botanical and geological

Fig. 282.—*Vallisneria spiralis.*

excursions round Montpellier, in the flower-decked wood of La Valette, or on the volcanic summit of Monferier. Thirty years have elapsed since those happy youthful days, and the recollection is just as vivid, just as present to my mind, as if I still heard the burning words of our then young teacher ringing in my ears, telling us, under our radiant skies, of the wonders of nature, and of the power of God.

Germination.

In order that a seed should germinate, three conditions are requisite—heat, air, and moisture; temperature, varying in different species, must not be much less than 50° or 59° (Fahrenheit), and it must not reach higher than 104° or 113°.

Moisture penetrating the seed beneath the ground softens it, swells all its parts, and allows their gradual unfolding.

Air is also as indispensable to the germination of seeds as it is to animal life. Seeds which are buried too deeply in the ground, and are thus cut off from the air, will never germinate.

What, then, is the important part that atmospheric air performs in the act of germination? It is just the same as that which it fulfils in the respiration of animals. Air acts on the seeds by means of its oxygen. The germinating seed, like the animal, breathes out carbonic acid. A portion of the carbon contained in the nutrient substances of the seed is burnt off, as it were, in the chemical changes which accompany germination, and combines with the oxygen of the air to form carbonic acid; but from the instant the young plant has produced small green leaves, the chemical phenomenon is, so to speak, reversed. In the day-time, and under the influence of light, the young plant absorbs carbonic acid from the air, and replaces it with oxygen; its respiration takes place just as we described it when speaking of this physiological function in the green-coloured portion of vegetables.

We will now follow the series of phenomena presented to the observer by the germination of a seed.

The first apparent effect of germination is the swelling of the seed, and the softening of the coverings that envelop it. If the seed contains an albumen, the embryo, which is in contact with the albumen either over its whole surface or the greater part of it, absorbs the nutritive matters which it contains, and increases in size in the same proportion as the albumen gets less, being developed at the expense of the substance stored up for this end by a provident Nature. If the seed is destitute of an albumen, and the embryo at the time of dissemination fills up the whole cavity of the seed, then

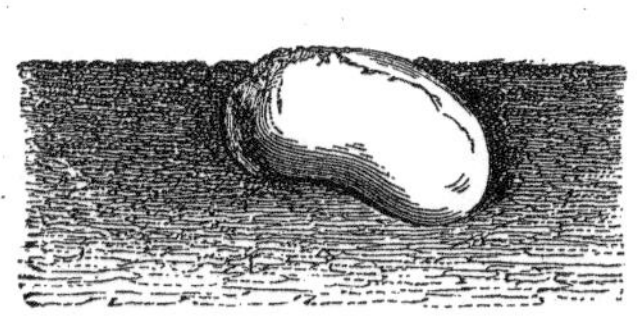

Fig. 283.— Haricot Bean germinating.

the cotyledons—which are farinaceous in the Pea, or fleshy in the Nut or Cole-seed—forming the greater portion of the embryonic mass, will perform the part of an albumen, as regards the rest of the embryo. Fig. 283 represents the first effort of germination in the Bean, whose seeds are exalbuminous.

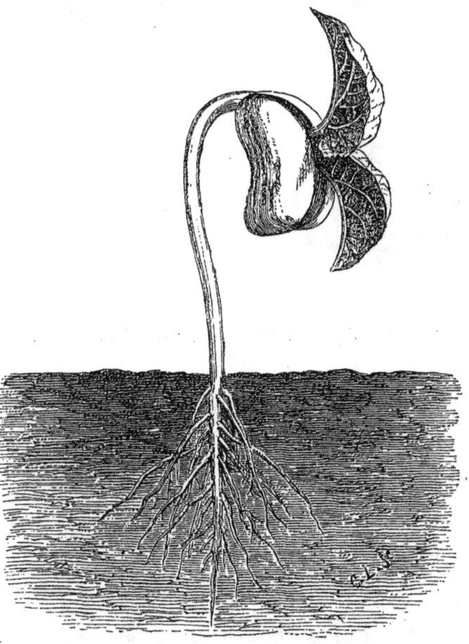

Fig. 284.—Germination of the Haricot Bean.

It was long a mystery how the starch of which the albumen of Wheat is almost entirely constituted can be absorbed by the young embryo, since the radicles of plants absorb soluble matters only, and starch is completely insoluble in cold water. But the interesting discovery has been made, that the insoluble starch becomes soluble under the influence of an energetic agent, which is developed near the germs at the time of the seed germinating; this dissolving agent has received the name of *diastase*. The starchy matter transformed by diastase into a soluble substance bears the name of *dextrine*.

Dextrine is modified in its turn under the influence of diastase, and losing part of its carbon, becomes sugar. We shall be right, then, in saying that the first nourishment of the young plant is sugared water.

Efforts have been made to discover if a grain of starch, while being transformed into dextrine, shows any visible trace of so complete a molecular change; whether it disappeared suddenly under the action of diastase, or is only gradually changed into a somewhat similar substance, so that one could follow out with the microscope all the phases of this change. It has been proved that this change is

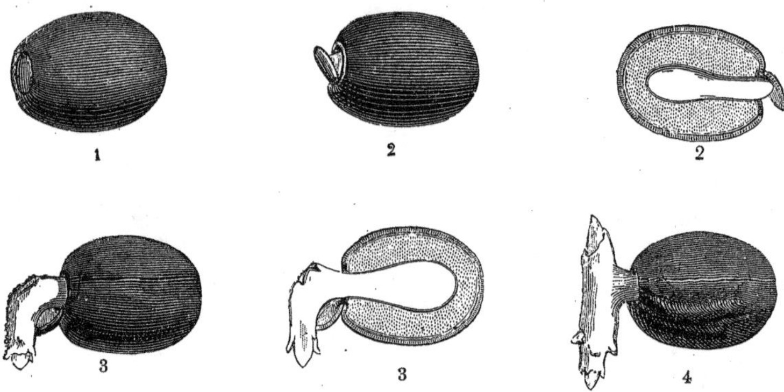

Fig. 285.—Germination of the Indian Shot (*Canna indica*), magnified.

only brought about by successive steps, and we are enabled to follow the progress of this alteration of the starch granules in the germination of several plants.

To return to the evolution of the embryo. However nourished and strengthened, either at the expense of the albumen or of its own cotyledons, the embryo quickly presses the integuments covering it on all sides, which in the end are broken, thus giving it a passage through. This rupture takes place sometimes in an irregular manner, as in the Spanish and other Beans (Fig. 284); sometimes in a very regular manner, as in the Virginian Spider-wort, the Date, and the species of *Canna*. In the last case the embryo appears outside, through an opening very regularly cut out in the integument covering the seed. This opening is hidden at first by a sort of disk, or lid, which the little root of the embryo lifts up in order to make its way out and bury itself in the ground.

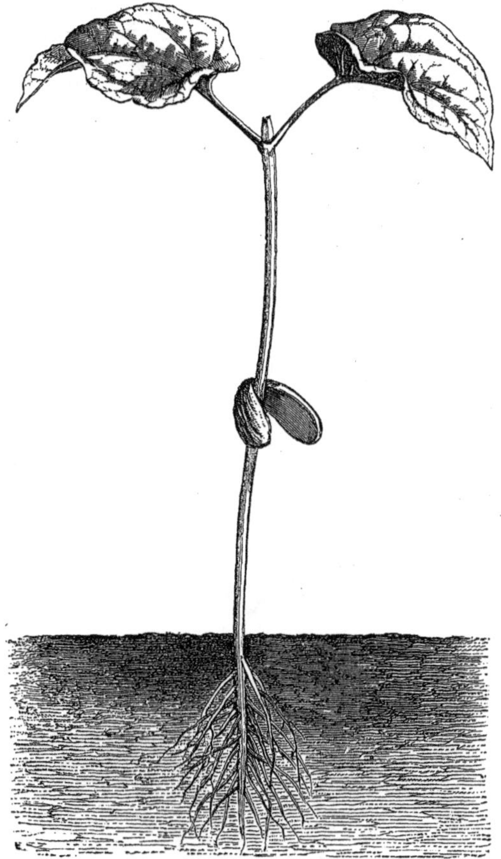

Fig. 286.—Germination of an exalbuminous seed, showing the Cotyledons which ascend above ground with the stem.

Fig. 285 shows the successive stages through which a germinating seed (1) of the Indian Shot passes. The little lid is lifted up and cast aside (2); the cotyledon is developed, elongated horizontally, and the radicle pushes its way out, and is quickly pointed towards the ground (3); the plumule, or bud, emerges from the opening of the cotyledon, which is transformed into a sheath; the radicle is increased in size, and the rudiment of the stalklet appears; lastly, the stalklet is formed (4). The seeds of most of the monocotyledonous plants are provided with albumen, and at the time of germination the limb of the cotyledon remains shut up in the seed, as we see in the Palms, Indian Shot, and Virginian Spider-wort.

PART II.

CLASSIFICATION OF PLANTS.

EVERY plant which grows on the surface of the earth or in the waters constitutes a distinct individuality. The careful examination and comparison of a certain number of these individuals of the Vegetable World will lead to the admission that a great many are quite identical in some of their characteristics, while others possess no character in common. Examine the individual plants, for instance, which compose a field of Oats; in each the root, the stem, the flowers, the fruit, present the same identical characters. The seed of any one whatever of these plants will yield other plants like those of the field. Every individual in the field belongs therefore to the same *species*—to the species *Avena sativa*.

The species, then, is a collection of all the individuals which resemble each other, and which will reproduce other individuals like themselves.

These species may present, as the result of diverse influences, such as change of climate or cultivation, differences more or less marked, more or less persistent, which withdraw them from the original type. To these, according to their importance, botanists give the name of *varieties* and *sub-varieties*. The Wheat-plant, the Vine, the Pear, the Apple, and most of our cultivated legumes, all yield, under the influence of culture extending over a long series of years, plants altogether different from the original in their exterior; but they preserve, one and all, the essential characters of the species. They are *varieties* of the Wheat-plant, of the Vine, of the Pear, of the Apple.

The assemblage of a certain number of distinct species presenting the same general characteristics, the same disposition of organs, the same structure of flower and fruit, constitutes a group to which the name of *genus* is applied. *Rosa canina*, *R. villosa*, and *R. Sabini* are three different species of the same group—the genus *Rosa*. The words *oak*, *poplar*, *barley*, are collective common names, which served, long before botanical science existed, to designate certain groups of plants. These are true generic names of popular creation, which botanists have accepted because they were the result of exact observation. "A man of observant eye and quick intelligence," says Auguste Pyramus de Candolle, "would observe certain groups in the vegetable kingdom which we call genera, before discerning the species."

The germs of botanical science are to be sought for in the rudimentary state in very remote antiquity. In the sacred writings we meet with constant allusions to the vegetable world. The cultivators of the science among the early Greeks and Romans were not botanists, but *Rhizotomæ*, or root-cutters, since they directed their attention to the roots in search of medicinal properties. Aristotle of Stagira, who lived in the fourth century before our era, may be regarded as the founder of botany; Mithridates, and the younger Juba, King of Mauritania, were among its cultivators. They established Botanic Gardens, some probably from love of the science, others of them in order to cultivate the deadly plants from which poisonous juices were obtained. Nicander of Colophon, Cato, Varro, Columella, Virgil, Pedanius Dioscorides of Cilicia, and lastly, the elder Pliny, all dwell upon the wonders of vegetation; and war, notwithstanding its desolating tendencies, was made to promote the interests of science.

To the Arabians of the twelfth century we are next indebted for our knowledge of botany. After them the darkness of the Middle Ages set in, and it is only since the illustrious Venetian, Marco Polo, came to examine and describe the wonders of the East, that the darkness has been dispelled. He examined the treasures of Asia and the East Coast of Africa, described many plants of India and the Indian Ocean, and from his day to the present our knowledge of the names of plants, as well as of their structure and physiology, has been continually on the increase.

The science of botany, as now understood, cannot be held, however, to date farther back than two centuries. In the year 1682 Nehemiah Grew, the Secretary to the Society of London, afterwards the Royal Society, published his "Anatomy of Plants." In 1684 the French botanist Tournefort, then Professor of Botany at the Jardin des Plantes, published his "Elements of Botany," being the first attempt to define the exact limits of genera in vegetables. Most of the genera established by Tournefort remain, proving the correctness of the formula from which he deduced their common characters. Tournefort succeeded to a large extent in unravelling the chaos into which the science of botany had been plunged from the days of Theophrastus and Dioscorides. Separating genera and species according to their characteristics, he described no less than 698 genera, and 10,146 species. He published, at the same time, a

Fig. 287.—Tournefort.

system for the classification of plants, eminently attractive, especially if we connect it with the times in which it appeared. The French botanist directed the attention of observers, probably for the first time, to those parts of plants most likely to excite admiration, namely, the different forms of the corolla.

In selecting the form of the corolla as the basis of his classifi-tion, Tournefort has, perhaps, contributed more to the progress of botany than any other *savant* of any age. The task of instruction was rendered a pleasure by thus taking, as a subject of scientific inquiry, the most attractive part of the plant. He soon made adepts of those who had hitherto only contemplated flowers as the source of an agreeable sensation.

Tournefort however fell into error in insisting upon the untenable division of plants into *Herbs* and *Trees*. In his next divisions he pointed out that the flowers of herbaceous plants are, or are not, furnished with a corolla; they are simple or compound; the corolla is *monopetalous* or *polypetalous;* it is regular or irregular. Such were the considerations on which Tournefort founded his classification of herbaceous plants.

As to trees, the flower is, or is not, provided with a corolla; that is to say, it is *apetalous* or *petalous*. The apetalous trees have the flowers disposed in catkins, or they have not; the petalous trees have the corolla regular or irregular.

Arranged and tabulated according to the system of Tournefort, beginning with trees, the scheme of the vegetable world will stand as follows:—

FLOWER-BEARING TREES.

Apetalous			APETALOUS plants, properly so called.
			AMENTACEÆ.
Petalous	Monopetalous		MONOPETALOUS.
	Polypetalous	Regular	ROSACEÆ.
		Irregular	PAPILIONACEÆ.

Trees, then, form five classes.

In the class of *Apetalous* plants are ranged the Box-tree and the Pistachio; in the class of *Amentaceæ* are the Oak, the Walnut-tree, and the Willows.

The Lilac, the Elder, and the Catalpa belong to the *Monopetalous* division; the Apple, the Pear, and the Cherry to the *Rosaceæ;* the Acacia, the Laburnum, to the *Papilionaceæ*.

The herbaceous flowering plants without corolla are subdivided into three classes: (1) plants provided with stamens; (2) flowerless plants provided with seeds; (3) plants in which the flowers and fruits are not apparent. Wheat, Barley, and Rice belong to apetalous herbaceous plants with stamens; the Ferns and Lichens to flowerless apetalous herbaceous plants provided with fruits; and the Mosses and Mushrooms to apetalous herbaceous plants without flowers, and having no apparent fruit.

Tournefort formed fourteen classes of flowering herbaceous plants provided with a corolla. The first eleven classes include the herbaceous plants with isolated and distinct flowers; the three others include the flowering herbs, which constitute the *Compositæ*, namely, the *Flosculosæ*, or flowers with funnel-shaped petals; the *Semi-flosculosæ;* and the *Radiatæ*—plants such as the Sunflower and the Daisies.

The following is a tabular arrangement of simple-flowering herbaceous plants according to the grouping of Tournefort:—

With Monopetalous corolla	Regular	*Campaniformes* . .	Campanula.
		Infundibuliformes .	Tobacco.
	Irregular	*Personatæ*	Snapdragon.
		Labiatæ	Salvia.
Polypetalous corolla	Regular	*Cruciformes* . . .	Stock Gilly Flower.
		Rosaceæ	The Rose.
		Umbelliferæ . . .	Angelica.
		Caryophyllæ . . .	Pink.
		Liliaceæ	Lily.
	Irregular	*Papilionaceæ* . . .	Pea.
		Anomalæ	Violet.

In addition, Tournefort has subdivided each class into sections, more or less considerable, based upon the composition and consistence of its fruit, and upon some particular modifications of the form of the corolla.

Such is the first known system for the classification of plants. This scientific conception met with great favour among his contemporaries, on account of its simplicity. Nevertheless, in its application, this system presented many difficulties. The form of the corolla is not always so exactly appreciable that the class to which that plant belongs can be settled from that character alone. But the gravest defect of the system is, that by it the vegetable world is divided into two classes, namely, Herbaceous Plants and Trees—a division which has no existence in Nature. The division destroys the natural analogies, for the size of a plant has no bearing upon its organisation and structure. In conclusion, the continually increasing number of new species, which were unknown in Tournefort's time, tests, in the strongest manner, the defects of his system of distribution. The greater number of vegetable species discovered since Tournefort's time could not be placed in either of his classes. This defect soon became very apparent, and the system fell by degrees out of favour with botanists even among his own countrymen, with whom it had found most admirers.

In England the study of plants had taken a more philosophical direction. About the middle of the seventeenth century the microscope was first applied to the study of the organs of plants; and in 1661 spiral vessels were detected by Henshaw in the Walnut-tree, and shortly afterwards the cellular tissues were examined by Hooke. These discoveries were followed by the publication of two works on the minute anatomy of plants by Malpighi and Grew. They examined the various forms of cellular tissues and intercellular passages in their minutest details, and with an exactness which causes their works still to be recognised as the groundwork of all physiological botany. The real nature of the sexual organs in plants was demonstrated by Grew; the important difference between the seeds with one and those with two cotyledons was first pointed out by him. Clear and distinct ideas of the causes of vegetable phenomena were gradually developed, and a solid foundation laid on which the best theories of vegetation have been formed by subsequent botanists.

About the time when Tournefort was engaged in arranging his system of plants, and when Grew had completed his microscopical observations, John Ray was driven from his collegiate employments at Cambridge by differences of opinion with the ruling powers of his University. He sought and found consolation in the study of natural history, to which he was ardently attached, and for which his powers of observation, capacious mind, and extensive learning so highly qualified him. Profiting by the discoveries of Grew and other vegetable anatomists, in 1686 he published the first volume of his "Historia Plantarum," in which are embodied all the facts connected with the structure and organs of plants, with an exposition of the philosophy of classification, the merits of which are better appreciated now than they were in his own days.

Ray was careful to guard his readers against the supposition that classification was other than a means of identification. He argued that there was no line of demarcation in Nature between one group or order, or even genus, and another, or that any system could be perfect. "What, indeed, I said before, I now repeat and insist on," he says, "that a system is not to be expected from me which shall be in every respect perfect and complete in all its parts; which shall so distribute plants into genera that every species shall be included, no one hitherto regarded as anomalous and exceptional being omitted; and which shall so mark out every genus by its peculiar indications and characteristics, that no species shall be found of uncertain family, so to speak, and referable to many genera. Nor by the very nature of things could this be done. For Nature (as is sometimes said) makes no leaps, passing from one extreme to the other, but takes a middle course, between the highest and the lowest, producing a certain order of things of a neutral and ambiguous character, partaking of the qualities of the objects which most resemble them on either side, as if to connect them, leaving it sometimes doubtful to which of the two they belong. Besides, Nature objects to be coerced by the narrowness of any system; and as if to show that her liberty and independence is perfect, she is in the habit, in every part of creation, of producing singular and anomalous species, which form exceptions to the general rule."

While he thus enumerated the true uses of classification, Ray also laid the foundations of the natural system, which has since been universally adopted by botanists. He separated Flowerless from Flowering plants, and he divided these again into Monocotyledonous and Dicotyledonous plants.

Forty years after the publication of Tournefort's system, and while Ray was yet pursuing his philosophical investigations, the Linnean system appeared. This new mode of distributing vegetable species was hailed with admiration. Its author, Charles von Linnæus, reigned supreme and without a rival till the end of the eighteenth century, and even in our days his partisans are neither few nor powerless. In Germany, for instance, more than one botanical work of character has for foundation the system of Linnæus, and many school-gardens are arranged after his classification.

The system of Linnæus rests upon the consideration of the organs of fecundation—organs almost overlooked until then, but whose physiological functions have since been ably demonstrated. He introduced in 1736 a salutary and much-wanted reform into botanical language and nomenclature, defining most rigorously the terms used to express the various modifications and characters of the organs, and reducing the name of each plant to two words, the first designating the genus, the second designating a species of the genus. Before his time, in fact, it was necessary to follow the name of the genus through a whole sentence in order to characterise the species, and in proportion as the number of species increased, the sentences were lengthened until it seemed as if they would never come to an end. It was like the confusion which would arise in society if, in place of using the baptismal name and surname, we were to suppress the baptismal name, and substitute for it an enumeration of many qualities distinctive of the individual; as if, for example, in place of saying Pierre Durand or Louis Durand, we said Durand the great sportsman, or any other phraseology applicable to the qualities of the individual. Nevertheless the Linnean or binary momenclature is one of the great titles to that glory which has been awarded to its immortal author. In the

Fig. 288.—Linnæus.

scheme of the Linnean system it has been found possible to describe all plants discovered since his time—an irrefragable proof of the great merits of this artificial classification of species.

At first Linnæus divided all known vegetables into two great groups: those in which the stamens and pistils are visible, which he called *Phanerogamous*; and those in which these organs are hidden, which he called *Cryptogamous*. These last form only a single class, namely, the twenty-fourth of his system.

Among the plants whose assemblage constituted the twenty-three classes, one portion have *hermaphrodite* flowers; the others are *unisexual*.

Plants with true sexual flowers have the male and female organs brought together on the same plant. They have a common habitation; they are *Monœcious*, as the name of the class to which the Oak, the Box-tree, Maize, and Castor-oil plant belong indicates. They are numerous, and form the twenty-first of the Linnean classes.

The male and female flowers are found upon two distinct individuals. There is a duality of habitation, as the name of the class to which *Mercurialis*, the Date, and the Willow belong, indicates. This is the class *Diœcia*, and the twenty-second.

A class which is only a combination of the two preceding groups includes the plants which present upon one or many individuals male, female, and hermaphrodite flowers. This is the twenty-third class, said to be *Polygamous*, in which we find ranged the Ash, the Pellitory, and the Nettle-tree (*Celtis*).

Plants with hermaphrodite flowers which have the stamens and the pistils combined the one with the other, as in the *Orchidaceæ* and *Aristolochiaceæ*, form the twentieth, or *Gynandrous* class.

If the stamens are equal in size and not coherent, their number determines the first twelve classes in the system. The twelfth and the thirteenth classes are founded upon the number of the stamens and their mode of insertion. The following are the classes:—

LINNEAN CLASSIFICATION.

One Stamen in each flower	1st class.	MONANDRIA (*Hippuris, Canna*).
Two Stamens	2nd class.	DIANDRIA (Jasmine, Lilac).
Three Stamens	3rd class.	TRIANDRIA (Wheat, Barley, Iris).
Four Stamens	4th class.	TETRANDRIA (Madder, Bedstraw).
Five Stamens	5th class.	PENTANDRIA (Borage, Hemlock).
Six Stamens	6th class.	HEXANDRIA (Lily of the Valley, Lily).
Seven Stamens	7th class.	HEPTANDRIA (Horse-Chestnut).
Eight Stamens	8th class.	OCTANDRIA (Heaths).
Nine Stamens	9th class.	ENNEANDRIA (Laurel).
Ten Stamens	10th class.	DECANDRIA (Pink, Lychnis).
Eleven to Nineteen Stamens	11th class.	DODECANDRIA (Purple Loosestrife).
Twenty Stamens or more, inserted upon the Calyx	12th class.	ICOSANDRIA (Myrtle, Rose).
Twenty Stamens or more, inserted upon the Receptacle.	13th class.	POLYANDRIA (Anemone, Poppy).

Linnæus founded two other classes upon the inequality of their free stamens: the *Didynamia* (fourteenth class), which comprises Thyme, Lavender, Foxglove, and Figwort, plants having four stamens, of which two are short and two long; the *Tetradynamia*, which comprises the Gillyflower, Cress, and Cabbage, has six stamens, of which four are longer than the others. When the stamens are coherent, the cohesion takes place either by their anthers or filaments. Plants in which the stamens cohere by the anthers, such as the Corn-flower, Dandelion, and Ox-eye, belong to the nineteenth class (*Syngenesia*). Those which unite by the filaments form three classes: the *Monadelphia* (sixteenth), in which all the filaments are united in one bundle, as in the Mallow; the *Diadelphia* (seventeenth), in which the filaments are united in two bundles, as in the Pea and the Bean; the *Polyadelphia* (eighteenth), in which the filaments are united in several bundles, as in the Orange.

The twenty-four classes being thus fixed, Linnæus, after some consideration, subdivided each of them—the first thirteen classes according to the number of their styles or distinct stigmata; the fourteenth (*Didynamia*) by the disposition of their seeds, sometimes bare (or at least what he considered as such), sometimes enclosed in a pericarp; the fifteenth (*Tetradynamia*) according to the form of the pod; the sixteenth, seventeenth, eighteenth, and twentieth, according to the absolute number of their stamens; the two following from the absolute number of their stamens, or from the manner of their cohesion; the twenty-third class (*Polygamia*) from the distribution of the hermaphrodite and unisexual flowers upon the same plant, or upon two or three different individuals. The nineteenth class (*Syngenesia*) is divided as follows: *—

I. Flowers all hermaphrodite and fertile (*Polygamia æqualis*), Goat's-beard, Lettuce, Thistle.

II. Hermaphrodite fertile flowers on the disk; fertile female flowers at the circumference (*Polygamia superflua*), Tansy, Wormwood, Groundsel.

III. Hermaphrodite fertile flowers on the disk; barren flowers at the circumference (*Polygamia frustranea*), Centaury, Sunflower.

IV. Hermaphrodite sterile flowers on the disk; female fertile flowers at the circumference (*Polygamia necessaria*), Marigold.

* Those syngenesious plants which have *solitary*, not aggregated flowers, are placed in the order to which the number of stamens refers them. The Linnean class *Syngenesia* corresponds, therefore, exactly to the natural family *Compositæ*.

V. Flowers provided severally with a "calyx" and clustered within a common involucre *(Polygamia segregata)*, Globe-thistle.

This classification of plants has received the name of the artificial system, because it groups the species according to a small number and not from the whole of their characteristics; in short, it rather permits one class to be distinguished from another, than makes each known in an intimate manner. It insists much upon their differences, little upon their resemblances. Between species thus compared, only one essential analogy may exist. The Rush takes place beside the Barberry, because each of these plants has six stamens and only one style. The Vine is ranged beside the Periwinkle, because they each have five stamens and one style. The Carrot is allied to the Gooseberry, &c. There may not be between the plants thus compared any natural bond, but only some trace of resemblance in a particular part of the organisation, which may be found also in a number of very different plants.

Linnæus was endowed with too sound a judgment, with a tact too exquisite, not to feel the defects of this artificial mode of classification. He detected by the force of his genius the existence of vegetable groups superior to genera, and connected them by a large number of characteristics. He called this group a *Natural order*, and it has since his time been called a "natural family." He also tried to distribute plants after a natural classification—that is to say, into families. After the death, and during the life, of Linnæus, botanists endeavoured to discover upon what principle he had founded his *natural orders*—that is to say, they sought to find the key to the hidden principle of his orders; but no one has succeeded. Linnæus himself does not appear to have had very fixed views on the subject. He created his orders by a sort of instinct which belongs only to the man of genius; by that kind of semi-divination which the man of learning acquires who possesses vast and profound knowledge of the objects which he passes his life in observing.

Linnæus created his natural orders, then, without any well-premeditated plan, and without having compared any well-defined assemblage of organs; this is sufficiently proved by the following conversation with one of his pupils, named Giseke, which has been perserved, and which we consider sufficiently interesting to repeat here, leaving the interlocutors to speak each for himself:—

LINNÆUS. Do you think, my dear Giseke, you are able to give the character of any one of my orders?

GISEKE. Yes, without doubt; for example, that of the Umbelliferæ.

LINNÆUS. Well, what is it?

GISEKE. Just this, to be umbelliferous, that is to say, it must bear flowers disposed in an umbel.

LINNÆUS. Very good; but you will readily recollect some plants whose flowers form an umbel, which nevertheless do not belong to the order.

GISEKE. That is true, I recollect some. I will add the two naked seeds.

LINNÆUS. Then, the *Echinophora* will not be of the order, for it has only one seed in the centre of the peduncle. Nevertheless it is an umbelliferous plant. And where do you place the *Eryngium?*

GISEKE. Among the *Aggregatæ*.

LINNÆUS. Not at all; it is most certainly an Umbellifera, for it has an involucre, five stamens, two pistils. What then shall be its character?

GISEKE. Such plants ought to be placed at the end of an order where they may bridge over the passage from one to another. The *Eryngium* would thus connect the *Umbelliferæ* with the *Aggregatæ*.

LINNÆUS. Oh, no, that is quite another thing; it is one thing to know the *passages* between, and another to describe the characters of two groups. I know them very well, and how the one ought to be joined to the other. One of our former pupils, named Fagraux, who is now at St. Petersburg, a most industrious young man, was quite wild upon the project of discovering the key to my orders. He laboured nearly three years, and sent me his notions. I could only laugh. In short, I can tell you one thing—if I publish a second edition of my book, I shall give a second arrangement of my orders.

In a letter to the same botanist we find the following passage: "You ask me for the characters of my orders. My dear Giseke, I assure you that I know not how to give them."

Magnol, Professor of Botany to the School of Medicine, in his work entitled "Prodromus Historiæ Generalis Plantarum" (1689), is the first author who uses the happy term "family," to designate natural groups of vegetable genera. M. Flourens speaks of the preface to this little book of a hundred pages as calculated to immortalise the author, as in it was first solved a very difficult problem. The following lines are taken from this much-admired preface: "Having examined the methods most in use," says Magnol, "and found that of Morison insufficient and very defective, and that of Ray much too difficult, I think I can perceive in plants a certain affinity between them, so that they might be ranged in divers *families*, as we class animals. This apparent analogy between animals and plants has induced me to arrange them in certain families, and, as it appeared to me impossible to draw the characters of these families from the single organ of fructification, I have selected principally the most noted characteristics I have met with, such as the root, the stem, the flower, the seeds. There is also found among plants *a certain similitude*, a certain affinity, as it were, which does not exist in any of the parts considered separately, but only as a whole. I have no doubt, for instance, but that the characters of families might be taken from the first leaf of the germ as it issued from the seed. I have followed the order that the parts of plants follow in which are

Fig. 289.—Magnol.

found the principal and distinctive characters of families, but without limiting myself to any one single part, for I have often considered many of them together."

Magnol established seventy-six families, but without giving their characters. His principles of classification are vague and uncertain; they only serve to announce the dawn of a new day which was soon to rise on the science. The few lines which we have quoted from the preface of the "Prodromus" reveal, as through a fog, the mere idea of a natural system. It is Bernard de Jussieu, Demonstrator of Botany in the Jardin des Plantes at Paris, to whom belongs the glory of working out the true natural system which was first established in principle by Ray, although it does not appear that Jussieu was acquainted with the works of the English philosopher.

Bernard de Jussieu, as his nephew Laurent de Jussieu tells us, regarded botany not as a science of memory or nomenclature but as a science of combination, founded on a profound knowledge of the characteristics of each plant. He would every day get together the materials out of which he had to form his natural orders, which he regarded as the "philosopher's stone" of botanists. He deferred the publication of his first Essay in his zeal and desire to perfect his work. He wrote little, but observed much, and the fruits of his labour would perhaps have been lost to science but for a favourable circumstance which obliged him to give his method of arrangement of plants to the world. Louis XIV., having seen the gardens of Saint Germains, in which the Marshal Duke de Noailles cultivated exotic trees and shrubs, formed the design of creating a School of Botany at Trianon. By the advice of Lemonnier, chief physician to the Dauphin of France, afterwards King, he selected Bernard de Jussieu to arrange the gardens. Thus forced to adopt some mode of classification, he thought it his duty to substitute his new arrangement for the old methods. These he considered to be mere schemes in which the plants were arranged in a convenient order for studying them. Science confined to such narrow limits is, however, very artificial, and remote from a natural system, which consists in the knowledge of the true connection of plants and their organisation. "When a man has so combined the characteristics of plants," says Laurent de Jussieu, "that he can in one species unknown determine the existence of many by the presence of a single character; when he can at once point to the order to which it belongs; when he has succeeded in destroying the prejudice so withering to Botany, that it is only a science of memory and nomenclature; when he has, in short, founded a science of combinations which furnish food to the mind and to the imagination, that man may surely be called a creator, or at least a restorer, of science.

"Others may perhaps have extended the limits, but he was the first to show the way, to trace the method, to establish the principles.

Fig. 290.—Bernard de Jussieu.

Jussieu consigned his discoveries to no book, but in the Gardens of Trianon the mind of the author is recognised. In examining the characters, he remarked that some were more general than others, and these furnished the first division. He recognised that the germination of the seed and the respective disposition of the sexual organs were the two principal and most persistent characteristics. He adopted them, and made them the basis of the arrangement which he established at the Trianon in 1759."

Four years later, another French botanist, Michel Adanson, a naturalist remarkable for the originality of his views and the extent of his conceptions, published a book upon the families of plants. He proposed a particular course for arriving at the true natural method. But what was that course? He proposed classing all the plants known according to a great number of artificial systems; and after considering them from all possible points of view, he proposed to arrange in the same group those plants which were classed as allies in the greatest number of systems. In this manner Adanson created sixty-five artificial systems, and by their comparison he formed fifty-eight families. He was the first to trace the precise characters and details of all these families; his work in this respect is far superior to those of his predecessors. Nevertheless, if Michel Adanson was right in employing all their characters in classifying these plants, he was wrong, on the other hand, in giving the same importance to all. He reckoned up the characters without considering that they were not all of equal value. The results of his calculations were frequently found to be false, as would inevitably happen in reckoning any sum where no regard was paid to the quality of the metal, but only to the shape and volume of the coin.

The year 1789 was the date of the real establishment of natural families among vegetables. It was in this year that Laurent de Jussieu published his celebrated "Genera Plantarum," which marked a new era in the science of botany, and hastened the advent of a natural system of zoological classification as well.

The catalogues of the Gardens of the Trianon, prepared by Bernard de Jussieu, and his conversations with his nephew, were the source whence the latter drew his inspirations. We shall, however, leave his son, Adrien de Jussieu, to state the true basis of the Natural System, and the considerations which guided his father. "Like Adanson," says Adrien, "Antoine-Laurent de Jussieu admits that the examination of all parts of a plant is necessary in order to its classification; but in pursuing this examination it is not sought to deduce from it theoretically the combination of the genera; in grouping them into families, he imitated the proceeding followed in the formation of the genera themselves. The botanist, struck

Fig. 291.—Adanson.

by the complete and constant resemblance of certain individual plants, has formed them into species; then, observing a resemblance equally constant, but much less perfect, he has formed species into genera. The characters which may vary even in the same species, ought to depend on causes exterior to the plant, and not on the plant itself: for example, its size, consistence, certain modifications of form and colour, which we see change with the sun, the climate, and under other influences purely circumstantial. The specific characters, on the contrary, which every individual should possess, in order to be referred to a particular species, whatever be the circumstances in which it is found, ought to depend on the nature of the plant. Among these characters some are more important than others, and less subject to variation from one plant to another. Finding these characters in a certain number of species impresses upon them a sufficiently striking resemblance to constitute them a genus. These would, from their general nature, have more value than the specific, as the specific is of more value than the individual characteristics.

"But how are we to discover and estimate these different values? Nature herself indicates to the observer species and many genera, by the features of resemblance with which she marks certain vegetables. All botanists are nearly agreed up to this point, although they separate further on to follow each a different road. Nevertheless, there are many great groups of vegetables connected together by features so strong in their resemblance that they could scarcely escape observation; it requires no botanist to recognise them. Besides these features common to all the species constituting one of these groups, there are others which are only common to a certain number of them, so that it may be subdivided again by a great number of secondary groups. These had been recognised as genera by botanists. We had already, then, some collections of genera evidently possessing more resemblance to each other than they had to any other group; in other words, there were some families incontestably natural. Jussieu thought that the key to the natural method was there; since, in comparing the character of one of those families with the genera composing it, he would obtain a knowledge of the relation of one to the other; and in comparing many families together, he would see how the characters common to all plants of the same family varied from one family to another; he would thus arrive at a true appreciation of the value of each character, and this value, once determined by means of groups so clearly designed by Nature, could be applied in its turn to the determination of other groups which were not so strongly impressed with the seal of family resemblance, and which were the unknown quantities in the great problem. He selected for this purpose seven families already universally recognised; those, namely, known as the Graminaceæ, the

Fig. 292.—Laurent de Jussieu.

Liliaceæ, the Labiatæ, the Compositæ, the Umbelliferæ, the Cruciferæ, and the Leguminosæ.

"It is recognised that the embryo is identical in all plants of any of these families; it is monocotyledonous in the Graminaceæ and Liliacæ, and dicotyledonous in the other five. The structure of the seed is also identical, the monocotyledonous embryo being placed in the axis of a fleshy albumen in the Liliaceæ, and upon the side of a farinaceous albumen in the Graminaceæ, while the dicotyledonous embryo is placed at the summit of a hard and horny albumen in the Umbelliferæ, and is destitute of albumen in the three others. That the stamens which vary in their number in the same family—the Graminaceæ, for example—do not vary in general in their mode of insertion, being hypogynous in the Graminaceæ and Cruciferæ, upon the corolla in the Labiatæ and Compositæ, and upon an epigynous disk in the Umbelliferæ. He ascertained thus the value of certain characters which ought never to vary in the same family; but along with these he found others more variable, which he sought to appreciate also, either to assist him in the study of other families analogous to them, or in explaining those which he had formed, by applying his first rules along with many others founded on observation. We cannot, however, follow him here into these details, which resulted, after immense labour, in the establishment of a hundred families, comprehending all the vegetables then known.

"We see in these remarks a principle employed which had altogether escaped the notice of Adanson—namely, that subordination of characters which in the method of Jussieu are, to use his own expression, 'weighed but not counted.'"

When the families were constituted, Laurent de Jussieu grouped them into fifteen classes, as in the following table:—

Acotyledones		Class I.	1. Fungi. 2. Algæ. 3. Hepaticæ. 4. Musci. 5. Filices. 6. Naiades.
Monocotyledones	Stamens Hypogynous	Class II.	7. Aroideæ. 8. Typhæ. 9. Cyperoideæ. 10. Gramineæ.
	Stamens Perigynous	Class III.	11. Palmæ. 12. Asparagi. 13. Junci. 14. Lilia. 15. Bromeliæ. 16. Asphodeli. 17. Narcissi. 18. Irides.
	Stamens Epigynous	Class IV.	19. Musæ. 20. Cannæ. 21. Orchideæ. 22. Hydrocharides.

Dicotyledones.	Apetalous	Stamens Epigynous ..	Class V. ..	23. Aristolochiæ.
		Stamens Perigynous ..	Class VI. ..	24. Elæagni. 25. Thymeleæ. 26. Proteæ. 27. Lauri. 28. Polygoneæ. 29. Atriplices.
		Stamens Hypogynous ..	Class VII. ..	30. Amaranthi. 31. Plantagines. 32. Nyctagines. 33. Plumbagines.
	Monopetalous.	Corolla Hypogynous ..	Class VIII. ..	34. Lysimachiæ. 35. Pediculares. 36. Acanthi. 37. Jasmineæ. 38. Vitices. 39. Labiatæ. 40. Scrophulariæ. 41. Solaneæ. 42. Borragineæ. 43. Convolvuli. 44. Polemonia. 45. Bignoniæ. 46. Gentianeæ. 47. Apocyneæ. 48. Sapotæ.
		Corolla Perigynous ..	Class IX. ..	49. Guaiacanæ. 50. Rhododendra. 51. Ericæ. 52. Campanulaceæ.
		Corolla Epigynous. Anthers Coherent ..	Class X. ..	53. Cichoraceæ. 54. Cinarocephalæ. 55. Corymbiferæ.
		Corolla Epigynous. Anthers Free ..	Class XI. ..	56. Dipsaceæ. 57. Rubiaceæ. 58. Caprifolia.
	Polypetalous.	Stamens Epigynous ..	Class XII. ..	59. Araliæ. 60. Umbelliferæ.
		Stamens Hypogynous ..	Class XIII. ..	61. Ranunculaceæ. 62. Papaveraceæ. 63. Cruciferæ. 64. Capparides. 65. Sapindi. 66. Acera. 67. Malpighiæ. 68. Hyperica. 69. Guttiferæ. 70. Aurantia. 71. Meliæ. 72. Vites. 73. Gerania. 74. Malvaceæ. 75. Magnoliæ. 76. Anonæ. 77. Menisperma. 78. Berberides. 79. Tiliaceæ. 80. Cisti. 81. Rutaceæ 82. Caryophylleæ.

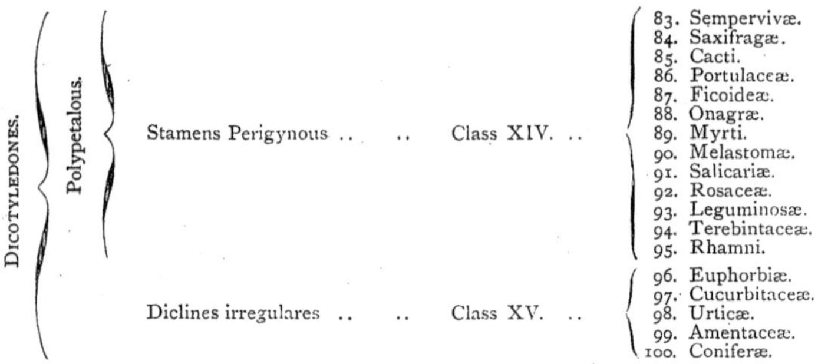

DICOTYLEDONES.	Polypetalous.	Stamens Perigynous	Class XIV. ..	83. Sempervivæ. 84. Saxifragæ. 85. Cacti. 86. Portulaceæ. 87. Ficoideæ. 88. Onagræ. 89. Myrti. 90. Melastomæ. 91. Salicariæ. 92. Rosaceæ. 93. Leguminosæ. 94. Terebintaceæ. 95. Rhamni.
		Diclines irregulares	Class XV. ..	96. Euphorbiæ. 97. Cucurbitaceæ. 98. Urticæ. 99. Amentaceæ. 100. Coniferæ.

Such, then, was the arrangement into which Antoine-Laurent de Jussieu distributed the twenty thousand plants known to botanists in 1789. The hundred orders or families he further subdivided into 1,754 genera. That the French botanist had acquainted himself with the principles of Ray's classification is unquestionable; in fact, Jussieu possessed the happy art of adapting the labours of others to perfecting his own conceptions. He made use of the simple language and accurate descriptions of Linnæus, divested of his pedantry. Ray had demonstrated that rigorous definitions in natural history are impossible, and, accepting the decision, Jussieu does not attempt to found his family orders or genera on any single character belonging to objects so various in their habits and organisation as plants.

During the last forty or fifty years other botanists have attempted various systems of classification. In those of De Candolle, Endlicher, Lindley, and of Brongniart, the distribution of plants into groups is founded, as in those of Ray and Jussieu, on the consideration of the cotyledons; of the polypetalous, monopetalous, and apetalous flowers; finally, upon the mode of insertion of the stamens. Names have changed; things remain the same; and if in their details the series of families or orders present certain differences, it only arises from the fact that a linear series is incompatible with the natural system, and that the connection of the intermediate groups may be expressed in various ways without affecting the general principles of the system. "The formation of natural orders by Jussieu," says Ad. Brongniart, "is even now a model which directs botanists in their studies to the affinity which connects the various forms of vegetation. Many of these orders have doubtless been subjected to important modifications, both in extending and limiting them; the numbers have been more than doubled; but the number of species now known is increased more than sixfold. Since the publication of the 'Genera Plantarum,' many points in the organisation of plants which were either scarcely touched upon or were altogether unsuspected, have now been considered, and it is found that they do not destroy, but confirm and perfect the work of Jussieu. One is even astonished to find that the numerous discoveries in the anatomy and organography of plants since the beginning of the century have not introduced greater modifications into the constitution of the natural groups admitted by the author of the 'Genera Plantarum.' It is here that we recognise the sagacity of the *savant* who established them, and the soundness of the principle which guided him."

The natural classification of plants, their distribution into families, well defined, and founded upon affinities, have been perfected and placed upon a basis more and more certain in our own days. Botanists have set themselves the task of unravelling and establishing the characters which dominate, and those which are subordinate, in each family; numbers have spread themselves over the globe, exploring the most distant regions, interrogating the solitudes of forests and plains which no European had hitherto visited, and have studied in their native wilds many exotic plants, comparing them with already known species, thus giving us a means of pointing out more precisely the tribes, genera, and species of each natural family. Monographs of a great number of such families have thus been written with great research. The study of the formation and evolution of organs; the discovery of the true mode of reproduction in Cryptogams, still unknown in Jussieu's time; the investigation of the inflorescence, of the fruits, of the ovules, of the embryos, have furnished elements for perfecting the limits of families and advancing natural classification.

Auguste Pyramus de Candolle is one of the botanists of this century who has most contributed to the general adoption of natural families. His "Essai sur les Propriétés des Plantes" is celebrated for the knowledge which it displays of the comparative physiological and medicinal action of vegetables, and the physical organisation which naturally connects certain plants as a group. His "Prodromus Systematis Naturalis Regni Vegetabilis," still continued towards completion by his pupils and his son, is a wonderful work for the extent and precision of its details.

In our own country, from the days of Ray, we have always had zealous followers of the science of botany, more especially in the class which may be called field botanists. Withering, Sir James Edward Smith, and hundreds of followers more or less eminent, employed their leisure in the fascinating and healthy pursuit of plants, and perhaps the most valuable contributions to science are the

Fig. 293.—De Candolle.

detailed descriptions of species, with their habits and habitats, with which they have enriched our botanical literature. Nor was the study of the physiology of plants—a science which may be said to owe its existence to the researches of Grew and Malpighi—neglected. To the former belongs the merit of having pointed out the difference between seeds with one and seeds with two cotyledons, on which Ray founded the first division of his system of classification.

The German botanists have always been distinguished for their patient and laborious investigations; and it was reserved for the first of Germans, the poet Goethe, to effect the last great revolution that the ideas of botanists have undergone. In 1790, shortly after the appearance of De Jussieu's "Genera," he published a pamphlet on the "Metamorphoses of Plants." At this time the functions of the organs of plants were supposed to be pretty well understood. The notion had, however, existed in a form more or less vague, from the times of Theophrastus, that the various parts of the flower were mere modifications of leaves, although their appearance was very different —a doctrine which Linnæus seems to have entertained at one time, as he speaks, in his "Prolepsis Plantarum," of the parts of a flower being mere modifications of leaves whose period of development was anticipated. Goethe's mind was, as he himself tells us, one more adapted to see agreements in things than to mark their distinctions. We are not surprised to find, therefore, that he takes up this theory, and demonstrates that the organs to which so many different names are applied, namely, the bracts, calyx, corolla, stamens, and pistil, are all modifications of the leaf: the bract being a contracted leaf; the calyx and corolla a collection or whorl of several; the stamens contracted and coloured leaves; and the pistil leaves rolled up upon themselves and variously coherent.

These views of the poet met at first with little attention from botanists, and we are chiefly indebted to Robert Brown for the elucidation of Goethe's theory. In his "Prodromus of the Plants of New Holland," and in many papers in the "Linnean Transactions," he demonstrates its truth as well as its practical value; showing, by the use of the microscope, that the law was applicable not only to the external parts of plants, but that it was followed in their development also. Robert Brown contributed largely to perfecting the natural method of classification. His great work upon the Flora of Australia has greatly extended the circle of our studies for that comparison of characters which is the basis of botanical genera and tribes.

The number of families of flowering plants admitted in the present day as the result of the investigations of the eminent men whose names have been mentioned, and many others which could not be quoted here without swelling our pages to undue proportions, number three hundred and three; and many of these are again

Fig. 294.—Robert Brown.

subdivided by botanists who have made certain families their special study.

We have had a tabular view of the vegetable world as arranged by Adrien de Jussieu. According to the division introduced (1828) by Brongniart, plants are divided into two great classes, *Cryptogams* and *Phanerogams*.

The CRYPTOGAMS, from κρυπτός, "hidden," and γάμος, "nuptials," are destitute of pistils and stamens; they are reproduced by means of organs of various kinds, which seem to have no analogy, except by their functions, with the reproductive organs in other plants. They possess no cotyledons, since their spores contain no embryo; hence, by Jussieu and others they were called Acotyledons.

The PHANEROGAMS, from φανερός, "visible," and γάμος, have perceptible reproductive organs formed of stamens, and ovules naked or enclosed in a pistil.

According, then, as Phanerogams have an embryo furnished with one or two cotyledons, they are divided into two great natural groups, the Monocotyledons or Dicotyledons.

Adrien de Jussieu (1844) divided the Cryptogams (which was a Linnean group), into two classes: *Cellular Cryptogamia*, including those composed of cellular tissue only, not traversed by *vessels*; and *Vascular Cryptogamia*, which are provided with vessels. As regards *Phanerogams*, he recognised the two great divisions of *Monocotyledonous* and *Dicotyledonous Phanerogams*, distributing the latter into two classes—(1) *Gymnosperms*, or *naked-seeded*, from γυμνος, "naked," and σπέρμα, "seed;" and (2) *Angiosperms* (*plants with enclosed seeds*), from ἀγγεῖον, "a vessel," and σπέρμα, "seed." The Dicotyledonous Gymnosperms of De Jussieu form only five families, and comprehend coniferous trees, a great number of which are what are called Evergreens; the Dicotyledonous Angiosperms were divided into many secondary groups, whose distinctive characters are drawn from peculiarities of the reproductive organs.

The *primary* groups into which flowering plants are divided, and in which therefore the families or orders are themselves comprised in the classification at present accepted, being founded upon the degree of cohesion and adhesion in the petals and stamens, are undoubtedly somewhat artificial. The problem of how the orders are themselves to be combined into natural groups is one which still engages the attention of systematic botanists. At the present time it is almost by general consent deferred till the enormous accumulations of plants, which are now available for study have been properly classified, and the genera and orders comprised in them accurately determined. When the vegetable kingdom is better known in detail, the formation of a classificatory system, which will be as truly natural in its higher divisions as in its lower, will be a task which will then be approached with some degree of success. Robert Brown saw the need of such an undertaking, and Dr. Lindley boldly attempted its accomplishment. From the nature of the case, as indicated above, however, his work has only met with partial acceptance. It was undoubtedly a move in the right direction.

We can only give in the present work an outline of his scheme, following it up with a short notice of the orders, and filling up the sketch with a few details of the species most interesting to man.

The vegetable kingdom is divided by Dr. Lindley into seven classes :—

FLOWERLESS PLANTS (CRYPTOGAMS).

I. THALLOGENS	Stems and leaves imperceptible.	A Thallus is a fusion of root, stem, and leaves into one general mass, and Thallogens are destitue of breathing pores, and multiply by the formation of spores, in their interior or upon their surface.
II. ACROGENS	Stems and leaves quite perceptible.	Beyond Thallogens are multitudes of species, flowerless like them, but approximating to more complex structures, sometimes acquiring the stature of lofty trees with breathing pores; their leaves and stems distinctly separated; they multiply by reproductive spores like the Thallogens. Their stem, however, does not increase in diameter, but at their summit, as the name of the class indicates.

FLOWERING PLANTS (PHANEROGAMS).

III. RHIZOGENS	Fructification springing from a Thallus.	The Rhizogens are a collection of anomalous plants, mostly leafless and parasitical, having the loose cellular organisation of Fungi, although traces of a spiral structure are usually found among their tissues. Some of them spring directly from the shapeless cellular mass which serves at once for stem and root, and seems to be analogous to the Thallus of the Fungi. Their flowers resemble those of more perfect plants; their sexual organs are complete, but their embryo, which is without any visible radicle or cotyledon, simply appears to be a spherical or oblong homogeneous mass.
IV. ENDOGENS	Cotyledon single. Permanent woody stem confused. Leaves parallel-veined.	In Endogens the embryo has but one cotyledon; the leaves have parallel veins; the trunk contains bundles of spiral and dotted vessels, surrounded by wood cells, arranged in a confused manner.
V. DICTYOGENS	Cotyledon single. Wood of the stem, when perennial, arranged in rings concentric with the veined pith. Leaves netted.	Dictyogens are distinguished from Endogens by the stems, which have concentric circles, and the leaves which fall off the stem by a clean fracture.

VI. GYMNOGENS .	Cotyledons, two or more. Wood of the stem in concentric rings, and youngest at the circumference. Seeds quite naked.	Gymnogens are Exogens which have no style or stigma, the reproductive organs being so constructed that the pollen falls immediately upon the ovules.
VII. EXOGENS .	Cotyledons two. Wood with concentric rings. Leaves netted-veined. Seeds enclosed in seed vessels.	Exogens have an embryo with two or more cotyledons; leaves with netted veins; the trunk consisting of woody bundles, composed of dotted vessels and woody fibres, arranged round a central pith, either in concentric rings or in a homogeneous mass, but always having medullary plates forming rays from the centre to the circumference

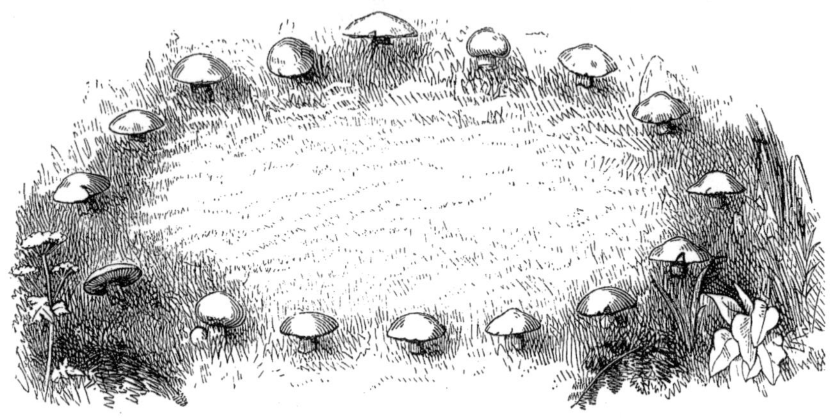

Fig. 295. –A Fairy Ring.

FLOWERLESS PLANTS.

CLASS I.—THALLOGENS.

The Thallogens of recent botanists correspond to a part of the Cryptogamia of older botanists—a name given to this division of the vegetable world, not from their reproductive organs being invisible, as had long been supposed, but from their being inconspicuous, and requiring close observation, and some exact knowledge of their organisation, in order to discover them. If we enter on a somewhat minute investigation of this important group, we are led to do so by the consideration that their organisation, structure, and development are not usually attempted in elementary botanical works; and we feel assured that the interest and novelty of the observations we have to offer will be our best apology for the relative extent of the descriptions.

The division includes vegetables destitute of stamens and of pistils, whose embryo is simple, homogeneous, and without distinct organs. A great number of them are of minute, microscopic dimensions; their reproductive organs can only be distinguished by the help of a lens or a microscope. Nevertheless, these, being so small, so humble, and to all appearance all but forgotten in more important creations, fulfil an important part in the economy of Nature. They constituted the first origin, and were even the source of all vegetation. Disintegrating rocks, they produced the earth which they further enrich with the products of their own destruction. The enriched soil nourished plants of more complex organisation, and these inferior creatures were gradually replaced by vegetable species of more perfect structure. All soil primitively sterile, all land recently emerged from the bosom of the waters, served first as the asylum of crustaceous and foliaceous Lichens; at a later period Mosses and Ferns made their appearance there; finally, a superior vegetation—namely, the Phanerogams, or Flowering Plants—presented themselves. Everything leads us to conclude that such has been the successive series of creations upon our globe, when it was sufficiently cooled on the surface to admit of organic life, and when the islands and continents were sufficiently elevated above the universal ocean of the ancient world to permit them to live.

Thus the higher orders of vegetables have only appeared, and will only continue to make their appearance, upon the *débris* of vegetation of a lower order.

But, on the other hand, by one of those striking contrasts of which Nature offers more than one example, vegetables of a superior order, when they are struck by death—sometimes even during their existence—are often the prey of humbler Thallogens, which attach themselves as parasites to these princes of the vegetable world, and devour them in the end; their destructive action is all-powerful and everywhere; it respects the works of man no more than the works of Nature.

To produce and to destroy life is, then, the double and providential mission which devolves on the Thallogens. Nevertheless, this multiplied work of creation and death is only bestowed on them on two conditions: the first is an evanescent and short existence, the second is to multiply themselves to infinity and with prodigious rapidity. There are some Mushrooms which produce 60,0000 cells per minute! The capsules of certain Mosses enclose spores of which it would require many thousands to make a pin's head in size. These spores float free and invisible in the air, which is, in a sense, filled with them.

In the Cryptogamia the reproductive organs differ fundamentally from the same organs in Phanerogams. Here, without pistil or stamens, ovary or flower, in the ordinary sense of the word, the reproductive organs, which are designated spores, are placed in a most varied manner, sometimes in the whole extent, sometimes in certain parts of the vegetable. These spores are sometimes enclosed in special receptacles, named sporanges, in other cases they are quite destitute of any envelope. In short, the reproduction of Thallogens is often the result of organic dispositions, quite special, which admit of no general description, and which can only be made intelligible by describing each individual case.

To examine all the families which constitute the class would be an immense undertaking. We shall confine ourselves to examine attentively certain types of the Algæ, Fungi, and Lichens.

ALLIANCES.	Class I.—THALLOGENS.	NATURAL ORDERS.
1. ALGALES	Cellular flowerless plants, nourished by the medium in which they vegetate; propagated by zoospores, spores, or tetraspores.	I. Diatomaceæ, or brittleworts. II. Confervaceæ. III. Fucaceæ, or seaweeds. IV. Ceramiaceæ. V. Characeæ.
2. FUNGALES	Cellular flowerless plants, nourished through their thallus, living in air, propagated by spores, sometimes enclosed in asci, destitute of green gonidia.	VI. Hymenomycetes, toadstools. VII. Gasteromycetes, or puff-balls. VIII. Coniomycetes, or blights. IX. Hyphomycetes, or mildews. X. Ascomycetes, or morels. XI. Physomycetes, or moulds.
3. LICHENALES	Cellular flowerless plants, nourished through their whole surface by the medium in which they vegetate, living in air, propagated by spores generally enclosed in asci, with green gonidia in their thallus.	XII. Graphidaceæ, cell-lichens. XIII. Collemaceæ, or jelly-lichens. XIV. Parmeliaceæ, or leaf-lichens.

ALLIANCE I.—ALGALES.

The DIATOMACEÆ are minute brittle bodies, which attach themselves to stones constantly under water, and are so obscure in their organisation that it was at one time a grave question whether they belong more to the animal or vegetable kingdom. Ehrenberg inclined decidedly to the former opinion, assigning as reasons that they exhibit a peculiar spontaneous movement, produced by locomotive organs; that many of them have a lateral opening, round which are corpuscles, which become blue when placed in water coloured with indigo, like many *Infusoria*. Their shells, or shields, are in structure silicious. Schleiden considers that such an artificial and complex structure as exists in the *Diatomaceæ* is without explanation, of great significance, and without "analogy in the vegetable world."

On the other hand, the discovery by Mr. Thwaites of the conjugation of the *Diatomaceæ*, discloses operations strictly in accordance with what occurs in the *Desmidieæ*. In *Epithemia turgida*, two individuals closely approximated *dehisce* in the middle of their long diameter; two protuberances now rise, which meet two similar ones in the opposite frustule; the two frustules become united by two channels, in each of which a sporangial mass is formed, which gradually shapes itself into the form of the parent frustule.

CONFERVACEÆ.—Towards the close of the year, in autumn, on moist, humid days, or after a heavy fall of rain, we frequently meet on roadsides, or in garden alleys, small gelatinous, greenish masses, more or less globular and wrinkled. These are a species of *Nostoc*. In studying the organisation of these curious plants, we shall follow M. Thuret in the observations he has made on *Nostoc verrucosum*, which he found growing in the brooks in the neighbourhood of Paris, attached to submerged stones, upon which many individuals had massed themselves together, forming on the stone a carpet of blackish green. Each *Nostoc* is a sort of irregular bag (Fig. 296), folded, rounded, and closed, filled with greenish gelatinous matter, whose appearance and consistence forcibly remind one of the pulp of the grape. In the very abundant mass of this matter we find numerous filaments composed of spherical globules placed end to end like the beads of a chaplet, and formed of a bluish-green granular matter. Fig. 297 represents the kind of necklaces which occupy the interior of the *N. verrucosum*, and accompany the mucilaginous matter. When the plant is examined in its full development, the pellicle formed by thickened mucilage is ruptured, and the green gelatinous substance and the necklaces escape. This matter is dispersed in the water with so much the more facility that it seems gifted at this stage of its existence with a very perceptible spontaneous movement. "In order to observe this phenomenon properly," says M. Thuret, " the most simple means is to arrange some gathered specimens in a plate filled with water. In two or three days the external pellicle will burst, and the necklaces will spread themselves through the water. If recourse is then had to a microscope, it will be seen that—originally very long and contorted in a thousand ways—the necklaces are divided into numerous fragments of unequal

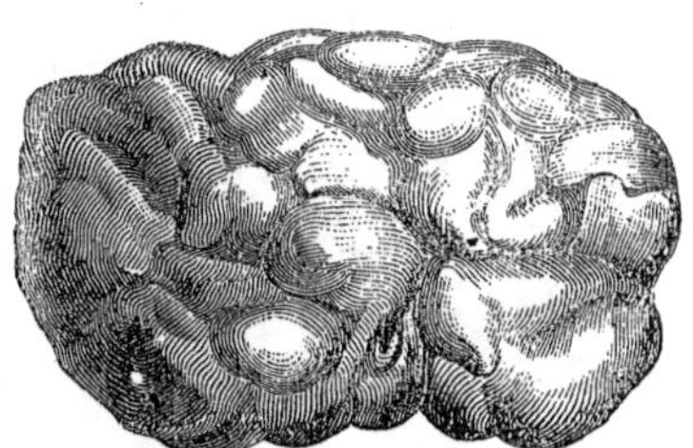

Fig. 296.—Mass of *Nostoc*.

length, nearly all straight, or only slightly flexuose, moving themselves in the direction of their length, and seeming to creep upon the plane of the field of vision. Their progress is slow, but very perceptible. If the observations are continued during some days, the necklaces will become immovable, and increase in size; at the same time a mucilage is developed by which they are surrounded as in a transparent sheath. Sometimes the cells enlarge considerably and divide themselves, so as to form two others, but laterally; this formation repeats itself many times, and it would seem natural to seek there for the origin of new chaplets. Unfortunately, the increase of the cells in number, by diminishing the transparency, no longer permits us to follow their increase with the same facility."

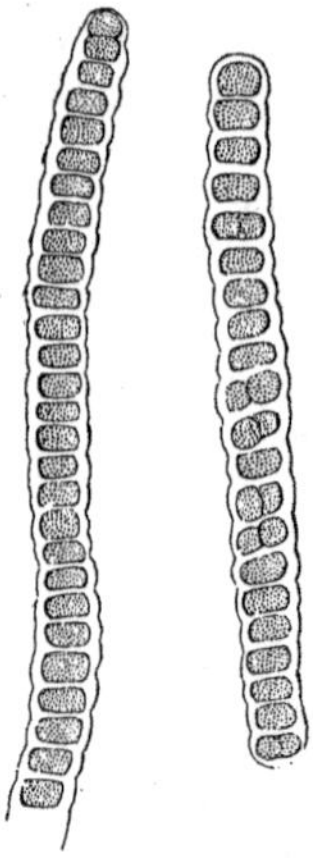

Fig. 297.
Necklace of *Nostoc.*

It is obvious, then, that these plants present an organisation quite rudimentary, and that their mode of reproduction consists of a species of *segmentation*, namely the division of the individual into new individuals.

The *Nostoc*—in consequence, no doubt, of the extreme rapidity of its vegetation—greatly attracted the attention of the alchemists, who often mention it; and it enters into many of their recipes for the pretended transmutation of metals.

Protococcus is a plant consisting of one cell only. The *Oscillatoria spiralis* is a simple vegetable thread formed by cells cohering at their extremities. The *Oscillatoria nigro-viridis*, shown in its natural state in Fig. 299, is

Fig. 298.—Two plants of *Sphærozyga Berkeleyana.*

Fig. 299.—Three plants of *Oscillatoria nigro-viridis.*

a simple vegetable structure, consisting of cells united at their extremities like threads. *Sphærozyga Berkeleyana* consists of threads which exhibit at intervals large swollen connecting joints (Fig. 298).

Sphæroplea annulina, belonging to the sub-family *Oscillatoridæ*, is a fresh-water species, composed of long filaments, formed of cellules more or less elongated, and associated end to end like the Nostocs. These cells contain in their adult state chlorophyll, a watery liquid, and some starchy granules; the whole distributed in such a manner that the liquid element forms coarse vacuoles (Fig. 300, 1).

In the month of April the contents of certain small cells are modified so as to give a spongy appearance, first by the multiplication of these vacuoles (Fig. 300, 2), then by the condensation of the green matter and grains of starch, as in figure 3, *a*, and by the disappearance of most of the vacuoles, as in figure 3, *b*, where the great flattened vacuoles represent superposed cavities. Soon the same cells contain a great number of free globular masses (Fig. 300, 4). These masses are young and soft spores, elastic, and destitute of membrane.

Long before the contents of these cells have submitted to the transformations indicated, the membrane proper to the cells presents small openings at various points, whose diameter varies from 1-300th to 1-500th part of a line (Fig. 300, 4 and 5, *o*).

But all the cells of the same filament of *Sphæroplea* do not present the same modifications which we have described, and which are finally converted into sporanges, filled with a multitude of spores. Phenomena of a very different nature occur in the meantime. The rings interposed between the uncoloured vacuoles become reddish, and the grains of starch which they contain disappear (Fig. 300, 5, *a*). Next the orange-coloured matter is organised into an infinity of short corpuscles, arranged in a confused mass. The rings break up, and suddenly one of the corpuscles is observed to disengage itself, and move in the cellular cavity; then other corpuscles in increasing numbers repeat the same phenomena. The movement which agitates them becomes excessively animated, and in a few minutes the whole substance of the ring under observation is an innumerable multitude of moving corpuscles. A second and a third ring of the same cell undergo the same process, until the whole mass is filled with elongated corpuscles, which move and swarm in all directions (Fig. 300, 5, *b*). Under a very strong magnifier these moving corpuscles appear as in Fig. 300, 6.

"It is," says Cohn—the botanical professor at the University of Breslau, to whom we are indebted for these observations—"a spectacle truly surprising to see all these movements of incredible activity in the bosom of the mother cell. The membrane of the cell is pierced at a given moment with one or many openings corresponding in form and dimensions to those we have already seen

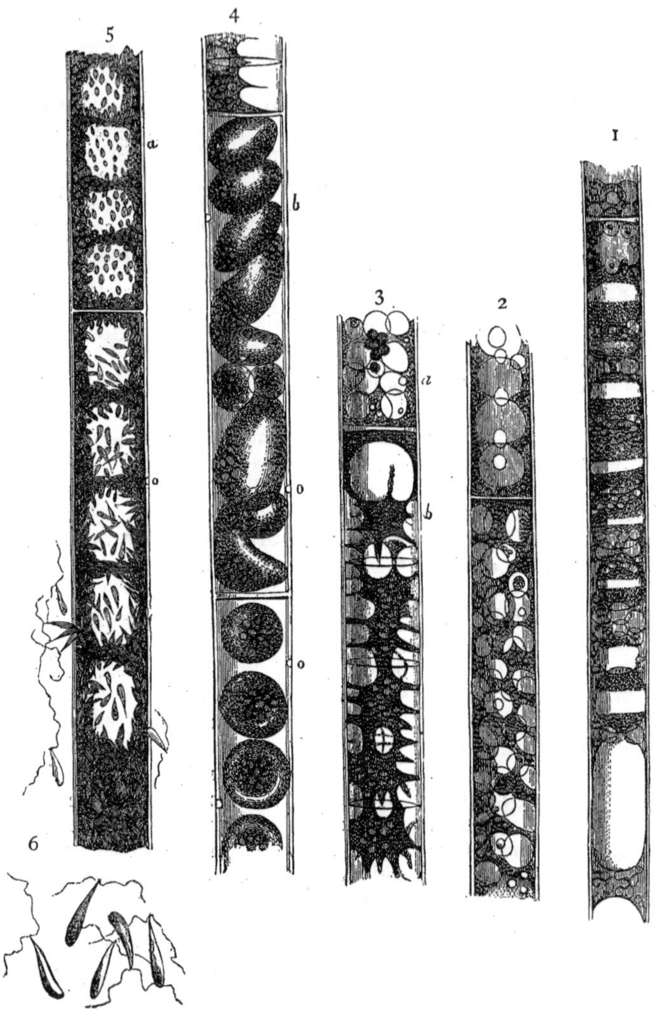

Fig. 300.—Reproduction of *Sphæroplea* (magnified).

in the sporanges. A corpuscle first escapes by one of these perforations, others follow, and they soon flow in multitudes. Their movement in the water is at first slow; their exit is often obstructed by the interposition of the mucilaginous envelope of a vacuole, against which the corpuscles press in vain. I have seen them, after struggling for twelve hours, still agitating themselves tumultuously within their prison-house, return finally to a state of repose, and become transformed into yellowish vesicles. The active corpuscles, of which we are speaking, measure about the 200th part of a line in length; their form is elongated and cylindrical, and reminds us of certain small Coleoptera. Their posterior extremity is slightly swollen, sometimes flattened and enlarged at the same time, of a yellowish tint, and granules can frequently be traced in its interior. The anterior extremity, on the contrary, is elongated into a species of straight and glassy beak, having at its extremity two long cilia, which become very visible in a solution of iodine, which seems to destroy the movements of the corpuscles. These movements of the ciliate corpuscles are very characteristic: they are gifted with little vital energy, they only oscillate with their beak as if groping; if they move more rapidly, they turn transversely round their median axis, like a tip-cat fixed by its middle and spun round; one sees them also move round upon themselves like a cat which runs round after its own tail, without changing its place. But for the most part they describe a cycloid by a movement of progression by jerks and leaps, as it were; more rarely they advance in a right line; their natural tendency towards the light being indicated by the fact that in the drop of water in which I observed them they massed themselves voluntarily towards the edge of the nearest window."

The exterior resemblance of these corpuscles to the antherozoids of *Vaucheria*, justify us in attributing to them analogous functions. When these antherozoids become free and diffused through the water, they reunite after a time round the cells, whose contents are organised into spores. They agitate themselves violently near these cells; they attach themselves to its walls, leaving it for an instant, then returning immediately. Finally, one of the corpuscles, approaching one of the openings existing in the membrane of the sporanges, holds itself there, and introduces its delicate beak. Sometimes the posterior part of its body is too large to pass with impunity; then we see it pushing itself on, but without relaxing the hold with its beak, contracting and making itself smaller. In short, it forces itself a passage, and penetrates into the cavity of the sporange. At the same time other antherozoids penetrate in the same manner or by analogous methods. Three or four of these are often engaged at once in the same opening; the smallest pass without difficulty at the first attempt, and in their movement of translation in the liquid in which they swim in the interior of the sporange, describe large circles, and constitute extremely curious phenomena. After a few moments the sporange may contain more than twenty of these antherozoids, which agitate themselves round the young spores. These are, as I have stated above, small soft spheroids, more or less completely filled with chlorophyll, and enveloped in a mucous coat, having, however, none of the characters of a cellulose membrane. The spermatozoids throw themselves from one spore to another, as if some electric force attracted and repelled them alternately, and that so rapidly that the eye can scarcely follow their movements. They are often observed to move with the same agility from one end of the sporange to the other, while the agitation of their vibratile cilia

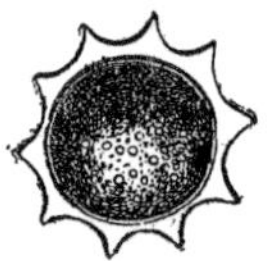

Fig. 301.
Spore of *Sphæroplea* (magnified).

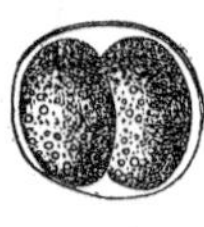

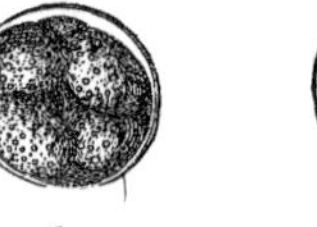

a b c

Fig. 302.
Spores of *Sphæroplea* in course of germination (magnified).

impresses upon the spores a slow movement of rotation. I have seen the antherozoids agitate themselves confusedly in the cavity of the sporange during more than two hours. This movement becomes slower and slower by degrees, and they finish by attaching themselves to the surface of the young spores. We can then see one or two fix themselves by the cilia and the beak upon each of these bodies, remaining as if implanted there, oscillating for some time; finally they become quite immovable, and attach themselves in all their length upon the spore. Their bodies lose their form; and are soon nothing but mucous droplets, of which a part seems to be absorbed by the spore. The primordial spore, now fecundated, soon becomes covered with a true cellular membrane, as represented in Fig. 301.

When these spores are prepared to germinate, their contents are subjected to many modifications. They become granulated, assuming a sombre tint of reddish brown, and a transparent circle shows itself in the centre. The reddish matter frequently assumes a greenish tint before germination. This change of colour proceeds slowly and step by step from the exterior of the spore towards the centre of its cavity, and the plastic contents finally separate, first into two, then into four, and ultimately into a much greater number of parts (Fig. 302, *a*, *b*, *c*), which rupture their double envelopes, and spread themselves freely in the water, as so many zoospores.

The form of these zoospores is as uncertain as their size and colour. During more than an hour these corpuscles, furnished with two cilia to their beak, agitate themselves by a slow jerking movement, interrupted from time to time by long pauses, until it almost seems as if they had lost their power of movement, when, after many hours of immobility, they suddenly resume their motion.

When these zoospores begin to germinate they elongate themselves more and more, in a spindle-like form, swelling in the middle, as in Fig. 303, *a*, *b*, *c*, *d*, *e*, *f*. In a short time the little plant, hitherto formed of a single cell, separates into two equal compartments; then, successively, into a great number of cells, in proportion as they grow; becoming finally a young *Sphæroplea.*

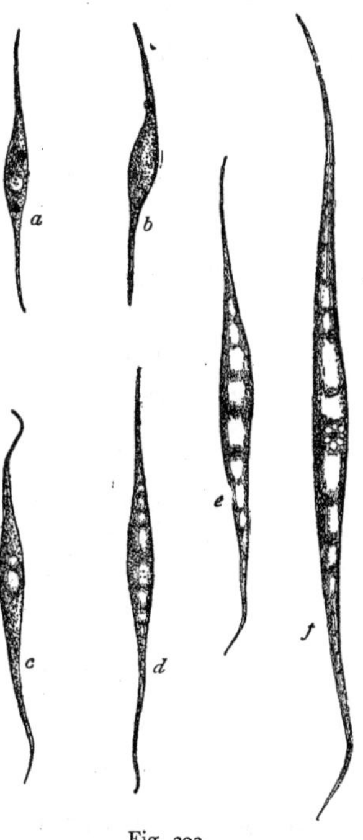

Fig. 303.
Germination of *Sphæroplea* (magnified).

Such is a brief history of *Sphæroplea annulina*, in which the descriptions of M. Cohn are only slightly abbreviated. The strange details give birth to feelings of profound admiration in the naturalist and thinker. Here are plants placed at the bottom of the vegetable scale which are reproduced by the emission of germs gifted with a movement of their own, and seemingly guided in their evolutions by a true kind of instinct.

FUCACEÆ are seaweeds which closely resemble Confervaceæ both in structure and local habitat. They differ in their mode of reproduction, the antheridia and sporanges being contained in small spherical cavities which raise the epidermis of the frond in a wart-like manner, and finally open upon the surface by a pore or minute mouth, to allow of their escape. These cavities, or conceptacles, also contain antheridia filled with antherozoids. Some of them are used as food, as *Alaria esculenta* and *Fucus vesiculosus*, which are eaten by the inhabitants of Scotland and Ireland, and the natives of the shores of the Pacific. They have also been extensively employed for industrial purposes, both in our own and other countries, in the manufacture of kelp, glass, and especially iodine, which is extracted from many species.

Dr. Lindley divides the order of Fucaceæ into the three following sub-orders, each distinguished by its own peculiarities:—

I. VAUCHERIÆ.	Frond mono- or pleio-siphonous, without bark, the utricles forming a lateral branchlet, proceeding from the upper joint, sometimes from the lowest.	Including Hydrogastrum and Vaucheria Dasycladus, Ectocarpus, Batrachospermum, and Chordaria.
II. HALYSEREÆ.	Frond polysiphonous, barked and jointed, vesicles scattered over the surface of the frond, or in heaps.	Including Sphacelaria, Dictyosiphon, Laminaria, and Sporochnus.
III. FUCEÆ. . .	Frond polysiphonous, often bladdery, vesicles seated in hollow conceptacles formed of a folding-in of the frond, pierced by a pore and surrounded by flocks; conceptacles scattered or collected upon a receptacle.	Including Lemanea, Fucus, Cystoseira, Sargassum, Turbinaria, and Seirococcus.

Two species of *Sargassum*, *S. vulgare* and *S. bacciferum*, are frequently found on our shores as far north as the mouth of the Clyde and the west coast of Scotland; but they are mere waifs and strays cast with other tropical productions on our shores. They are the "Gulf-weeds" which form floating meadows in the midst of the ocean. "Midway the Atlantic," says Maury, in his "Physical Geography of the Sea," p. 88, "in the triangular space between the Azores, Canaries, and the Cape de Verde Islands, is the great Sargasso Sea, covering an area equal in extent to the Mississippi valley; it is so thickly matted over with the Gulf-weed that the speed of vessels passing through it is often retarded. When the companions of Columbus saw it, they thought it marked the limits of navigation, and became alarmed. To the eye, at a little distance, it seems substantial enough to walk upon."

Tufts of *Vaucheria* are formed of a network of cylindrical filaments, branching and continuous, enclosing green granules and a colourless mucilage. This small plant, common enough in marshy places, is rendered very remarkable by its diverse modes of reproduction. It has been the subject of most interesting and minute investigations made by Thuret and Pringsheim. Its reproductive spores are, as we shall see, gifted at one period of their existence with a true movement, one might almost believe them to be mobile animals. This very remarkable fact shows how difficult it sometimes is to establish precise differences between plants and animals, and to define absolutely the limits of what are sometimes called the kingdoms of Nature.

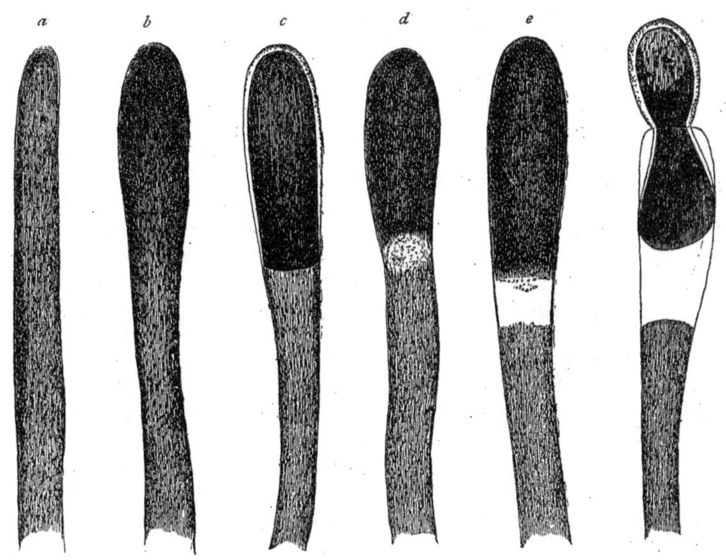

Fig. 304.—Transformation of *Vaucheria* (magnified).

Fig. 305.—Spore of *Vaucheria* escaping

The extremity of the filaments of this plant are swollen into a club-like form, and the green matter is condensed there until it assumes a blackish tint. In Fig. 304 are represented the successive alterations which the extremity of Vaucheria presents at the moment when the work of reproduction is in progress. The letters *a b c d e* indicate the progressive modifications which takes place. According to M. Thuret, the granules separate themselves slowly, the one from the other, towards the base of the club-like swelling, leaving there a vacant space. The granules now approach each other again, and once more form a junction. But then a great change takes place, for this singular operation seems to determine the separation of the mother plant from the reproductive body or *spore.* Henceforth the spore is invested with a proper membrane, and possesses a distinct organisation. This is the moment when the crisis approaches. The upper extremity of the spore begins to protrude (Fig. 305); at the same time it begins to turn on its axis so completely that the granules contained may be observed passing rapidly from right to left, and from left to right, as if they moved in the interior of a transparent cylinder. The narrow opening by which the spore seeks its egress produces a very marked state of strangulation. In a few moments it succeeds in disengaging itself and throws itself into the water. Once detached from the mother, the individual spore now, as represented in Fig. 306, continues to turn upon itself unceasingly, but with a very irregular motion, its movement being now quick or slow in one direction, and now in another. Generally it gains the edge of the object-glass on which the observation is being made, as if it sought to escape from its prison-house. Sometimes its motion is arrested for an instant, and immediately it resumes its more active career.

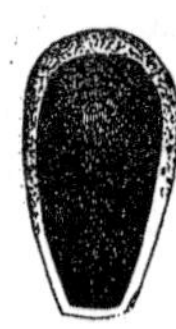

Fig. 306 (magnified).

The whole surface of the spore is covered with vibratile cilia (Fig. 307), which, however, are quite invisible from the rapidity of their motion. In order to see them properly it is necessary to stop them by means of some reagent, such as opium or iodine. The effect of this application is very remarkable. The opium reduces the movements of the spores of the *Vaucheria* so as to admit of the play of the organs being quite perceptible. The iodine arrests them suddenly, and makes their motion more perceptible from the suddenness of their arrest. The solution of iodine employed by Thuret contained only the one seven-thousandth part of that substance.

This observer has succeeded in following the movements of a spore of *Vaucheria* in water during more than two hours. When the cilia finally cease to move, the spores remain immovable, and soon begin to germinate (Fig. 308), giving birth to a new *Vaucheria.*

Unger's remarks upon the clubbed *Vaucheria, V. clavata,* are highly interesting. While examining this plant it seemed to him that the vesicular summits had the power of contraction, and that by this process they expelled the contained spores, which, after expulsion, ascended to the surface of the water. Happening to look at the surface of this water, he was surprised to find it covered with small globules of unequal size, swimming freely here and there, and gliding round globules that remained motionless, stopping and again setting themselves in motion, like animated beings. The next day a great number of these globules aggregated round the bubbles of gas disengaged from the *Confervæ*, and floated at the surface—some dark green, others round or elongated, but evidently in a state of germina-

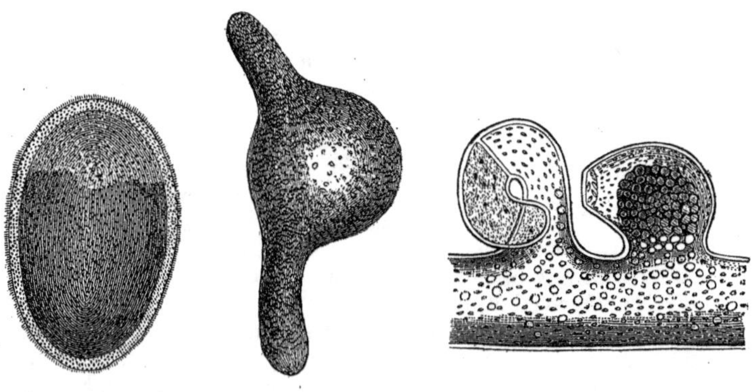

Fig. 307. Spore with its Cilia (magnified).

Fig. 308. Young *Vaucheria* (magnified).

Fig. 309. Antheridium and Sporange of the *Vaucheria* after fecundation (magnified).

tion. Other globules were oval, dark at one extremity, and almost transparent at the other, which swam freely about; within the space of an hour he succeeded in tracing not only the diminution of apparent vitality in these mobile globules, but their subsequent development into germinating plants.

The French botanists are more cautious in expressing their views; they dare not pronounce upon the animality of these beings; and the researches of M. Decaisne prove beyond a doubt the vegetable nature of the so-called zoospores, a doctrine to which the more eminent English physiologists had long given their adhesion.

When the spore has become fixed, it develops itself regularly; its progress is easily followed under the microscope. The elongation of the filaments takes place, so to speak, under the eye.

Besides these non-sexual multiplications by zoospores, there has been recently discovered in the same plant a true sexual reproduction, produced by the assistance of two distinct organs, borne at a little distance the one from the other on the filaments. The *antheridium* (Fig. 309, B) is a sort of short branch, re-curving as it were upon itself, like a snail or small horn. The other, A, is a sort of shell, light and thin, and fashioned like a beak, which is the *sporange.* These two organs are separated the one from the other upon the tube which carries them by a transverse chamber. In the interior of the sporange, and towards its base, certain green granules are found, while towards the beak is a colourless matter, very finely granulated. At the extremity of the antheridium, which is bounded by a thin partition, a great number of small spindle-shaped corpuscles are found, more or less surrounded by a colourless mucilage.

Such is the state of things when fecundation takes place. At this moment the membrane of the sporange is broken at its tip, and some of the matter contained in this sac is poured out through the opening (Fig. 310). Immediately after the sporange opens, the antheridium, by a marvellous coincidence, opens also at its extremity, and pours out its contents. The innumerable spindle-like corpuscles of minute size, that is to say, *antherozoids*, now issue from the antheridia. They penetrate the adjacent opening of the sporange (Fig. 311), filling it almost entirely. Arrived at the surface of the mucous and granular layer, which prevents them, by its consistence, from penetrating further, they advance and retire, continuing this backward and forward motion during the next half-hour, presenting a most singular spectacle to the observer. A partition is then formed in advance of the mucous layer, which prevents the further action of locomotive corpuscles from exercising themselves upon it. These movements will endure for another hour; but they become gradually slower and slower, ceasing entirely at the end of a few hours.

It is after the introduction of the antherozoids into the sporange that a large cell or spore forms itself in the interior of the sporange, which completely fills it. At first greenish, this cellule becomes paler by degrees, and presents in its interior many larger bodies of a dusky brown (Fig. 311). Sometimes it is isolated from the tube in consequence of the membrane of the sporange beginning to decompose. At a certain time—that is, in about three months—this spore begins to recover its green colour; it becomes elongated slowly, and soon assumes the form of a young tube of Vaucheria, and in a short time it is a perfect reproduction of the mother plant (Figs. 313 and 314).

Such is the double and singular mode of fecundation in the *Vaucheria*. It was foreseen by Vaucher, who recognised the first, and suspected the importance of the antheridia; but we are indebted

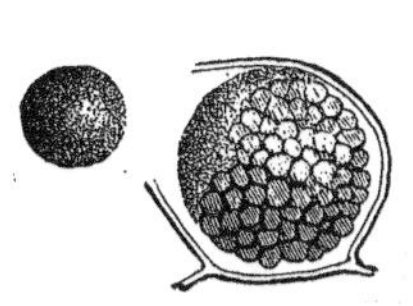

Fig. 310.
Antherozoid of the *Vaucheria* (magnified).

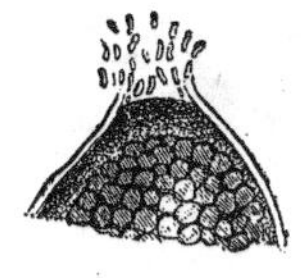

Fig. 311.
Antherozoids penetrating the Sporange (magnified).

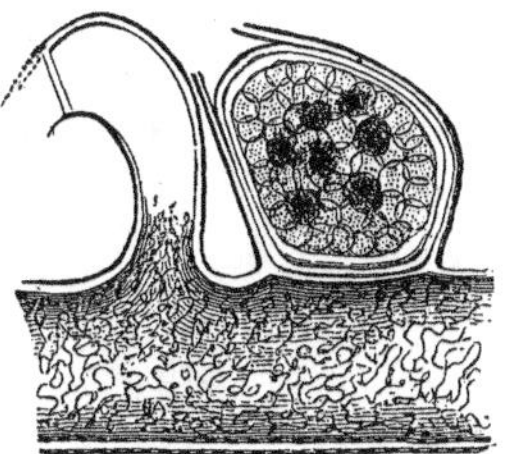

Fig. 312.
Formation of Spore magnified).

to Pringsheim, an able German anatomist, for the complete and circumstantial relation which is here presented.

The *Fucus vesiculosus* (Fig. 315) is the commonest and best known of all the marine algæ. It is found on all the shores of

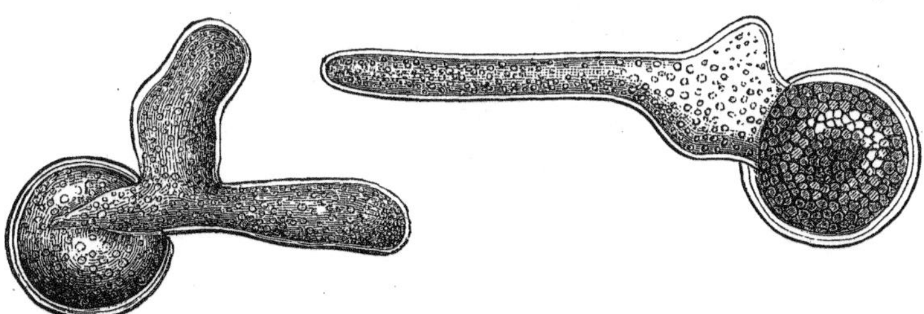

Figs. 313, 314.—Spores of *Vaucheria* germinating (magnified).

Europe; and in the North Sea it grows so abundantly that it is used for domestic purposes, such as roofing houses. It is also cut twice a year in some places, in order to extract carbonate of soda from its ashes. The plane part of its bifurcated frond is studded with globular vesicles, or air-bags, which are probably intended to sustain the plant in the water and perform the functions of the swimming bladder in fishes. Certain minute tubercles invest the extremity of the bifurcations of the frond.

If portions of the plant are taken out of water at the time when

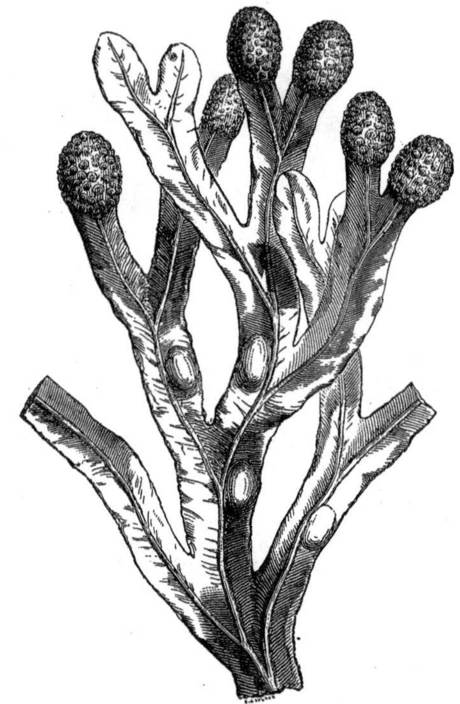

Fig. 315.—*Fucus vesiculosus.*

the tubercles are fully developed, it is sometimes observed that their orifice is closed by a drop of reddish liquid, while other portions of the same plant present in similar circumstances a sort of secretion, no longer red, but olive.

This change of colour seems at first glance to indicate a physiological difference in the tubercles borne upon the different fronds.

The fact seems to be that each of these tubercles is only a cavity or conceptacle, enclosing in the one case a fecundating apparatus, in the other a fruit-bearing organ; and these organs are borne upon separate tubercles. *Fucus vesiculosus* may therefore be considered as *diœcious.* "The fructification of the Fuceæ," says M. Thuret, "is contained in small spherical cavities situated beneath the epiderm, called conceptacles. These are completely closed at first, but open eventually; they open at the surface of the frond by a small pore, or mouth, through which the reproductive bodies escape, assisted by

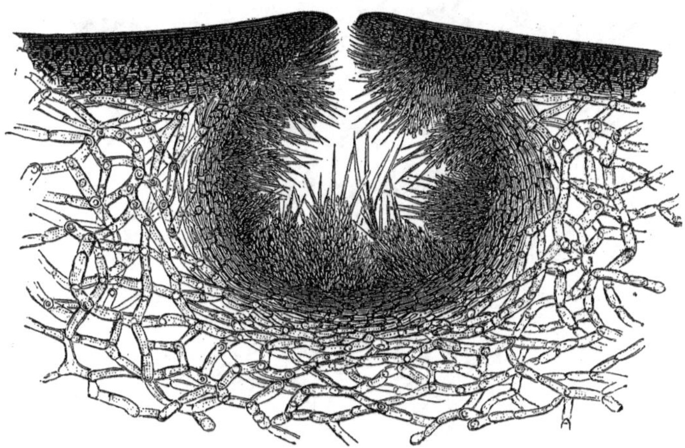

Fig. 316.—Transverse section of the Conceptacle (magnified).

jointed and branched hairs, or cilia, which line the conceptacle which supports the antheridia, and it is at their base that the spores are fixed. In certain species, sporanges and antheridia are found in the same conceptacle; in others, on the contrary, these organs are produced in different conceptacles, and on different individuals."

The male conceptacle of *F. vesiculosus* (Fig. 316) is found to contain small ovoid, transparent sacs. These are the *antheridia* which are inserted or supported upon the branched reticulated hairs which line the conceptacle. They enclose at first granular and colourless matter, but in time it condenses into little bodies forming a greyish mass, sprinkled with orange or reddish spots. These corpuscles are the antherozoids, which are so packed that neither form nor structure can be recognised. The antheridia of Fucus, Ozothallia, Pelvetia, and Himanthalia, have a double envelope ; that is, the sac in which they are contained is itself inclosed in another of the same description, which remains fixed to the hair on which it was produced, while the inner one is expelled through its summit and falls into the conceptacle, whence it glides as far as the mouth. The

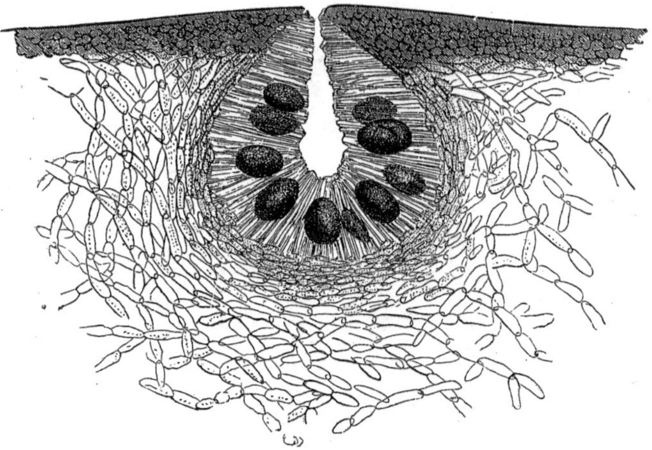

Fig. 317.—Transverse section of the Conceptacle enclosing Spores (magnified).

antherozoids which fill it soon become violently agitated ; the sac opens at one or both ends, through which they force their way into the water, and disperse. In Halidrys, Pycnophycus, and Cystoseira, the second envelope is absent, the outer sac only is found attached to the jointed hairs, and the antherozoids are compelled directly and in a mass, remaining for a time in ceaseless struggling, and turning upon one another before dispersing in the liquid, where they move with great vivacity. Their locomotive organs consist of two cilia, greatly attenuated, the shorter of which appears to be inserted at the smallest extremity of the body, which is always in advance during its progress. The second cilium is drawn behind the corpuscle.

In what botanists designate the female conceptacle of *Fucus vesiculosus* (Fig. 317), certain membranous sacs, more or less spherical or ovoid, are found, enclosing a rounded opaque mass, of a greyish brown, divided into eight parts. These sacs or sporanges, are borne upon a short peduncle, and surrounded by articulated filaments. When the sporange opens, as the antherozoid does at a given moment, the mass which it contains is set at liberty, still preserving its original form, so long as an inner sac which exists is sufficiently strong to hold it together. But matters do not remain long in this state; the spores isolate themselves more and more in the membranous envelope which confines them, and finally they become free. They are then perfectly round, of an olive yellow, and absolutely destitute of skin.

M. Thuret, to whom we are indebted for some excellent observations on the structure of these vegetables, has established by his experiments what becomes of the spores when disengaged from their envelope. "If the male fronds, which are easily recognised by the yellowish colour of their receptacles," he says, "are placed for a short time in a humid atmosphere, an effect is produced analogous to what has been described in the female. The antheridia, expelled in immense quantities from the conceptacles, form on the surface of the frond at each *ostiole*, or mouth, small viscous protuberances of an orange colour. When a portion of the viscous matter is detached with a needle's point, and examined with the microscope in a drop of sea-water, it is seen that it is entirely composed of antheridia, which almost immediately discharge the antherozoids which they contain. These move about with the utmost activity, and their movements are prolonged sometimes till the next day, diminishing in their intensity, however, and by the third day at latest they have become decomposed.

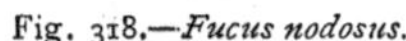
Fig. 318.—*Fucus nodosus.*

"In order to fecundate the spores and prepare them for germination, it is necessary to mix them with the water which contains some antheridia. If the experiment is made upon a glass plate, and the antherozoids are present in considerable quantities, one of the most curious spectacles which the study of these interesting plants can produce will be presented to the observer. The antherozoids, attaching themselves in great numbers to the spores, communicate to them, by means of their vibratile cilia, a movement of rotation, sometimes very rapid. Sometimes the entire field of the microscope is covered with these thick brownish spheres, bristling with antherozoids, which roll themselves about in all directions, surrounded with swarms of these corpuscles.

"In about half an hour, rarely longer, this movement of the spores ceases; the antherozoids continue, however, to be agitated for some time, but with diminished activity, until all movement is finally arrested. From the day after the spores are thrown into contact with the antherozoids they are already invested with a membrane."

M. Thuret remarks on this movement of rotation among the spores, that the phenomenon, however curious, does not perhaps merit much consideration. He does not think it necessary to the fecundation of the spores, and does not admit that the movement takes place in Nature.

CERAMIACEÆ are seaweeds distinguished by their rose and purplish, rarely olive or violet, colour, hence called Rose-tangles. They are cellular, or rather tubular, unsymmetrical bodies, their cells being occasionally only polygonal. They are multiplied by means of tetraspores and sphærospores formed in fours, or within a transparent perispore, or mother cell. The variety in their fructification, as Dr. Lindley remarks, "seems to indicate their being the highest form of Algales." They are entirely marine species, and, according to Endlicher, they chiefly abound between the 35° and 48° N. lat., diminishing in number towards the pole and the equator, and being rare in the southern hemisphere. It is among the genera of this order that the more gelatinous Seaweeds occur: the material out of which the swallows construct the edible nests so valued by the Chinese has been supposed to be *Gelidium*. *Plocaria compressa* and *Chondrus crispus* possess similar qualities. *Rhodomenia palmata*, the Dulse of the Scottish coast, the Dillesk of Ireland, and the Saccharine Fucus of the Icelanders, is eagerly sought on the coasts throughout the maritime countries of Europe.

The CHARACEÆ are aquatic plants, composed of an axis consisting of parallel tubes, which are either transparent or encrusted with carbonate of lime, and of whorls of symmetrical tubular branches; they are multiplied by spiral-coated *nucules* filled with starch. Among the most obscure creations of the vegetable world in regard to their reproductive organs, their transparency makes them interesting objects of investigation, and they were the first in which an actual circulation of sap was observed. They inhabit stagnant, fresh, or sea water, and live always immersed, giving out a fetid odour, and having a dull greenish colour. Their stems are regularly branched, brittle, and surrounded here and there by smaller branches or whorls, the axils of the uppermost whorls concealing the reproductive organs, consisting of a *nucule*, or sporange, and a *globule* representing the antheridium.

The order consists of three genera, *Chara*, *Nitella*, and *Charopsis;* the first containing on the outsides of its central tube a thick layer of calcareous matter, which is the result, according to Greville, of the peculiar economy of the plant itself, and according to Brewster, analogous to the silicious deposit on *Equisetum*. *Nitella*, on the contrary, is transparent, and free from all foreign matter. This property seems to have recommended it as a subject for experiment. If the stem of any transparent *Chara* is examined with a good microscope, a distinct current will be seen to take place in every tube of which the plant is composed, setting out from the base of the tubes to the apex, and returning in *C. vulgaris* at the rate of about two lines per minute; this movement is destroyed by the application of a few drops of spirits, by pressure, or by lacerating the tube. M. Thuret's account of the structure and action of these plants is interesting:—"The antheridia in all the species are in the form of orange-red globules, immediately below the spore-cases. These globules consist of eight slightly concave triangular valves, crenulate at the edge, the crenulatures dove-tailing together so as to form a sphere when united, each crenulature corresponding to a partition directed towards the centre of the cell, occupying about a third of the breadth of each valve; that part which is turned towards the centre of the antheridium is clothed with a lining of red granules, the rest of the cell containing a colourless liquid which gives it transparency, and produces the appearance of the antheridium being surrounded by a whitish ring. To the centre of each valve is attached an oblong vesicle filled with orange-coloured granules arranged in lines, and presenting a very remarkable circulation. The eight vesicles emanating from the eight valves converge in the centre of the antheridium, where they are united by a small cellular mass; while a ninth vesicle of the same nature, but larger and bottle-shaped, fixes the antheridium, its broad base being attached to a branch of the plant, while its other extremity, penetrating the four lower valves, is fixed in the central cellular mass of the antherid. From this point proceed a number of wavy transparent tubes divided by partitions, in each joint of which is borne a thread-like antherozoid rolled up on itself several times. When these tubes are young their joints contain only a small granular mass forming a nucleus of oval form and greyish colour. This nucleus subsequently disappears, leaving a brilliant point encircled with black on each side of the joint. This is the first indication of the appearance of an antherozoid, and it is produced by the circumvolution of their thread-like body. By degrees the antheridium dehisces; the valves or cells turn back on the branch, dragging with them the oblong central vesicle, to the extremity of which a portion of the cellular mass adheres, bearing the tubes filled with antherozoids. These now present themselves, under the microscope, twisting and turning in all directions in the cavity which contains them. They eventually escape by a sudden movement which resembles the action of a spring. When free they resemble a thread twisted up into a corkscrew form with three or four turns, like the fragments of spiral vessels. The field of the microscope is quickly covered with little thread-like bodies swimming with a singular tremulous motion. They turn upon their axis, always preserving the screw form, for their spiral seems to be somewhat stiff. Their motion seems to be caused by the continual agitation of two long cilia of excessive fineness, which spring from a little behind the anterior extremity of the spiral on which they seem to fold themselves. The posterior extremity—namely, that which is dragged along by the advancing antherozoid—is granular, thicker, and less defined than the rest of the body. When the cilia diminish in activity, it is easy to see that motion originates at their base, and extends by waves in the direction of their length. Iodine, alcohol, ammonia, and the acids stop these movements, the cilia resisting the action of ammonia longer than the other parts of the antherozoids." The phenomena which he has described M. Thuret considers unquestionably of the same nature as in the Mosses, their function being, as he believes, impregnation; to which purpose the sporangium seems to be specially adapted, being surmounted by five cells surrounding a small canal, forming, when young, a sort of stigmatic coronet, which disappears at a later period, when the reproductive body has arrived at a certain stage of growth.

Alliance II.—Fungales.

The Fungi, or Mushrooms, form an extensive section of Thallogens. Their best-known form is the Mushroom. But the Alliance includes forms of infinite diversity, the least noticeable of which perhaps are the minute kinds which appear in and upon decomposing fluids, and as vegetable parasites upon many living animals and plants. All alike, however, consist exclusively of cellular tissue, although they differ widely in their outward appearance, and especially in the nature of their reproductive organs. Fungi, in fact, may be divided into two great sections, characterised by the mode in which their reproductive bodies are formed. In one section minute bodies, called sporidia, are contained in enlarged cells, called asci. But in the section to which the Mushroom belongs the spores are simply the terminal joint or joints of the component threads or cells. Fungi vary widely in form, size, colour, and duration; but one of their most common characteristics is rapid growth and brief duration. The most conspicuous species are distinguished by elegance of shape and bright glossy colours; but nothing can exhibit greater extremes of development if the highest and lowest forms are contrasted; the large fleshy *Boletus*, for example, which grows on the trunks of trees, and the microscopic Mould, composed of threads much too delicate to be distinguished by the naked eye. Viewed in their whole extent, the Fungi may be described as cellular flowerless plants, nourished through their thallus or *mycelium*, living in air, destitute of gonidia, propagated by spores, colourless or brown, and sometimes enclosed in asci.

The Fungi are very extensive as regards genera and species. New forms are being constantly added. "In their simplest forms," says Dr. Lindley, "Fungi are little articulated filaments, composed of simple cellules placed end to end; such is the mouldiness found upon various substances, the mildew of the Rosebush, and, in short, all the tribes of *Mucor* and *Mucedo;* in some of these the joints disarticulate and appear to be capable of reproduction; in others, spores collect in the terminal joints, and are finally dispersed by the rupture of the cellule that contained them. In a higher state of composition they are masses of cellular tissue of a determinate figure, the whole centre of which consists of spores attached, often four together, to the cellular tissue, which at length dries up, leaving a dust-like mass, intermixed more or less with flocci, as in the puff-balls; or sporidia, contained in membranous tubes, or asci, like the thecæ of Lichens, as in *Sphæria*. In their most complete state they consist of two surfaces, one of which is even and imperforate, like the cortical layer of Lichens; the other, separated into plates or cells, and called the *hymenium*, to whose component cells, which form a stratum resembling the pile of velvet, the spores are attached by means of little processes, and generally in fours, though occasionally the number is either less or greater.

A most formidable array of Fungi which are parasites on man and living animals, can be adduced. Still more numerous are those which infest fruits and vegetables. Smut in wheat and other cereals is produced by *Ustilago carbo;* rust is produced by *Puccinia graminis.* The potato disease, which caused such distress in many quarters of the globe a few years ago, is due to the attack of *Peronospora infestans;* the spores of which, according to Berkeley, are supposed to attack the leaves, causing disease in them, which afterwards extends to the tubers. The dry rot in timber is another form of Fungus.

Fries, whose work is the foundation of most of the modern arrangements of this immense family, separates them into six classes.

The following arrangement taken from Berkeley, is essentially that of Fries with slight modifications :—

DIVISION 1.—SPORIFERA. Spores naked.

Class 1. *Hymenomycetes.*—Hymenium free, mostly naked, or if enclosed at first, soon exposed.

Class 2. *Gasteromycetes.* — Hymenium enclosed in a peridium, seldom ruptured before maturity.

Class 3. *Coniomycetes.*—Spores mostly terminal, seated on inconspicuous threads, free or enclosed in a perithecium.

Class 4. *Hyphomycetes.* —Spores naked, variously seated on conspicuous threads which are variously compacted; mostly small in proportion to the threads.

DIVISION 2.—SPORIDIIFERA. Sporida in asci.

Class 5. *Physomycetes.*—Fertile cells seated on threads, not compacted into an hymenium.

Class 6. *Ascomycetes.*—Asci formed from the fertile cells of an hymenium.

Fungi are leafless, stemless, and rootless; they respire, producing thereby carbonic acid, like the flowers of plants and animals.

The organs of vegetation and those of reproduction are quite distinct. The first are composed of a felt-like substance, consisting of very thin interlacing filaments, termed *mycelium.* This mycelium is subterranean, not very apparent, and often doomed to immediate destruction. It is upon the mycelium that the apparatus of reproduction, large in size compared with the organs of vegetation, and often different in kind for the same species, develops itself. This multiplicity of reproductive organs is recognised in certain species of which we shall have occasion to speak; the *Oidium Tuckeri*, which attacks the Vine, is supposed to be a particular state of an *Erysiphe*, a genus in which five distinct kinds of reproductive apparatus is noted.

Fungi exist under the most opposite conditions, and in every kind of locality. Some appear on the surface of the earth, as the cultivated Mushroom, the edible Boletus, the Morel, the Puff-ball, &c. Some grow upon the trunks of trees, upon branches, and upon leaves; others, as the Truffle, are found buried in the earth at a considerable depth. Thousands of small species live as parasites upon other plants; the *Oidium Tuckeri* on the Vine, *Peronospora infestans* on the Potato. Others attack animals. No one is ignorant that the malady which destroyed so many silkworms in the nurseries in the south of France was produced by a Fungus which developed itself in the interior of the living larvæ. In short, these microscopic and encroaching objects attack even the skin and mucous membrane of man and animals, producing new and dangerous diseases.

Mushrooms in many countries are a source of nourishment to the poor, who look for their return as a providential manna. But others conceal a mortal poison. Animals, such as worms, insects, and snails, feed upon them. It is not, therefore, without use that Nature has scattered them with so much profusion over the globe. It would be out of place to enter here upon an elaborate consideration of Fungales in general. We must therefore limit our remarks to a few types selected from the best known species among those which interest us from their utility, or from the dangerous maladies to which they give birth.

The common MUSHROOM (*Agaricus campestris*) will serve as an example of the class Hymenomycetes. This species, commonly raised for the table (Fig. 319), consists of a footstalk, or *stipes*, ranging from an inch and a half to two inches and a half in height. When young, they resemble little balls, usually called Button Mushrooms. Afterwards, when the stalk appears, the cap separates, and it becomes convex or slightly conical, of a white or palish yellow, with rose-coloured gills, and a thick white fleshy cap. At a more advanced age the cap becomes concave, the colour grey, and the gills nearly black. A whitish membrane, like a species of veil, at first entirely covers the young gills, forming afterwards a sort of collar, more or less perfect, round the stalk.

The Field Mushroom grows naturally upon the grass sward where it is exposed to the sun. It is also obtained by culture in dark places where there are heat and moisture, as in caves and quarries. But it is necessary to guard against its being confounded with other species which are dangerous poisons; such are the Fly Mushroom (*A. muscarius*) and *A. vernus*, which is distinguished by the footstalk

Fig. 319.—*Agaricus campestris.*

being bulbous at the base, enveloped as by a purse (*volva*), and by the colour of its gills, which are not flesh-coloured, as in *A. campestris*, but of a pale whitish colour. That we may have some exact idea of the structure of the Mushrooms in general, let us consider the structure of the comestible species.

Let us detach one of the laminæ, or gills, which occupy the lower face of the cap. We shall readily observe, by looking at it through a lens, that the two surfaces are of a velvety texture; but it is only through the microscope that their true organisation can be appreciated.

If a transverse incision is made in the thinner parts of these laminæ perpendicular to their surface, we may assure ourselves that each presents three very distinct layers: the middle layer being in connection with the substance of the cap, and bearing the perpendicular elements of the two others. These elements consist of cells of three distinct kinds (Fig. 319, 4). The first are shorter than the others, carrying nothing at their free extremity; the next are a little longer, terminating in four points, which each bear a small spherical sac at their summit (Fig. 319, 5); the third are much larger, but have neither point nor sac at their extremity.

We are assured by experiment that the little sacs, disposed in fours at the summit of the centre cells, are the reproductive organs, which germinate and reproduce the mother plant. We call them *spores;* the cellules which support them are called *basidia.* The result of the germination of these spores is that *mycelium* of which we have already spoken as being the vegetative apparatus of the Fungi, which is seen in the form of filaments at the foot of the Mushroom in Fig. 319, 1. Fragments of this mycelium can multiply the plant much as a fragment of rhizome of a phanerogamous vegetable does. It is on this principle that gardeners sow the mycelium, which they term *mushroom spawn,* and which may be preserved for many years without losing its germinating properties. In cultivating Mushrooms a hotbed is prepared, consisting of horse-dung, covered with a bed of earth of about three feet thick, in which the mycelium is planted, watering it from time to time in order to maintain a certain degree of humidity. In a short time small tubercles will appear, which at a later period become young mushrooms.

The TRUFFLE, belonging to the class of Ascomycetes, has its membranous sporangia, or asci scattered on a serpentine vein-like hymenium, and enclosed in a stout coat or peridium. The species are very generally diffused over the temperate parts of the globe, growing ten or twelve inches beneath the surface of the soil.

The Common Truffle (*T. æstivum*) is of irregular form, nearly black in colour, and warty in appearance. It seems to affect the soil covered with woods, especially oak and beech woods, but there is no reason to suppose that there is anything approaching a parental bond between the Truffle and the roots of trees among which it grows by preference. It develops itself by *sporidia,* which make their appearance in the matured plant. They are singularly small, something less than the one-thousandth part of an inch in length. When the truffle is left after maturity to decompose in the sun, these sporidia produce whitish filaments, analogous to the mycelium of the Hymenomycetes.

If we examine the soil of a truffle bed in Poitou, in the month of September for instance, we find that it is traversed by great numbers of these white cylindrical threads, composed of microscopical filaments. These threads are divided into cells by internal transverse partitions, and are in continuation with a floccose mycelium of the same nature, which surrounds the young truffles, forming round them a sort of white packing, some twentieth part of an inch thick. These filaments connect themselves directly with the external layer of the young truffle. But this enveloping network is soon destroyed; at first slowly and partially, then entirely, and the truffle appears completely isolated in the soil. The structure of the Truffle is much more complicated than was formerly imagined; and we are indebted to the works of the brothers Tulasne for much that we now know respecting the organisation of this singular vegetable.

The young Truffles present very irregular sinuous cavities, partially communicating with each other, which end sometimes on a single opening corresponding with an exterior depression, sometimes on many points of the surface. When more advanced in age they are traversed by a double system of veins—the one white, the other coloured. The coloured veins are continuous with the exterior tissue which composes the envelope. In their middle parts they are formed of a network of filaments running also in the same direction, whence issue shorter filaments perpendicular to them, whose swelling extremities become the asci. The white veins seemed to be formed by the elongation of sterile filaments intermingling with the asci, between which air is found to interpose itself. They come out on the surface in one or many points. The sporidia contained in the asci, whose forms are much varied, though constant for the same species, are limited in number, which rises from four to eight. Their external membrane is soft, downy, or reticulated.

The *Tuber brumale*, *melanosporum*, *æstivum*, and *mesentericum* are the only species sought for in France. *T. æstivum* grows abundantly in various parts of England, where it is gathered about the size of a large walnut, having a peculiar smell, and something of the flavour of the mushroom. In Algeria *Terfesia leonis* occupies the place of the Truffles of Western Europe. The large *Mylitta Australis*, which attains a weight of more than two pounds, is known by the natives of those regions as native bread, and is supposed to be allied to the Truffles.

The Truffles especially affect a calcareous soil, or a mixture of clay and chalk. In France they abound in Poitou, Touraine, the Vivarais, the Comtat-Venaissin, Provence, at Brives, and at Cahors. They require for their full development a shaded soil, rendered fertile by the decomposed leaves and fruit which annually fall from the sheltering trees; at the same time it must be disintegrated by the subterranean network of the roots. The oak and elm are the trees most favourable to its growth. The truffle is sought for by hogs trained for the purpose. Having indicated the spot by slightly scratching it, they generally leave to man the labour of digging it out of the earth; but if the soil is sufficiently permeable, they do not pause in their scratching until they have seized the truffle. In Burgundy the sheep-dog, and in Italy the water-spaniel, are trained and employed for the purpose. But dogs only seem to hunt the truffle from love of or obedience to man. The hog is more given to egoism; he loves the truffle, and hunts it out for his own use. The trained hog, when on the traces of one, is immovable, with nose over his prize, waiting till it is dug out. Nor does he wait long, but, unless prevented, seizes and devours the odorous prey himself, with the least possible delay. In Upper Provence a hog trained to truffle-hunting is worth 200 francs.

The Morel (*Morchella esculenta*), like the Truffle, belongs to the class Ascomycetes. This Fungus has a stalk from one to three inches long; a spherical hollow cap, the size and shape of an egg-cup, of a pale brown or even grey colour, and deep pitted over its whole surface. It grows in orchards and woods, springing up in early spring and summer, and is generally believed to abound in woods where fires have been made. The country people in Germany were so persuaded of this, that they made a practice—until the custom was put down by law—of firing the woods in order to obtain a crop of Morels. The plant has a slight smell and an agreeable taste; it is employed for various purposes in cookery, both fresh and dried.

When the vast numbers and universal dissemination of Fungi are taken into consideration, together with their diversity of form and size, it is not surprising that botanists have been much puzzled over them. Fries discovered no fewer than two thousand species within the compass of a square furlong in Sweden. Of the Agarics alone above a thousand species are described. In size Fungi range from the minute species (*Torula sporendonema*) which is found to produce death in the common house-fly, and the *Mould*, which M. Deslongchamps found in the air-cells of the Eider duck while alive, and which Professor Owen found in the lungs of a flamingo, up to the

Great Puff-ball, which is said to attain the diameter of a foot in a single night.

"Some writers have questioned the propriety of classing Fungi as plants at all; and it has been proposed to establish them as an independent kingdom, equally distinct from plants and animals. Others have adopted the unphilosophical notion that they are mere fortuitous developments of vegetable matter, called into action by special conditions of light, heat, earth, and air." Fries thus argues against these notions. "The sporules," he says, "are infinite, for in a single individual of *Reticularia maxima* I have reckoned 10,000,000, so subtle as to resemble thin smoke, as light as if raised by evaporation and dispersed in so many ways—by the sun's attraction, by insects, by adhesion, and elasticity—that it is difficult to conceive a place from which they could be excluded."

The BUNT or fungus which attacks the wheat-plant is *Tilletia caries*. It belongs to the class Coniomycetes. This must not be confounded with *Ergot*, which is produced in Rye and some other grasses by one of the Ascomycetes (*Claviceps purpurea*). From the investigations of Quekett it appears that the great mass of *Ergot* consists of the albumen of the grain in a diseased state. After the outside is scraped off, the interior, under the microscope, is found to be composed of cells filled with globules of fatty oil. Outside of the ergotised grain are found a number of small oval or elliptical bodies about the one five-hundredth part of an inch long, containing smaller granules. These are sporidia of a fungoid plant, and are attached to filaments which developed themselves early in the growth of the grain, producing this diseased state. In this form Quekett gave the fungus the name of *Oidium abortifaciens*. When the Ergot is sown in a damp soil the perfect fructification makes

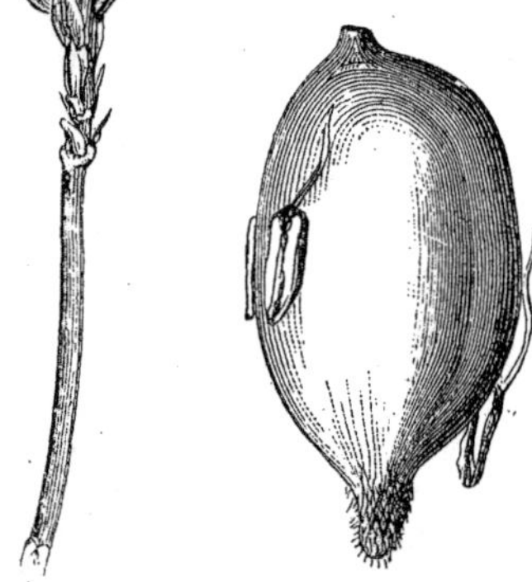

Fig. 320. Bunt on Wheat.

Fig. 321. Bunted Grain of Wheat (magnified).

its appearance as a short purple fleshy stalk, with a small globular head.

Bunt grows in the interior of the ovary of the cultivated wheat (Fig. 320). At the maturity of the plant which it has invaded, the diseased grain is nearly of the size and form of the healthy grain (Fig. 321), differing chiefly in its brownish colour and unequal division. The fungus would seem to originate in the flower of the

Figs. 322, 323, 324.—Germinating Spores of Bunt (magnified).

wheat-plants; it involves the wasting of the stigmata and stamens of the parent plant. "Having," says M. Tulasne, "subjected the pulverulent matter which fills the ovary to the microscope, and especially the parts adjacent to the periphery, which seemed to ripen more slowly, we recognise that spores attach themselves to it in great numbers by short peduncles, to a sort of trunk, or common, thin, colourless branches of a fragile nature, which seem to be re-absorbed, or at least to disappear, as the spores they engender approach maturity. The tissue constituted by them developes with the ovary. This proceeds until the whole ovary is crammed with spores of the vegetable parasite. When a spore germinates," says M. Tulasne, in continuation, "its reticulated integument is broken at some point of its surface, but without regularity (Fig. 322), and a thick and flexible tube issues from the opening, which continues to grow till it attains the length of about fifteen times the diameter of the spore. This rarely fails to be crowned with a sheaf and a bundle of secondary spores (Figs. 323 and 324). These are very slender linear bodies united into pairs in their lower part by a short and rigid band, which gives to the pair the form of the letter H. This bouquet of spores being matured, the primary spores soon perish. The reproductive paired-spores are then isolated from each other, dispersing,

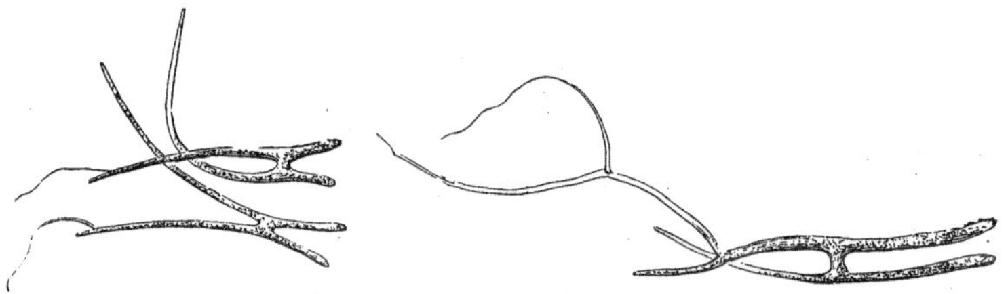

Figs. 325, 326.—Bunt—germinating spores of second generation (magnified).

without dissociating themselves, on the surface of subjacent bodies. Some germinate quickly, and emit, especially towards their summit, very fine filaments, which quickly branch (Figs. 325 and 326). Others, and those in much greater numbers, give birth to tertiary spores, thickish oblong or arched bodies, which appear to be the most important agents in the multiplication of these plants. These germinate by emitting one or many very fine filaments at certain points of their surface."

Smuts are another group of parasitic Fungi belonging to the Coniomycetes. The spores are simple, deeply seated in the tissues of the affected plant, and ultimately forming an irregular powdery mass. The Smut, properly so called (*Ustilago carbo*), particularly attacks oats and barley. It develops itself in the parenchyma of the floral envelopes, in the axes of the spikelet, and in the rachis of these Gramineæ. When the wind has dispersed the spores of the parasite, the plant only remains a blackened skeleton, and scarcely recognisable. The presence of the Fungus involves the abortion more or less complete of the organs of the flower which it has attacked, the sterility of the spike, and a notable alteration of the normal structure.

Another species, *Ustilago Maydis*, with black spores, is equally

Fig. 327.—Smut on Maize.

Fig. 328.—Section of an Ovary of Maize attacked by Smut.

disastrous to the cultivators of Maize or Indian Corn. Fig. 327 represents a spike of maize with white grains. Fig. 328 presents the vertical section of an ovary surrounded with bracts, tumefied by the presence of the Fungus. The black spots indicate the formation at these points of the black powder of the *Ustilago Maydis.*

This Fungus attacks also the stem, producing upon it excrescences more or less voluminous and deformed. "In dissecting the ordinary excrescences while they are still gorged with juices," says M. Tulasne, "we find them to be formed of a parenchyma of large cells, frequently with hollows, and traversed by a small number of fibro-vascular bundles; the bracts and ovary infested by the entophyte present an analogous structure. The chasms in this parenchyma, and frequently even the interior of its constituting cells, are filled at whatever instant they are examined before the final pulverescence of the Ustilago takes place, with the matter of the Fungus. It is a mucous, gelatinous, and perfectly colourless substance, which separates little by little into small polyhedral rounded masses, which, clothing themselves with a tegumentary system, become spores."

The Vine Fungus, which is the *Oidium Tuckeri* of Berkeley, is supposed to be a state of a species of *Erysiphe*—a genus of small Fungi which the world has a great interest in being acquainted with. The elegant structure and varied form of some of these minute Fungi had fixed the attention of mycologists upon them long before the unforeseen result of M. Tulasne's investigations became known. These microscopic plants possess, according to M. Tulasne, no less than three kinds of reproductive apparatus, which make their appearance successively, and the Fungus destructive to the vine is only a species of *Erysiphe*, which runs through the first two phases only of the evolution of its reproductive organs.

The vegetative organs in *Erysiphe* consist in a *mycelium* formed of fine thread-like filaments, furnished with suckers, the form and functions of which remind us in many respects of the suckers of the Dodder-plant; this leads to the inference that in these Fungi we see parasites which live upon the green or living parts of vegetables, particularly upon the leaves. Certain filaments of the mycelium bear straight branches more or less numerous, which swell at the extremities into ellipsoidal cells, and constitute small bodies, often in the form of a chaplet of beads, formed of reproductive cells, analogous to the deciduous buds produced by some flowering plants. To this first reproductive system M. Tulasne gives the name of *Conidia* (Fig. 329). Another class of organs consists of spherical or ovoid vesicles, generally pedicellate, and filled with innumerable small oval or oblong corpuscles. This second system, represented in Fig. 330, he calls *Pycnidia*.

Such are the two sorts of reproductive organs exhibited by the *Oidium Tuckeri;* it is, however, probably only a state of an unknown species of *Erysiphe*, the last and perfect form of which has not yet developed itself. This, the latest and most important form, consists in globular conceptacles, sessile, at first colourless, then yellow

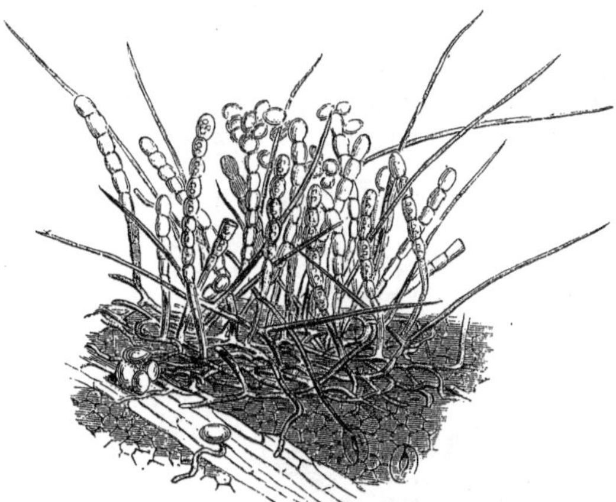

Fig. 329.—Reproductive Organs (Conidia) of *Erysiphe*, magnified.

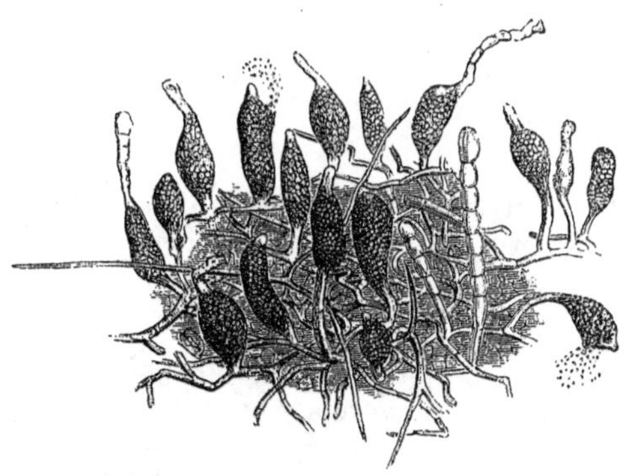

Fig. 330.—Reproductive Organs (Pycnides) of *Erysiphe*, magnified.

brown, and finally black and more or less spotted, which bear, like the first two sets of organs, certain filaments of the mycelium. They are all accompanied at maturity with a variable number of filiform appendages, whose form, dimensions, and position, vary with the species under consideration (Fig. 331). They are simple or branching, and frequently terminate in arms divided into pairs. In the bosom of the conceptacles are found sacs or *sporangia*, variable also in number, generally ovoid, attached by a short pedicel to the base of the conceptacle. The number of spores, which are constant to each species, varies from two to eight. The conceptacles open irregularly in order to permit the spores or the *sporangia* to issue.

MOULDS were long supposed to have a very simple organisation,

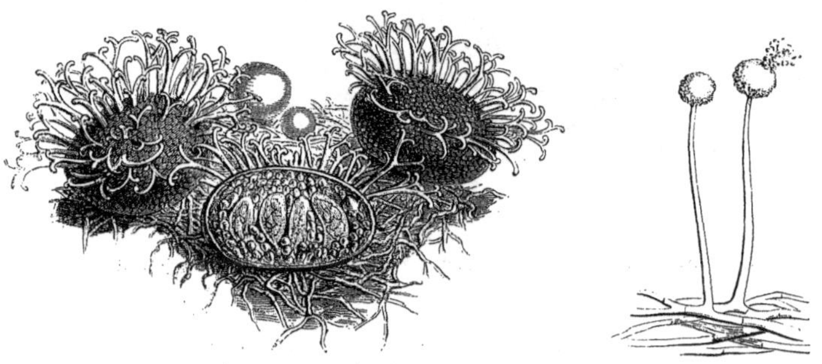

Fig. 331.
Reproductive Organ (Conceptacles) of *Erysiphe*, magnified.

Fig. 332.
Mucor (magnified).

because they had been imperfectly observed. Even in our day they are very imperfectly known. It is now known, however, that some of them are endowed with multiple reproductive apparatus. The species of *Mucor*, which are the commonest of the Moulds, grow on organic substances in a state of decomposition, forming large cottony tufts, with vesicles full of greenish spores on the summit of long slender pedicels (Fig. 332). It has only recently been ascertained that the genera which had been named respectively *Aspergillus* and *Eurotium* are only two different and successive modes of fructification of the same plant; *Aspergillus* being the young, *Eurotium* the adult state.

The parasite which has proved so destructive to the potato belongs to the family of Fungi now under consideration; and here also we note two modes of fructification; one, namely, in which the spores are produced naked at the extremity of the filaments; while in the other the spores are contained in vesicles. In a recent memoir, M. de Bary, professor in the University of Friburg in Breisgau, has directed attention to the very curious phenomena attending the germination of the naked spores, and it may be useful to present a *résumé* of these interesting researches.

The spores, or rather the so-called naked spores of the potato parasite, or the *Peronospora infestans*, present us with three distinct modes of germination. In the first process germination is indicated by the emission of simple or ramified filaments, which possess the power of penetrating the tissues of the potato by piercing the walls of its superficial cells. The second form of germination is characterised by the formation of a secondary spore, from the summit of which

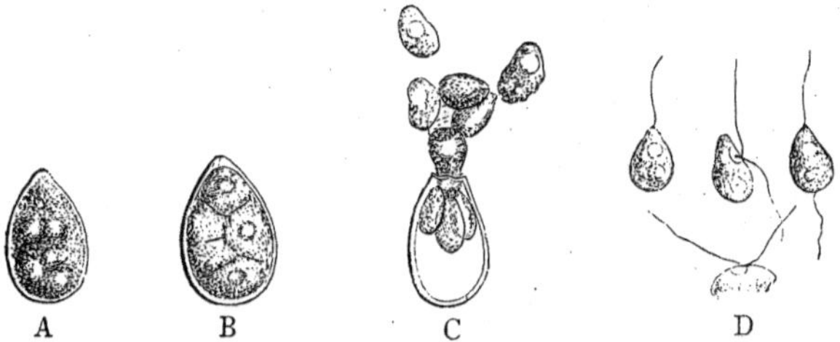

Fig. 333.—Germination of the Spores in *Peronospora* (magnified).

issues a simple tube, which in due course attains the length of two or three times its greatest diameter, expanding like a vesicle at the extremity. When all the plastic contents of the spore are enclosed in this term inalvesicle, it is isolated from the *filamentous portion* by a partition, and thus constitutes a distinct cell. But this spore of the second order is a phenomenon of rare occurrence, and of secondary importance only, according to M. de Bary. In the third mode of germination the spores (Fig. 333, A) divide themselves into a certain number of polyhedric portions (Fig. 333, B), which in a short time begin to issue, one after the other, by a round opening (C), thus constituting egg-shaped zoospores furnished with two unequal cilia, the shorter directed forward, in advance of the corpuscle, and the other dragging after it (Fig. 333, D). The movement of these small bodies lasts for about half an hour, describing a circle, and becoming slower and slower until it enters a state of perfect repose. Now that it has become immovable, the zoospore takes a regularly rounded form, giving birth on one side to a germ-tube, slender and curved, which lengthens rapidly in water.

If the zoosporangia are sown upon portions of the nurse plant, and other circumstances are favourable, the zoospores which proceed from it attach themselves to the epidermis of these fragments, produce their usual germs, which, after spreading a short time on the outside, begin to enter the epidermic cells. Their extremities thus attached soon acquire a considerable thickness, which afterwards increases in a tubular form, which perfectly resembles the filaments of the adult mycelium in the *Peronospora*, and it soon insinuates itself into the depth of the tissues of the nurse plant.

ALLIANCE III.—LICHENALES.

Lichens are cellular plants, and of the simplest structure. They form irregular patches, more or less dry, according to their exposure upon the surface of stones, trees, and other bodies which they cover, while decorating them with a thousand varied hues of colour. These Thallogens live in the air, never in water. Their existence may endure for hundreds of years; their growth and propagation are both excessively slow. They are found in all regions of the globe, from the tropics up to the poles. They grow on the summit of the loftiest mountains, in the plains and valleys, and at all intermediate heights. Near the limits of eternal snow, where all other signs of vegetation have disappeared, on the edge of glaciers, and up to the 70th degree of north latitude—that is to say, at the nearest point to the pole which has been approached by man—the Lichens still vegetate. Humboldt and Boussingault found them growing near the summit of Chimborazo, and they are the last vegetable which are found on the slopes of Mont Blanc.

In form these universally-disseminated plants are cellular, and formed of a lobed and foliaceous substance called a thallus, which consists of a cortical and medullary layer—the former being simply cellular, the latter both cellular and filamentous. Some species are crustaceous, when the cortical is coloured and the medullary layer chiefly differs in being colourless. "The reproductive matter is of two kinds:—1. Spores, naked, or lying in membranous and amylaceous tubes or thecæ immersed in nuclei of the medullary substance, which burst through the cortical layer, and colour and harden by exposure to the air in the form of little discs, called shields. 2. The separated cellules of the medullary layer of the thallus; these, called *gonidia*, are of a green colour, and lie singly or in clusters beneath the cortical layer of the thallus, or break out in clusters called *soredia*, or in cups called *cyphelia*" (Lindley).

Although Lichens, with one or two exceptions, are never immersed in water, they are said in all cases to be developed in humidity, and to be in that state mere Confervæ; but as soon as the humidity diminishes, the under part dies, and an inert leprous crust is formed, which becomes the basis of the plant. Hence the two sorts of tissues—living cellules forming the vegetating part, and dead cellules which have lost their cohesion; the latter have lost their

Fig. 334.—*Cetraria islandica.*

reproductive powers, while every part of the living stratum has been ascertained to possess reproductive properties.

Some Lichens are employed in medicine, others in domestic economy, some in the art of painting. The Iceland Lichen (*Cetraria islandica*), Fig. 334, is demulcent, and is employed in many affections of the chest; the great quantity of fecula which it contains renders it esculent. *Sticta pulmonaria* (the Lungwort Lichen), Fig. 335, is in Siberia a substitute for the hop in the preparation of beer. *Cladonia rangiferina* is an excellent substitute for pasture to the reindeer, whose instinct readily discovers it, even under the snow.

The Orchil, produced by a species of *Roccella*, is a valuable dye, and *Lecanora tartarea* is the cudbear of commerce, which is imported to the value of £80,000 annually.

One of the most curious of this family of Thallogens is *Lecanora esculenta*, a lichen which is frequently met with in the mountains of the arid deserts of Tartary. It is found in great abundance in the Kirghis deserts to the south of the river Jaïk. It seems to fall from the sky as a sort of miraculous manna. Men and beasts are nourished on it. But what is really remarkable is that it occurs in

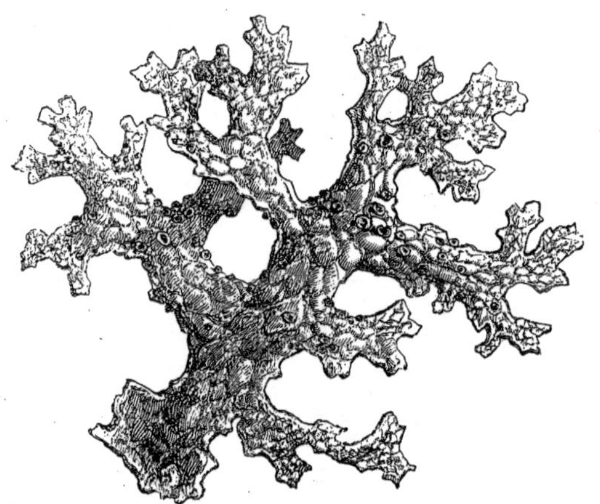

Fig. 335.—*Sticta pulmonaria.*

the form of small globules, the size of which may vary from the head of a pin to a hazel-nut. They are found free, being attached to no other body, except probably when young. It follows from all this that the Lichen spreads itself rapidly; it has vegetated and increased while the wind transported it from one place to another. The light mass which constitutes these Lichens is, in short, often transported by the air to great distances. The manna which supported the Israelites in the wilderness is supposed to have been a species of comestible Lichen of rapid growth, which the wind carried and spread out at their feet. These falls of the so-called manna are by no means rare in our days. In proof of this, one of the secretaries of the Turkish embassy, Fahri-Bey, wrote to me on the 22nd August, 1864, as follows :—"Last year, in the neighbourhood of Kutahia, in Asia Minor, after a great storm, accompanied with a heavy rain, the grains enclosed fell from the sky in great quantities. As a great dearth had prevailed there for some time, the inhabitants eagerly profited by the occasion, and made it into bread. In stating this fact for your information, I would beg of you to analyse them, and favour me with your opinion upon them." The grains enclosed in the letter were the grains of the comestible Lichen, the *Lecanora esculenta* now under consideration.

The thallus of lichens, which is sometimes imperceptible, may nevertheless attain the length of ten yards. The colours which it commonly presents are white, grey, yellow, citron, orange, green, brown, or black. As to its form, it is *foliaceous* in *Parmelia; fruticulose*, in *Usnea; crustaceous*, in *Squamaria; hypophlœodal*, or hidden under the epidermis of the trees, or under the woody fibres, as in *Verrucaria*, *Xylographa*, &c.

In order to give some idea of the anatomical structure of the thallus, it will suffice to mention *Parmelia parietina*, the thallus of which does not exceed the 300th of an inch in thickness.

Thin as it is, however, this organ, as already partly pointed out, presents four very distinct regions. Its upper part consists of a bed of thick, closely-consolidated cells of a yellowish colour at their surface only. In the lower part of the thallus is another cellular layer, like the first, but white. Between these two epiderms are confined, 1st, the green grains known as *gonidia*, and forming the *gonidial* layer; 2nd, a kind of *medulla*, or pith, formed of filamentous elements loosely interlaced or knitted, which is the medullary layer, and which encloses air in its meshes.

If we pass from the vegetative to the reproductive system, we find that it consists of the fructifying or female organs, and a fecundating or male apparatus; the first being represented by the *apothecia*, the second by the *spermogonia.*

The *apothecia*, or fruits of the Lichens, develop themselves on the upper face of the thallus, or upon that part of it which is turned towards the light. They strongly resemble small cups or discs or nuclei, black, brown, yellow, rose-coloured, red, and sometimes powdered with a white or glaucous dust. In size they are extremely variable; the smallest are the 250th of an inch in diameter, whilst the largest may be an inch.

The *spermogonia* are generally very small organs, rounded or oblong, lodged sometimes in particular tubercles, but more frequently immersed in the superficial layers of the thallus.

Many reasons lead to the conclusion that the *spermogonia* are the male organs of the Lichens. They present themselves simultaneously with the fruit in the same individual, and at other times only upon sterile individuals in such a manner that in the latter case the *apothecia* and the *spermogonia* develop themselves upon different individuals. The tenuity of the corpuscles contained in the *spermogonia*, their immense numbers relatively to the number of their spores, their solidity, their form, their equality as to size, the absence of all germinating faculty, are so many circumstances which seem to identify them as agents of fecundation analogous to the antherozoids of other Thallogens. But they possess, so far as is known, no organs of locomotion.

Class II.—Acrogens.

So called from their possessing stems growing in height and not merely consisting in diffused expansions of cellular tissue. They have stems and leaves which are distinguishable, and in other respects they approach closely to higher structural forms, even acquiring the stature of lofty trees in some of the orders. They have breathing pores, or stomata, on their surface; their leaves and stems are distinctly separated in most of the species; in some of them the spiral vessels, which form an important portion of the internal anatomy of the higher forms of vegetation, are found well developed, and they are propagated by spores. These give rise to a vegetative structure upon which reproductive organs (*archegonia* and *antheridia*) of distinct sexes are developed.

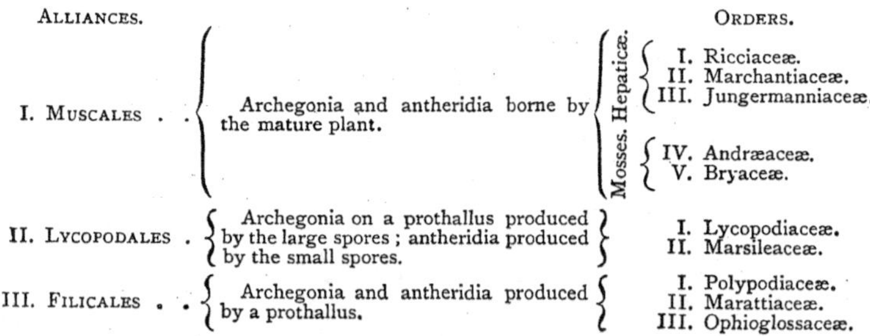

Alliances.			Orders.
I. Muscales . .	Archegonia and antheridia borne by the mature plant.	Hepaticæ.	I. Ricciaceæ. II. Marchantiaceæ. III. Jungermanniaceæ.
		Mosses.	IV. Andræaceæ. V. Bryaceæ.
II. Lycopodales .	Archegonia on a prothallus produced by the large spores; antheridia produced by the small spores.		I. Lycopodiaceæ. II. Marsileaceæ.
III. Filicales . .	Archegonia and antheridia produced by a prothallus.		I. Polypodiaceæ. II. Marattiaceæ. III. Ophioglossaceæ.

Alliance I.—Muscales.

Of the Muscales, Hepaticæ have a loose cellular texture, usually prostrate, and producing rootlets in their sides: they grow in damp places. Sometimes the stem and leaves unite and form a confluent expansion; in other cases the leaves are distinct from the stem. "The most remarkable point of structure in Hepaticæ," says Dr. Lindley, "is the elater, or spiral filament, lying (except in *Ricciaceæ*) among the sporules, within the theca or spore case, and having a strong elastic force. This consists of a single fibre, or two twisted spirally in different directions so as to cross each other, and contained within a very delicate transparent perishable tube. They have been supposed to be destined to aid in the dispersion of the sporules." The spore case is destitute of any definite lid, and the margins of the fissure by which the sporules are discharged is naked and without any fringe of teeth.

The Mosses are humble plants, but they have no insignificant part to play either in the economy of Nature or in the physiognomy of the landscape. Trees, walls, rocks, and ruins assume a smiling or picturesque aspect under their covering in its varied colours. The species of *Phascum*, growing in the gravelled alleys of woods and gardens, are so very minute that in some cases they scarcely attain the height of the hundredth part of an inch. Those of *Hypnum*, which often clothe the banks of brooks in shady places, or form small islets of verdure at the foot of willows and poplars, or are attached to the trunks of these trees, are vigorous vegetable organisms which do not readily decay. The genus *Fontinalis* consists of small grass-like mosses which float in the middle of running brooks. The various kinds of *Sphagnum* grow in marshy places, where they perform an important part in the formation of bog-turf. These aquatic mosses grow very rapidly, extending in such a manner as to occupy by degrees the whole interior of the pool which they inhabit. Their thin and delicate tissues, when they die, accumulate in the bottom of the water, and form with mud and the detritus of other plants a mixture which, when consolidated by time, is cut out in oblong squares, and, under the name of peat-turf, forms an important article of domestic fuel in some countries.

The *Polytrichum*, commonly called the Hair Moss, is one of the most elegant of its kind; it is larger than the Common Mosses, and grows generally among heaths, in fir-woods and bogs. Its principal stem creeps along the ground, throwing out from time to time adventitious roots, which penetrate the soil, and branches to the surface. These branches carry the leaves, which are narrow, lanceolate, or awl-shaped, finely dentate or serrate on their edge, and imbricate in a spiral round the stem; the leaves of the lower part assume a reddish colour when past maturity.

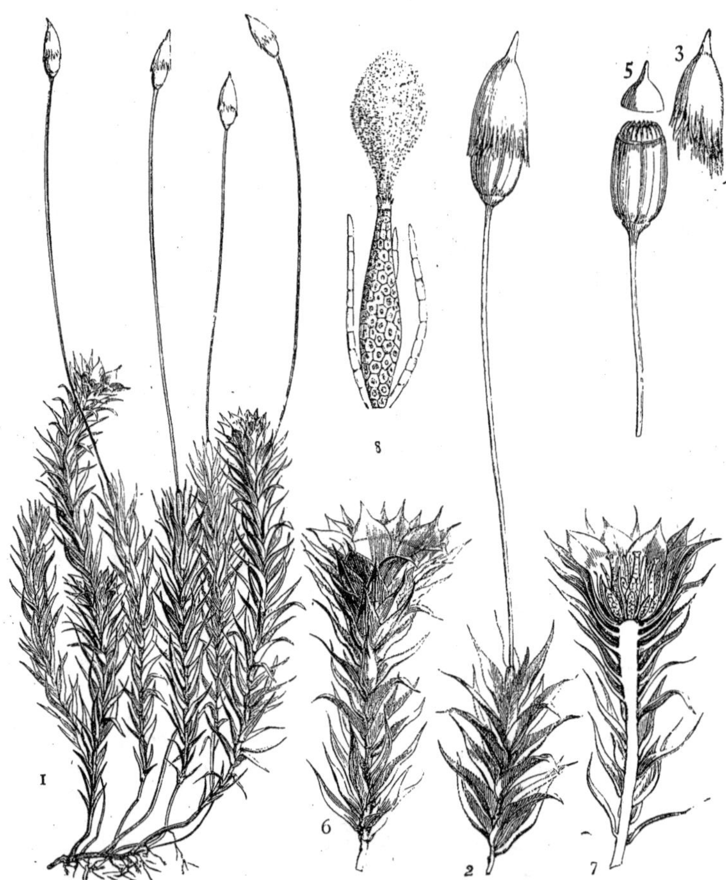

Fig. 336.—*Polytrichum.* (6, 2, 7, 8, 4, 5, are magnified.)

In Fig. 336 the stems terminate, as we see in 1 and 2, by a long reddish filament carrying a pointed or cone-shaped cap (3) composed of silky hairs, disposed longitudinally, and of a bright yellow colour. If this cap is raised, we see that it is the cover of an urn-shaped vessel (4), furnished with a species of covering (5), resting on a rim which circumscribes a thin greyish membrane, stretched horizontally like a drum. The rim is composed of small pointed teeth curving towards the interior, and connected by the horizontally-stretched skin. There are sixty-four of these teeth. As to the interior of the urn-shaped vessel, it is hollow, and encloses a multitude of small greenish granules, perfectly free, and which make their escape with great facility when the urn bursts.

There is every reason to believe that these granules reproduce the species by germination. Their organisation is so simple, and shows so little resemblance to the seeds of the higher order of plants, that they are called *spores*. These are enclosed in the interior of a membranous sac, which covers the walls of the urn, and adheres to a central axis called the *columella*. The free edge of the urn crowned by the teeth is the *peristome;* here the peristome has sixty-four teeth. The covering reposing on the peristome bears the name of *operculum*. The cap of yellow hair which shelters the urn almost entirely is the *calyptra*. Finally, the filament which contains the stem and supports the urn is the *seta*.

This urn, then, results from the development of a small apparatus bearing a strong resemblance to a long-necked bottle, which is traversed very perceptibly in its entire length by an open canal, expanding at the summit in such a manner as to present some analogy with the pistil of plants of higher organisation, and which has been called an *archegonium*. At first many archegonia are enclosed in the terminal rosette of the stem (as in 6 and 7, Fig. 335), but only one of these is ultimately developed into the urn borne at the summit of the seta.

The appearance of the archegonia is contemporaneous with that of the fertilising apparatus which makes its appearance in the centre of the terminal rosette of stems different to the urn-bearing. The Polytrichums are then *diœcious*. These supposed fertilising organs, or *antheridia*, consist of small greyish elongated bodies (as represented in 8, Fig. 336). They are more or less spindle-shaped, and are accompanied by cylindrical filaments called *paraphyses*. They are cellulose sacs, which open above, allowing their contents to escape by abrupt jerks, until the organ is completely emptied.

When the matter thus ejected is carefully examined, it is found to consist of cells, of which each encloses a small body rolled up, or antherozoid. These little bodies are in a continual state of rotatory motion; the tissue which contains them dissolves quickly on contact with water. The *antheridium* becomes flat and dry after the emission of the movable corpuscles which it contains. These are *antherozoids*.

We have said that the appearance of the archegonium is contemporaneous with that of the antheridia. Whatever difficulties may appear to oppose the idea that the antherozoids reach the archegonia, it is impossible to deny that such transposition takes place, for in the archegonia of certain mosses, living antherozoids have been found which had already traversed one-third of the length of the neck.

It follows, therefore, from the structure of the archegonium and antheridia, and from the curious observations which have been recorded, that there is now little doubt of the sexuality of these little plants. This is further confirmed by the fact on which Hedwig founds his principal argument, namely, that in the diœcious mosses the archegonium arrives at maturity only when individuals furnished with antheridia grow in the vicinity.

Alliance II.—Lycopodales.

To this group of Acrogens belong Club Mosses, and the beautiful cultivated species of *Selaginella*. They are closely related to Ferns, and play an inconspicuous part in vegetation.

Alliance III.—Filicales.

In their most graceful type—the Tree Ferns—this group of Acrogens rivals the most beautiful Palms. When they have attained a height of forty or fifty feet their stems form a noble column; from the summit of which droops a plume of pennate leaves, incised by a thousand dentations; these unfold themselves from the terminal bud as a sort of crosier, whose graceful curve adds greatly to the elegance of the plant. The chief anatomical peculiarities of the group are as follows:—

The leaves are termed *fronds*, and they bear the organs of fructification in little groups, termed *sori*, on their edges, or on their under surface. These consist of little membranous sacs termed *sporangia*. These are cellular in structure, and surrounded by a ring, or *annulus*. They contain a number of cells termed *spores*, from which the new plant is produced. The foot-stalk of the leaf or frond is called a *stipe*, and contains bundles of woody fibre and scalariform vessels, connected together by cellular tissue, which pass into the stem, forming the external fibro-vascular cylinder. In the Tree Fern the rind or bark consists of cellular tissue, and is marked from top to bottom by the cicatrices left by the fallen fronds (Fig. 337, *a*).

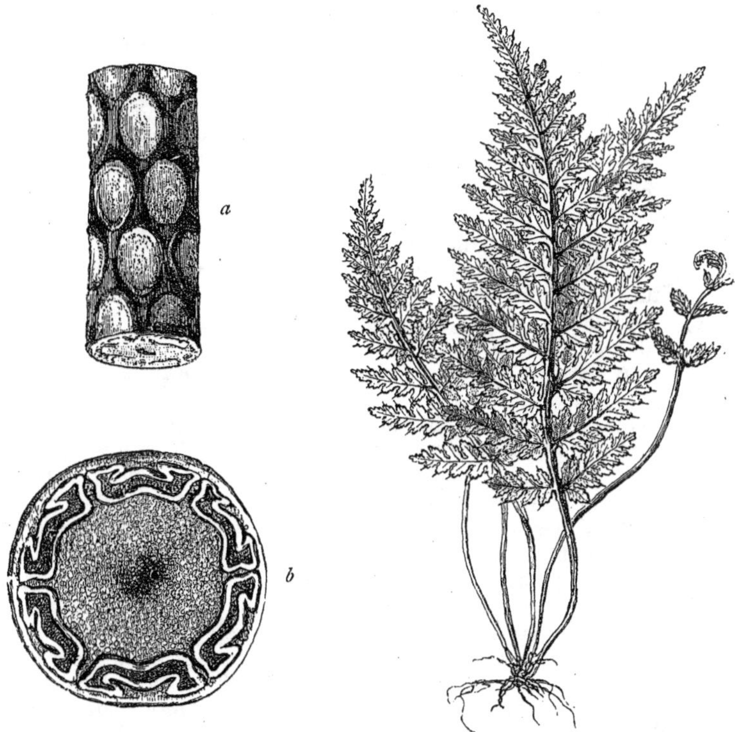

Fig. 337.—External surface and transverse section of a Stem of a Tree Fern.

Fig. 338.—Trichomanes.

These cicatrices occur irregularly and at considerable distances apart near the foot of the tree, but at regular distances and almost close together towards the summit of the stem, showing that its fronds are produced at the top and in regular succession, and that the trunk has increased in height after their fall. A large portion of the transverse section of the trunk is seen to consist of cellular tissue, the centre being occupied by a mass of it. The wood, composed largely of scalariform ducts, is arranged in plates with a wavy outline, forming a cylinder near to the bark. These plates seem to be continuous with the fibro-vascular bundles sent down from the fronds, and as the fronds surround the stem, the bundles sent down from them lie side by side and form a cylinder. The bundles pass from the fronds towards its inner surface; hence an opening in the cylinder corresponds to the organic basis of each frond.

PLATE IV. represents Arborescent Ferns of the Brazilian forests. In our climate these Acrogens are far from presenting the dimensions which they attain in the tropics. Our Ferns are only perennials, with a short or spreading rhizome, the fronds of which rarely exceed three to four feet in length. Even in the tropics and in the southern hemisphere the *Hymenophyllum* and *Trichomanes* (Fig. 337), which grow in humid places, at the foot of old trees, or upon rocks bathed in running brooks, are generally of small size. The delicate leaves are destitute of epidermis, and consist of a simple layer of cellular tissue, traversed by nerves formed of scalariform vessels.

In order to study more closely the structure of a Fern, let us examine the *Nephrodium Filix-mas*, commonly known as the Male Fern (Fig. 339).

This plant is common in the woods. It carries upon its rhizome, which creeps along somewhat horizontally, reddish scales. The fronds are large, petioled, and much divided. On the under surface of the fronds we find little rounded or rather kidney-shaped projections. Each of these projections is formed by groups of small bodies, yellowish green at an early age, brown at their maturity, and covered by a thin greyish pellicle. These groups of little bodies or sporangia bear the name of *sori;* the pellicle which covers them is called the *indusium.* Fig. 341 is a greatly-magnified representation of the organs which occur on the lower surface of the fronds of the Male Fern.

The sporangia (Fig. 342) are pedicelled cellular sacs, furnished on their circumference with an almost entire circle of cells, larger and thicker than those of the other parts of the wall. These cells form a sort of ring, which by enlargement, or by hygrometric changes, seems to determine the irregular rupture of the walls of the capsule (Fig. 343), and by these movements pour out a number of egg-like irregular globules, which were long considered to be the seeds of the plant, and were called *spores.* But this is ascertained to be a false analogy, and absolutely opposed to the facts. In different genera of Ferns, the apparatus under consideration has very different modes of arrangement.

IV.—Arborescent Ferns of Brazil.

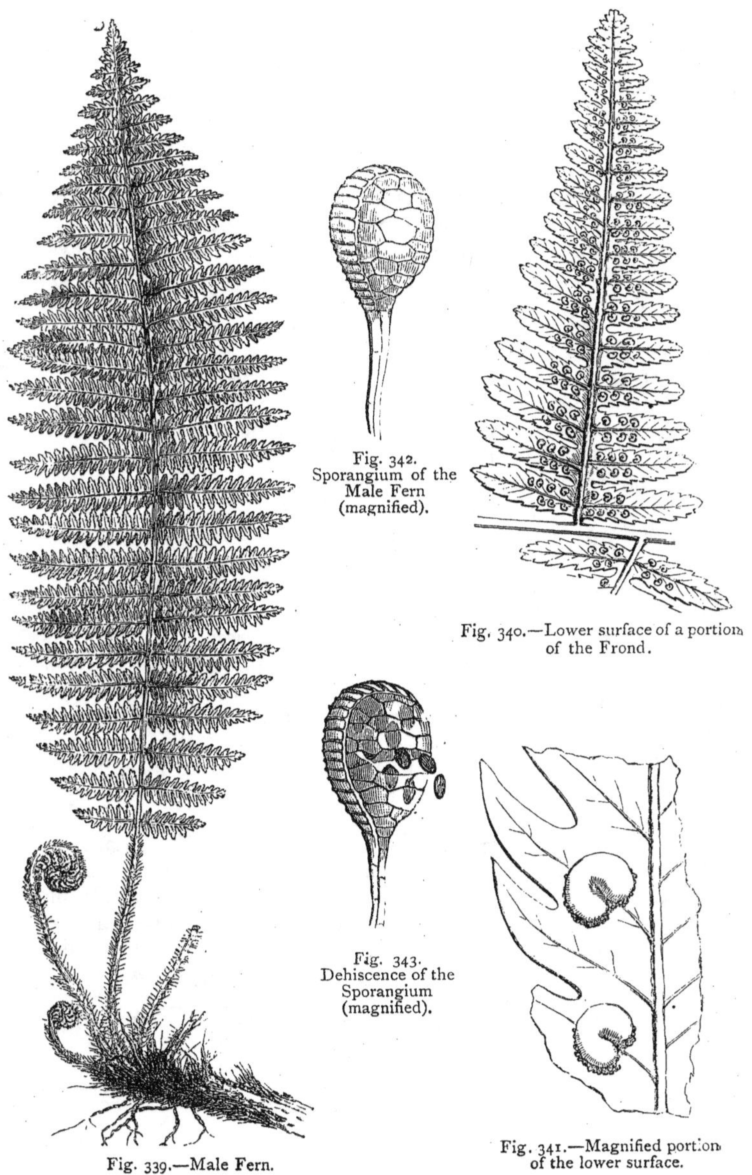

Fig. 342. Sporangium of the Male Fern (magnified).

Fig. 340.—Lower surface of a portion of the Frond.

Fig. 343. Dehiscence of the Sporangium (magnified).

Fig. 339.—Male Fern.

Fig. 341.—Magnified portion of the lower surface.

In our indigenous *Polypodium* the rounded sori are destitute of an indusium. In *Pteris* it extends along the edge of the frond, and opening on the inner side protects the sori. In *Scolopendrium*, the sori, approaching in pairs, are protected by an indusium, which is to appearance bivalve, and they are disposed in oblique lines. In *Osmunda* the capsules form terminal clusters upon the nerves of the upper contracted and modified parts of the frond, and are destitute of indusium, and have the annulus imperfect and contracted.

The reproduction of Ferns has been closely studied in our days by Nägeli, a distinguished German botanist, and still more recently by Leszczyc-Suminski.

It had long been known that the so-called spores of Ferns were susceptible, in favourable conditions, of germinating and reproducing the original plant. The mode of development of these plants seemed, therefore, fully understood; the capsules or sporanges were considered to be the female organs, and the male organs were supposed to be found in the hair-like glandular filaments in their vicinity. Some new and remarkable observations, however, showed that the phenomenon was not so simple as it was thought. The structure of the body which was supposed to be the male organ did not correspond with that of the antheridia of other cryptogams. Neither had the presence of antherozoids confirmed the propriety of the term assigned to it. In short, Nature has neither placed the antheridia of the ferns in the middle of the sorus, nor upon the pedicels of the sporangia. Contrary to the provisions demanded by theory, it is upon plants in process of germination that we find these organs; upon individuals which have only been in existence for a few weeks, and which still consist of only a small number of cells. For the important discovery of the antheridia we are indebted to Nägeli, and it was confirmed some years later by the observations of Leszczyc-Suminski, who detected the archegonia as well.

If we follow the germination of a fern-spore with Leszczyc-Suminski, we find that its external membrane, resistant and coloured, is broken, and by the opening thus formed, the external membrane issues in the form of a sort of tube; and cells are produced and multiply themselves at the extremity of the tube. From this there results a small foliaceous heart-shaped expansion (Fig. 344, *a*), whose dimensions in *Pteris serrulata* may be an eighth of an inch by a tenth. In the under part of this small organ or *prothallium* appear in due course the roots or radicles, then the antheridia, and finally the archegonia.

The *antheridia* are small cellular protuberances, formed, according to Thuret, of three cellules superimposed on each other, as in Fig. 345. In the young antheridia (*a*), says this botanist, the central cavity, surrounded by the second cell like a ring, is only filled with a greyish granular matter; by degrees, small spherical bodies are seen, which are the antherozoids. As these develop themselves the central cavity increases in volume, and presses strongly upon the walls of the peripheral cell. Finally, the time comes when the pressure is so great that the antheridium is suddenly burst; the uppermost cell, which had served as a cover or lid to the central cavity, falls off, or is torn in a stellate fashion, allowing (Fig. 346) the antherozoids to be expelled at the same time.

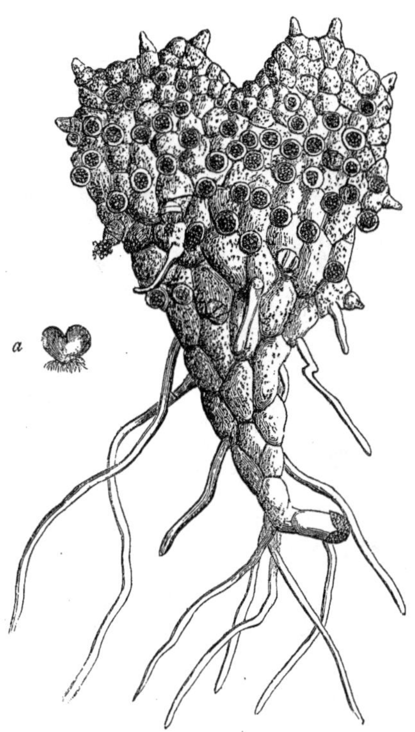

Fig. 344.—Prothallium of *Scolopendrium*, with Antheridia (magnified).

At the moment of their expulsion the antherozoids present themselves in the form of little greyish spherical vesicles, whose contents are very indistinct (Fig. 347). At first they are immovable; but after some minutes they begin to unrol themselves suddenly, and move in the ambient liquid with inconceivable rapidity. They now turn themselves with gyratory movements, which are sometimes continued without interruption during one, and even two hours. If a drop of iodine is added under the microscope, these movements are suddenly arrested. Their body, twisted and contorted, forms a sort of spiral ribbon; it is imperfectly defined, especially at the extremities. The locomotive organs of these curious bodies consist of bundles of numerous short cilia, forming a sort of crest from the anterior part of the body. The number of these cilia is sufficient to account for the extreme rapidity with which these antherozoids move.

These facts overturn all our notions as to the distinctions of

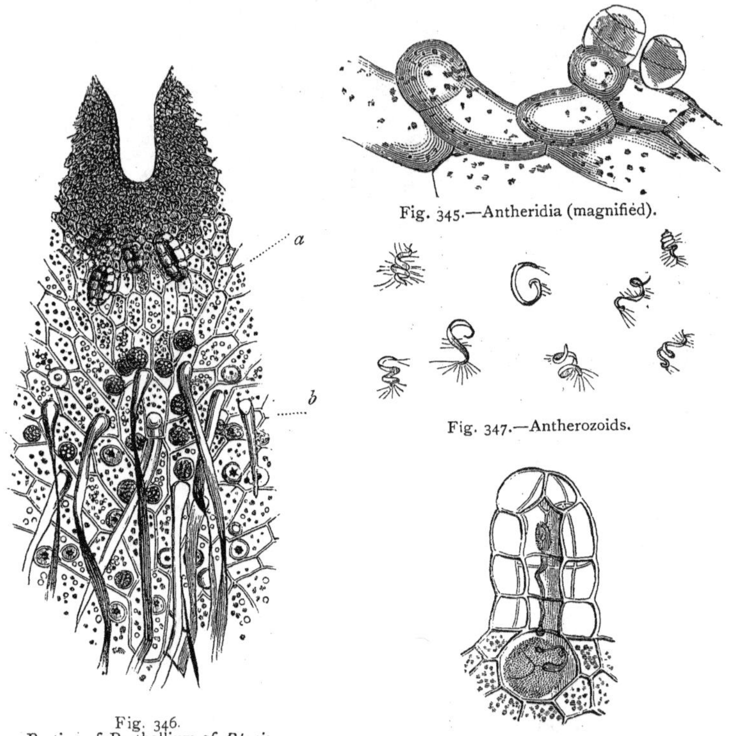

Fig. 345.—Antheridia (magnified).

Fig. 347.—Antherozoids.

Fig. 346. Portion of Prothallium of *Pteris serrulata,* showing Antheridia and Archegonia.

Fig. 348.—Isolated Archegonium, showing the action of the Antherozoids upon the Embryonal Cell.

animals and plants. Here are simple vegetable organs which seem to have the power of motion; and if we reflect on the other hand there are animals, as the sponge, corals, and adult oysters, which are altogether immovable, we may well ask which is the plant and which the animal? We can only reply that the distinctions which science is compelled to draw among living beings become indecisive when we reach the confines of what are usually designated the two kingdoms of Nature.

The female organs of the plants which occupy our attention are less numerous than in the preceding orders; a prothallium does not bear more than from four to twenty (Figs. 346 and 348). They occupy its lower surface, but in front of the notch in it; each of them presents itself as a rounded cavity, plunged in the interior of the parenchyma, and communicating with the exterior by a sort of tube, formed by sixteen transparent cells disposed in fours, the one above the others (Fig. 348).

We ought to remark here, that the two kinds of organs which have been described may exist at once in the same prothallium, as in Fig. 346, or they may be distributed upon several, as in Fig. 344. They are, then, monœcious or diœcious. As to the fact of the fecundation, it can no longer be contested. Leszczyc-Suminski has seen and figured the antherozoids in the interior of the cavity of the archegonium (Fig. 348), and his observations have been confirmed by other observers.

Without entering into details respecting the development of the embryonal cell in the interior of the cavity of the archegonium, we may remark that we only see a single plant issue from the prothallium, as if only a single archegonium had been fertilised, or at least one assumes such a development as to hinder the growth of all others.

To conclude, the capsules which develop themselves on the lower surface of the fronds of ferns are not fruit, as has been assumed until lately; nor are the spores enclosed in the capsules seeds. The male and female reproductive organs are developed on a small and transitory cellular apparatus resulting from the germination of the spores.

CLASS III.—RHIZOGENS.

The Rhizogens are a most anomalous collection of plants, at once leafless and parasitical. They have the loose cellular organisation of Fungi, traces of a spiral structure among their tissues, with a cellular mass serving at once for stem and root, and suggesting an analogy to the thallus of Fungi. Their flowers are like those of more perfect plants, and their sexual organs are complete; their embryo, without visible radicle or cotyledons, presents the appearance of a spherical or oblong homogeneous mass. Lindley placed in this class the following families, now considered to have little affinity with one another.

RHIZOGENS . .	Ovules, solitary, pendulous; fruit one-seeded.	1. Balanophoraceæ.
	Ovules numerous, parietal; fruit many-seeded; calyx 3; 4; 6 parted; anthers opening by slits.	2. Cytinaceæ.
	Ovules numerous, parietal; fruit many-seeded; calyx 5 parted; anthers opening by pores.	3. Rafflesiaceæ.

They are a singular class of parasitical plants, which have cellular scales instead of leaves, but true flowers. They agree with Exogens in having sexual organs, and with the Fungi in their parasitical habits and in their fungus-like consistence. The *Balanophoraceæ* are leafless root parasites, with flowers, brown, red, white, or yellow, but never green, having underground stems, rhizomes, or tubers, from which spring erect simple peduncles. They are found on the roots of the Vine, Maple, and Oak; abounding in the mountains of tropical countries, especially of America, the Himalayas, and Khasia. The *Cytinaceæ* are parasitical on the roots of the Cistus of the South of Europe, and (*Hydnora*) on the succulent Euphorbias of the Cape of Good Hope. *Rafflesiaceæ* are stemless plants of the East Indies, the flowers of which spring immediately from the surfaces of the branches upon which they are parasitic, and are immersed among the scales which represent leaves. The order takes its name from Sir Stamford Raffles, who discovered in Sumatra the species named by Robert Brown *Rafflesia Arnoldi*, which has flowers more than a yard across (Fig. 137).

The plants which Dr. Lindley grouped together as Rhizogens are now considered to have little real common affinity; their parasitic habit merely brings about a superficial resemblance. They have, however, been retained here, as their true position is uncertain.

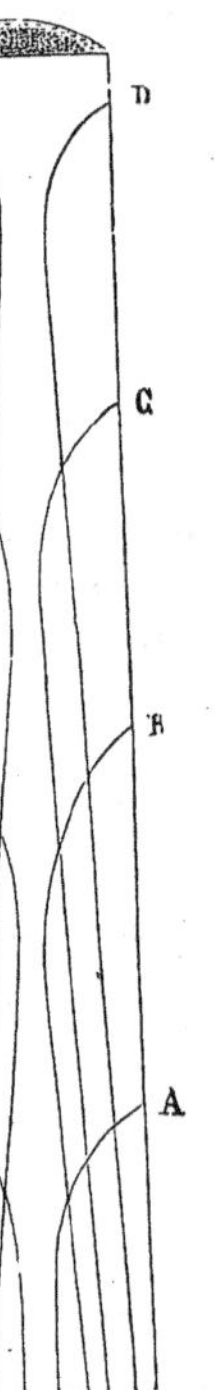

Fig. 349.—Diagram of course of fibro-vascular bundles in Endogens.

CLASS IV.—ENDOGENS, OR MONOCOTYLEDONS.

Endogenous or Monocotyledonous plants are herbaceous, rarely woody plants, *Ruscus aculeatus* being the only species with woody fibres indigenous to Britain. Schleiden, in describing the peculiarities of Endogens, and the manner in which they differ from Exogens, says that all plants whose development proceeds from the interior to the exterior are either limited or unlimited in their growth Woody fibre exists under two different physiological phases. It commences as an extremely delicate tissue (*cambium*), capable of rapid development, in which new cells are continually generated and

Fig. 350.—Transverse Section of the Stem of a Palm.

matured in two different directions, namely, next the circumference, as *liber* or *bast*, a tissue of a peculiar lengthened kind, with thick walls; and on the side next the centre in the form of woody fibres, proper or reticulate, or porous vessels. Up to a certain point the development of the fibro-vascular bundles is the same, but in Endogens the active and delicate formative cellular tissue suddenly changes, the partitions of the cells become thicker, the generating power ceases. Finally, they cease to convey any kind of formative sap; all further addition of new elements to the bundles becomes impossible, and therefore their dimensions are limited. In Exogens, on the contrary, this tissue retains its vital activity during the whole life of the plant, and additions are made to the fibro-vascular bundles during the whole of that period. They are, therefore, unlimited.

The general characteristics of Endogens are the single cotyledon;

the leaves parallel-veined; the wood of the stem always confused (Fig. 350), and containing fibro-vascular bundles, which are limited in their growth.

Lindley estimated Endogens to consist of 1,420 genera and 13,684 species. Dr. Lindley divides this class into—

SECTION 1.—FLOWERS GLUMACEOUS, THAT IS WITH IMBRICATED, COLOURLESS, OR HERBACEOUS SCALES.

ALLIANCES.		NATURAL ORDERS.
1. GLUMALES	In the first two orders the ovule is erect or ascending, the pistil compound; in the other three the ovule is pendulous, the pistil simple. The first approach the Palms, and the last the Rushes.	I. Graminaceæ. II. Cyperaceæ. III. Desvauxiaceæ. IV. Restiaceæ. V. Eriocaulaceæ.

SECTION 2.—FLOWERS PETALOID, WITH TRUE CALYX OR COROLLA, OR WITH BOTH, OR NAKED, USUALLY UNISEXUAL.

2. ARALES	Flowers naked, furnished with scales or corolla, or both, having the sexes on different flowers, with rudiments of the absent sexes present; embryo axile, albumen mealy or fleshy (some altogether without albumen). Here we have the simplest structure amongst flowering plants, gradually approaching the Palms in the Screw-pines.	I. Pistiaceæ. II. Typhaceæ. III. Araceæ. IV. Pandanaceæ.
3. PALMALES	Flowers perfect (having both calyx and corolla), sessile on a branched scaly spadix; embryo minute, lodged below the surface of a horny or fleshy albumen.	Palmaceæ.
4. HYDRALES	Floating water-plants; flowers perfect or imperfect, not arranged on a spadix; embryo axile, exalbuminous.	I. Hydrocharidaceæ. II. Naiadaceæ.

SECTION 3.—FLOWERS WITH A TRUE CALYX AND COROLLA ADHERENT TO THE OVARY, HERMAPHRODITE.

5. NARCISSALES	Flowers symmetrical; stamens, three to six, or more, all perfect; seeds albuminous. Some of the Bromeliaceæ have the calyx free, but so fleshy and permanent as to have all the appearance of being adherent to the ovary.	I. Bromeliaceæ. II. Taccaceæ. III. Hæmodoraceæ. IV. Hypoxidaceæ. V. Amaryllidaceæ. VI. Iridaceæ.
6. AMOMALES	Flowers unsymmetrical, with one to five stamens, some of which are abortive; seeds albuminous; differ from Narcissales in having the veins of the leaves diverging.	I. Musaceæ. II. Zingiberaceæ. III. Marantaceæ.
7. ORCHIDALES	Flowers unsymmetrical; stamens one to three; seeds exalbuminous; embryo, a solid homogeneous body, destitute of visible radicle or cotyledon.	I. Burmanniaceæ. II. Orchidaceæ. III. Apostasiaceæ.

SECTION 4.—FLOWERS WITH TRUE CALYX AND COROLLA FREE FROM THE OVARY, HERMAPHRODITE.

ALLIANCES.		NATURAL ORDERS.
8. XYRIDALES	Flowers half herbacous, 2-3-petaloid; embryo with a fleshy albumen, axile in Philydraceæ, minute on its outside in Xyridaceæ, half immersed in Commelynaceæ, minute and outside in Mayaceæ.	I. Philydraceæ. II. Xyridaceæ. III. Commelynaceæ. IV. Mayaceæ.
9. JUNCALES	Flowers herbaceous, dry, and permanent, bisexual, scaly, scarious, if coloured; albumen copious; embryo, minute and undivided in the Rushes axile with a cleft on one side in the Orontiaceæ.	I. Juncaceæ. II. Orontiaceæ.
10. LILIALES	Flowers hexapetaloid, succulent, withering; albumen copious; anthers turned outwards in Melanthaceæ, inwards in Liliaceæ.	I. Melanthaceæ. II. Liliaceæ. III. Pontederaceæ.
11. ALISMALES	Flowers, 3-6-petaloid; with separate carpels; without albumen; many-seeded in Butomaceæ; few-seeded, with solid embryo in Alismaceæ; flowers scaly; embryo with large plumula in Juncaginaceæ.	I. Butomaceæ. II. Alismaceæ. III. Juncaginaceæ.

ALLIANCE I.—GLUMALES.

So called from their flowers being composed of bracts, and not collected in whorls, but consisting of imbricated, colourless, or herbaceous scales. The grasses and sedges of which they consist constitute a very large proportion of the vegetation of the globe, covering our fields with verdure and furnishing food for man and beast. They are provided with stamens and pistils, which are indispensable to the production of seeds, but there is little trace of calyx or corolla. The *Graminaceæ* are distinguished by their round, hollow, and prominently-jointed stems, their slender, parallel ribbed leaves with slit sheaths; the *Cyperaceæ*, or Sedges, by their angular and solid stems, and inconspicuous joints, with leaf-sheaths which are nerve-slit; the *Juncaceæ*, by their round tapering stems, and many-seeded capsular fruits; the *Eriocaulaceæ*, by their angular stem and capitate inflorescence.

The important family of GRAMINACEÆ, or Grasses, supplies us with Wheat, Rice, Rye, Barley, Oats, Maize, and the Sugar-cane; and constitutes, besides, the grass of our meadows and our hill-sides. The oat is an annual, the lower stem of which forms a short rhizome, from which secondary stems emanate; these are interrupted by brown inflated nodes or joints, which become solid, whilst the parts intervening between the knots are hollow tubes. From these knots spring the leaves. Their petioles form a split sheath on one side, which embraces the stem for some distance before spreading out into a very long slender leaf, traversed by parallel and simple veins, converging towards the summit. At the point where the leaf separates from the sheath, we find a small, whitish, membranous scale (*ligule*), which appears to be a continuation of the inner lining of the petiole beyond the origin of the lamina of the leaf. In the accompanying engraving (Fig. 351) of an annual grass, *Poa annua*, *s* is the sheathing petiole, *l* the ligule, *a* the lamina, *n* the tumid node, at which the leaf originates.

The inflorescence of the Oat is a loose but ample panicle, extending its branches in all directions. If we examine more closely the little pendent bodies which it supports, and which from the delicacy of their peduncles oscillate freely to the breeze which skims over the surface of the field of oats, two pointed scales will be observed on the outside, almost equal in size, but one of which is inserted a little lower than the other; these constitute the envelopes of three distichous flowers, forming each little ear or spikelet; they are called the *outer glumes*. The lower flower is fully developed, the second is smaller, and the third rudimentary and sterile. If we examine the lower flower, it is composed essentially of three stamens and a pistil. The filaments of the stamens are delicate, and the anthers, which are in the shape of an X, are loosely attached by the back. The pistil is composed of a shaggy, hairy ovarium, which is surmounted by the two feathered styles. There is only one cell in the interior, containing a solitary and *anatropal* ovule. These essential organs are protected by two scales, of which the external or lowest, or *flowering glume*, bears upon its back a caducous and rigid filament, slightly bent, while the internal one (*pale*), which is smaller, is supplied with two lateral veins. In addition to this we find on the outside of the most external stamen two small collateral and fleshy bodies, which are designated *lodicules*.

When the ovule has been subjected to the fertilising influence of the pollen-tubes, it is developed into a seed, which presents this peculiarity, that it is blended firmly with the fruit by its integument in such a manner as to constitute what botanists call a *caryopsis*—a term used when the pericarp of a one-seeded indehiscent seed-vessel is membranous, and adheres firmly to the integument, as in Fig. 352, which represents a caryopsis of Wheat, which is often confounded with the seed itself. The greater part of this is composed of a farinaceous albumen. On the outside and beneath, a small distinct body is perceptible, sunk in the surface, and scarcely projecting from it. This is the embryo which rests upon the albumen by a cushion-like lateral expansion of the radicle.

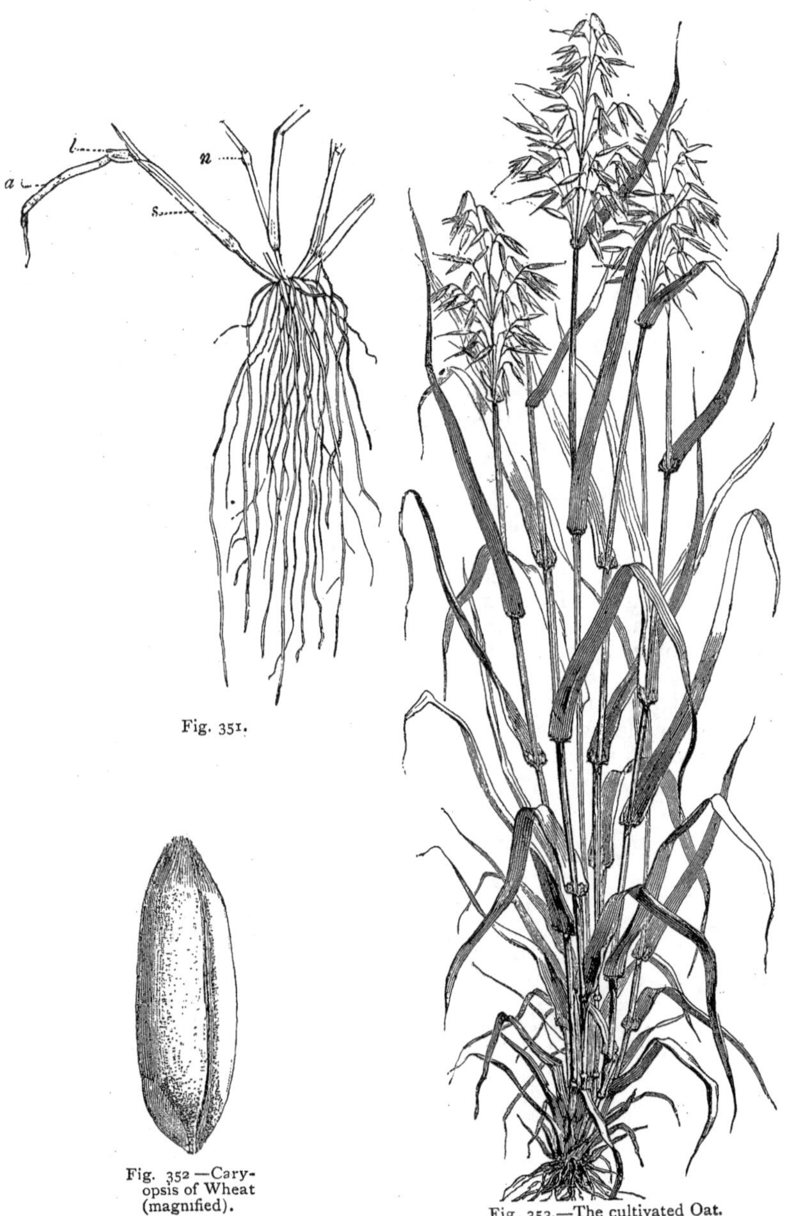

Fig. 351.

Fig. 352.—Caryopsis of Wheat (magnified).

Fig. 353.—The cultivated Oat.

Wheat (*Triticum vulgare*), originally from Persia, has three to five flowered spikelets; their sides opposite the axis, and disposed, as we know, in ears (Fig. 152).

Rice (*Oryza sativa*), originally from India, presents a panicle of rigid and erect branches, with unifloral spikelets, the flower presenting six stamens.

Maize (*Zea Mais*) is monœcious—that is to say, it presents both the sexual organs upon the same plant. The flowers with stamens are disposed in a terminal panicle. The flowers with pistils have their spikelets close together, in a lateral spike, enclosed in a large spathe, which is nothing but the sheathing petiole of a leaf deprived of its limb. The stigmas of these pistils are thread-like and very long, and the whole of them together is like a handful of long filaments, hung carelessly towards the earth like a tuft of hair. The names of Turkish, Spanish, and Guinea Wheat and Indian Corn, which have been given to Maize, are quite fallacious, for it is indigenous to the tropics of America. With the exception of Wheat and Rice, Maize is the most useful as well as the most universally cultivated of the grasses. To Asiatics, Africans, and South Americans, it is more important than wheat. The Sugar cane (*Saccharum officinarum*), another grass of the East, furnishes us with sugar for domestic purposes.

Endlicher divides the 229 genera of which the Graminaceæ consisted when he wrote (which is now increased to 245), into thirteen tribes, as follows:—

I. ORYZEÆ	Spikelets one-flowered with minute glumes, or two-three-flowered; lower floret with one pale, and neuter, the upper only fertile.	14 genera, containing the Rice-plant and some other interesting species.
II. PHALARIDEÆ	Spikelets hermaphrodite, one-two- or three-flowered; the terminal fertile; glumes equal; pale hardened and shining	Containing 19 genera, including *Zea*, the Maize-plant, and some pretty flowering plants.
III. PANICEÆ	Spikelets two-flowered; the lower incomplete; glumes thin, the lower often, both rarely, abortive; pales coriaceous, awnless; caryopsis compressed.	Containing 33 genera, varying in height from the minute to the tall arborescent Panicum, which rises to the height of a lofty tree in India, with a stem as slender as a goose-quill.
IV. STIPACEÆ	Spikelets one-flowered; lower pales rolled inwards; awn simple, articulated at the base.	10 genera, including the graceful *Stipa pennata*.
V. AGROSTIDEÆ	Spikelets one-flowered; glumes and pales two, membranous, herbaceous; stigma sessile.	14 genera, including the bent grasses and many curious genera of easy culture.

VI. ARUNDINACEÆ	Spikelet one with or without a rudimentary upper flower, or many-flowered; florets surrounded with long hairs; glumes as long as the florets.	11 genera, usually tall grasses of reed-like appearance. *Arundo donax* is grown in France and Italy for fencing, vine-poles, fishing-rods, &c.
VII. PAPPOPHOREÆ	Spikelets two- or many-flowered, the upper withering; glumes and pales two.	7 genera of obscure and valueless plants, including the curious bearded *Amphiphogon* of Australia.
VIII. CHLORIDEÆ	Spikelets unilateral, one- or many-flowered, upper florets withering; glumes and pales two, glumes permanent on the rachis.	22 genera of insignificant grasses, chiefly intertropical.
IX. AVENACEÆ	Oat grasses; spikelets two- or many-flowered, the terminal usually withering; panicle branching, or spike-like; glumes two, pales two, the lower awned, awn dorsal and twisted.	24 genera, including the cultivated Oat and many other species of less interest.
X. FESTUCACEÆ	Pasture grasses; spikelets many-flowered; glumes two; pales two; both awned; panicle branched and spreading.	44 genera of grasses, including *Poa*, celebrated for pasture, and *Eragrostis* for its graceful dancing spikelets, and *Bambusa* for its canes.
XI. HORDEACEÆ	Barley plants; spikelet one- three- and many-flowered; terminal floret withering; glumes two, sometimes absent; pales two; ovary hairy; inflorescence spiked.	10 genera, including Wheat (*Triticum*), Barley (*Hordeum*), Rye-grass (*Lolium*), Rye (*Secale*).
XII. ROTTBOELIACEÆ	Spikelets one- two- rarely three-flowered; inflorescence spiked, lodged in the hollows of the jointed rachis; glumes one or two.	13 genera, generally of insignificant grasses, including *Tripsacum*, the forage grasses of the West Indies.
XIII. ANDROPOGONEÆ	Spikelets two-flowered; lower floret incomplete; pales transparent, and thinner than the glumes.	24 genera, including the Sugar-cane (*Saccharum*), and many pretty grasses cultivated in hot-houses.

This vast family of Endogens is universally diffused. *Agrostis algida* was found by Phipps on Spitzbergen. On the mountains of the south of Europe, *Poa disticha* and other grasses ascend almost to the snow line; and this is also the case on the Andes with *P. dactyloides*, *Deyeuxia rigida*, and *Festuca dasyantha*. Their different dimensions are equally striking. Some species of *Bambusa*, true grasses, are fifty to sixty feet high; in our own country we are better acquainted with grasses as forming the compact grassy turf of our meadows, lawns, and hay-fields.

It would be impossible to exaggerate the importance of this great family of plants. Most of them contain abundance of nutritious starchy matter, and comparatively few of them are objectionable, although the cereal grasses only are cultivated for human food. Those reckoned deleterious may be briefly enumerated. *Lolium temulentum*, a weed in some parts of Britain, is injurious: persons have died from its effects. *Bromus purgans* and *catharticus* are emetic and purgative. *Bromus mollis* is also reported to be unwholesome; and *Festuca quadridentata* is said to be poisonous in Quito. *Molinia varia* is said to be injurious to cattle. The most esteemed pasture grasses are *Lolium perenne*, *Phleum pratense* and *Festuca pratensis*, *Cynosurus cristatus*, with several species of Poa and Dwarf Festuca, to which the fragrance of the sweet Vernal Grass (*Anthoxanthum odoratum*) adds a fine aroma.

The CYPERACEÆ, or Sedges, are grass-like herbs, with solid angular stems, sometimes enlarged at the base in corms or tubers, with narrow, tapering leaves wrapping round the stem, but without the slitting sheath. The flowers consist of imbricated solitary bracts, the lowermost of which are often empty, very rarely enclosing other opposite bracts at right angles with the first, and called glumes. There is no diaphragm at the articulations; the seed has its embryo, lying in the base of the albumen, within which it is enclosed.

The Sedges are found on the margins of ditches, marshes, and running streams, heaths and forests, on the sands of the seashore, and on the tops of mountains. In Lapland, according to Humboldt, they are equal to grasses in number; but from the temperate zone to the equator the proportion decreases. As we approach the equatorial regions the character of the order changes—*Carex*, *Scirpus*, and *Schœnus* give place to *Cyperus* and other analogous genera. On the banks of the Nile, in Egypt, the *Papyrus antiquorum*, of which boats, paper, and ropes were made, no longer grows. It is found, however, in Nubia, and lingers in two spots in Palestine. *P. corymbosus* is equally useful in India, where it is manufactured into matting for rooms; while *Cyperus textilis* makes a kind of rope.

Of the remaining orders constituting glumaceous plants the DESVAUXIACEÆ consist of genera of small tuft-like herbs, distinguished from the Sedges by their ovaries, which are variable in number and distinct from each other, ranged round a common axis, as in the Ranunculus. They are insignificant plants of the South Sea Islands and New Holland.

The RESTIACEÆ are herbaceous under-shrubs, with naked stems, or protected by slit sheaths, flowers in spikes, separated by bracts, and generally unisexual. They are distinguishable by their pendulous ovules and lenticular embryo at the extremity of the seed most remote from the hilum, by their stamens opposite the inner glumes, their simple unilocular anthers, and their glumaceous flowers. They are natives of the woods and marshes of New Holland and South Africa.

The ERIOCAULACEÆ are perennial marsh-plants with linear cellular spongy leaves sheathing at the base, very minute capitate and bracteate flowers. Glumes two, and unilateral, or three. A membranous tube, with two or three teeth or lobes, surrounds the ovary. They are chiefly aquatic, two-thirds of them natives of tropical America, the rest Australian and North American. One species is found in the Isle of Skye and Western Ireland.

ALLIANCE II.—ARALES.

These consist of Endogens of secondary interest, the several orders of which are enumerated in the brief summary we have given of them.

ALLIANCE III.—PALMALES.

The PALMACEÆ are for the most part trees of gigantic growth, always forming wood, and occasionally reaching dimensions altogether unknown among other plants. They are the culminating point of vegetative power among Endogens. The species of *Calamus*, or Rotangs, for instance, are sometimes 500 feet long, rising to the tops of the highest trees and falling again. The Palms occupy their place in the first ranks of vegetation as much by the majestic beauty and elegance of their appearance as for their services to the inhabitants of the tropics, to whom they furnish at once bread, oil, and wine.

Let us examine the Date Palm (*Phœnix dactylifera*). This beautiful tree (represented in Fig. 354), which has deservedly received the name of the Prince of Vegetables, raises its straight and column-like stem from ninety to one hundred feet. It is crowned by an ample tuft of from forty to fifty leaves, which sometimes attain the length of ten or twelve feet, floating from the summit with rigid, linear, sword-shaped leaflets, arranged like the fringe of a feather. From the axils of the leaves issue coriaceous spathes of a single piece, opening on one side and permitting the passage of the long branched spadix, which bears small flowers, and are generally male on one tree and female on another; for it is to be noted that the Date Palms are diœcious trees, and it is well known that in order to produce fruit from this tree, it is necessary to have recourse to artificial impregnation of the female flowers, a practice which has been carried on from the earliest times in the countries where the Date is cultivated. The male flower of the Date-tree has a calyx with very short sepals, a corolla of three petals much larger, with six stamens, furnished with long linear anthers, the two cells of which open themselves inwards by two longitudinal slits.

The female flowers present a double floral envelope, each whorl of which is formed of three pieces, and possesses three distinct pistils, each surmounted by a stigma in the form of a hook.

Of these three pistils one only develops itself, ripens, and becomes an elongated, ovoid berry, with a thin epicarp of a yellowish red, a solid and slightly viscous pulp, and an endocarp represented by a slight pellicle enveloping the stone, which is the seed. This seed is cylindrical, thinner at its two extremities, deeply grooved in its whole length on one side, and presenting in the middle of the other a small circular depression—an operculum, which is destined to fall out at the moment of germination to let out the radicle, in the manner described (p. 204) in the chapter on Germination, in speaking of the Indian Shot. In short, this operculum corresponds with a little hollow where the germ is placed in such a manner that its great axis (if one can speak of the great axis of such a little thing) is perpendicular to the surface of the seed. It will be seen, from Fig. 354, which gives a representation of the seed of a Date, that it is almost entirely composed of a hard horny albuminous substance, the thick-walled cells of which are filled with matters which are drawn upon by the embryo in the process of germination.

The Date-tree, indigenous to Arabia and the north of Africa, is pre-eminently the tree of the oasis of the desert; that which, according to the allegorical language of the Orientals, plunges its foot into the water and its head into the fires of heaven. It is planted as an ornamental tree in Corsica, Sardinia, the north of Italy, the Ionian Isles, and Northern Greece; but it does not fruit in those countries, or only imperfectly. By incisions in the trunk of the Date Palm a sweet liquid is obtained, which is called the milk of the Palm-tree, and which, after fermentation, takes a vinous flavour. When distilled, this liquid furnishes arrack. The stem of the tree supplies the natives with their wood for firing and construction, its leaves are employed to roof their houses, and from its leaflets they manufacture baskets, mats, hats, &c. The compass of this work will not permit us to give the history of more than a few of the different species of Palms, which are so numerous and so interesting, from their structure, beauty, and utility. We must confine ourselves to the mention of those whose form is the most remarkable.

Fig. 354.—The Date Palm.

The Cocoa-nut (*Cocos nucifera*) is an inhabitant of the whole torrid zones, chiefly in the neighbourhood of the seas. It rises to the height of a hundred feet, and is surmounted by a crest of pinnate leaves about twelve feet long; the fruit is a drupe as large as the head of a man, with a fibrous mesocarp and a bony endocarp; the seed is almost entirely formed of a fleshy white albumen; in the interior the centre of this albumen is occupied by a space filled by a clear liquid, which is an agreeable and refreshing beverage, a sort of vegetable milk. A fixed oil is obtained from the Cocoa-nut tree; every part of the tree, in short, is useful to man, either to clothe, feed, or shelter him.

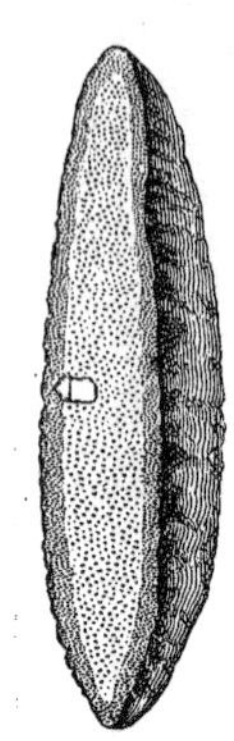

Fig. 355. Section of the seed of a Date.

We borrow from Bonifas-Guizot the following passage, which, whether imaginary or the experiences of an actual traveller, gives with some piquancy an idea of the infinitely varied advantages which the inhabitants of tropical countries draw from the Cocoa-nut tree and its products :—

"Imagine a traveller passing through one of these countries, situated under a burning sky, where coolness and shade are so rare, and where habitations, in which to take the repose so necessary to the traveller, are only to be found at considerable distances. Panting and dispirited, the poor traveller at length perceives a hut surrounded by some trees with straight, erect stems, surmounted by an immense tuft of great leaves, some being upright and the others pendent, giving an elegant and agreeable aspect to the scene. Nothing else near the cabin indicates cultivated land. At this sight the spirits of the traveller revive; he collects his strength, and is soon beneath the hospitable roof. His host offers him an acidulous drink, with which he slakes his thirst: it refreshes him. When he has taken some repose, the Indian invites him to share his repast. He produces various courses, served in a brown-looking vessel, smooth and glossy; he serves also some wine of an extremely agreeable flavour. Towards the end of the repast his host offers him sweetmeats, and he is made to taste some excellent spirits. The astonished traveller asks who in this desert country furnishes him with all these things. 'My Cocoa-nut tree,' was the reply. 'The drink I presented you with on your arrival was drawn from the fruit before it is ripe, and some of the nuts which contain it weigh three or four pounds. This kernel, so delicate in its flavour, is the fruit when ripe. This milk, which you find so agreeable, is drawn from the nut; this cabbage, whose flavour is so delicate, is the top of the cocoa-nut, but we rarely regale ourselves with this delicacy, for the tree from which the cabbage is cut dies soon after. This wine, with which you are so satisfied, is still furnished by the Cocoa-nut tree. In order to obtain it an incision is made into the spathe of the flowers. It flows from it in a white liquor, which is gathered in proper vessels, and we call it palm wine; exposed to the sun, it gets sour and turns to vinegar. By distillation we obtain this very good brandy which you have tasted. This sap has supplied the sugar with which these sweet-meats are made. These vessels and utensils have been made out of the shell of the nut. Nor is this all: this habitation itself I owe entirely to these invaluable trees; with their wood my cabin is constructed; their leaves, dried and plaited, form the roof; made into an umbrella, they shelter me from the sun in my walks; the clothes which cover me are woven out of the fibres of their leaves. These mats, which serve so many useful purposes, are produced by them also. The sifter which you see was ready made to my hand in that part of the tree whence the leaves issue; with these same leaves woven together we can make sails for ships. The species of fibre that envelops the nut is much preferable to tow for caulking ships, it does not rot in the water, and it swells in imbibing it; it makes excellent string, and all sorts of cable and cordage. Finally, the delicate oil that has seasoned many of our dishes, and that which burns in my lamp, are expressed from the fresh kernel.'

"The stranger would listen with astonishment to the poor Indian, who, having only his Cocoa-nut tree, had nearly everything which was necessary for his existence. When the traveller was disposed to take his departure, his host again addressed him: I am about to write to a friend I have in the city. May I ask you to charge yourself with my communication?' 'Yes'; "but will your Cocoa-nut tree supply you with what you want?' 'Certainly,' said the Indian; 'with the sawdust from the wood I made this ink, and with the leaves this parchment; in former times it was used to record all public and memorable acts.'"

In the great conservatories of Kew, of the Museum of Paris, and of St. Petersburg, magnificent specimens of Palms are cultivated, which flourish there and frequently fruit. One of these, shown in Fig. 8, is one of the two plants of *Chamærops humilis* which decorate the entrance of the great amphitheatre to the Jardin des Plantes at Paris. This species is indigenous to the South of Europe; the others belong almost exclusively to the torrid zone, and to the warmer

Fig. 356.—Palm Trees.

regions of the temperate zone. The species of *Chamærops* are numerous in India and the Indian Archipelago. They swarm in equatorial America, but are comparatively rare on the African continent, in consequence of the long periods of dry weather to which the climate is subject. Another species of Palm, widely dispersed in Central America, and which forms immense forests in Brazil, is the *Mauritia flexuosa*.

The Oil Palm (*Elæis guineensis*) is a magnificent tree, originally from Guinea, whence it has been transported to Asia and America. Its fruit, which is about the size of an olive, is of a golden yellow, and has an oily husk, which yields the liquid known under the name of palm oil, which serves for the manufacture of soap, and is imported to Europe for that purpose, being one of the principal objects of exportation from the east coast of Africa.

The Sago Palm (*Sagus Rumphii*), originally from the Malacca Isles, contains in its often voluminous stem a very nourishing starch; but the finest sago is said to be prepared from *S. lævis*.

The Areca Palm (*Areca Catechu*), cultivated in all the warmer parts of Asia, furnishes the highly valuable catechu. The albumen of its seed, cut up in slices, powdered with chalk, and enclosed in a leaf of the Betel-tree, is much used by the natives of the East for chewing as a stimulant. Another species of Areca (*Areca oleracea*) is particularly esteemed for the excellence of its large and tender terminal bud, and is commonly known under the name of the Cabbage Palm.

The Rotang (*Calamus*) has a slight stem, few or no leaves, but climbing, by means of which it sometimes extends itself along the whole length of a tree, passing from one branch to another; from *Calamus Scipionum* the polished and flexible canes are made, known by the name of Malacca canes.

The Wax Palm (*Ceroxylon andicola*) has a trunk which rises to a height of 200 feet, and is covered with a coating of resin-like wax.

ALLIANCE IV.—HYDRALES.

These are represented by the HYDROCHARIDACEÆ—floating water-plants; natives of fresh water in Europe, North America, and the East Indies; among indigenous plants they include the Frogbit and the American water-weed (*Anacharis Alsinastrum*). The NAIADACEÆ inhabit both fresh water and the ocean. To these belong the Grass-wracks, whose habitat is the bottom of the ocean from the North Sea to the Mediterranean and the Indian Ocean.

ALLIANCE V.—NARCISSALES.

The BROMELIACEÆ are distinguished by their short stem, rigid channelled spinous leaves, often covered with cuticular scales, and flowers of gay colour, borne in racemes or panicles; the most remarkable species is the well-known Pine-apple (*Ananassa sativa*), celebrated for the sweetness and aroma of its fruit.

The TACCACEÆ are large tuberous-rooted perennial herbs, found in damp woods in the hotter parts of India, in the South Sea Islands, and in the tropical parts of Africa. The HÆMODORACEÆ (Blood-roots) are herbaceous plants, with fibrous perennial roots, and permanent sword-shaped leaves and woolly flowers; natives chiefly of North America, the Cape of Good Hope, and Australia. The HYPOXIDACEÆ is an inconsiderable order of herbaceous plants, with tuberous or fibrous perennial roots, natives of the Cape of Good Hope, Australia, the East Indies, and tropical America.

The AMARYLLIDACEÆ are generally bulbous plants, occasionally fibrous-rooted, sometimes with a tall cylindrical woody stem, ensiform leaves with parallel veins—singularly elegant plants, which have long been the favourite inhabitants of the greenhouses. The order includes the Daffodil, the Belladonna, and Guernsey Lily, the showy Brunsvigias and Blood-flowers (*Hæmanthus*) of the Cape of Good Hope, and the American Aloe—all characterised by their six stamens, a brilliantly-coloured flower, and inferior ovary. With all their beauty, however, there is no family of plants possessed of more noxious properties. The viscid gum drawn from the bulb of *Hæmanthus toxicarius* is used by the natives of South Africa to poison their arrow-heads. The bulb of the common Daffodil and Snowdrop contain an acrid principle which renders them emetic. The flowers of *Narcissus pseudo-Narcissus* are not only emetic, but a dangerous poison. On the other hand, many of the species possess fine medicinal properties; and from the succulent root of *Alströmeria pallida* fine arrow-root is prepared. Others of the tribe, as *Bomarea Salsilla*, yield a substitute for sarsaparilla. The American Aloe (*Agave americana*), which according to a gardening fable only blooms once in a hundred years, forms an impenetrable fence with its hard spinous leaves, while its fibre forms excellent cordage after being steeped in water for some time, and the succulent substance beaten out of it. In Mexico, where the Aloe is extensively cultivated, sap of an agreeable sourish taste is drawn from it by cutting out the inner leaves just before the flower-scape is ready to burst forth. This sap, when fermented, forms a vinous beverage, resembling cider, called *pulque*, while by distillation a very intoxicating brandy is made from it. The juice yields a very considerable revenue to the State.

The IRIDACEÆ connect the Narcissales with the Amomales; the chief external distinction between the two being a singular change in the development of the foliage. In the Narcissales, and especially in *Iris* and *Gladiolus*, the leaves are long, slender, and sword-shaped, with the veins running in parallel lines converging to the apex; but in the Musaceæ and other Amomales, the veins run perpendicular

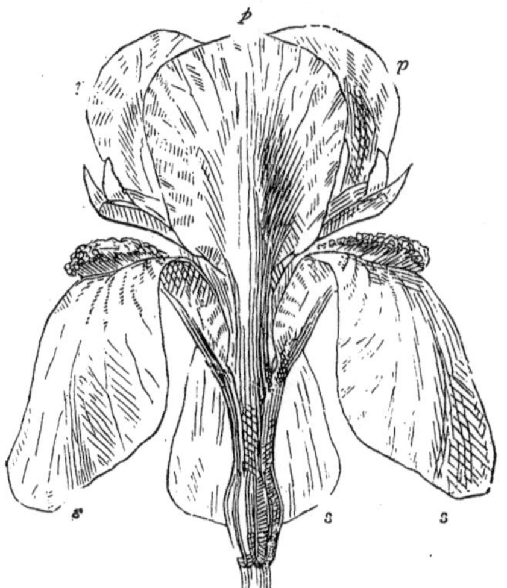

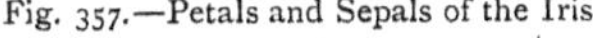
Fig. 357.—Petals and Sepals of the Iris.

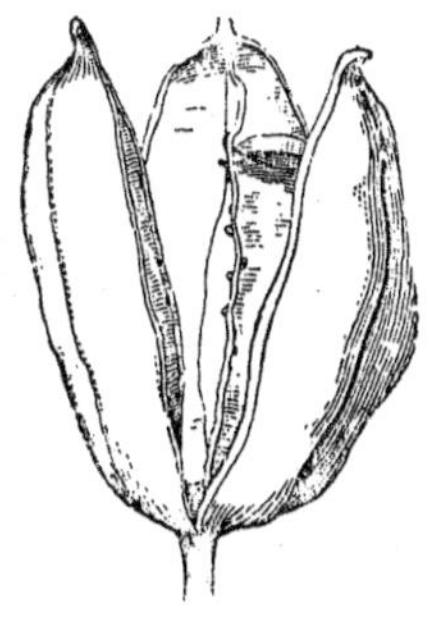
Fig. 358.—Capsule of the Iris.

to the midrib—an arrangement which gives to the foliage a totally different character.

The Iridaceæ are herbaceous Endogens, in which the exterior envelope of the flower or the calyx is composed of three richly-coloured sepals, re-curving outwardly, as shown in Fig. 357, where *s s s* represent the three reflexed or bent-back richly-coloured sepals, and *p p p* the erect petals at the summit of the flower. These six divisions, which are free in the young plant, and arranged in two rows, are afterwards united, and form a perianth of singular appearance, taking a tube-like form towards the base. On depressing the sepals, three stamens with broad flattened filaments and elongated

Fig. 359.—*Iris germanica.*

anthers, bifurcated in the form of an arrow-head, are observed opening with two longitudinal grooves and filled with voluminous pollen grains. These stamens, which are at first completely independent of the perianth, are united in the adult state to that organ. The pistil consists of an inferior ovary united to a style attached by its base to the foot of the tube of the perianth, and terminating in three petaloid stigmas. The ovary presents three cells, enclosing numerous ana-tropal ovules attached to its middle, or disposed in two series placed at the internal angle of each cell. The fruit is capsular, and opens into three distinct portions, divided by a partition at the centre. Fig. 358 shows this loculicidal dehiscence, the valves with the septa in the centre, each valve being formed of the half of each contiguous carpel. The seeds being horizontal and flattened, present a straight embryo placed in an axis of fleshy albumen.

The Iris has a thick, branching, fleshy, horizontal rhizome, a simple or branching stem. The leaves are for the most part in fascicles, radical folded longitudinally, and attached nearly in their whole length by the two halves of their internal face. The central vein corresponds with the outer edge. The cauline leaves are alter-nate and sheathing. The flowers form a sort of composite cluster, which are of great volume, and exhale a powerful odour.

The genera are numerous, inhabiting all the temperate parts of the world; *Iris* and *Crocus* representing the order in the northern, as *Gladiolus* and *Ixia* do in the southern hemisphere. Most of the genera are strikingly beautiful, the number and brilliancy of the varieties in *Gladiolus* produced in cultivation alone being almost unexampled.

Among the interesting species which constitute the genus *Iris* we may mention the *Iris Germanica* (Fig. 359); the *Iris florentina*, the rhizome of which produces the highly-scented orris root, which is used extensively in preparing perfumery. The cultivated Saffron, *Crocus sativus*, is a native species, the stigmas of which form a crest containing a very odorous volatile oil united with a bitter principle; it is employed in medicine, and also in painting. The species of *Gladiolus* have bilabiate flowers of great brilliancy; they are chiefly natives of Southern Africa.

Tigridia pavonina, so called from its spotted and brilliantly-coloured flowers, is another of the bulbous plants for which we are indebted to tropical America. It is a native of Mexico, remarkable at once for its large size, originality of form, and lively colours.

Alliance VI.—Amomales.

We may briefly pass over the Musaceæ, so celebrated for the

nutritive food yielded by their fruit, known in tropical countries as Plantains and Bananas; the ZINGIBERACEÆ, distinguished for the beauty of their floral appendages, as in *Hedychium coronarium*, and for the rich and glowing colours of the bracts in *Curcuma Roscoeana*—they are still more valued for the aromatic stimulating properties of their rhizome; and the MARANTACEÆ, valued for the starch which abounds in the rhizome and fleshy corms of *Canna* and some other genera.

ALLIANCE VII.—ORCHIDALES.

The ORCHIDACEÆ are a group of epigynous Endogens, with one to three stamens, consolidated with the style into a central column, and the seeds without albumen. They are herbaceous plants, and always perennial, and they occur in all parts of the world, but in the warmer latitudes in countless numbers. In the forests of tropical America, in the Indian Archipelago, and in India and other hot countries, they generally attach themselves to the branches of trees, stones, and rocks, to which they fix themselves by means of their long fleshy roots. In temperate countries epiphytes are rare, although it is believed by many (but apparently without reason) that we have an instance of them in our own climate, in the Bird's-nest Orchis (*Neottia Nidus-avis*) a brownish scaly plant springing up occasionally in woods, and deriving its nourishment from the roots of the tree upon which it grows.

"There is no order of plant," says Dr. Lindley, in his papers on Orchidaceæ in the "English Cyclopædia," "the structure of whose flowers is so anomalous as regards the relation borne to each other by the parts of reproduction, or so singular in respect to the form of the floral envelopes. Unlike other endogenous plants, the calyx and corolla are not similar to each other in form, texture, and colour; neither have they any similitude to the changes of outline that are met with in such irregular flowers as are produced in other parts of the vegetable creation. On the contrary, by an excessive development and singular conformation of one of the petals called the labellum or lip, and by irregularities either of form, size, or direction of the other sepals and petals, by the peculiar adhesion of these parts to each other, and by the occasional suppression of a portion of them, flowers are produced so grotesque in form that it is no longer with the vegetable kingdom that they can be compared, but their resemblance must be sought in the animal world. Hence we see such names among our native plants as the Bee, Fly, Man, Lizard, and Butterfly Orchis, and appellations of a like nature in foreign countries." Of these resemblances some idea may be formed by the annexed engraving, where 1 represents *Oncidium raniferum*, or

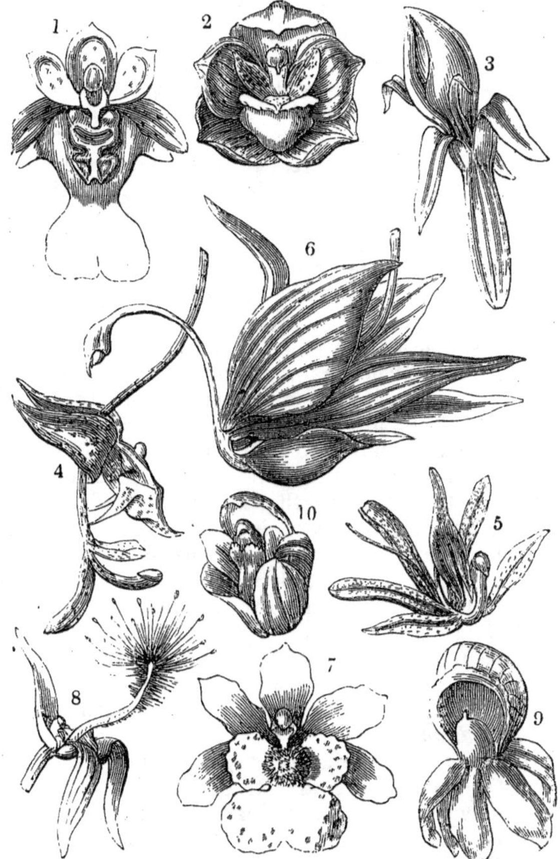

Fig. 360.—Forms of Orchidaceous Flowers.

the Frog Oncidium, so called because its lip bears at its base the figure of a frog couchant; 2, *Peristeria elata*, the Spirito Santo plant of Panama, in whose flower we find the likeness of a dove in the act of descending on the lip; 3 is *Prescottia colorans*, whose lip is a fleshy hood; 4, *Gongora fulva;* 5, *Cirrhœa tristis;* 6, *Cycnoches ventricosum*, singularly like a swan, the arched column forming the head and neck; 7, *Oncidium pulvinatum;* 8, *Bolbophyllum barbigerum*, 9, *Catasetum viride;* and 10, *Peristeria cerina.* The roots, according to the same authority, are—(1) slender, simple-branched fibres of a succulent nature, incapable of extension, but burrowing under ground, as in the genus *Orchis.* (2) Annual fleshy tubercles, having a bud at their extremities containing amylaceous granules for the nutrition of the plant. (3) Fleshy branching bodies, tortuous and irregular in form, as in *Neottia*, or resembling tubers, as in *Gastrodia.* (4) Simple or branched shoots, capable of extension, protruding from the stem into the air, and formed of a woody and vascular axis, covered with cellular tissue, the subcutaneous layer being often green, and composed of large reticulated cells, the points of the roots being usually green, but sometimes red or yellow.

The stem is in its simplest state in *Ophrydeæ*, where it is only a growing point surrounded by scales, and constituting a leaf-bud when at rest, which eventually grows into a secondary stem or branch on which the leaves and flowers are developed. This kind of stem develops every year a lateral bud with a tubercular root attached, which, after unfolding its flowers and ripening its fruit, perishes, to be succeeded by the stem belonging to the lateral bud; hence the species having this kind of stem has always a pair of tubercles attached, one shrivelling and in process of exhaustion, the other swelling and progressing towards completion. The leaves are uncertain in their appearance—usually they are sheathing at the base, and membranous; but in *Vanilleæ* they are hard, stalked, articulated at the base, and have no trace of a sheath. Sometimes they are leathery and veinless; frequently they are membranous and strongly ribbed, and both these conditions occur in the same genus, as in *Maxillaria* and *Cypripedium.*

The floral envelopes are constructed irregularly upon a ternary type, and consist of three exterior and three interior pieces, the exterior being nearly equal, but less brilliantly coloured than the interior, but the two lateral ones are often of a somewhat different form from the other, which is anterior when young, but becomes posterior when the flower expands, in consequence of the flower-stalk becoming twisted or curved.

The centre of the flower is occupied by a body called the Column, which is formed by consolidation of style and stamens, of which there is, in the greater part of the order, only one present, placed opposite the intermediate sepal, and, consequently, alternate with the petals; but in *Cypripedium* there are two stamens. In the greater part of the order a single anther terminates the column. This is usually two-celled, but often has its cells divided into two or four other cavities by the extension of the endothecium between the lobes of the pollen-masses, or it is occasionally more or less completely one-celled by the absorption of the connective.

The pollen consists of lenticular or spheroidal grains, either single or cohering in pairs, threes, or fours, or in larger masses in indefinite numbers; they are usually held together by an elastic filamentous substance, which forms an axis round which the pollen grains are arranged, and frequently assumes the appearance of a strap or *caudicula*. This body either contracts an adhesion with a gland originating on the margin of the stigma, as in *Ophrydeæ*, *Neottieæ*, and *Vandeæ*, or it is folded upon the pollen-masses, as in *Epidendreæ*, or it terminates in an amorphous dilatation, as in many *Malaxideæ*.

These differences in the structure of the column, anther, and pollen furnish botanists with the best means of classifying the order, and breaking it into sub-orders, thus:—

I. MALAXIDEÆ	Anther free, opercular or covered with a lid; pollen-masses waxy, with neither caudicula nor gland.	Including Pleurothallis, Liparis, Dendrobium, and Corallorhiza.
II. EPIDENDREÆ	Anther opercular; pollen waxy; caudicula folded back upon the pollen-grains, and no gland.	Including Epidendrum, Cœlogyne, Cattleya, Lælia, and Bletia.
III. VANDEÆ . .	Anther opercular; pollen waxy; with a membrano-cartilaginous caudicula and gland.	Including Vanda, Sarcanthus, Cryptochilus, Brassia, Oncidium, Maxillaria, and Calanthe.
IV. OPHRYDEÆ .	Anther erect; pollen sectile or granular.	Including Orchis, Serapias, Gymnadenia, Holothrix, Disa, and Corycium.
V. ARETHUSEÆ .	Anther opercular; pollen granular or powdery.	Including Limodorum, Acianthus, Pterostylis, Caleya, Pogonia, Gastrodia, Vanilla.
VI. NEOTTIEÆ .	Anther dorsal, pollen powdery.	Including Cranichis, Listera, Neottia, Spiranthes, Physurus, and Thelymitra.
VII. CYPRIPEDIEÆ	Anthers two, separated by a broad sterile lobe.	Including Cypripedium only.

Among the *Malaxideæ* we find many British species, as the Coral-root, *Corallorhiza innata*, whose root gives out the scent of the vanilla when drying; *Liparis Loeselii*, with its yellowish ten-flowered spike; but the more beautiful species of this sub-order are found in *Dendrobium*, an extensive genus of East Indian Epiphytes found in the moister parts of Asia and East Australia.

The *Vandeæ* have no representative among British Orchids.

Epidendrum, originally a name given to all Orchidaceous Epiphytes, is now restricted to a genus of the order having the labellum united to the column, and four pollen-masses adhering to as many caudicles bent back upon them—a group containing some showy and interesting plants, but many of them inconspicuous and unimportant. Among the other genera, however, some of the most brilliant ornaments of the conservatory are found. The colours of some of the species of *Cœlogyne* are rich and delicate in hue—*C. cristata*, a dwarf evergreen species from Nepaul, throws out leaves six inches long, and six or eight drooping spikes of flowers of a delicate white, with a blotch of yellow on the lip, each flower being three or four inches wide.

The species of *Cattleya* are, however, the most striking of all the Orchids. Their dark evergreen foliage and compact habit recommend them especially to the cultivator. The flowers are large, elegant in form, and their prevailing colours, of violet, rose, crimson, and purple, are unsurpassed for depth and brilliancy. *C. granulosa*, from Brazil, produces large olive-coloured flowers, with rich brown spots; the lip whitish, spotted with crimson in *C. guttata*. *C. Leopoldii*, another Brazilian species, grows about twenty inches high, with a dark green foliage. The sepals and petals are dark brown, spotted with crimson, and purple lip. Others are rose-coloured, margined with white, or the sepals and petals pure white, with beautifully-fringed lip of richest crimson.

Lælia rivals while it resembles the *Cattleya;* the species are compact in growth, with evergreen foliage, producing their flowers in spikes from the top of the bulbs. In *L. acuminata*, from Mexico, the sepals and petals are white, and lip white with a dark blotch on the upper lip. In *L. anceps* the sepals and petals are rose lilac, and lip a beautiful purple, the flowers being three or four inches across; others purple, with a crimson lip, or delicate rose colour, the lip white with lilac spots, with flowers four inches across, as in *L. majalis*, In short, so far as graceful foliage, brilliancy of colouring, form and size of flowers are concerned, the Orchids of this division are the gems of the Vegetable World.

The *Vandeæ* include many of the most curious and interesting among Orchids, chiefly Epiphytes, or air plants.

The species of *Aërides* combine with their rich, evergreen, and gracefully-curving stem and opposite leaves, flowers of peculiar elegance, proceeding from the axils of the leaves, and extending their rich and waxy petals in delicate racemes, sometimes two feet in length, and yielding a most agreeable fragrance. They are natives of the hottest parts of India and other tropical countries, attaching themselves to trees, generally such as overhang running streams of water. *A. affine* throws out its pink and white flowers in branching spikes two feet long. *A. crispum*, another Indian species, with purple-coloured stem and dark green foliage, throws out long spikes of flowers of pure white tipped with pink. *Saccolabium* closely resembles *Aërides* in habit; the flowers are produced in graceful racemes, often a foot and a half long, springing from the axils of the leaves.

Oncidium is another extensive genus belonging to the Vandeæ; the species are chiefly natives of tropical America. The prevailing colour of their flowers is yellow spotted with rich reddish brown, and they are known by their broad labellum, more or less lobed, continuous with the column, and furnished at the base with a tuberculated disc, spreading sepals and petals, with a membranous ear on each side of the column. In their native forests these epiphytes wholly overrun the trees, clasping them round, and covering them from top to bottom with their brilliant and grotesque flowers.

The *Ophrydeæ* have fleshy, bulbous roots, with radical fibres, leafy stem, anther continuous with the column, and the pollen-mass agglutinated and attenuated into a pedicel. In *Orchis* and *Gymnadenia* the lip has a spur; in *Ophrys* it is thick and spurless; in *Habenaria* the spur is very long; in *Aceras* the outer and inner divisions converging, form a hood, lip in three linear divisions, and spurless. This and the following group are natives of temperate Europe, many of them of the British Islands.

As a representative of the sub-family we may take the plant so well known in this country and in the north of France, commonly called the Flower of Pentecost, or the Spotted Orchis (*Orchis maculata*), Fig. 360. The floral envelope of this species is composed of six petal-shaped pieces, disposed alternately in two whorls (Fig. 361). Of the three exterior pieces, two are slightly lateral; the middle one is curved forward in such a manner as to form, with two divisions of the internal whorl, a sort of casque or helmet. The third division has, on the contrary, a shape peculiar to itself; on the upper side it presents the appearance of a large hanging apron, prolonging itself below into a sort of spur. This is the *labellum*, or lip, of the flower. The corolla is then essentially irregular. When the six pieces of the floral envelope are removed a central column becomes visible, having in front two cells, the longitudinal openings of which face the apron. Below, an almost square cup, glossy and viscous, is observable. If we open these cells with a point of a

Fig. 361.—*Orchis maculata.*

needle, it will be found that each of them contains a pyriform body, the upper part of which is swollen, and composed of little angular masses, bound together by a sort of elastic network, while the lower part is lengthened into a kind of pedicel, or footstalk. These two pedicels are inserted at their base in the contiguous compartments of a little pocket. If we depress one of these pyriform bodies towards

Fig. 362.—Flower of *Orchis maculata*. Fig. 363.—Tubers and Root of *Orchis mascula*.

the cup, it adheres there firmly. We may easily satisfy ourselves, when this phenomenon is produced by insect agency in Nature, that the pollen-tubes are emitted upon the viscous surface, and that they quickly penetrate its tissues. This cup, then, is the *stigma*, the pyriform bodies the pollen, and the two cells which enclose them constitute an anther. Thus, in this curious flower the style and the andrœcium are united to form the central column, and it has only one stamen.

Beneath the point of insertion of the floral divisions, the column is continued by a sort of greenish appendage, with six longitudinal ribs, and twisted upon itself. This is the ovary, which, as we see, is inferior. This ovary presents one single cell, and encloses a great number of very small ovules inserted upon three placentas attached to the internal walls. The fruit is capsular, and opens by three valves, which bear the placentas in their axes, whilst the midribs remain in their place, united at the base as well as at the summit. Let us now examine the organs of vegetation in the Purple Orchis (*O. mascula*). The parts which are under the ground consist of two unequal tubers (Fig. 363), one of which is wrinkled, soft, and apparently devoid of sap, whilst the other is whiter, larger, and much firmer. The substance of the first, in short, has been exhausted in the development of the aërial stem, which now bears flowers; whilst the other is reserved for the development of the stem in the following year. These two tubers, which are ovoid or egg-shaped, are stores of nutriment. Above these may even be perceived another little shoot, which will not be fully developed till two years later. In other indigenous species—*Orchis maculata*, for instance—the tubers, instead of being ovoid, are palmate or hand-shaped. Both forms are accompanied by the ordinary cylindrical root-fibres, whose principal function is absorption. The leaves of the *Orchis maculata* are sheathed and arranged spirally upon the stem; their lanceolate blades are generally sprinkled with black spots. This study of an Orchid, so frequently met with, will give a sufficient general idea of them all. But to realise thoroughly the appearance presented by the plants of this remarkable order, they must be seen or imagined as they appear in the tropical forests. Many of the tropical Orchids are epiphytes, but not parasites—that is they grow in the clefts and angles of branches, either erect or gracefully suspended, but without drawing their nourishment from them. Their flowers are disposed in spikes, branches, or tufts of different sizes, and their colours are often most rich and varied, frequently yielding a sweet perfume. They always present an original and somewhat fantastic appearance; some resembling a fly, some a spider, others a butterfly, and some a man suspended by the head. The diversity in the size and appearance of their flowers, and their strange beauty, cause this group of plants to be one of the most cherished ornaments of our hothouses.

The *Arethuseæ* are among the least interesting of Orchids. Dr. Lindley once thought *Vanilla* of sufficient importance to constitute a sub-order, which he has since withdrawn. The *Vanillaceæ* are climbing Orchids, but not epiphytes; the leaves are fleshy, subcordate at the base, and articulated with the stem, which is square, and climbs to the height of twenty or thirty feet. The flowers are fleshy, the perianth articulated with the ovary; the sepals and petals nearly equal, and free at the base; the labellum is entire and united with the column; the anthers terminal and opercular; the pollen-masses two, bilobed, and granular. There are eight species, two of them found in Asia and six in America. The fruits of most of them are aromatic, and there is still some doubt which of these species yields the Vanilla of commerce. It is supposed to be the product of several species, probably the fruit of *V. aromatica* and *V. planifolia*, which contains a large proportion of essential oil and benzoic acid.

Vanilla, by far the most expensive of spices, is the seed-vessel of some of these plants. They flower very abundantly, but impregnation can only be effected by extraneous aid, which, in South America, is performed by a species of bee, which, in robbing the nectar, carries the pollen from flower to flower. The non-success of *Vanilla* in India is to be attributed to the absence of this bee, for the plant blooms abundantly. The few pods which ripen are the product of flowers which have been operated on by man.

The *Neotteæ* include many of our British Orchids; they are fibrous-rooted, rarely having a fleshy bulb; the anthers are distinct from the column; the pollen-grains loosely coherent and nearly powdery. The species are generally small, being six to fifteen inches high in the Bird's-nest *Neottia*, flowers and stem pale yellowish brown; column notched with two short beaks, and lobes of the lip divergent; it is found in chalky, shady places. Lady's Tresses (*Spiranthes*) bears small white flowers on a one-sided spike; one rare indigenous species is found in bogs in the New Forest.

The *Cypripedeæ* are singularly beautiful in their foliage. They differ from all other Orchids in having *two* lateral anthers. The form of the flower is curious, being slipper-shaped; hence their name of Lady's Slipper; we have one native species of *Cypripedium*, found in the woods of the north of England, but now nearly extinct. The sepals are ribbed, of a rich dark-brown colour, the two lower ones united. Lip yellow, and marbled, about an inch long, reticulated with veins, and spotted internally.

The exotic species are also dwarf species, but compact and evergreen, the leaves of many of them being spotted. *C. barbatum* is a pretty species, with beautifully spotted foliage; the colour of its solitary flowers brownish-purple and white. In the variety *grandiflorum* the foliage is finely variegated, and the flowers considerably larger. *C. venustum* is an Indian species, with variegated foliage, four inches high. The blossoms are white, variegated with a purplish brown.

It will have been observed from the preceding remarks, that, besides the physiological differences in accordance with which botanists have arranged orchids, there is a distinct difference of habit; that a portion of them root in the soil and draw their support from the earth, while others attach themselves to trees, stones, and rocks, where they receive little or no support through their roots. The first, including such genera as *Phaius*, *Calanthe*, *Bletia*, *Cypripedium*, are known as *Terrestrial Orchids*. The others are termed *epiphytes*. They grow chiefly upon other plants, adhering to their bark or rooting among the scanty soil that occupies their surface, not as parasites, by striking adventitious roots into the wood, and nourishing themselves upon the sap of the individual to which they are attached, but using the tree apparently as a means of attaining a height where they can obtain the air and light, or the heat, moisture, and shade, as the case may be, necessary to their existence.

Orchidaceous epiphytes are much the most numerous and interesting; and now that our great cultivators have been enabled to study their natural habits, they are grown in a state of perfection which it is doubtful if they ever attain in a state of Nature. In the tropical forests they establish themselves upon the branches, and either vegetate in the midst of decayed vegetable and animal matter, or cling to the naked branches by their long, succulent, grasping roots; they attach themselves also to rocks and stones in moist places, where they grow luxuriantly.

On the confines of the Orchidales plants we find the family APOSTASIACEÆ, herbaceous perennials of the Indian woods, in many respects resembling the Orchids; differing from them chiefly in having a three-celled fruit, and a style altogether separated from the stamens for the greater part of its length.

ALLIANCE VIII.—XYRIDALES.

The PHILYDRACEÆ are grassy-looking herbaceous plants of Australia and Asia, exhibiting the great spathaceous bracts of the Musaceæ with the habit of Sedges.

The XYRIDACEÆ are herbaceous fibrous-rooted plants, with sword-shaped radical leaves enlarged and equitant at the base, the flowers having imbricated scaly heads; sepals three, glumaceous; petals three, coloured; stamens six. These plants join to the habit of

Sedges and other glumaceous plants, some approach to the peculiarities of Liliaceous plants.

ALLIANCE IX.—JUNCALES.

The JUNCACEÆ, or Rushes, approaching the Grasses in the glumaceous character of the calyx and corolla, partake of the characteristics of the *Xyridales* in their calyx and bracts. They are herbaceous plants, with tufted or creeping roots and tapering stem, often with a distinct pith.

The ORONTIACEÆ are herbaceous plants, with broad, entire, or deeply-divided, sometimes sword-shaped, leaves. They occupy woodland stations, chiefly within the tropics of both hemispheres, but are found also in colder regions; one of them, *Symplocarpus*, being com mon in the swamps of the United States; another, *Calla palustris*, in the deep muddy marshes of South Lapland, in 64° north.

ALLIANCE X.—LILIALES.

The beautiful plants which constitute this alliance have been cultivated and admired for ages. The LILIACEÆ is by far the most important order included in it. In Fig. 364 the petaloid perianth of the Lily is represented; and in Fig. 366 the group of Lilies represented will give a general idea of the habit of the type.

The protecting envelope of the flower of the Lily (Fig. 364) is composed of six petaloid leaves, which, as a whole, form the delicate white and odorous flowers. Of these six leaves, the three exterior ones may be held to constitute a calyx, the three inner ones, which are placed alternately with those of the outer circle, and differ slightly in form and colour from them, to constitute the corolla.

The androecium is composed of six stamens, disposed in two verticils with white filaments, elongate two-celled anthers attached by their backs, filled with a yellowish pollen, and opening longitudinally. The pistil is composed of three carpels, as may be ascertained by an examination of the constituent parts. These three carpels are united, constituting one whorl, thus appearing as one organ standing in the centre of the flower, *o* being the ovary, *s* the style, and *st* the stigma (Fig. 365). The ovary, which is free or superior, presents three swelling sides externally, with three inner cells, the walls of which correspond to the three deep external grooves formed by three carpellary leaves. Numerous ovules are inserted in two series at the central angle of the cells. The style, which is thickest at the summit, is crowned by a three-lobed stigma. The matured fruit forms a capsule which opens of itself by an opening in the dorsal suture of each cell; that is to say, the dehiscence is *loculicidal.* The seed presents an embryo in a direct axis in fleshy albumen.

The Lily is a deep-rooted perennial plant, with bulbous root. The bulb is scaly (Fig. 367); the stems of the large proportion of

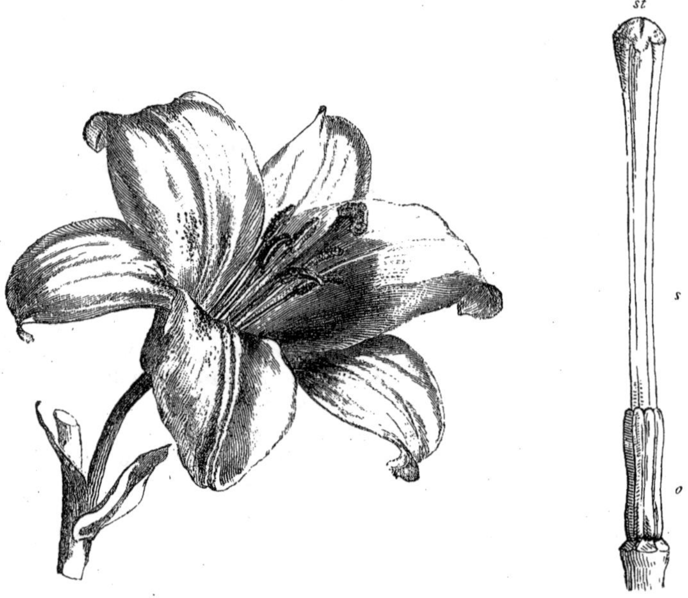

Fig. 364.—Petaloid Perianth of the Lily.

Fig. 365.—Pistil of the Lily.

those which are natives of cold countries perish after ripening their leaves, flowers, and fruit. The leaves generally are lanceolate in their lower parts and often linear above. The flowers form a white raceme.

The Liliaceous plants are generally large and showy, especially in those with annual stems, as the Lily itself, the Fritillary, the Hyacinth, the Star of Bethlehem, the Day Lilies, and the Tulip, which, combines all that is rich and beautiful in colour and form. But there are Liliales of arborescent size and stem, as the Dragon-tree (*Dracæna Draco*), in which the flower is less in size, so that the largest trees have the smallest flowers.

Fig. 366.—Group of White Lilies (*Lilium candidum*

The sub-order *Tulipeæ* consist of bulb-producing plants, with with annual stems, bearing cup-shaped flowers remarkable for their colours, without spathes, and the anthers lightly attached to a stiff filament. This division includes the Lilies, Fritillaries, and the Dog's-tooth Violet. *Lilium chalcedonicum* covers the plains of Syria with its scarlet flowers.

Fig. 367.—Bulb of Lily.

The *Hemerocalleæ*, or Day Lilies, have the calyx and corolla joined together, so as to form a tube of considerable length. The fragrant Tuberose and *Agapanthus* belong to this division, and the Aloes resemble them in almost all their parts, except the thick succulent foliage.

The *Asparageæ* includes the Common Asparagus and the Lily of the Valley, *Dracæna* and *Ruscus*.

The geographical limits of the order are as wide as its differences. Aloes abound in the southern parts of Africa. The Dragon-trees, the most gigantic of their order, attain the greatest size in the Canaries, where the Dragon-tree of Orotava (PLATE V.) is described as having been between seventy and seventy-five feet high, and forty-six feet in circumference at the base. All travellers to Teneriffe were accustomed to visit this gigantic plant, which was, according to tradition, an object of adoration to the Guanches, who are the primitive people of these islands. It was probably long anterior to historic times. At the conquest of Teneriffe by the Spaniards, it was already as large and as hollow as it is to-day.

"This gigantic tree," says Von Humboldt, in his "Pictures of Nature," "grows in the garden of M. Franqui, in the little villa of Orotava, called Taoro, one of the most beautiful spots in the civilised world. In 1799, when we ascended the peak of Teneriffe, we found

V.—Dragon Tree (*Dracæna Draco*).

that this enormous tree was forty-eight feet in circumference a little above the root." Sir George Staunton asserts that at the height of ten feet the tree is twelve feet in diameter. Tradition reports that this tree was an object of veneration to the Guanches, as the Ash of Ephesus was to the ancient Greeks; and that in 1402, when Bethencourt first visited the island, it was as large and as hollow as it is now. "When I saw the tree it seemed to enjoy an eternal youth, and still bore flowers and fruits."

"When Bethencourt, the French adventurer, conquered the Canary Isles in the beginning of the fifteenth century, the Dragon-tree of Orotava was found to be as sacred in the estimation of the natives as was the Olive of the Athenian Acropolis in the eyes of its inhabitants. In the torrid zone, a forest of *Cæsalpinia* and of *Hymenæa* is perhaps a monument a thousand years old; and remembering that the Dragon-tree of Orotava is of very slow growth—that its appearance now differs very slightly from the same tree described 400 years ago—we may conclude that it is extremely aged. It is, without contradiction, with the *Baobab*, perhaps the most ancient inhabitant of our planet."

"It is very singular," he adds, "that the Dragon-tree has been cultivated from very remote times in the Canaries, in the islands of Madeira and Porto Santo; it could not have come, therefore, from the East Indies, a fact which does not contradict the assertion of those who would represent the Guanches as an Atlantic race, entirely isolated, and as having no connection with the Asiatic or African races."

A few years ago this grand old plant was destroyed by a storm.

The species of *Dracæna* are evergreens, either shrubby or arborescent, having long slender stems, often columnar after the manner of the Palms; their trunks present marks, cicatrices produced by fallen leaves; they are soft and cellular at the centre, with a circle of stringy fibres towards the exterior. The leaves are simple; but in some of the species, instead of the veins running parallel with the midrib, they are perpendicular to it, after the character of the leaves of *Musaceæ*. They are usually clustered together at the end of the branches, like the inflorescence, which is terminal.

Alliance XI.—Alismales.

The Alismales, including the Butomaceæ, Alismaceæ, and Juncaginaceæ, conclude the important class of Endogens in Dr. Lindley's arrangement of the vegetable kingdom.

The BUTOMACEÆ are made interesting to us by the Flowering Rush, perhaps our handsomest indigenous flowering plant. They are all water-plants, with erect and leafless stems, narrow leaves dilated at the base, and pedicelled perfect flowers, forming a terminal umbel, subtended by three membranous bracts, a perianth with six divisions, the three outer segments slightly coloured and distinct from the inner, which are larger and more highly coloured. Stamens nine or more, with a free ovary, consisting of six carpels (in *Butomus*) more or less united by the ventral suture. The style is short, terminating in a stigma.

The ALISMACEÆ are aquatic plants, floating on ponds or growing in swampy places, distinguished from the other orders of the same alliance by the sepals and petals being perfectly distinct from each other both in colour and position. The root is usually a perennial creeping rhizome. The flowers form umbels, racemes, or panicles. The leaves expand into a broad blade with parallel veins. They are chiefly natives of northern regions, but several species of *Sagittaria* are found in the tropics of both hemispheres.

CLASS V.—DICTYOGENS.

Among the Monocotyledons of Jussieu and Endogens of later botanists there is a small group of plants which are referable to Endogens in the structure of the embryo, but which more resemble Exogens in their broad net-veined foliage, which usually disarticulates with the stem, their small green flowers nearly resembling those of *Menispermaceæ*. This class Dr. Lindley considers to be distinct from Endogens on the one hand, and Exogens on the other. He calls them Dictyogens, from the netted structure of their foliage. They are distinguished as having leaves net-veined, deciduous; wood of the stem, when perennial, arranged in a circle round the central pith, the wood of the rhizome being exogenous, that is, with concentric circles in the herbaceous species.

"For these reasons," says Dr. Lindley, "I have endeavoured to show that they ought to be regarded as a transition class, partaking somewhat of the nature of Endogens and also of that of Exogens. In the rhizome of the whole genus" (of Sarsaparilla), he adds, "the wood is disposed in a compact circle, below a cortical integument, and surrounding a true pith; in *Smilax aspera* the woody matter is disposed in the form of a cylinder, enclosing a centre of soft cellular matter, the vessels of the cylinder having an evident tendency to arrange themselves in lines forming rays from the centre."

Flowers unisexual: perianth adherent; carpels consolidated, and several seeded.	I. Dioscoreaceæ.
Flowers hermaphrodite, or polygamous; carpels several, placentas axile	II. Smilaceæ.
Flowers hermaphrodite; carpels several; placentas parietal	III. Philesiaceæ.
Flowers tripetaloid, hermaphrodite; carpels half-consolidated; placenta axile.	IV. Trilliaceæ
Flowers hermaphrodite; carpels solitary, simple, many-seeded; placenta basal.	V. Roxburghiaceæ.

The DIOSCOREACEÆ are distinguished by their diœcious flowers, adherent calyx and corolla, six stamens and three-celled ovary. They are all twining shrubs. *Dioscorea* is the typical genus of the order. Upwards of 150 species are known, mostly belonging to tropical countries, and principally to America. They are all characterised by fleshy tuberous roots and herbaceous stems, twining to the left; the leaves are almost always entire, and usually alternate, very rarely being opposite. The various species of *Dioscorea* and *Testudinaria* produce edible farinaceous tubers, but *Tamus* exhibits a dangerous acridity.

The tropical esculent, the Yams, forms a genus of fleshy-rooted diœcious plants, with annual twining stems, broad alternate leaves with a netted arrangement of veins, and bearing small green flowers in clusters. The seed vessel is a thin, compressed, three-winged capsule, containing one or two membranous seeds.

The SMILACEÆ, a small order, consist of the *Smilax*, or Sarsaparilla, or *Ripogonum;* the former evergreen climbing shrubs (Fig. 369), a few of which are found in temperate, but the majority in the warmer regions of both hemispheres. They are fibrous or tuberous rooted plants, with stems often prickly, leaves alternate and petiolate, and with a tendril on either side of the petioles. The botanical name *Smilax* occurs in Greek authors, as Theophrastus and Dioscorides, but applied to some poisonous plant. *Smilax aspera*, a native of the south of France, Italy, &c., yields Italian Sarsaparilla. The different kinds of *Sarza*, or Sarsaparilla, are now obtained from South America. Though the genera are limited, the species are numerous and important.

The PHILESIACEÆ are chiefly twining, sometimes upright shrubs, with large and showy flowers. *Lapageria rosea* is a beautiful greenhouse climber. The TRILLIACEÆ are simple-stemmed, herbaceous plants, with tubers or rhizomes, resembling the Sarsaparillas in many respects. The ROXBURGHIACEÆ are twining shrubs, with tuberous roots; large, showy, and somewhat fetid flowers; reticulated and coriaceous leaves. They are all natives of the hotter parts of India.

CLASS VI.—GYMNOGENS.

An important class of plants, which mark a transition between

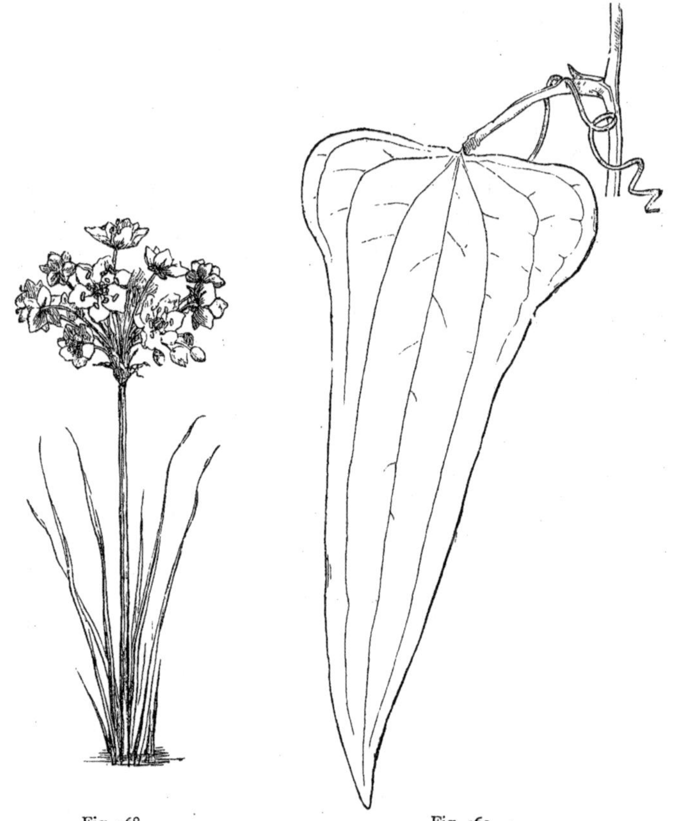

Fig. 368.
Flowering Rush (*Butomus umbellatus*).

Fig. 369.
Leaf and stalk of *Smilax*.

the simpler forms of vegetation and plants of more complicated structure. They differ from most *Vasculares* by the fibres of their wood having large perforations or discs; but their habits of growth are essentially exogenous, while their tissues are peculiar to themselves. The stem consists of concentric zones; they have a vascular system, with spiral vessels and a central pith. Some Cycads resemble Acrogens in their growth, which is continued by a terminal bud. They are without style or stigma, but so constructed that the pollen falls immediately upon the ovules.

Stem simple, continuous; leaves parallel-veined, pinnate; scales of the cone antheriferous.	I. Cycadeaceæ.
Stem branched, continuous; leaves simple, acerose; female flowers in cones.	II. Pinaceæ.
Stem repeatedly branched, continuous; leaves simple, sometimes fork-veined; female flowers solitary; anthers two-celled, opening longitudinally.	III. Taxaceæ.
Stem repeatedly branched, jointed; leaves simple, net-veined; anthers one-celled, opening by pores.	IV. Gnetaceæ.

The GYMNOGENS (from the Greek γυμνὸς, "naked," the ovules being uncovered.)—The plants comprehended in this great class agree with the flowering plants in having their vascular tissues complete, approaching the higher forms of vegetation in the Joint-firs (*Gnetaceæ*) which combine the habit of growth of *Chloranthus* with the structure of their own class. With the Ferns and Club-mosses, some of the *Cycadeaceæ* agree in habit in the peculiar gyrate vernation of the leaves, and in the less perfect structure of the spiral vessels and reproductive organs.

The CYCADEACEÆ are characterised by the cylindrical and unbranched growth of the trunk, and the continuous development of one terminal bud, and by its diœcious flowers, the male flower generally growing in cones composed of scales antheriferous below. In *Zamia* the female flowers are disposed in cones consisting of peltate scales; in *Cycas* they are placed on the toothings of abortive leaves occupying the centre of the terminal bud. The leaves are pinnate, having some resemblance to those of Ferns and Palms; their wood is arranged in numerous consecutive circles in *Cycas*, and also in a confused manner in the central pith, thus partaking of the peculiarities both of the Exogens and Endogens. Robert Brown demonstrated the similarity of conformation between the flowers of *Cycas* and *Pinaceæ*; and Adolphe Brongniart determined the resemblance between them in the structure of the vessels of their wood, and confirmed also the proximity of the former to Ferns. Their relation to the *Pinaceæ* is established by both being dicotyledonous, and both their seeds having naked ovules, "constructed," says Dr. Lindley, "in a similar remarkable manner, and borne in both cases not upon an ordinary axis of growth, but upon the margin or face of metamorphosed leaves;

the same peculiar form of inflorescence, the same kind of male flowers, the same constant separation of the sexes, and a like imperfect formation of spiral vessels; and both agree in having the vessels of their wood marked with circular discs—a character which, if not confined to them, is uncommon elsewhere." They are all natives of the tropics, or of temperate America, and Asia; in the eastern part of the colony of the Cape of Good Hope they characterise the thickets along the Caffre frontier.

Fig. 370.

The PINACEÆ are noble trees or evergreen shrubs, with a branching trunk abounding in resin. The leaves are elongated, linear, lanceolate, acicular, or pointed, sometimes fasciculated in bundles of two, three, or five, each bundle being a little branch with very short axis; when thus fasciculated the primordial leaf to which they are then axillary is membranous and enwraps them like a sheath, as in *Pinus strobus*, in Fig. 370. The plants are monœcious, that is, they bear male and female flowers upon the same stem. The male flowers are composed of a floral axis, along which are inserted a considerable number of stamens having a short filament, and an anther which opens in two longitudinal clefts; this anther is surmounted by a dilated connective like a tongue. The female flowers are disposed in a catkin, and are each composed of a scale-like ovary destitute of style or stigma, and bearing on its internal surface two suspended ovules. Fig. 371 represents the male flower with an enlarged view of an anther; Fig. 371, the female flower with carpellary scales, showing on the right hand a pair of inverted ovules. When these flowers have ripened, the scales become hard, ligneous, and thickened at the summit; they then form collectively the composite fruit we call a Cone, which has given a name to the family in botanical systems. The cone is represented in Fig. 373. It is formed, besides the scale-shaped ovaries, now enlarged and hardened, sometimes of bracts also, which are occasionally obliterated, and sometimes extend beyond the scales in the form of a lobed appendage. These scales finally drop from the tree, become disintegrated and scattered, and buried in the soil, thus completing the end of their existence, namely, the propagation of the species.

The *Pinaceæ* are resinous, mostly evergreen and hard-leaved trees or shrubs, all but universally diffused over the globe. Gigantic in size, rapid in growth, noble in aspect, and robust in constitution, these trees form a considerable portion of the woods and plantations

VI.—*Pinus sylvestris.*

in cultivated countries, as well as of primeval forests in all temperate countries. In Europe, Siberia, China, and North America the species are abundant; the timber trees being exceedingly valuable in commerce, being well known as Deal, Fir, Pine, and Cedar woods, while their juices yield oil of turpentine, Canada balsam, Burgundy pitch, all equally well known. The common Larch yields Venice

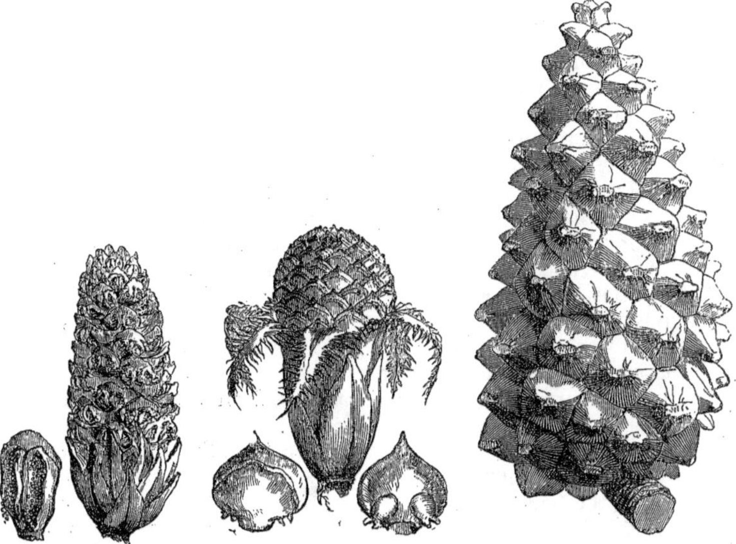

Fig. 371.—Male flower of *Pinus sylvestris*.

Fig. 372.—Female inflorescence of *Pinus sylvestris*

Fig. 373.—Cone of *Pinus sylvestris*.

turpentine; spruce beer is made from branches of the Hemlock Spruce; and oil of savin, a well-known irritant, is made from the *Juniperus Sabina*.

Conifers are usually divided into three sub-orders:

I. *Abietineæ*, comprehending the Firs, Pines, Spruce, Larch, and Wellingtonia, all of which bear cones with one or two inverted ovules at the base of each scale of the cone; pollen oval, and curved.

II. *Cupressineæ*, or Cypress tribe, bearing an indurated globular cone with erect ovules, and spheroidal pollen, including the Cypresses and Junipers.

III. *Taxineæ*, or Yew tribe, of which Dr. Lindley forms a distinct order. The fruit is a species of drupe, with a solitary ovule in the centre.

Among the Pines with leaves in pairs worthy of consideration we may mention the Scotch Pine (*Pinus sylvestris*) represented in PLATE VI.; the Maritime Pine (*P. Pinaster*); the Corsican Pine (*P. Laricio*), a noble tree spread over the mountains of Corsica, Greece, and Turkey, celebrated for its rapid growth and excellent timber, and amongst the species with leaves in fives, the Weymouth Pine (*P. strobus*).

Firs (*Abies*) differ from the Pines (*Pinus*) in their cones, which are furnished with thinnish scales slightly rounded at the apex, and scattered distichous leaves; such is *Abies pectinata*, the Silver Fir, from which Strasburg turpentine is extracted; while Burgundy pitch and oil of turpentine are obtained by incision from *Pinus sylvestris*, which is also a valuable timber (red deal) for building purposes. The Larches (*Larix*) again differ from the Firs in this: their leaves spring from a bundle of scaly buds, and become scattered or solitary in consequence of the lengthening of the stem; the imbrication of the scales of the cone is very loose; the leaves of the Larch are deciduous during winter. The Larch of Europe attains a height of from ninety to a hundred feet; the wood is of a reddish colour, its tissues closer and considerably harder than those of the Fir-trees, and a very pure turpentine, which is used in arts and medicine, oozes out from incisions made in its bark.

The Cedars (*Cedrus*) are distinguished from the Larches by their leaves being persistent during several years after the elongation of the bud, and by the scales of the cone being more closely imbricated. The Cedars of Lebanon (PLATE VII.) are trees having an aspect full of grandeur, spreading their vast horizontal arms thirty or forty feet from the stem, which rises forty or fifty feet above the soil. Upon the back of Mount Atlas, in the north of Africa, and in the temperate countries of Asia, the Cedar forms immense forests of a most majestic and imposing aspect. There is indeed no nobler object than the Cedar. "The Lebanon," say the Arabian poets, "bears winter on his head, spring on his shoulders, and autumn in his bosom, while summer sleeps at his feet;" and in confirmation of the truth of the sentiment a few venerable Cedars still remain; they form a beautiful grove on the line of route from Baalbec to the coast. They are large and massy, rearing their heads to an enormous height, and spreading their branches afar; but they have a strangely wild aspect, as if wrestling with some invisible person bent on their

VII.—Cedar of Lebanon.

destruction while life is still strong in them; but they are gradually disappearing. The grove near the Kadisha has been described by every traveller. In 1575, of the largest trees there were twenty-four standing in a circle; in 1630 Fermanil counted twenty-two; there are now seven standing near each other, and a few more almost in a line with them. According to Tristram the Cedar is also to be

Fig. 374.—Group of Cypress Trees (*Cupressus fastigiata*).

found scattered through the northern and inaccessible parts of the Lebanon.

Sequoia Wellingtonia (FRONTISPIECE), the Wellingtonia of our gardens, was originally discovered in Upper California, and introduced into Europe in 1853. The tallest tree was 327 feet high, and 90 feet in circumference at the base.

The *Cupressineæ*, of which we have now to speak, including

Thuja and the Juniper, differ in many respects from the *Abietineæ*. The Thujas have their male flowers composed of a filiform floral axis, upon which are inserted numerous stamens, which may be likened to nails with which old-fashioned doors are sometimes studded, supporting under their heads four unilocular one-celled anthers. The female flowers are disposed in catkins, each scale of which bears two erect ovules. These soon become fleshy and consolidated, but when at maturity, they dry up, and in doing so detach themselves and separate, thus setting the seeds free. Thujas are evergreen trees with flattened branches with very small imbricated and compact leaves; they are well known in gardens under the name of Arbor Vitæ, a name of uncertain origin. *Thuja occidentalis* is a native of America; *T. orientalis* of Japan and China.

The Cypress (*Cupressus*) very much resembles the Thujas; it is essentially distinguished from them by its seeds, several in number, and pressed into the base of each scale. In the common Cypress (*Cupressus sempervirens*) there are two very distinct varieties, in one of which (*C. fastigiata*), the branches leave the parent stem at a very acute angle, giving to the tree its very peculiar physiognomy (Figs. 77 and 374).

The common Juniper (*Juniperus communis*) is a native shrub, with long, narrow, sharp-pointed leaves, in whorls of three, spreading, rigid, and almost prickly. The scales of the female catkin, which are not more than six in number, present this curious fact, that they become fleshy, and constitute a sort of spherical berry, black or blue, containing ordinarily three bony seeds called botanically a *galbulus*. In some countries in the north of Europe these fruits are used to flavour a malt spirit known under the name of Gin or Hollands, sometimes Geneva, from the French name *Genièvre*. The Virginian Juniper, also called Red Cedar, furnishes a light odorous wood, with which the cylinders are made in which we enclose the lead of our pencils. The savin, of stimulating, diuretic, and obstetric powers, is *J. Sabina*.

The *Taxineæ*, or Yews, are trees with leaves very close together, entirely veinless, almost distichous, linear, and sharp-pointed, of a deep green colour; flowers diœcious. The male flowers are composed of an elongated floral axis, upon the whole length of which are inserted a variable number of stamens, which may be said to resemble studs or nails, the connective being the head. On the lower side of this connective, six or eight bilocular anthers are disposed circularly round the *filament*. The female flowers are solitary and surrounded with imbricated bracts; they consist of a sessile ovule at the centre of a highly-developed disc. When arrived at maturity this disc becomes fleshy, and forms a little cup of a lively red, which loosely envelops the seed. The tree then appears as if studded with coral drops. The leaves of the Yew are highly poisonous to some animals, as the sheep; but the berries, although insipid and uninviting to the palate, may be eaten with comparative impunity. These plants occur in milder climates all over the world, and in elevated situations in the tropics. They are resinous, like Conifers, and possess excellent medicinal qualities. They include *Salisburia*, a tree of great beauty and elegance.

The *Gnetaceæ*, or Joint Firs, are small trees or sarmentose twiggy shrubs of the temperate parts of Asia, South America, and Europe, with opposite leaves or clustered branches, and thickened separable articulations.

In the Gymnosperms, which connect the lower with the higher forms of organisation, the transition is very distinctly marked. In Cycads the stem is simple and cylindrical, the departure from its terminal mode of development being exceptional and accidental, while the Conifers exhibit a constant tendency to a rapid evolution of leaf-buds in every axil. An increasing value in their products is also observable. The Cycads, for instance, yield a mucilaginous juice, mixed with starch, from which common articles of food are prepared. At the Cape of Good Hope the fruit of the various species of *Encephalartos* are called "Caffer bread," and a kind of arrowroot is prepared in Mexico from the seeds of *Dion edule.* In Japan a sago is procured from the cellular substance occupying the stem of *Cycas revoluta*, and also from *Cycas circinalis* in the Moluccas. Other species are also utilised in the countries of which they are natives. The Pines and Fir-trees are chiefly valuable for their timber, and the Yews and their allies are valuable for their resinous products, and also for timber, which is unsurpassed for elasticity and durability; and in Amboyna the seeds of *Gnetum Gnemon* are eaten roasted, boiled, or fried, and the green leaves are a favourite vegetable cooked and eaten as spinach.

CLASS VII.—EXOGENS.

The Dicotyledons of Jussieu and more recent botanists include the more highly organised plants, which are moreover endowed with proportionate vitality, for "while a century or two terminate the existence of most endogenous trees, some existing Exogens were possibly monarchs of the forests at the beginning of the Christian era." As already explained, Exogens add their new wood to the outside between the bark and wood of last year's growth.

The class of Exogens numbers over 66,000. Their germination is *exorhizal;* the embryo has two cotyledons; the leaves have a network of veins; the trunk is formed of woody bundles of ducts and fibres, arranged round a central pith, forming either concentric rings, or in a homogeneous mass, but always having medullary plates radiating from the centre to the circumference. Genera, 6,191; species, 66,225 (in 1853).

All botanists are agreed that the organs of reproduction may be expected to furnish the best characters for classification after those necessary for nutrition. Linnæus was of this opinion, and he made them to a considerable extent the basis of his system, but he mainly relied upon their number. The importance of the stamens and pistil did not escape the observation of Jussieu, who separated from all other Exogens those having the stamens in one flower and the pistil in another, and he called them DICLINOUS. By this means he brought together a collection of natural orders corresponding with the monœcious and diœcious plants of Linnæus. But, in carrying out his system, he excluded a vast number of truly diclinous plants. Some of these anomalies have been corrected by recent observers; and Dr. Lindley has divided the whole of the vast class of Exogens into those which are (1) DICLINOUS, (2) HYPOGYNOUS, (3) PERIGYNOUS, and (4) EPIGYNOUS.

SUB-CLASS I.—DICLINOUS EXOGENS.

Having male and female without any tendency to bisexual flowers.

ALLIANCE I.—AMENTALES.

Flowers forming catkins, without or with only one floral envelope; carpels superior; embyro small, with little or no albumen.

Ovary one-celled; ovules one or two, ascending; radicle superior	I. Casuarinaceæ.
Ovary two-celled; ovule single, pendulous; radicle superior	II. Betulaceæ.
Ovary two-celled; ovules indefinite; seeds winged	III. Altingiaceæ.
Ovary one-celled; ovules indefinite; seeds cottony	IV. Salicaceæ.
Ovary one-celled; ovule single, erect; radicle superior	V. Myricaceæ.
Ovary one-celled; ovule single, ascending; radicle inferior	VI. Elæagnaceæ.

The BETULACEÆ are trees or shrubs, with simple alternate leaves, with primary veins running nearly straight from the midrib to the margin; stamens usually distinct; ovary two-celled; fruit

VIII.—Alder Tree (*Alnus glutinosa*).

membranous, indehiscent, and one-celled; seed pendulous. The Birch (*Betula*) is monœcious, with alternate leaves, ovate, and petiolate, acuminate and dentate or double dentate, green and glossy above, of a pale glabrous green below. Its straight upright stem, smooth silvery bark, long, round, slender, flexible, and pendulous branches, render the Birch a graceful ornament in the landscape. The Birch is in flower in the month of April, when the white silvery bark produces the happiest effects contrasted with the deeper and more sombre tints which the trunks of the Elms and Oaks present.

The Alder (*Alnus glutinosa*, PLATE VIII.) inhabits the margin of

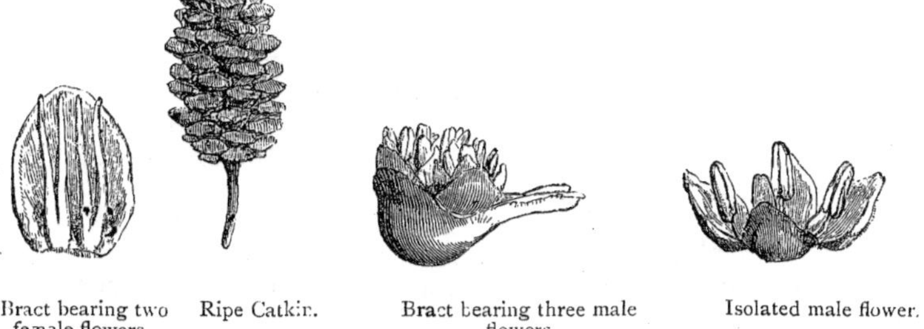

Bract bearing two female flowers. Ripe Catkin. Bract bearing three male flowers. Isolated male flower.

Fig. 375.—Details of Inflorescence of the Alder.

rivers and marshy places in woods; its round, nearly orbicular leaves are of a dark green on the upper surface, and a pale glabrous green beneath, covered when young with a varnished investment. This tree, like the Birch, is monœcious. In the month of February, before the leaves have appeared, the cylindrical elongated catkins composed of male flowers hang from the tree, while the egg-shaped catkins composed of female flowers are erect and directed towards the sky.

In the axils of each scale of the male catkin may be noted three flowers, one in the centre and two placed laterally; these three flowers are composed each of a perianth of four divisions (Fig. 375), and of four stamens opposite to these divisions, with bilocular anthers. They are surrounded, independently of the scale of the catkin, by four other secondary scales, two of which are to the right and two to the left.

In the axil of each scale of the female catkin, four secondary scales and two flowers may be observed; each of these flowers consists of a pistil only, the free ovary of which is surmounted by a short style divided into two stigmatic branches. The ovary presents two cells, in each of which is suspended an anatropal ovule. The fruit catkins are in the form of a Pine cone, with horizontal persistent scales, closely approximated, and rendered coherent by a resinous substance, but sporating ultimately to permit the fruit to escape. These fruits are compressed, surrounded on all sides by a coriaceous cork-like edging. They are unilocular, and enclose only one seed.

The SALICACEÆ consist of the Willows and Poplars, trees or shrubs with alternate simple leaves with deliquescent primary veins, frequently with glands on the edges, and deciduous or persistent stipules; flowers amentaceous, ovary one-celled, and numerous cottony seeds. These downy seeds of Willows and Poplars, adhering to the base of a leathery two-valved capsule, cannot be mistaken.

The Willows (*Salix*) are very numerous in species, varying in size from the dwarf (*S. herbacea*), with creeping underground stem and ascending branches, to *S. alba*, the White Willow, thirty feet high; they consist of Willows, Sallows, and Osiers, generally with round, slender, flexible branches, with simple entire stipulate leaves, diœcious inflorescence, and male and female flowerets in long cylindrical catkins. The natural habitat of the Willow is on the banks of sluggish rivers, and in low marshy places; the Osier beds are generally found where they can be under water the greater part of the year; while the Sallows, as *S. caprea*, rather affect dry woods and hedges; others, as *S. lanata*, are beautiful mountain shrubs, the fertile catkins of which are sometimes found a span long in Glen Dole and Glen Callater.

Willows abound in temperate regions, but decrease sensibly in number towards the South of Europe and in Algeria. They serve to consolidate the borders of pieces of water and rivers. The Willow furnishes means for the basket-maker's work.

The White Willow (PLATE IX.) is the most important of its species, on account of the large dimensions which it acquires: it is very productive in osiers as a Pollard.

The Weeping Willow (*Salix Babylonica*), of which we have already given a representation at page 74, is particularly remarkable for the length, flexibility, and graceful drooping habit of its branches, which give it an appearance of melancholy grace.

Salix reticulata, the Wrinkled Willow, is a very small shrub of

IX.—The White Willow (*Salix alba*).

from one to two feet high, which grows in the Alps, the Pyrenees, the Welsh and Scottish mountains, and the whole Arctic region.

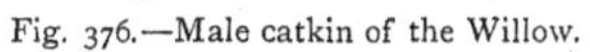

Fig. 376.—Male catkin of the Willow.

Fig. 377.—Female catkin of the Willow.

S. herbacea, the Herbaceous Willow, is the smallest British shrub, with a stem creeping under ground and emitting branches almost

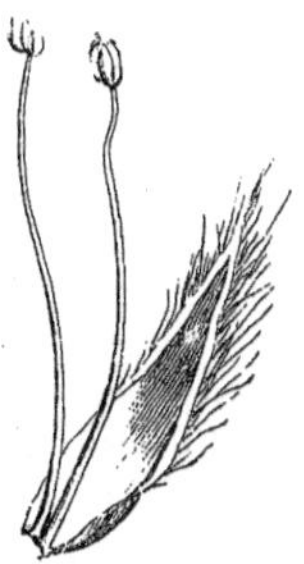

Fig. 378.—Male Flowers of *Salix alba* (magnified)

Fig. 379.—Female Flower of *Salix alba* (magnified).

completely herbaceous; its distribution is the same as that of *S. reticulata*. Willows are diœcious; their flowers are in catkins, and solitary in the axil of each scale of the catkin. They have no

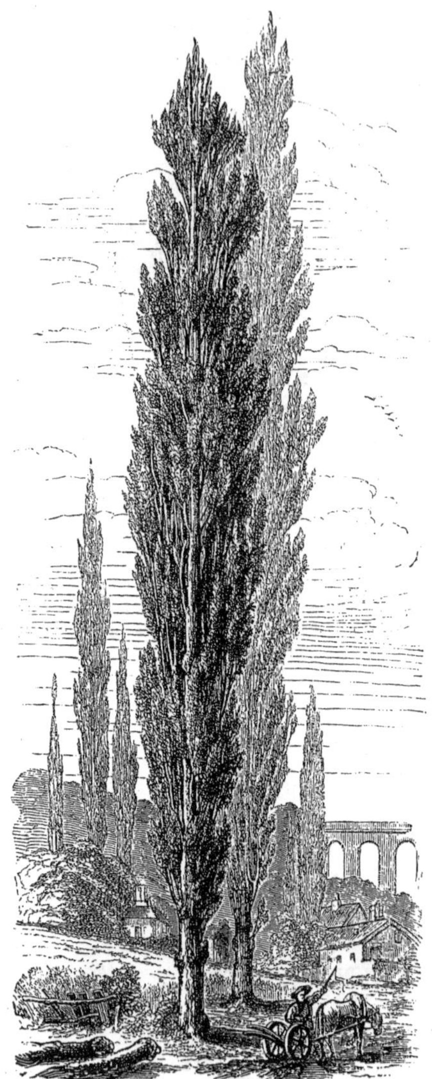
Fig. 380.—Lombardy Poplar.

floral envelopes. Figs. 376 and 377 represent the male and female catkin of the White Willow; Figs. 378 and 379 show the isolated flowers.

The POPLARS (*Populus*) are nearly allied to the Willows, and in the flowers they are only distinguished from them by the greater number of stamens, which are inserted on the internal face of a kind of cup.

The Black Poplar (*Populus nigra*) is among the larger of these trees, commonly known as the Swiss Poplar. The White Poplar (*Populus alba*), a fine tree with an ample crown, is furnished with leaves remarkable for their extreme whiteness underneath.

The Aspen (*Populus tremula*) is the only real forest tree of this genus; it is of middle height, with leaves which are very mobile in consequence of the length and the vertical compression of their petioles, which leads to the horizontal tremulous motion which distinguishes the tree and procured its name.

The Lombardy Poplar (*Populus dilatata*, Fig. 380) originally from the Himalayas and Persia, was introduced in France from Italy in 1749, is remarkable for its erect branches, which spring from nearly the base of the trunk, and altogether form a long, straight, pyramid: only the male plant of this species is known.

The CASUARINACEÆ are for the most part Australian trees, or scrubby bushes of little value to man.

The ALTINGIACEÆ are tropical plants of India, North America; and the Levant-Storax is yielded by several species of *Liquidambar.*

The MYRICACEÆ, or fragrant Gales, are leafy shrubs, or small trees, having resinous glands; natives of the temperate parts of America, the Cape of Good Hope, and India.

The ELÆAGNACEÆ, or Oleasters, are trees or shrubs of the northern hemisphere down to the equator. *Elæagnus hortensis* bears a fruit about the size of an olive, which is brought to market in Persia. The red drupes of *E. conferta* and several others are eaten in India. The only species growing wild in Britain is *Hippophaë rhamnoides*, a spiny shrub with diœcious flowers, and small, round, orange-coloured acid berries; it grows on the cliffs in some places near the sea; its fruit becomes rather a pleasant preserve when sufficiently sweetened. *Elæagnus angustifolia* is a fragrant garden plant, filling the air with its perfume, while the dull yellow flowers which exhale the delicious fragrance attract little attentton.

ALLIANCE II.—URTICALES.

Flowers scattered, with a single floral envelope; carpel single, superior; embryo, large, lying in a small quantity of albumen.

Radicle superior; ovules twin, suspended; embryo straight, albuminous; anthers two-lobed, with vertical fissures.	I. Stilaginaceæ.
Radicle superior; ovule erect solitary; embryo straight, albuminous; juice limpid; stipules small, flat.	II. Urticaceæ.
Radicle inferior; embryo exalbuminous; plumule many-leaved, large.	III. Ceratophyllaceæ.
Radicle superior; ovule solitary, suspended; embryo hooked, exalbuminous.	IV. Cannabinaceæ.
Radicle superior; ovules solitary, suspended; embryo hooked, albuminous.	V. Moraceæ.
Radicle superior; ovule solitary, erect or suspended; embryo straight, exalbuminous; juice milky; stipules large, convolute.	VI. Artocarpaceæ.
Radicle inferior; embryo exalbuminous; plumule minute; juice limpid; stipules large, deciduous.	VII. Platanaceæ.

The URTICACEÆ contains the Nettles, Figs, the Hop, the Mulberry, the Hemp, and the celebrated Upas tree; the species are widely diffused over every part of the world—in the frozen North, and in the hottest tropical countries. The order, as formerly constituted, was nearly synonymous with the alliance, but it is now limited to a few genera characterised by the causticity of the limpid juice they yield. The stinging effects of the Nettles, *Urtica dioica* and *U. urens*, will be familiar to most readers; but these are not to be compared for a moment with some of the East Indian species. Listen, for instance, to De la Tour's description of the effects of the sting of *U. crenulata.* "One of the leaves," he says, "slightly touched the first three fingers of my left hand; at the time I only perceived a slight pricking, to which I paid no attention. This was at seven in the morning. The pain continued to increase. In an hour it had become intolerable; it seemed as if some one was rubbing my hand with a red-hot iron. Still there was no remarkable appearance, neither swelling, nor pustule, nor inflammation. The pain spread rapidly along the arm as far as the armpit. I was then seized with frequent sneezings, and with a copious running at the nose. About noon I experienced a painful contraction of the back of the jaws, which made me fear an attack of tetanus. I went to bed, hoping that repose would alleviate my suffering, but it did not abate; on the contrary, it continued nearly the whole of the following night; but I lost the contraction of the jaws about seven in the evening. The next day the pain left me. I continued to suffer for two days, and the pain returned when I put my hand into water; and I did not finally lose it for nine days."

The CANNABINACEÆ, represented by the Hemp (*Cannabis sativa*), Figs. 381 and 382, originally came from Persia, but has since become acclimatised to all parts of Europe; its bast fibres make this plant eminently precious to man. The Hemp is a diœcious herbaceous annual, with opposite lower leaves. The upper leaves are often alternate and deeply partite, with from five to seven acuminate linear-lanceolate divisions, strongly dentate, rough, and of a pale green colour underneath; two lateral stipules accompany them. The male flowers are disposed in racemes, and are composed of a calyx with five divisions and five stamens opposite the divisions, with bilocular anthers opening inwards by two longitudinal clefts, opposite to them. The female flowers are disposed in axillary leafy glomerules, and present a calyx formed by two divisions and a pistil composed of a superior ovary, surmounted by a short style, with very long filiform stigmata. The unilocular ovary encloses a single ovule. The fruit is an achene; the seed, without albumen, encloses an embryo folded upon itself.

It is from this species of Hemp (*C. sativa*) that the Arabians make the intoxicating liquor which is commonly known under the name of *haschisch;* the Orientals make a deplorable abuse of its intoxicating powers.

The Hop, Fig. 28, (*Humulus Lupulus*), a perennial plant, with twining stems and opposite lobate leaves in the shape of those of a

Fig. 381.—Male Plant of *Cannabis sativa.*

Palm, belong to the same family as the Hemp. It is found wild in Europe in hedges and upon the banks of rivers. The Hop is cultivated in England, Germany, France, and Belgium. The female flowers are disposed in compact ovoid spikes, forming cones when at maturity by the development of their sepals and bracts. The fruit or achenes are covered with a granular powder of a greenish or golden yellow colour; they are very odorous, and contain an active principle to which chemists have given the name of *lupuline.* The cones of the Hop are used in the manufacture of beer; they are tonic, and slightly narcotic.

The MORACEÆ are trees and shrubs, sometimes climbing plants, mostly yielding a milky juice. They are natives of warm countries, where they form vast forests, the thick trunks and strong boughs of

Fig. 382.—Female Plant of *Cannabis sativa.*

the Fig, with its large head, being conspicuous. Travellers speak of the noble aspect of the Wild Figs of hot countries, of their gigantic dimensions, and the thick, delightful shade cast by their leafy heads. Fraser, speaking of their habits at Moreton Bay, says, "I observed several species of *Ficus*, upwards of 150 feet high, enclosing immense Iron-bark trees, on which the seeds of those Fig-trees had been originally deposited by birds. Here they had vegetated and thrown out their parasitical and rapacious roots, which adhering close to the Iron-bark tree, had followed the course of the stem downwards to the earth, where, once arrived, their progress and growth are truly astonishing. The roots increase rapidly in number, envelop the stem, and send out at the same time such gigantic branches, that it is not unusual to see the original tree, at the height of seventy or eighty feet, peeping through the Fig as if *it* were a parasite on the real intruder." But the Banyan (*Ficus indica*) excels all others in its magnitude, one tree being capable of giving shelter to a regiment of cavalry. This tree is a native of India and the islands of the Indian Ocean, reaching its greatest perfection in the villages on the skirts of the Circar mountains. (PLATE I.) The branches cover a vast extent of ground, dropping their roots here and there, which as they reach the ground rapidly increase in size till they become as large as the parent trunk. Roxburgh says he has seen such trees fully 500 yards round the circumference of the branches, and a hundred feet high, the principal trunk being twenty-five feet to the first branches, and eight or nine feet in diameter.

All the species of *Ficus* abound in a milky juice containing caoutchouc, the best-known quality of that valuable product being obtained from *F. elastica.* The leaves of *F. indica* are ovate and entire; when young, downy on both sides, smooth when more matured, and from five to six inches long, and three to four broad, having a broad, smooth, greasy-looking gland on the under side of the leaf-stalk at the top. The figs grow in pairs from the axils of the leaves; they are downy, and about the size and colour of a ripe cherry at maturity.

F. elastica, the Indian Caoutchouc-tree, will be known to most readers; it is now common in all the hothouses in the country, and numbers of fine plants may be seen in the Palm House at Kew. It has large glossy leaves, thick, oval, and pointed; small axillary uneatable fruit of the size of an olive, and long reddish terminal buds contained in the rolled-up stipule. In its native fields it grows to the size of the European Sycamore, chiefly among decomposed rocks and vegetable matter over the declivities of mountains, growing with great rapidity as a young tree, attaining the height of twenty-five feet in four years, and with a trunk a foot in diameter. The milk is extracted by making incisions through the bark to the wood, at the distance of a foot from each other, all round the tree, and up to the top. After one course of tapping the tree requires to rest a fortnight, when the process may be repeated. When the liquid is exposed to the air it becomes a firm and elastic substance, fifty ounces of pure milky juice yielding about fifteen ounces of clean, washed caoutchouc. The

Pippul, or Sacred Fig of India (*F. religiosa*), is known by its rootless branches and heart-shaped foliage, with long attenuated points. It is common in every part of India, where it is planted for the sake of its grateful shade. It is held in superstitious veneration by the Hindus, because, according to tradition, Vishnu was born under its shade. The long pointed leaf has a wavy edge, and long, slender, and flat footstalks, which produce a tremulous motion in the air, like that caused by the Aspen-tree (*Populus tremula*). Silk-worms seem to prefer this leaf to the Mulberry; and they are used by the natives of Arabia for tanning leather. The Sycamore Fig (*F. Sycamorus*) is a large tree which grows in Egypt round the villages near the coast, and gives grateful shelter to the villagers under its widely-spreading head. The leaves are broad, ovate, and angular, and the fruit is produced in clusters upon the twigs of the trunk and old limbs. The figs are sweet, but insipid and woody.

The Common Fig (*Ficus Carica*) was originally found in the eastern and western regions of the Mediterranean. It was introduced and has been cultivated in Europe from the most ancient times. They are frequently found growing almost spontaneously in the South of France. Generally a shrub, the Fig is sometimes a tree of four or five yards in height. The leaves vary in form on the same plant. They generally present from three to seven unequal and obtuse lobes. The flowers are unisexual, and placed upon the internal walls of a common receptacle, pierced at the vertex with a small orifice, protected by a large number of imbricated bracts. The male flowers have a calyx composed of three sepals, with three stamens opposite to them, and two-celled anthers that open inwards by two longitudinal clefts. The female flowers have a calyx formed of five sepals, and a pistil composed of a superior ovary, surmounted by a style, that divides itself into two stigmatic branches. This ovary is one-celled, and encloses only one ovule. The fruit, as botanists regard it, is an achene (what is generally known to the world as such is the thick, fleshy, and succulent *receptacle*); the seed contains under its integument a fleshy albumen, in which is a re-curved embryo.

The ARTOCARPACEÆ abound in the tropics, and in the tropics only. They bear so close a resemblance to the Nettle tribe, that botanists find it difficult to separate them by any well-defined characteristics. Their chief characters are leaves with conspicuous stipules, a rough foliage, and an acrid milky juice which often contains caoutchouc; the flowers are collected over a head or fleshy receptacle. The milk of the Upas-tree (*Antiaris*) is intensely poisonous. The *Galactodendron utile*, or Cow-tree of Demerara, belongs to this family. When incisions are made in it a milky sap exudes, which, if left to stand, is covered by a creamy scum; both are said to be wholesome.

The Bread-fruit reminds us of a reversed fig. The trees have stems of considerable size, large rough leaves, stipules like the Fig, the stamen-bearing flowers disposed in long club-shaped spikes, the pistil-bearing ones in round heads. The female flowers soon grow together, and form, with the receptacle, a large solid globular mass, in contradistinction to the Fig, in which the receptacle bears the female flowers in its interior. *Artocarpus incisa*, the Bread-fruit of the South Sea Islands, is green, and equal in size to the larger melons. One variety produces the fruit free from spines on the surface and without seeds; others split into deep lobes, or are covered all over with the sharp-pointed fleshy tops of the calyx. The seeds when roasted are said to taste like chestnuts; but it is principally for the fleshy receptacle that it is valued, and this when roasted becomes soft, tender, and white, and not unlike the crumb of bread when eaten new.

The Upas-tree (*Antiaris toxicaria*), the half-fabulous poison-tree of Java, was said to be a large tree growing in the midst of a desert produced by its own pestiferous qualities, and causing death to every other plant and animal which came under its influence. To approach the tree for the purpose of wounding its stem and carrying off its juice was said to be the task of criminals condemned to death. There is a measure of truth in the fable. There is the Upas-tree in Java, and its juice, taken internally, is speedy death to any animal; and there is a tract of land where no animal can exist; but the two circumstances have no connection. The poisoned tract is the crater of a volcano, which emits carbonic acid and other gases continually—a spot where not even the Upas-tree can grow. The Upas-tree is one of the *Artocarpaceæ*, which abounds in milky juice, and this juice, as we have said, is like many of its congeners, a deadly poison when mingled with the blood.

The PLATANACEÆ are exogenous trees or shrubs, with palmate deciduous leaves, toothed and stipulate, unisexual naked flowers in globose catkins, the barren flowers with single stamens mixed with scales; the fertile flowers with one-celled ovary, terminated by a thick and awl-shaped style. The Oriental Plane (*P. orientalis*) is a tree of noble growth, in many respects resembling the Sycamore, with large palmate leaves. Its wood is fine-grained and hard, and when old it acquires dark veins somewhat resembling the Walnut. *P. occidentalis* ranges from Mexico to Canada.

The Plane-tree is one of the largest trees of temperate regions. Pliny relates that in his time there existed a celebrated Plane-tree in Lycia, the hollow trunk of which formed a kind of grotto, measuring ninety feet in circumference. Its branching crown resembled a little forest; the branches composing it covered with their shade an immense space of ground. The hollow of the trunk was carpeted with moss, which gave it still more the appearance of a natural grotto. Licinius Mutianus, the Roman governor of Lycia, gave a feast in this grotto to eighteen guests. Pliny mentions another Plane-tree which the Emperor Caligula found in the neighbourhood of Velitræ, the branches of which were so disposed as to form a grotto of natural verdure, in which the Emperor dined with fifteen persons. Although the Emperor occupied a part of the tree alone, the guests were all quite at their ease, and the slaves were able to perform their offices with perfect convenience.

At Caphyæ, in Arcadia, eight hundred years after the Trojan War, an old Plane-tree was shown bearing the name of Menelaus carved on its bark. It was then said that this prince planted it himself before his departure for the seat of war. It is also related that Agamemnon planted a Plane-tree at Delphos, which was seen many centuries after the death of the hero. These assertions are probably fabulous; but what makes recitals of this kind somewhat credible, is the fact, that at the present time, Plane-trees of an age and dimensions quite extraordinary still exist in the East. De Candolle, in his "Physiologie Végétale," records the statement of a modern traveller in the East, to the effect, that in the valley of Bujukdéré, three leagues from Constantinople, there exists a Plane-tree 100 feet in height, the trunk of which was 165 feet in circumference. The trunk presented an excavation of eighty feet in circumference. Its shadow extended over 500 square feet. PLATE X. is a representation of the Plane-tree of Bujukdéré, a celebrated tree all over the East, although the documents which would determine its exact age are wanting.

ALLIANCE III.—EUPHORBIALES.

Flowers scattered, monodichlamydeous; carpels, superior, consolidated; placentas axile; embryo large, with abundant albumen.

Ovules definite, suspended, anatropal; flowers scattered; fruit tricoccous.	I. Euphorbiaceæ.
Flowers amentaceous; ovules definite, anatropal, suspended; radicle superior.	II. Scepaceæ.
Aquatic; ovules definite, suspended, amphitropal; radicle superior . .	III. Callitrichaceæ.
Ovules definite, ascending, anatropal; radicle inferior	IV. Empetraceæ.
Ovules indefinite; radicle inferior.	V. Nepenthaceæ.

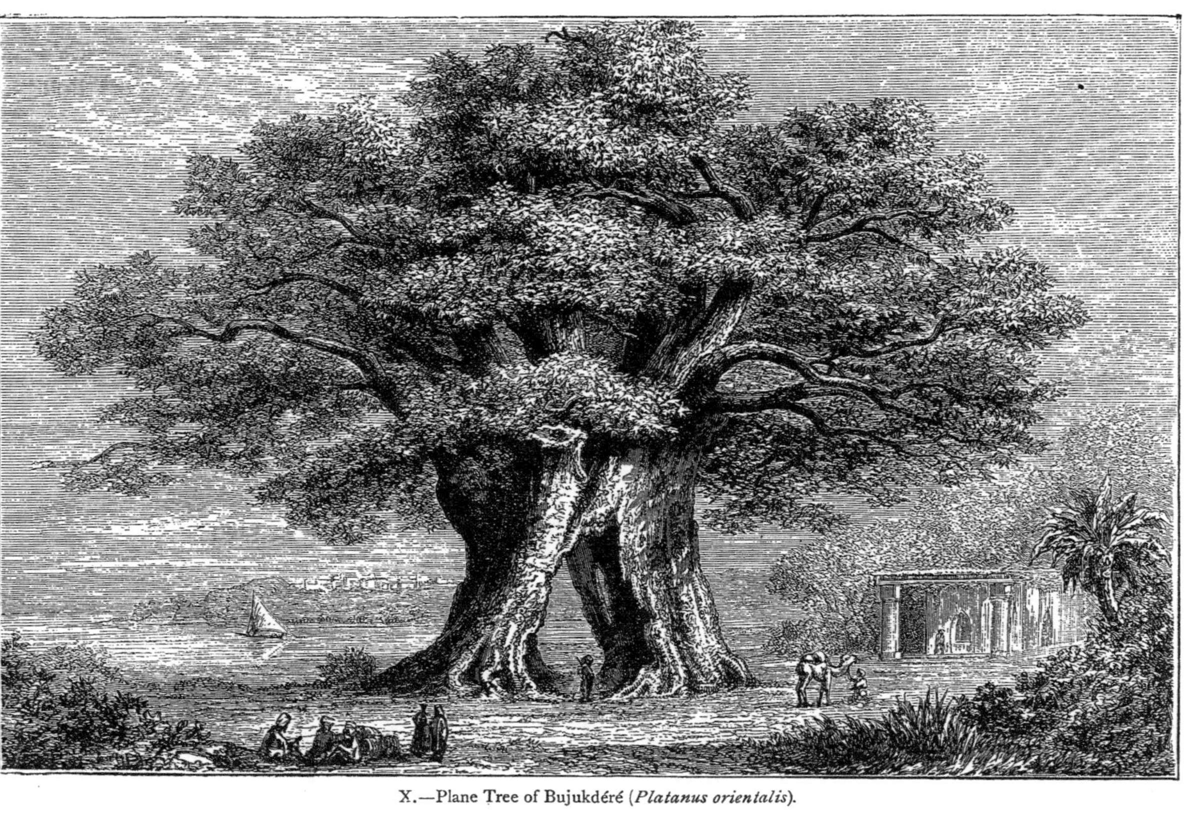

X.—Plane Tree of Bujukdéré (*Platanus orientalis*).

The EUPHORBIACEÆ number no less than 2,000 species, three-

Fig. 383.—Pitcher-plants.

eighths of which are natives of intertropical America, sometimes forming trees, bushes, and even weeds; occasionally they are deformed, leafless, succulent plants, resembling the Indian Figs in appearance, but altogether different from them in properties; many of them are medicinal, as *Ricinus communis*, the Castor-oil plant, and the species of *Croton*.

The Pitcher-plants (NEPENTHACEÆ) are placed here for the excellent reason that no place more suitable could be found for them. With certain insignificant resemblances to the tribe, they possess other characteristics which are without parallel. The pitcher-shaped termination of the leaf, and its articulated lid, are among the curiosities of the vegetable world.

Nepenthes khasiana is illustrated in Fig. 383. The pitchers are believed to be dilatations of a gland at the apex of the leaf.

ALLIANCE IV.—QUERNALES.

Flowers in catkins, monochlamydeous; carpels inferior; embryo fleshy, exalbuminous.

Ovary with two or more cells; ovules pendulous or peltate . . . . . I. Corylaceæ.
Ovary one-celled; ovule solitary, erect . . . . II. Juglandaceæ.

The CORYLACEÆ include the most important trees and shrubs of the European Flora.

The Hazel (*Corylus Avellana*) is a shrub common in woods, thickets, and copses, and often planted in hedges. The branches are erect, slender, and flexible, with simple alternate leaves, doubly dentate, sometimes superficially lobed, accompanied by two caducous stipules. The male flowers are disposed in from one to three pendent catkins, at the extremity of the branches, or upon the short lateral shoots. These catkins begin to make their appearance towards the end of autumn, before the fall of the leaf, and they flower at the end of winter, before the development of the new leaves takes place. The male flowers, contained between two little scales, have eight stamens, with one-celled anthers. The female flowers are composed of a calyx, with a very small denticulate limb, and an inferior ovary with two cells, each containing a suspended anatropal ovule. This ovary is surmounted by two long styles of a lively red. At the period of fructification the involucre has undergone great development; it has become foliaceous, a little fleshy, and slightly bell-shaped at its base, open at the summit and containing a fruit, or nut, which is an achene, in consequence of the abortion of one of the cells, and of the ovule, which it encloses. The seed with a thin membranous skin, contains an embryo destitute of albumen ; the cotyledons are smooth on the surface of one side, and convex on the other.

The common Yoke Elm, or Hornbeam (*Carpinus Betulus*), when allowed to retain its natural dimensions, is a graceful tree, rising sometimes to the height of seventy feet, with a smooth and ashy grey

Fig 384.—Male Catkin of the Oak.

bark. The male flowers are disposed in cylindrical catkins, the imbricated scales of which directly protect six to twenty stamens, with short bifurcated filaments and one-celled anthers barbed at the summit. The female flowers are disposed in racemes; each of the caducous bracts bears two one-flowered involucres, presenting a structure very analogous to that of the Hazel. The fruit has a foliaceous, veined, reticulated cupule, with three lobes, the middle one of which is much larger than the other two. The Hornbeam produces an excellent fire-wood. It is much used in the fabrication of certain tools, and in parts of machines which are subjected to much friction, and where great durability is necessary. The Hornbeam is an indigenous British tree, very common in copses, and is frequently pollarded by the farmer. When checked and stunted in this way it retains its withered leaves all the winter, and is useful where winter shelter is required for tender plants.

The Oaks (*Quercus*) are monœcious trees with simple alternate leaves, each having two caducous stipules. The male flowers are disposed in filiform, slender, interrupted, and pendent catkins (Fig. 384). Each flower presents a calyx with six or eight free, unequal, fringed divisions, and an equal number of opposite stamens,

Fig. 385.—Female Flowers of the Oak.

Fig. 386.—Fruit of the Oak.

with two-celled anthers. The female flower is composed of an inferior ovary, surmounted by a perianth with three or six divisions, and a short style, which is divided into three stigmatic branches, added to which it is surrounded by a sort of little cup, formed by a fold of the peduncle, upon which a large number of small imbricated bracts are inserted. The ovary is three-celled, each having two anatropal ovules. At maturity, two of these three cells, with their contents, are abortive. The fruit, designated under the name of an acorn (Fig. 386), is of an ovoid or oblong shape, umbilicated at its summit with a leathery and shiny pericarp, becomes, therefore, unilocular and monospermous. Under its covering the seed presents an embryo destitute of albumen, the cotyledons of which are convex on the outside, and flat on the inside; they are also fleshy and farinaceous.

XI.—The Oak (*Quercus sessiliflora*).

The fruit is enveloped at its base by the indurated and ligneous cupule of which we have spoken.

The Oaks belong almost exclusively to the northern hemisphere, where they inhabit the temperate regions, or the high mountains of equatorial countries. The species which they include are numerous and difficult to discriminate; they are scarcely known in a wild state in the southern hemisphere, their southern limits being the islands of the Indian Archipelago, whence they spread westward along the Himalayan range until they reach Europe. The Oaks are the most majestic trees of our forests, with robust, hardy trunks, and powerful, far-spreading branches.

The Common Oak (*Q. pedunculata*) is much influenced by the soil on which it grows; and for the Oak, the soil of Sussex seems to surpass any other. It is recognised by the very short stalks of the leaves, while the acorns are borne on long peduncles.

Quercus sessiliflora, represented in PLATE XI., is a tree of variable size, with petiolate, oblong, almost oval, sinuate leaves, with a fruit peduncle much shorter than the petioles; the acorns arrive at maturity the same year with the flowers which produce them. It is of more rapid growth than the last, and attains much larger dimensions, and its timber is also excellent. It is of a much darker colour than the above, and is known as brown oak or "chestnut."

Quercus Ilex is an Evergreen Oak, native of the European continent, of some fifty to seventy feet in height; the leaves are shiny above, grey or whitish and tomentose on the lower surface; the acorns are sessile, or borne by the short, downy, or rather hairy peduncle, with tubercular, scaly cupule. It grows in arid places, and is common in the South of France. The wood is very combustible; besides which it is largely employed in naval construction, and carpenter's and cabinet work.

The Cork-tree (*Quercus Suber*), of which we have already spoken in the chapter on Bark, is closely allied to the Evergreen Oaks; its leaves are persistent till the end of the second and even the third year. It is, as already stated, the external portion of the bark, largely developed, which produces the substance known under the name of cork. It grows upon the sides of mountains of slight elevation, a little removed from the basin of the Mediterranean. Limited to some parts of the South of France and to Spain, the Cork Oak is the predominant inhabitant of the forests of Algeria, where it constitutes woods of great extent, occasionally mixed, however, with other denizens of the forest.

The Kermes Oak (*Quercus coccifera*), is a tufty bush, of from seven to twelve feet high, with small, oblong, cordate, strongly dentate, smooth persistent leaves; common in dry, sandy, and stony places in the regions of the Mediterranean. It is upon this little Oak that the hemipterous Kermes insect lives, from which a beautiful colour is obtained. It is at present used for dyeing the cloths from which the red fez caps are made. This colour has been greatly superseded by that of the Coccus of the Indian Fig (*Opuntia cochinellifera*), which is infested by the cochineal insect (*Coccus cacti*), the females of which—many times larger than the males—alone afford the dye.

The Turkey Oak (*Quercus Cerris*), widely disseminated in south-eastern Europe, is remarkable for the long loose hairy scales of its cup.

The Spanish Oak (*Q. hispanica*) bears its branches erect; leaves nearly evergreen, lanceolate, acute, finely serrate, dark green on the upper side, glaucous green on the under, cups top-shaped, with shaggy, prickly, spreading scales. This tree grows in Spain and Algeria, and is found in some of our nurseries under several synonyms.

Besides these species, there are the Austrian Oak (*Q. austriaca*), found in Hungary and Lower Austria; the Prickly-cupped Oak or Valonia (*Q. Ægilops*), which grows in the Morea, valuable for its acorns, which are largely imported for tanning purposes; and various Oaks, the produce of the mountains intervening between India and Asia Minor. Of the species which Dr. Royle found in the Himalayas, most of them are too tender for acclimatisation with us, but some of them are beautiful trees.

The American Oaks are numerous in species, but their timber is by no means of the same value as that of their European congeners. The White Oak (*Q. alba*) produces sweet acorns and excellent timber, some specimens in the American forest attaining the height of seventy or eighty feet. The Chestnut-leaved Oak (*Q. Prinus*) is cultivated in all the nurseries, and under eight or ten synonyms. It is a handsome tree, but its timber is light and porous. The Black Oak (*Q. tinctoria*) is a native of the Carolinas, Georgia, and Pennsylvania, where it attains a great size, with large, handsome leaves, downy beneath, which become dull red, or yellow in the autumn. The tree is more appreciated for its dyeing properties than for its timber, the latter being coarse-grained; but its inner bark abounds in a yellow dye of great brilliancy, known as Quercitron, which is much sought after. The Live Oak (*Q. virens*) is a valuable timber tree, which grows in the Southern States of the Union, on the shores of the creeks and bays. It is a heavy, compact, fine-grained wood, with coriaceous, elliptic, oblong leaves, obtuse at the base, clothed with starry down beneath; acorns oblong, and said to be very sweet-tasted.

Many species of Oak are found on the high lands of Mexico and the adjoining States, growing at a height of 5,000 or 6,000 feet above the sea; some of them, as the Iron-Wood Oak (*Q. sideroxyla*) and the Large-leaved Oak (*Q. macrophylla*), trees either yielding valuable timber, or of great beauty. The Large-leaved Oak is, perhaps, the finest Oak in the world; its leaves, which are downy beneath, tapering at the point, and heart-shaped at the base, being from twelve to eighteen inches long, and broad in proportion; and its acorns are as large as French walnuts.

The Beech (*Fagus sylvatica*) is one of our best known and most important forest trees. It attains great dimensions, sometimes rising to the height of 120 feet; its smooth, strong stem, which becomes ashy grey by exposure to the weather, is visible to the top of its crown. It is sometimes free from branches to the height of sixty feet. Its leaves, petioled, oval, or oval-oblong, are generally pointed or acuminate, loosely dentate, waving, coriaceous, ciliate with prominent veins; they are alternate, and accompanied by two brownish stipules. The flowers, which are unisexual, appear at the same time as the leaves. The male flowers are disposed in globular catkins, with long pendent peduncles. The female flowers are enveloped, to the number of two or three, in an involucre, consisting of four bracts covered exteriorly with a number of filaments; the fruit is Beech-mast. The seed contains an embryo without albumen, the cotyledons of which are irregularly folded up inside. The oil obtained from this seed is both adapted for cooking and is a good lamp oil.

Among the Beeches many handsome varieties, well adapted for ornamental purposes, from their variously-coloured foliage, have originated. The Purple Beech (*Var. purpurea*) has the young buds and shoots of a rich rose colour. In the Copper Beech (*Var. cuprea*) they are pale copper colour. In some, the leaves are curled up; in others, as *Var. pendula*, the branches are pendulous, or weeping.

The smooth thin bark of the Beech is apt to develop the knobs called embryo-buds, or abortive branches, which are sometimes used by cabinet-makers. Its branches are numerous, and its foliage dense and shady; the Bird's-nest Orchis is often found under its shade, as some believe, though perhaps wrongly, parasitical on its roots; among which also, but not upon them, the Common Morel flourishes in the Beech forests of France and Germany.

The Chestnut (*Castanea vulgaris*) is a tree of rapid vegetation, and endowed with great longevity. It attains a height of 100 feet, occasionally presenting an enormous circumference. Its leaves are large, petioled, oblong, acute, lanceolate, deeply dentate, coriaceous, smooth and shining, with prominent secondary parallel nerves, accompanied by two caducous stipules.

The flowers are unisexual, and appear after the leaves. The male flowers are in very small catkins, each flower being composed of a calyx with five or six divisions, with as many or more stamens, having bilocular anthers. The female flowers are, to the number of five or six, enveloped in a common four-lobed involucre consolidated externally with numerous unequal linear bracteoles. Each female flower consists of twelve abortive stamens and an inferior ovary surmounted by a calycine limb with five to six lobes and an equal number of styles. It encloses a like number of cells, each containing two ovules. When arrived at maturity, which is in the month of September or October, the involucre is thick and coriaceous, prickly on the outside, and enclosing from one to five fruits unilocular by abortion, known under the name of Chestnuts. The pericarp of each is coriaceous, fibrous, and hairy on its internal surface. The seed under a membranous covering contains an embryo without albumen; the cotyledons are voluminous, and plicated with fissures of greater or less depth, and, as is said, farinaceous. The nut is the principal produce obtained from this useful tree; it forms a great part of the food of the poor populations of the central plains of France, and of the valleys of the Alps.

Improved by culture, the Chestnut-tree has given place to the variety called *Marronier* by the French cultivators, of which several varieties are known. They yield the large chestnuts which sometimes come into our markets.

The native country of the Chestnut is not very clearly ascertained. It was introduced into Europe from Asia Minor. The name *Castanea* (whence Chestnut) was derived by the Romans from Castanum, a town of Thessaly, where it grew in great abundance. In the same way we have Currant corrupted from Corinth Grape.

The famous Chestnut-tree of Mount Etna, said in Sicily to be the "Chestnut of a Hundred Horses" (*Castagno de cento Cavalli*), is reported to be 170 feet in circumference. (PLATE XII.) Jean Houel gives the history and dimensions of this gigantic tree. "We started," he says, "from Aci-Reale, in order to visit the so-called Chestnut of 'the hundred horses.' We passed through Saint Alfio and Piraino, where these trees are common, and where we found some superb

XII.—The Chestnut Tree of Mount Etna (*Castanea vulgaris*).

woods of Chestnuts. They grow very well in this part of Etna, and they are cultivated with great care. Night not having yet come, we went at once to see the famous Chestnut which was the object of our journey. Its size is so much beyond all others that we found it impossible to express in words the sensation we experienced on first seeing it. Having examined it carefully, I proceeded to sketch it from Nature. I continued to sketch the next day, finishing it on the spot, and I can now say that it is a faithful portrait. I demonstrated to my own satisfaction that the tree was 160 feet in circumference, and heard its history related by the *savants* of the hamlet. It obtained its name of the 'Chestnut of a Hundred Horses' in consequence of the vast extent of ground it covers. They told me that Jeanne of Aragon, while journeying from Spain to Naples, stopped in Sicily and visited Mount Etna, accompanied by all the noblesse of Catania on horseback. A storm came on, and the Queen and her *cortége* took shelter under this tree, whose vast foliage served to protect her and all these cavaliers from the rain. It is true that out of the hamlet the tradition of the Queen's visit is looked upon as fabulous; but however that may be, the tree itself seems very capable of doing the office assigned to it. This tree is entirely hollow. It is supported chiefly by its bark, having lost its interior entirely by age; but is not the less crowned with verdure. The people of the country have erected in it a house, with a kiln for drying the chestnuts and other fruits which they wish to preserve. They are so indifferent to the preservation of this wonderful natural curiosity that they do not hesitate to cut off branches to burn in the furnace. Some persons think that this mass of vegetation is formed of many trees which have united their trunks; but a careful examination disposes of this notion. All the parts which have been destroyed by time or the hand of man have evidently belonged to a single trunk. I have measured them carefully, and found the one trunk, as I have said, 160 feet in circumference."

We should be inclined to adopt the opinion that this monster tree was the union of several, but M. Houel's sketch and description seem conclusive; and his opinion is further confirmed by the fact that many Chestnuts in the neighbourhood of Mount Etna are twelve yards in diameter, while one actually measures eighty-three feet in circumference.

What age can be assigned to the Mount Etna Chestnut? It is difficult to say. If we are to suppose that each year its concentric layers have only been a line in thickness, this venerable tree would be not less than 3,600 to 4,000 years old.

At Neuve-Celle, on the Lake of Geneva, there exists another Chestnut of gigantic proportions.

The type of the JUGLANDACEÆ, among whose prevailing qualities are astringency, is the Common Walnut (*Juglans regia*). (PLATE XIII). It is a large tree, with whitish bark, more or less fissured according to its age. It has a cylindrical stem, rising to a considerable height without branches; the branches are large and spreading, forming an ample and rotund head. The leaves are of a dull greenish colour. The Walnut is indigenous to the Caucasus, Persia, and India. This tree only prospers and is abundantly fruitful when it is completely isolated. The leaves are compound, alternate, smooth, and coriaceous; they are composed of seven or eight elliptical, acute leaflets. The flowers are monœcious and disposed in catkins; but in the female catkins the flowers are less numerous. The male catkins are pendent, with loosely imbricated scales, cylindrical, very caducous, and placed in the axil of the leaves which have fallen the preceding year. At the axil of each scale is a flower with a perianth of five to six divisions, and a valuable number of stamens ranging from eighteen to thirty-six. The female flowers are in clusters of drooping catkins, of from one to four, borne on the summit of the young shoots, each presenting a very short, scarcely-toothed exterior involucre, and an interior calyx with four divisions. A short style rises from the centre of the flower, which soon divides itself into two stigmatic scaly plates. The ovary is inferior, one-celled, with one erect ovule. It is subdivided by partial dissepiments or partitions,

Fig. 387.—Male catkin of the Walnut (*Juglans regia*).

XIII.—The Walnut Tree (*Juglans regia*).

starting from the placenta, into four imperfect cells. The fruit is a drupe; the external covering a fleshy husk of one piece, separating into irregular segments; the woody shell is two-valved, and very hard, furrowed, and wrinkled. The seed—single, erect, and wrinkled by the furrows in the shell—is four-lobed at the summit and at the base, by the dissepiments. The exterior envelope is at first whitish, then yellowish green, more or less deep, and is remarkable for its astringent properties. The embryo is destitute of albumen, and straight; the cotyledons thick, fleshy, oily, bilobed, resemble in figure the convolutions presented by the sinuosities of the brain of a vertebrated animal. It is these cotyledons which form the kernel of the nut.

The Walnut-tree was known to the Greeks, and cultivated by the Romans, by whom it was much valued for its wood as well as for its nut. There is no record of its introduction into Britain; but Gerarde tells us that "the green and tender nuts, boyled in sugar and eaten as suckade, are a most pleasant and delectable meat, comforting to the stomach, and expell poyson." Before the introduction of mahogany and rosewood, walnut was in great estimation, and within the last few years it has been restored to its old pre-eminence, its favourite purpose, however, being for gun-stocks, for which its lightness is its qualification. In many parts of Spain, France, Italy, and Germany the nut forms a great article of food to the people. In all these countries the Walnut-tree is extensively cultivated; the district of the Bergstrasse, between Heidelberg and Darmstadt, is almost entirely planted with them, and in some places, according to Evelyn, in his days, "no young farmer is permitted to marry a wife until he bring proof that he is father of a stated number of Walnut-trees." We need not enlarge on the well-known fruit, although we may quote Cowley on its virtues:—

"On barren scalps she makes fresh honours grow.
Her timber is for various uses good:
The carver she supplies with useful wood:
She makes the painter's fading colours last.
A table she affords us, and repast;
E'en while we feast, her oil our lamp supplies.
The rankest poison by her virtues dies."

Alliance V.—Garryales.

Flowers with a single floral envelope, sometimes amentaceous; carpels inferior; embryo minute in a large mass of albumen.

Flowers amentaceous ; leaves opposite, exstipulate	I. Garryaceæ.
Flowers fascicled ; leaves alternate, stipulate	II. Helwingiaceæ.

Alliance VI.—Menispermales.

Flowers monodichlamydeous ; carpels superior, disunited ; embryo surrounded with abundant albumen.

Albumen copious, solid ; seeds pendulous ; embryo small, external ; stamens perigynous.	I. Monimiaceæ.
Albumen copious, solid ; seeds erect ; anthers opening by re-curved valves.	II. Atherospermaceæ.
Albumen copious, ruminated ; sepals united into a valvate cup . .	III. Myristicaceæ.
Albumen copious, solid ; seeds parietal ; embryo minute	IV. Lardizabalaceæ.
Albumen copious, solid ; seeds pendulous ; embryo minute, internal, stamens hypogynous.	V. Schizandraceæ.
Albumen sparing, solid ; seeds amphitropal ; embryo large	VI. Menispermaceæ.

The Myristicaceæ are an order of tropical trees, amongst which is the Nutmeg.

The Menispermaceæ, while possessing many features in common, present very extreme differences of structural arrangement, yet they are very distinct, as a whole, from any other class of plants. The plants of this order are common within the tropics in Asia and America ; climbing among the forest trees to a great height, and remarkably tenacious of life—broken branches throwing out slender thread-like shoots, which rapidly establish a connection with the soil. Thirty-one genera, and 300 species.

Alliance VII.—Cucurbitales.

Flowers monodichlamydeous ; ovary inferior ; placentas more or less parietal ; embryo without albumen.

Fruit pulpy ; placentas strictly parietal ; monopetalous	I. Cucurbitaceæ.
Fruit dry ; placentas strictly parietal ; apetalous	II. Datiscaceæ.
Fruit dry ; placentas projecting and meeting in the axis ; monodichlamydeous	III. Begoniaceæ.

The Cucurbitaceæ are plants, either climbing by means of tendrils, or trailing, with unisexual flowers, and have scabrous stems and leaves, with a lobed foliage. The order comprehends the Melon, Cucumber, Colocynth, Bryony, and other genera, of which Dr. Royle remarks that "they afford large and juicy fruit in the midst of the Indian desert and on the sandy islands of Indian rivers ; nor does excessive moisture appear to be injurious, as the majority of them are cultivated in the rainy season, and even on the bed of weeds floating on the Cashmere lakes." Two principles pervade them ; the one saccharine and nutritious, the other bitter, acrid, and purgative.

In the Melon, the Gourd, and their allies, the first exists almost exclusively, but even here accompanied in some degree by a laxative principle. In the Colocynth, Bottle Gourd, and in some species of *Luffa* and Bryony, the bitter principle is strongly concentrated.

The native country of the Melon is doubtful, although Linnæus places it in Tartary, which is probably erroneous. To the Cucumber the same nativity is ascribed, but without corroborative authority. The Colocynth Gourd, yielding the well-known medicine, grows wild in Egypt and the countries of the Grecian Archipelago. The Water Melon (*Cucumis Citrullus*) is extensively cultivated all over India and the tropical countries of Africa and America.

The characteristics of the genus *Cucumis* are:—The flowers monœcious, the male flowers solitary, growing at the axils of the leaves, or more often, fascicled by the contraction of the common peduncle; the calyx campanulate, with five teeth; the petals five, ovate, acute, and spreading. There are three free stamens, two being entire, bilocular, the other unilocular, with flexuose anther cells, a connective prolonged above the anthers into an appendage, simple in the unilocular stamen, bifid in the others. The female flowers are solitary, and composed of a five-toothed calyx, a corolla similar to that of the male flowers, with an inferior three-celled ovary, surmounted by a short style with three thick stigmata. The ovary was originally one-celled, with three parietal placentas, each with two series of ovules, and are advanced towards the centre of the cavity, where they become united, and where they soon become fleshy. The fruit is a smooth or warted fleshy berry. The seeds are oval, more or less compressed, and containing a straight embryo destitute of albumen.

It is principally upon the modifications of the fruit that the classification of the Melons into several tribes has been adopted. They are again sub-divided into secondary groups, of which we will give a rapid glance.

The Cantaloups form a rather characteristic group. In the principal varieties the fruit is of large dimensions, varying in shape from that of a very depressed sphere to an ovoid oblong, more or less prominent, with smooth or warty skin. The flesh of the Melon proper is thick, of a reddish-orange colour, delicate, melting, and succulent. All these Melons turn yellow when ripening, and then exhale an agreeable odour.

Another group is that of the *Melons brodés*, which comprises the Market Melon, the *Melon de Coulommiers*, and some others cultivated largely in the gardens round London and Paris.

The *Sucrins* are ranked in the third class. Their flesh is white or greenish, with a sweeter and more penetrating perfume than the Cantaloups; the flesh is delicate, melting, and sugary.

The Winter Melons form a fourth group, of which the finest

Fig. 388.—Male and Female Flowers of the Melon.

European representative is the Winter Melon of Provence, or *Melon de Cavaillon*. The skin of this variety is thin, and its flesh very thick and firm, and of a white, pale, yellowish-green colour, according to the variety, without perfume, but melting and very sweet. It is highly esteemed in the South of France and Europe, where it is cultivated on a large scale, and fills the markets during a greater part of the summer and autumn. It is also being introduced into Paris. But we must pause in our enumerations of these esculent fruits, or we should pass the limits of this work.

Many species of *Cucurbita* are worthy of an attentive examination. Among these we may mention the Melon Pumpkin (*C. maxima*), the Pumpkin (*C. Pepo*), and the Vegetable Marrow.

Bryonia dioica, a very common and widely known species, decorates our hedges with its charming little round, red, sometimes yellow, berries. It is a perennial, with large succulent roots, from which springs a slender, pale green, hairy stem, which climbs among bushes by means of its tendrils, after the manner of the order. The leaves are palmate, and rough on both sides, with callous points. The stamens and pistils are on different plants, the male flowers being largest, whitish, with pale green veins.

ALLIANCE VIII.—PAPAYALES.

Flowers dichlamydeous; carpels superior, consolidated; placentas parietal; embryo, surrounded by abundant albumen.

Corolla monopetalous; throat of female flowers without scales	I. Papayaceæ.
Corolla polypetalous; throat of female flowers with scales.	II. Pangiaceæ.

The best known representative of the PAPAYACEÆ is the Papaw-tree (*Carica Papaya*), of tropical America and the West Indies, which is said to have the singular property of rendering the toughest animal substances tender; newly-killed meat suspended among its branches becomes tender in a wonderfully short time. Old hogs and poultry soon fatten when fed upon its leaves; the leaves are also used by the negroes as a substitute for soap. The fruit of the plant has an insipid flavour, but when candied resembles citron; the seeds have a sharp and biting taste. Some species of the *Carica* are deadly poisons.

The PANGIACEÆ are mostly natives of the hotter parts of India. All are poisonous plants, but contain medicinal properties.

SUB-CLASS II.—HYPOGYNOUS EXOGENS.

Flowers with the stamens entirely free from the calyx or corolla.

ALLIANCE IX.—VIOLALES.

Flowers monodichlamydeous; placentas parietal or sutural; embryo straight, with little or no albumen.

Characters	Order
Flowers scattered, apetalous or polypetalous ; petals and stamens both hypogynous ; leaves dotless, or with round dots only.	I. Flacourtiaceæ.
Flowers in catkins, apetalous, scaly, polygamous ; stamens unilateral.	II. Lacistemaceæ.
Flowers scattered, apetalous, tubular, hermaphrodite ; leaves marked with both round and linear transparent dots ; stamens perigynous.	III. Samydaceæ.
Flowers polypetalous, or apetalous, coronetted ; petals perigynous, imbricated ; stamens on the stalk of the ovary ; styles simple, terminal ; seeds arillate ; leaves stipulate.	IV. Passifloraceæ.
Flowers polypetalous, coronetted ; petals perigynous, imbricated ; stamens on the stalk of the ovary ; styles simple, dorsal ; seeds without aril ; leaves without stipules.	V. Malesherbiaceæ.
Flowers polypetalous ; calyx many-leaved ; petals perigynous ; anthers one-celled ; fruit stipitate, consolidated, siliquose ; seeds without albumen ; stamens perigynous.	VI. Moringaceæ.
Flowers polypetalous ; calyx many-leaved ; petals hypogynous ; stamens all perfect ; anthers crested and turned inwards ; fruit consolidated ; seeds albuminous.	VII. Violaceæ.
Flowers polypetalous ; calyx tubular, furrowed ; petals hypogynous, unguiculate.	VIII. Frankeniaceæ.
Flowers polypetalous ; calyx many-leaved ; petals hypogynous ; styles distinct ; fruit consolidated ; seeds indefinite, basal, comose, exalbuminous.	IX. Tamaricaceæ.
Flowers polypetalous ; calyx many-leaved ; petals hypogynous ; stamens partly sterile and petaloid ; anthers opposite the petals, naked, turned outwards ; fruit consolidated ; seeds albuminous.	X. Sauvagesiaceæ.
Flowers polypetalous or monopetalous ; calyx many-leaved ; petals hypogynous ; fruit follicular, apocarpous.	XI. Crassulaceæ.
Flowers polypetalous ; petals perigynous, contorted ; styles forked ; leaves exstipulate.	XII. Turneraceæ.

This important group of Orders, with three exceptions, forms, as Dr. Lindley informs us, a perfectly natural group, the exceptions being the *Moringaceæ*, the *Tamaricaceæ*, and the *Crassulaceæ*. The FLACOURTIACEÆ are small trees or shrubs, natives of the hottest parts of the East and West Indies, and Africa. The LACISTEMACEÆ grow in low places in equinoctial America. The SAMYDACEÆ are all tropical, and chiefly American.

"The PASSIFLORACEÆ," says Dr. Lindley, "are the pride of South America and the West Indies, where the woods are filled with their species, which climb from tree to tree, bearing at one time flowers of the most striking beauty, and of so singular an appearance, that the zealous Catholics who first discovered them, adapted Christian traditions to those inhabitants of the South American wilderness; and at other times, fruit tempting to the eye and refreshing to the palate."

The name is derived from a fancied resemblance to the emblems of our Saviour's crucifixion. In the five anthers, the Spanish monks saw his wounds; in the triple style, the three nails by which he was fixed to the cross; and in the column on which the ovary is elevated, the pillar to which he was bound; while a number of filaments which spread from the cup within the flower were finally likened to the crown of thorns. In reality, the flower consists of a calyx and corolla, each of five divisions, consolidated into a cup, from within the rim of which spread several rows of filamentous processes, regarded by some as barren stamens. From the sides of the cup, and within these, there proceeds one or more raised rings, notched or undivided, and in various degrees of development, and evidently of the same nature as the filamental processes themselves. In the centre of the flower stands a column, to the sides of which the five stamens are united, but spread freely beyond its apex, and bear five oblong horizontal anthers. The apex of the columns bears the one-celled ovary, with three parietal polyspermous placentas, having three club-shaped styles at its vertex. The plant produces a gourd-like fruit, containing many seeds, each having its own fleshy aril, usually enveloped in a subacid mucilage.

The MALESHERBIACEÆ are herbaceous plants of Chili and Peru, of little known interest. The MORINGACEÆ are a small group of trees of the East Indies and Arabia, which De Candolle placed with the *Leguminosæ* erroneously, as Dr. Lindley thinks.

The VIOLACEÆ contain many plants, like the Violet, with irregular flowers; but this is not an essential character of the order. The Violet (Fig. 389), is a stemless or very short-stemmed plant. Its leaves, which are radical, or growing upon runners, are heart-shaped. The stipules are ovate, acuminate, or lanceolate. The flowers have a sweet aroma, a reddish-blue colour, and each is borne upon a slender peduncle, which is re-curved at the summit. The calyx has five sepals, each of them having a small appendix at its base, which descends beyond the point of its insertion. The corolla is composed of five petals; the inferior and largest is hollowed out, and terminated at its base by a short and obtuse spur, and directed downwards with the two lateral ones, which are entire and bearded. The two upper petals, which are equally entire, are directed upwards. There are five stamens alternating with the petals; they are nearly sessile, and slightly joined by their anthers. Each anther is surmounted by a little slender yellow tongue, which is a prolongation of the connective. In addition to this, the two anterior stamens are provided at their base with a kind of tail, which lodges in the hollow spur of the lower petal. The pistil is composed of a free ovary, surmounted by an ascending style, swelling a little above the base, and terminating in a slender hooked stigma. In the interior of the ovary, which is unilocular, are three parietal placentas, with straight anatropal ovules. The fruit is a capsule, which opens in three valves, each having a placenta in its midst. The seeds contain a straight embryo in the axis of a fleshy albumen.

Such is the Violet (*Viola odorata*, Fig. 389) to the botanist; the poet would give a very different description of it.

It is well known that there are many other species of this plant which, to the disappointment of many, are inodorous; such are the *Wood Violet* and the *Dog Violet*. The pretty little plant called the Pansy belongs to the genus *Viola*. In the section to which it belongs the upper and lateral petals are directed upwards, and only

Fig. 389.—The Sweet Violet (*Viola odorata*).

the lower one is directed downwards, and the stigma is capitate. There are many varieties of the Pansy, or *Viola tricolor;* amongst them the Wild Pansy, the corolla of which does not exceed the calyx, and the Garden Pansy, the petals of which extend far beyond it. The size and colour of the Pansies have been greatly varied by cultivation.

The TAMARICACEÆ include a very familiar sea-coast shrub, the Tamarisk. A closely allied species in Arabia, when punctured by a species of *coccus*, exudes a saccharine substance, which is called, like the substance mentioned in Scripture, manna.

The CRASSULACEÆ comprehend many of the plants commonly known as succulents, on account of the quantity of water which they enclose in their tissues, and the general thickness of their leaves. The Biting Stonecrop (*Sedum acre*), Fig. 390, will serve as a type of the family. It is a little fleshy plant, common upon old walls,

Fig. 390.—*Sedum acre.*

thatched roofs, and stony places which are exposed to the sun. It has a slender stem, prostrate and creeping, throwing up branches here and there, covered with short, sessile, and fleshy leaves resembling little eggs slightly flattened above, and bearing five or six flowers disposed in a kind of scorpioid cyme. The flowers have a calyx composed of five fleshy pieces, five free petals, and double the number of stamens, with flattened filaments pointed at the summit, and anthers bilocular; lastly, a pistil composed of five free unilocular carpels, enclosing many anatropal horizontal ovules inserted on the ventral suture of each carpel. At maturity these carpels become dry like follicles, and enclose extremely small seeds. These singular plants grow and keep fresh in the most arid places, from the store of liquid held in reserve in their fleshy tissues, and from the almost total absence of all exhalation. The genus *Crassula*, which has given its name to the family, is remarkable for the structure of its flowers, which have been taken as a type of floral symmetry. This flower has five sepals, with five petals alternating with them; five stamens alternating with the petals, and five carpels alternating with the stamens.

The Houseleeks belong to the genus *Sempervivum*, the calyx of which has from six to twenty divisions, the corolla from six to twenty petals, the andrœcium twelve to forty stamens, the pistil six to twenty carpels. All of us have seen this beautiful plant creeping along the thatch of cottages, with its succulent leaves disposed in a rosette, from the centre of which rises a straight cylindrical stem, garnished with thick fleshy leaves, and terminating by a scorpioid cyme of purplish flowers.

Among the numerous exotic plants with which this family furnishes horticulture, we may mention the different species of *Crassula Echeveria*, *Cotyledon*, and *Rochea*.

Alliance X.—Cistales.

Flowers monodichlamydeous, placentas parietal or sutural; embryo curved or spiral, with little or no albumen.

Stamens not tetradynamous, generally indefinite; flowers trimerous or pentamerous; fruit closed; seeds albuminous.	I. Cistaceæ.
Stamens tetradynamous; flowers tetramerous.	II. Cruciferæ.
Stamens not tetradynamous, definite; flowers not tetramerous; fruit open at apex; seeds exalbuminous.	III. Resedaceæ.
Stamens not tetradynamous; flowers tetramerous; seeds exalbuminous; fruit closed.	IV. Capparidaceæ.

The Cruciferæ is one of the most natural groups of the whole vegetable world, and in studying one of the species which compose it we may study all. Let us take as a type one flower, the Wallflower (Fig. 391). This flower is regular. The calyx is composed of four sepals, free and straight, the two lateral bulging out at the base. The corolla has four unguiculate petals alternating with the sepals, with limbs which are spreading and entire. The stamens are tedradynamous, that is to say, they number six, of which four are large and two smaller; they are also hypogynous. Their anthers are bilocular, and open inwards by two longitudinal clefts. At their base

Fig. 391.—The Wallflower.

is a disc composed of glands, two of which are of a greenish colour, enclosing, so to speak, the base of the small stamens, whilst the other four are smaller, and are placed on the outside of the long stamens. The pistil is composed of a very elongated ovary, surmounted by a rather short style, of which the stigma is bilobate. When at maturity this ovary presents two cells. It is, however, unilocular when young, with two parietal placentæ, with ovules. It is only at a certain period of its development covered; that, a partition starting from one of these two parietal placentæ, and advancing towards the interior, it meets the partition of the opposite placenta and is joined to it, and this constitutes a dissepiment which separates the ovarian cavity into two compartments. When arrived at maturity this ovary becomes a *siliqua*, as the long pod of the *Cruciferæ* is termed; of which the seeds are destitute of albumen, and enclose an embryo with flat cotyledons and lateral radicle, that is to say, folded upon the commissure of the cotyledons.

We have already said that the family of the *Cruciferæ* is constructed upon one common type. There are nevertheless some secondary characters, arising from the regularity or irregularity of the corolla, the shape of the fruit and of the embryo, which serve to distinguish the different genera; for example, *Iberis* has two petals much larger than the others. Then the fruit, which in the Wallflower is a siliqua, is a silicula, or as broad as long, in *Thlaspi*. It becomes Lomentaceous, *i.e.*, an indehiscent legume in *Raphanus*; monospermous and indehiscent in *Isatis*, &c. With regard to the embryo, the radicle does not always fold upon the commissure of the two cotyledons, as it does in the Wallflower; it is sometimes folded upon their back, as in *Isatis*. The cotyledons themselves are not always level, but may be folded longitudinally, or rolled upon themselves. Among the more remarkable plants belonging to this vast vegetable family, we will mention *Cochlearia officinalis*, which is a most powerful antiscorbutic; the Garden Cress (*Lepidium sativum*); the Watercress (*Nasturtium officinale*); the Horse-radish (*Armoracia rusticana*); the Radish (*Raphanus sativus*), of which two varieties (the Turnip and Long Radish) appear at our tables; Turnip (*Brassica Rapa*); *Brassica campestris*, the seeds of which furnish the colza oil, so excellent for lamps; the Cabbage (*Brassica oleracea*), which furnishes us with the Cabbage, the Cauliflower, the Broccoli; the Black Mustard (*Sinapis nigra*), the seeds of which afford a volatile oil, to which they owe their exciting virtue (the pungent oil of mustard seed does not exist in it until it is moistened with water, being, in fact, the product of the chemical decomposition of substances contained in the seed; this is proved by the expressed oil of mustard having no pungent flavour, being, in fact, mild and emollient); the Woad (*Isatis tinctoria*), the roots of which furnish the blue colouring matter called *woad*, used formerly for dyeing, now superseded by indigo, a substance resembling it in colour and properties. The plants of this family are very little cultivated for ornamental purposes, although some remarkably pretty flowers are found within its limits. For fragrant aroma few plants excel the Wallflower. Who does not remember

> "Some rude stone fence, with fragrant Wallflower gay,"

or the bright green shining *Cardamine pratensis*—the

> "Ladies' smock, all silver white,"

whose praises Shakespeare sang?

The RESEDACEÆ, of which the Common Mignonette (*Reseda odorata*) may be considered the type, are soft herbaceous plants, or, in a few instances, small shrubs, with alternate entire or pinnately divided leaves and minute gland-like stipules; flowers in racemes or spikes; calyx many-parted; petals broad fleshy plates, with lacerated appendages at the back. They are all natives of Europe and adjoining parts of Asia, and of the shores of the Mediterranean. A few species occur at the Cape, North India, and California. The Common Mignonette, whose delicious scent has caused it to be universally cultivated, is a native of Northern Africa. It is known by its lanceolate entire or trifid leaves, a six-parted calyx, equal in length to the petals, which are finely cleft into many club-shaped divisions, the two lowest simple; the capsules three-toothed. The Cultivated Mignonette (*Reseda odorata*), as commonly grown in this country, is an annual. If it is prevented from flowering the first two years, it becomes perennial, and may attain the size of a shrub; but this in the English climate requires the protection of a conservatory.

The CAPPARIDACEÆ are chiefly found in the tropics and the countries bordering on them. The flower buds of the Caper are stimulant, antiscorbutic, and aperient; they are a favourite as a pickle. *C. ægyptiaca* is thought by Sprengel to be the hyssop of the Old Testament.

ALLIANCE XI.—MALVALES.

Flowers monodichlamydeous; placentas axile; calyx valvate in æstivation; corolla imbricated or twisted; embryo with little or no albumen.

Stamens columnar, all perfect; anthers two-celled, turned outwards. .	I. Sterculiaceæ.
Stamens monadelphous, in most cases partly sterile; anthers two-celled, turned inwards.	II. Byttneriaceæ.
Stamens free; disk none; seeds albuminous; embryo curved; petals permanent; calyx ribbed.	III. Vivianaceæ.
Stamens free; disk none; seeds exalbuminous; embryo amygdaloid. .	IV. Tropæolaceæ.
Stamens columnar, all perfect; anthers one-celled, turned inwards. .	V. Malvaceæ.
Stamens free, on the outside of a disk; seeds albuminous; embryo straight.	VI. Tiliaceæ.

Among the STERCULIACEÆ, which are large trees or shrubs, all natives of the tropics, having the columnar stamens of the Mallows, none excited greater surprise than the Baobab-trees of Africa, or *Adansonia*, when they were first introduced to the botanical world; their enormous size, and their prodigious longevity—estimated in some instances at some thousands of years—took the world by surprise.

The Baobab (*Adansonia digitata*) is a tree of tropical Africa, which has been transplanted by man into Asia and America. Its trunk does not exceed fifteen or seventeen feet in height, but its girth is enormous, attaining, as it sometimes does, the circumference of thirty to forty feet. This trunk separates at the summit into branches fifty to sixty feet long, which bend towards the earth at their extremities. The trunk being short, and the branches thus curving towards the earth, it follows that the Baobab presents at a distance the appearance of a dome, or rather a ball, of verdure, over a circuit of 160 feet. Adanson concluded, from the observations he made, and from his calculations upon their growth, that some of the specimens which he studied could not have been less than 6,000 years old. But it is the general opinion of botanists that this estimate was enormously overrated. One of these monstrous trees is represented in PLATE XIV. This colossal vegetable was first observed by Adanson in Senegambia, and, after him, the genus was named *Adansonia*. The Baobabs have since been discovered along the Mozambique coast, and in Abyssinia.

The leaves of *Adansonia digitata* are of a deep green, and divided into five unequal parts, each of which forms a narrow lanceolate figure, radiating from a common centre, the outermost being smallest. The flowers, which grow singly in a pendulous position, before the appearance of the leaves, are large and white, crumpled at the edge, the petals being much reflexed; the stamens numerous, and collected into a tube, which spreads at the top into an umbrella-like head, from which rises a slender curved style, terminating in a rayed stigma.

The bark and leaves of this tree possess considerable emollient properties, of which the natives take advantage. The natives make a daily use of the pounded leaves of the Baobab, which they call *lalo*,

XIV.—Baobab Tree (*Adansonia gigantea*), from a Photograph.

to mix with their food, for the purpose of inducing perspiration. Its flowers are proportioned to the gigantic trunk, their breadth being from five to six inches. The fruit, called by the French settlers on the Senegal *Monkey Bread*, is ovoid, pointed at one of its extremities, and from eight to eighteen inches long by six to seven broad. It encloses in its interior from ten to forty cells, containing kidney-shaped seeds, surrounded by mucilaginous pulp, which is sweet, and of an agreeable flavour; the juice, when extracted and mixed with sugar, forms a beverage very useful in the putrid and pestilential fevers of the country. The fruit is transported into the eastern and southern parts of Africa; and the Arabs carry it to the countries round Morocco, whence it finds its way into Egypt. The negroes take part of the damaged fruit and the ligneous bark, and burn them for the sake of the ashes, from which they manufacture soap by means of palm oil. They make a still more singular use of the trunk of the Baobab; they deposit in it the bodies of those among them whom they consider unworthy of the honours of sepulture. They select the trunk of some Baobab already attacked and hollowed out by insects or decay; they increase the cavity, and make a kind of chamber, in which they suspend the body. This done, they close up the entrance of this natural tomb with a plank. The body becomes perfectly dry in the interior of this cavity, and becomes a perfect mummy without further preparation. This kind of sepulture is especially reserved for the *Gueriots*; they are the musicians and poets, who preside at all fêtes and dances at the courts of the negro kings. During their life this kind of talent gives them influence, and makes them respected by other negroes, who look upon and honour them as sorcerers; but after death this respect is succeeded by a kind of horror. These superstitious people imagine that if they consigned the body of one o' these sorcerers to the earth, as they would the bodies of other men, they would draw upon themselves the celestial malediction. Hence the monstrous Baobab serves as their resting place. It is a strange sentiment which leads barbarous people to bury their poets between heaven and earth in the heart of this vegetable king.

The various species of *Bombax*, which also belong to the order, are gigantic American forest trees, with huge buttresses projecting from their colossal trunks. The Silk Cotton-tree (*Bombax ceiba*) yields a substance resembling thistle-down, which will not card, but forms a good stuffing for pillows.

The BYTTNERIACEÆ are trees, shrubs, sometimes climbers. They are natives of Australia, New Zealand, South Africa, Asia, and tropical America, the most remarkable species being *Theobroma Cacao*, a small tree, of which Demerara has whole forests. The seeds of this plant, when dried, roasted, and ground, form chocolate; and an ardent spirit is distilled from the pulp of the fruit.

The VIVIANACEÆ are herbaceous or half-shrubby plants of Chili and Brazil, of little interest.

The TROPÆOLACEÆ are smooth, herbaceous, trailing, or twining plants, of tender texture and acrid taste, better known for the strange form and rich colours of their flowers. They are all natives of temperate North and South America. The flower is distinguished by its irregular spurred calyx (Fig. 392), the spur *s* being a backward development springing from the base of the upper sepal.

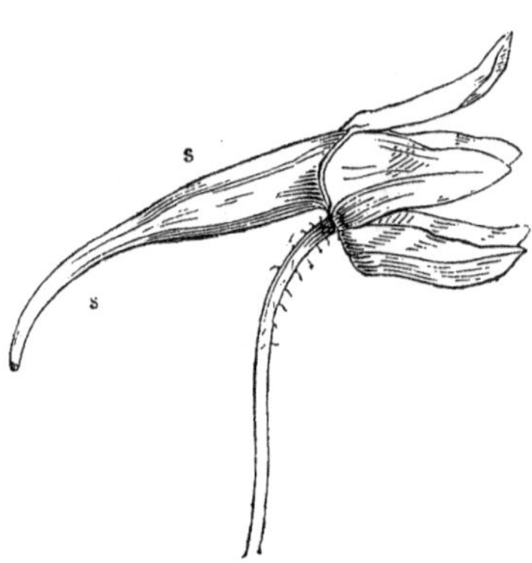

Fig. 392.—Indian Cress (*Tropæolum majus*).

The MALVACEÆ are herbaceous plants, trees, and shrubs, with showy flowers, often enclosed in an involucre of various forms. They are abundant in the tropics, diminishing as we approach the north. In our climate the Mallows are the well-known representatives of the order. The Common Mallow (*Malva sylvestris*), Fig. 393, has a stem ascending or patulous, branching, and hairy; the inferior leaves somewhat orbicular, heart-shaped, or truncate at the base, having from five to seven shallow and obtuse lobes; the upper leaves present three to five lobes, usually much deeper. The flowers, the corolla of which is veined with purple, passing to violet, are disposed in axillary fascicles. The calyx is in five divisions, and is furnished exteriorly with an involucre of three bracteoles (Fig. 394 *b*.) Five alternate petals, coherent at the base of their claw, constitute the corolla. The stamens are numerous and monadelphous; that is, they appear as if their unequal filaments, free only in their upper part, were united by their lower parts into a tube covering the ovary. These filaments are surmounted by a one-lobed anther, opening by a semicircular cleft. The pistil is composed of a multilocular ovary, surrounded by as many styles as there are cells. These styles are filiform, consolidated in their lower part, and forming a sort of brush. An ascending ovule is inserted at the central angle of each of these cells.

Fig. 393.—Common Mallow (*Malva sylvestris*).

The fruit is composed of a great number of small carpels, each with a single seed connected circularly round a common central axis. Under their integuments the seeds contain a curved embryo, in a rather abundant mucilaginous albumen, the cotyledons folded upon and fitting into one another.

The Marsh-mallow (*Althœa officinalis*) has stems from two to three feet in height. It grows in marshes near the sea, but is rather local. The whole plant is covered with hairy down. The leaves are ovate, cordiform, thick, entire or 3-5-lobed, and toothed. The flowers are one to two inches in diameter, and rosy; the involucre is 6-9-lobed.

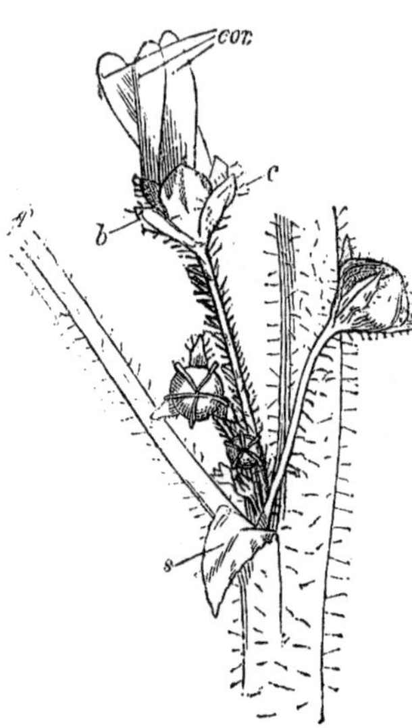

Fig. 394.
Common Mallow (*Malva sylvestris*).

Amongst the most remarkable of the Malvaceæ is the Cotton (*Gossypium*), of which several species are largely cultivated in many parts of Asia, America, and the North of Africa, within the tropics, for the sake of the hair which covers the testa of their seeds. This forms the textile substance known under the name of cotton. *Hibiscus* has a five-valved capsular fruit; several species form an ornament to our gardens. The young mucilaginous capsule of *Hibiscus esculentus* furnishes a stew, very much liked in America. The Hollyhock (*Althœa rosea*), when well grown, is a noble flowering plant. *Malope*, *Sida*, and the finely-pencilled-flowered *Abutilon*, are members of the large family of the *Malvaceæ*, which abound throughout the tropics, and also in Europe; they are interesting in many respects. The uniform properties of the order are, an abundant mucilage, and total absence of deleterious properties. The Mallows and Marsh-mallows of Europe yield a tasteless, colourless decoction, which is salutary in coughs and other local causes of irritation. The flowers of the Hollyhock are used in Greece for the same purpose; various species of *Sida* are used as emollients. The bark of many, as *Malva crispa*, and several species of *Hibiscus*, yield a strong fibre suitable for cordage. The leaves of the Hollyhock (*Althœa rosea*), is said to yield a dye not inferior to indigo. There are four species of Cotton-plant cultivated in America, namely, the Nankeen Cotton, supposed to be derived from *G. herbaceum*, and possessing naturally the yellowish colour which distinguishes it; the Green-seed Cotton, producing white cotton and green seeds, appears to be a variety of *G. peruvianum*. The variety cultivated on the low sandy islands lying between Charleston and Savannah, and known as the Sea Island Cotton,

Fig. 395.—Inflorescence of the Lime tree (*Tilia*).

celebrated for its long staple and the high price it produces in the market, belongs to *G. barbadense*. *G. arboreum*, the Tree-cotton, cultivated in India and Africa, produces some of the varieties cultivated in America.

The TILIACEÆ are trees or shrubs, rarely herbaceous. The Limes, the types of the order, are large trees with light white wood; their leaves are broadly ovate, oblique, acuminate, dentate, pubescent, or glabrous. They are alternate, distichous, and furnished with caducous stipules. The flowers have the remarkable character, that they are disposed in axillary few-flowered corymbs with a peduncle, which in its lower half is consolidated with a whitish membranous bract.

The Common Lime (*Tilia europæa*), PLATE XV., usually called the Linden, is widely disseminated through Europe; it is rarely found beyond the altitude of 350 feet above the level of the sea. It is planted, for its elegant and graceful form and delicious aroma, in parks and public promenades. Its buds are velvety, and its mature leaves are slightly pubescent on their lower surface. The calyx consists of five lanceolate sepals, the corolla of five petals longer than the sepals. The stamens are hypogynous and very numerous, free or irregularly polyadelphous at their base. The ovary is free, and generally has five cells, each with two anatropal ovules. The style is simple, and the stigma five-lobed. The fruit is coriaceous, indehiscent, unilocular by the disappearance of the dissepiments, and only contains one or two seeds by abortion. The seeds each enclose a fleshy albumen and an embryo, with foliaceous rolled-up cotyledons. The flowers of the Lime contain a volatile oil, with sugar, mucilage, gum, and tannin. They are employed in medicine as antispasmodics. Of all indigenous plants the liber of the Lime yields the strongest fibre; it supplies the bast used by gardeners. The wood, which is easily wrought, is employed chiefly by joiners and carvers.

Sparmannia africana, a pretty shrub from the Cape of Good Hope, has evergreen leaves, white flowers disposed in umbels. The leaves of *Corchorus oltorius* are used in Egypt as a pot-herb. Fishing lines, nets, and a coarse bagging are made from the fibre of Jute (*Corchorus capsularis*) in India. In general terms, the order may be said to abound in plants both useful and ornamental.

ALLIANCE XII.—SAPINDALES.

Flowers monodichlamydeous, unsymmetrical; placentas axile; calyx and corolla imbricated; stamens definite; embryo with little or no albumen.

Flowers complete, partially symmetrical; calyx valvate; anthers 2-4-celled, opening by pores.	I. Tremandraceæ.
Flowers complete (irregular), unsymmetrical; petals naked; anthers 1-celled, opening by pores; seeds caruneulate.	II. Polygalaceæ.
Flowers complete, unsymmetrical, very irregular; petals naked; anthers opening longitudinally; carpels 3; seeds winged.	III. Vochyaceæ.
Flowers complete, partially symmetrical; calyx imbricated; ovules ascending; stigmas simple; leaves opposite with stipules.	IV. Staphyleaceæ.
Flowers complete, unsymmetrical; petals usually apendiculate, or wanting; anthers opening longitudinally; carpels 3; seeds usually arillate, wingless.	V. Sapindaceæ.
Flowers apetalous; carpels solitary.	VI. Petiveriaceæ.
Flowers complete, unsymmetrical; petals naked or wanting; anthers opening longitudinally; carpels 2; seeds without an aril.	VII. Aceraceæ.

XV.—The Lime Tree (*Tilia europæa*).

Flowers complete, partially symmetrical; calyx imbricated; petals naked, stalked; ovules hanging by cords; stigmas simple; embryo usually convolute.	VIII. Malpighiaceæ.
Flowers complete, partially symmetrical; calyx imbricated; petals apendiculate; ovules sessile, pendulous; stigmas capitate; embryo straight.	IX. Erythroxylaceæ.

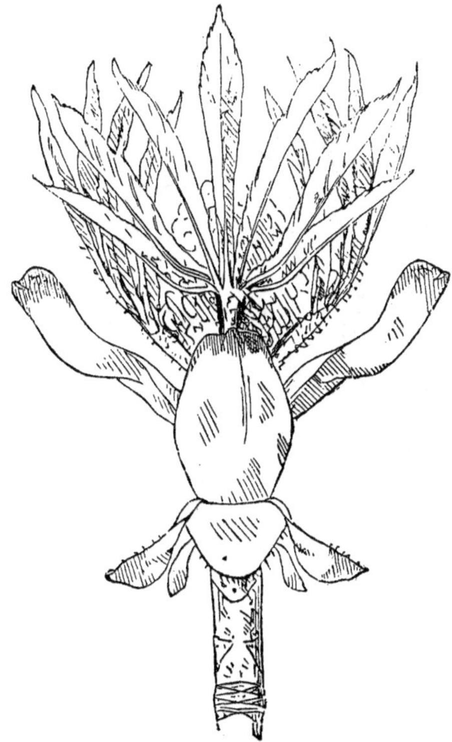

Fig. 396.—Opening Bud of Horse Chestnut.

In skimming over the surface of the vegetable world, as our limited space compels us to do, we necessarily leave unnoticed a vast number of genera interesting alike for their properties, for their beauties, and for the physiological peculiarities they present.

The SAPINDACEÆ are natives, for the most part, of the tropics, especially of South America, India, and Africa. While many species yield most delicious fruit, the leaves of others are deadly poisons, even where the fruit is safe and delicious. Of *Sapindus senegalensis*, the seeds are known to be poisonous to man and beast.

Fig. 397.—Panicle of the Sycamore.

The ACERACEÆ are trees of temperate countries in the Northern Hemisphere. The Sycamore (*Acer pseudo-platanus*), is a species of Maple (Fig. 397.) Different species of *Nephelium* produce fruits of great delicacy in the Indian Archipelago. The fruits of the Buckeye, or American Horse-Chestnut (*Æsculus ohiotensis*), which belongs to this order, are said to be poisonous; and the seeds of the Common Horse-Chestnut (*Æsculus Hippocastanum*), although they are said to be excellent food for sheep and deer, are not wholesome when raw for man, although when roasted they have been used as a substitute for coffee, in common with those of the yellow Iris of our marshes.

ALLIANCE XIII.—GUTTIFERALES.

Flowers monodichlamydeous; placentas axile; calyx imbricated; corolla imbricated or twisted; stamens numerous; embryo with little or no albumen.

Leaves simple, alternate, with large convulute stipules; flowers symmetrical; petals equilateral; calyx unequal, permanent, winged; anthers beaked; fruit one-celled, one-seeded.	I. Dipteraceæ.
Leaves simple, alternate, without stipules, or with very small ones; flowers symmetrical; petals equilateral; anthers versatile; seeds few or single; stigmas on a long style.	II. Ternströmiaceæ.
Leaves digitate, opposite; flowers symmetrical; petals equilateral; stigmas sessile; seeds solitary; embryo with an enormous radicle.	III. Rhizobolaceæ.
Leaves simple, opposite, exstipulate; flowers symmetrical; petals equilateral; anthers adnate, beakless; seeds solitary or few; stigmas sessile, radiating.	IV. Clusiaceæ.
Leaves simple, alternate, exstipulate; flowers unsymmetrical; petals equilateral; anthers versatile; seeds indefinite, minute; stigmas sessile.	V. Marcgraviaceæ
Petals oblique, glandular; seeds indefinite, naked; styles long, distinct.	VI. Hypericaceæ.
Petals oblique, glandless; seeds few, shaggy; styles long, distinct.	VII. Reaumuriaceæ.

Dr. Wight describes the DIPTERACEÆ as trees of majestic size and handsome form, and deserving of cultivation for the beauty of their clustered flowers, and the richly-coloured wings of their curious fruit. They are natives only of India and the Indian archipelago.

The CLUSIACEÆ or Guttifers are often valuable for their succulent juicy fruit, in many cases large, that of the Mangosteen (*Garcinia Mangostana*) being the most delicious. They are chiefly natives of tropical America, a few of Madagascar and Africa, and similar regions where great heat and moisture are combined. Gamboge is the secretion of a variety of plants, the kinds usually met with are Siam and Ceylon gamboge; the last is the inspissated juice of *Garcinia Cambogia*. The Coor gamboge is produced by *Garcinia elliptica*. Gamboge is one of the chief ingredients in Morison's pills. The name is derived from Cambodja, a part of Siam.

ALLIANCE XIV.—NYMPHALES.

Flowers dichlamydeous; placentas axile or sutural; stamens

indefinite; embryo on the outside of a large mass of mealy albu men (sometimes exalbuminous).

Carpels united into a many-celled fruit	I. Nymphæceæ.
Carpels distinct; albumens copious	II. Cabombaceæ.
Carpels distinct; albumens absent	III. Nelumbiaceæ.

The NYMPHÆACEÆ are floating plants, with peltate or fleshy cordate leaves, arising from the prostrate trunks growing in quiet waters. Their flowers are large, showy, and of bright white, yellow, red, or blue colours. They inhabit the whole of the northern hemisphere; occasionally they are met with on the South African coast, but generally they are rare in the southern hemisphere. In the South American Continent they are represented by the Victoria Lily. Among the tribe *Euryalidæ*, so named after one of the Gorgons, the tube of the calyx is adherent to the disc. *Euryale ferox* is an elegant aquatic, covered with prickles, with peltate orbicular leaves, and bluish purple or violet flowers. This species emulates *Victoria regia* in the size of its leaves, but has an insignificant flower. The *Victoria regia*, as we have seen in an earlier chapter, produces leaves six feet and a half in diameter, and flowers fifteen inches across. These inhabit the cool translucent lake-like rivers of Demerara, as illustrated in PLATE III. The tribe *Nupharidæ* have calyx and petals both distinct. The species are about twenty. The Blue Water Lily (*Nymphæa cærulea*), the sacred plant of the ancient Egyptians, is very fragrant. *N. edulis* contains abundance of starch in its root, and is an article of diet in India. *N. lotus*, the Egyptian Lotus, grows in slow-running streams and in the rice-fields in Egypt. It has large white flowers, with sepals, red at the margins: the seeds and roots were dried and made into bread by the ancient Egyptians. In *N. alba*, the common White Water Lily of our ponds and ditches, the flowers, according to Linnæus, open in the morning about seven o'clock, and close again on the approach of evening. The leaves of Water Lilies are large, succulent, and floating; the sepals and petals numerous, imbricated, and passing gradually into each other; the leaf is rounded oval, usually purplish beneath, the lobes at the base almost parallel, and the leaf-stalk cylindrical. In *Nuphar*, on the contrary, they are ovate, pointed, the basal lobes slightly divergent, and the leaf-stalk angular, especially on the upper part.

The NELUMBIACEÆ, or Water-beans, are natives of stagnant waters in the temperate and tropical regions of both hemispheres; chiefly remarkable for the beauty of their flowers. The fruit of *Nelumbium speciosum* is supposed to have been the Egyptian Bean of Pythagoras, and the flower that mythic Lotus so common upon the monuments of Egypt and India.

ALLIANCE XV.—RANALES.

Flowers monodichlamydeous; placentas sutural or axile; stamens indefinite; embryo minute, enclosed in copious fleshy or horny albumen.

Carpels distinct; stipules large, convolute; corolla imbricated; albumen uniform.	I. Magnoliaceæ.
Carpels distinct; stipules absent; corolla valvate; albumen ruminate.	II. Anonaceæ.
Carpels distinct; stipules absent; corolla imbricated; albumen uniform; seeds arillate.	III. Dilleniaceæ.
Carpels distinct; stipules absent; corolla imbricated; albumen uniform; seeds without an aril.	IV. Ranunculaceæ.
Carpels united; calyx persistent; placentas axile	V. Sarraceniaceæ.
Carpels united; calyx caducous; placentas usually parietal. . . .	VI. Papaveraceæ.

Ranales—so called from *rana*, "a frog," from many of the species inhabiting humid places usually the haunt of that animal—are characterised by the presence of a distinct calyx and corolla, sometimes so blended together, however, as to be indistinguishable; while in other instances there is no corolla, and occasionally both are absent. In general there is an indefinite number of stamens, but there are exceptions to that rule, as in the *Bocagea*, among the Anonaceæ, in which the stamens and carpels are definite.

The MAGNOLIACEÆ include some of the finest trees and shrubs in the world. The typical *Magnolia grandiflora* is an evergreen tree of North Carolina, which sometimes attains the height of seventy feet. Its ovate oblong coriaceous leaves—the upper surface of a pale green, shining and glossy, the under rusty, with white and erect flowers, with nine to twelve expanding petals—constitute it one of the noblest trees of the American forest. There are many species of the genus, of which this is the grandest example.

Other plants of the order are remarkable for the beauty of their flowers. *Liriodendron tulipifera*, the Tulip-tree, with half a dozen popular synonyms, is a handsome tree, with four-lobed truncate saddle-shaped leaves, and large elegant flowers, coloured green, yellow, and orange, sometimes attaining the height of 120 feet, and a circumference of twenty feet. Nor is the order deficient in officinal properties or fragrancy. Their general character is bitter and tonic in taste, with fragrant odorous flowers. *M. glauca* is so stimulating as to produce fever and even inflammatory gout, according to Barton. A species of *Michelia*, called Tsjampac, is the delight of the people of India for its fragrant properties.

The ANONACEÆ are trees and shrubs of the tropics in both hemispheres, whence they spread to the north and south within certain limits. Their general properties are their powerful aromatic taste and smell. The flowers of some are sweet and fragrant. Others, as *Anona squamosa*, have a heavy, disagreeable odour.

The DILLENIACEÆ are trees and shrubs of Australia, India, and tropical America, of comparatively little interest, though some of the Indian species are of great beauty. Dr. Wight speaks of them as equally remarkable for the grandeur of their foliage and the magnificence of their flowers. The plants generally are astringent.

The RANUNCULACEÆ, the typical order of the Ranales, are herbaceous, rarely shrubby, plants, and they are chiefly natives of Europe, with a sprinkling of North American, Indian, and African plants, on the shores of the Mediterranean; acridity, causticity, and poison are the general characteristics of the order, which includes a powerful sudorific in *Ranunculus glacialis*, a strong diuretic in *Aconitum Napellus* and *A. Cammarum*, drastic purgatives in the *Hellebores*, a virulent poison in some of the seeds of *Aconitum* and *Ranunculus Thora*, while many of the order are vermifugal and tonic.

To give any sufficient idea of this important family, it will be necessary to study successively the Columbine, the Hellebore, the Larkspur, Aconite, Ranunculus, Clematis, and Peony. Country people, struck with the form and elegance of the Columbine (*Aquilegia vulgaris*), have bestowed the name of "Our Lady's Gloves" upon its flowers (Fig. 398). Its petals are furnished with spurs, the shape of a hollow horn, hooked at their extremities; there are five of them, which alternate with as many flat and petaloidal sepals. The stamens are numerous, disposed in ten phalanges, five of which alternate with the petals and five with the sepals, which they exceed in length. The whole of the anterior face of the anthers is attached to the filaments, and they dehisce by two lateral clefts. Of these stamens ten are only represented by filaments dilated in the form of membranous scales of a silvery white, folded upon their edges and applied to the pistil. This organ is composed of five free unilocular ovaries, containing many anatropal ovules inserted in two contiguous vertical series. The ovaries, when at maturity, change into five free follicles. The seeds which they enclose contain a very small embryo, placed at the base of a horny and very abundant albumen. With regard to the organs of vegetation, the stems are erect, solitary, or more or less numerous, bearing numerous flowers, and branching in the upper portion. The radical leaves for the most part stalked; base of the stalk dilated; leaflets broad as long, and stacked with stalked divisions

Fig 398.—The Columbine (*Aquilegia vulgaris*).

of the first order; irregularly three-lobed stem leaves, a few with short stalks, the upper ones sessile; bracts narrow and three-lobed.

The Columbine is found growing in its primitive simplicity in hilly woods, and upon the borders of the forests of Bondy, Montmorency, St. Germain, Versailles, and in woods and copses where the soil is calcareous, in various parts of England, Scotland, and Ireland, though often naturalised. In the chalky copses of Kent and Surrey there is, perhaps, no doubt of its being indigenous.

In the *Black Hellebore*, or Christmas Rose, the sepals, five in number, are large, and the petals small and numerous, situated at the base of the andrœcium. The corolla of the Columbine (Fig. 399) may be said to be a calcarate polypetalous corolla; but culture has worked many curious modifications in the Columbine. We frequently see the five horn-like petals enclosing others, fitted in series, whilst five similar series are placed opposite the sepals. It seems, in fact, as if the flower was entirely composed of these series of horns contained one within the other. In other varieties, on the contrary, in place of the hollow spur or horn, we find only oval and almost flat petals, often in considerable numbers. As in all these cases, the stamens become fewer as the supernumerary petals increase in number, one is led to think that the formation of these petals arises in some way from the metamorphosis of the stamens.

Fig. 399.—Corolla of the Columbine. *s*, sepals; *p*, petals.

The Hellebores have a calyx with five sepals, a corolla having from five to ten petals, short, tubulous, and bilabiate, an indefinite number of stamens and pistils, varying from two to ten. They are slightly coherent at the base; and the ovary, like that of the Columbine, contains two series of ovules. The fruit also is a follicle. To give the reader a better idea of the aspect of plants of this tribe, we will take the Bearsfoot (*Helleborus fœtidus*), common enough in gardens, and occurring rarely in South of England, in stony places by the roadside, and in the glades of woods. It is a plant with a poisonous odour, with a thick, generally erect, vertical stem, terminating in a tap-root. The stem, which is evergreen during the winter, and ranging from one to three feet in height, is strong and erect, bare in its lower parts, leafy towards the summit, and divided into flower-bearing branches. The leaves are coriaceous, and of a deep green colour, are stalked, with lanceolate segments, narrow, dentate, and generally free to the base. The flowers are drooping, and disposed in branching corymbs. The sepals are concave, erect, and greenish, often edged with purple. The follicles are oblong, and terminate in a long beak or spur.

The Hellebores are all interesting to lovers of gardening, because for the most part they flourish during the winter; such, in particular, are the Black Hellebore, or Christmas Rose, and the little Yellow Hellebore, or Winter Aconite, known to botanists under the name of *Eranthis hyemalis*, which flourishes when the snow begins to melt.

Fig. 400.—Larkspur (*Delphinium*).

The Larkspurs (*Delphinium*) have a calyx with five unequal petaloid sepals, the upper with a straight, pointed, horn-like spur. The petals, which in some species are four in number, in others are reduced to only one by some process of abortion and consolidation—for originally there are always eight petals, six of which are developed in pairs opposite three of the sepals, while two are developed singly opposite the two other sepals. However that may be, the two superior petals in one case, and the single and superior petal in the other case, are prolonged into a pointed horn included in the spur of the calyx. The stamens, which are very numerous, are disposed in eight series, opposite to the eight original petals. The carpels, from one to five in number, are free, sessile, and verticillate, changing into follicles at a later period, and occupying the centre of the flower. The *Delphinium Consolida*, commonly known under the name of

Wild Larkspur, is frequently found at harvest time in cultivated fields round Paris, and other places, but in our own Flora it has only been found in corn-fields in the Channel Islands. Its stem is slight and straight, with numerous branches, its leaves are slender, cut into thong-like divisions, its short bunches of blue flowers form a panicle. The seeds of this species partake of the acrid and poisonous properties of the genus. The expressed juice of the petals, mixed with alum, makes a blue fluid ink. The form of the flower, with the spurred calyx, is represented in Fig. 400. Many beautiful species of Larkspur are cultivated in gardens, such as the *Delphinium elatum*, *Delphinium grandiflorum*, &c., species originally from Siberia; one of the most elegant is the *Delphinium Ajacis*, originally from the East and from Algeria; it is often met with in gardens, and has also been found in our corn-fields, where its seeds have been disseminated.

The Aconites (*Aconitum*) have five unequal petaloid sepals, the superior of which is formed like a helmet, or *galea*, lapping over the corolla. This latter organ is composed of from two to eight petals, of which the two upper form an elongated claw terminating in a reversed hood; whilst the lower, which are very small and filiform, are often absent altogether. The numerous stamens are disposed in series, as in the Larkspurs; and in the centre of the flower we find from three to five pistils, which become follicles. The Aconites are very poisonous narcotic herbs; but when applied with discrimination, they become eminently useful in medicine, being employed in cases of neuralgia, rheumatism, paralysis, and purulent infections. The most poisonous species is the *Aconitum ferox*, which is acrid in the highest degree. The *Aconitum Napellus* (Fig. 401) is another officinal species; it is rarely met with, and doubtfully native in the west of England, but is often found by tourists in the mountains of Switzerland. On the Jura range it grows to the height of three feet, having straight simple stems, the upper part slightly branching, furnished with leaves, shiny, of a deep green colour above, and of a pale green below, with five or seven segments, having oblong incised lobes. Its flowers, which are blue and of an elegant aspect, form elongated racemes, with two little bracts below each flower.

The *Ranunculus* (Fig. 402) has a green calyx, composed of five sepals; the corolla has five petals, furnished at the internal base of their claw with a nectariferous pore, which is covered by a scale. The stamens and pistils are very numerous; the former are of the ordinary structure, but the latter are disposed in a globular or oblong head; they have a short beak, and enclose a single ascending and anatropal ovule; after a time they become achænes. The British species have all white or yellow flowers. Many of them are submerged or floating plants, with slender stems, some of them with handsome showy flowers, sometimes covering the whole surface of the ponds or ditches. This is the case with *R. aquatilis*, the Water Crowfoot, whose dark green and rather rigid foliage is very conspicuous, and whose leaves, according to Dr. Pulteney, are not only innocuous, but nutritive to cattle, the cottagers on the banks of the Avon gathering them into boats, from which their cows eat them with avidity. Many species of this tribe possess blistering properties, that is to say, they produce an irritation which goes far towards the destruc-

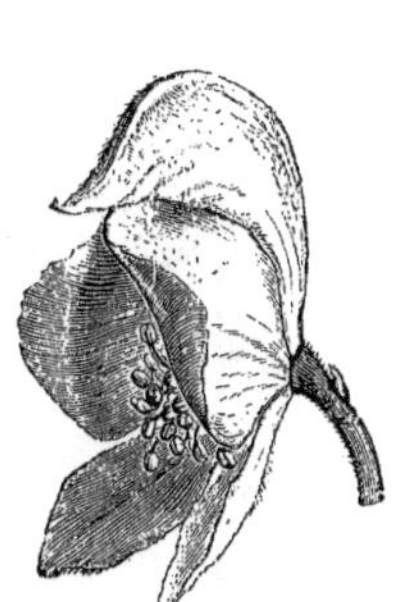

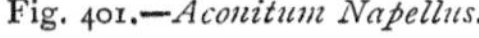

Fig. 401.—*Aconitum Napellus.*

Fig. 402.—*Ranunculus acris.*

tion of the epidermis, and the formation of a sore: the species used for these purposes are the *Ranunculus Flammula* (Lesser Spearwort); *R. Lingua* (Great Spearwort); *R. repens* (Creeping Crowfoot, or Golden Crowfoot). When the juice of these flowers is distilled, the liquid drawn from it contains a very acrid principle. Animals will not touch the Ranunculi when fresh, but when dried and used as hay their taste is lost. The Sweet or Wood Crowfoot (*R. auricomus*), or Golden-locks, differs remarkably from most of the tribe; it grows in tufts with numerous stems, very slightly branching above, its habitat, woods and moist shady places, and it is without the acridity so common to the tribe. The Buttercup, Kingcup, or Meadow Crowfoot (*R. acris*), in spite of its known acridity, has been a favourite with the poets. One sings of

"The kingcup of gold brimming over with dew,
To be kissed by a lip just as fresh as its own."

One variety of this species having become double, was a favourite in old gardens, and probably is still, under the name of Bachelors Buttons.

The Creeping Crowfoot (*R. repens*) is the Cuckoo-bud of Shakespeare.

> "When daisies pied and violets blue,
> And cuckoo-buds of yellow hue,
> Do paint the meadows with delight."

Next to *Ranunculus* we may also mention *Clematis*, *Anemone*, *Hepatica*, and *Adonis*.

The *Clematis* (Travellers' Joy, with several other popular names) has a calyx with four petaloidal divisions, and no corolla. The

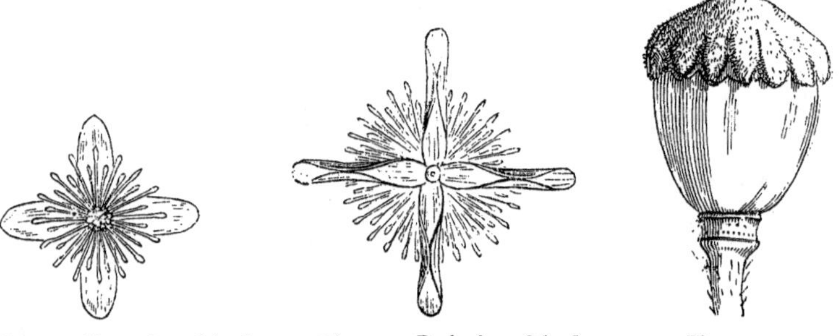

Fig. 403.—Front view of the flower of *Clematis vitalba*.

Fig. 404.—Back view of the flower of *Clematis vitalba*.

Fig. 405. Capsule of the Poppy.

stamens and carpels are numerous as in *Ranunculus;* their carpels are unilocular; after inflorescence they become achænes, with seeds reversed, and surmounted by a sort of plumose tail, resulting from the growth of the style. They are climbing shrubs. The leaves, when pulled and laid upon the skin, produce inflammation, vesication, and ulceration, properties which mendicants sometimes use to produce artificial ulcers—hence one of their common names, Beggar's Herb. Its flowers are white, disposed in axillary panicles. The name of "Virgin's Bower" was given to this species by old Gerarde in 1597, "by reason," as he tells us, "of the goodly shadowe which they make with their thick bushing and climbing; as also for the beauty of the flowers, and the pleasant savor or scent of the same." And, indeed, there is no finer ornament in our English hedges than this pretty bush, with its copious clusters of white blossom. Fig. 403 is a front view of the flower, Fig. 404 a back view of the same flower. We meet some times on lawns and gardens with *Clematis Flammula* (Sweet Clematis), where it is used to ornament palisades and bowers, and where it is distinguished from the preceding species by its sepals, which are covered with a hairy down only at the edges. We frequently see clumps of *Clematis viticella*, where the sepals are violet, purple, or rose colour, the flowers of which are produced double by cultivation. We sometimes see beside this species the *Atragene* of the Alps, remarkable for the beauty of its large flowers of violet blue. The genus *Atragene* is distinguished from *Clematis* by the existence of a corolla, composed of numerous petals shorter than the sepals.

Peonies (*Pæonia*) have the calyx foliaceous, coriaceous, persistent, with unequal sepals; the corolla is composed of five, six, or ten nearly equal petals; the pistils are variable in number, and enclose a great number of ovules; the fruit is a coriaceous follicle; the receptacle is swollen into a fleshy disc, which forms a sort of enveloping sack. The seeds are furnished with a slightly-developed aril rising out of the placenta which surrounds it. The Peonies are herbaceous or somewhat shrubby perennial plants, with alternate leaves; they are among the earliest ornaments of our gardens in the spring-time. In the Chinese Tree Peony (*P. Moutan*) the petals are sometimes white, marked at the base with a purple shade, and sometimes rose-coloured. This Peony has become double under culture; by cultivating this species for fifteen centuries the Chinese have obtained 200 varieties. It has been introduced into France since the commencement of the present century. We may also note the Common Peony *Pœonia officinalis*) with red, rose-coloured, or variegated petals, the flowers of which are easily grown double.

PAPAVERACEÆ.—The Poppies have a calyx with two caducous sepals, a corolla of four petals, with numerous stamens, provided with a long filament, whose anthers open laterally by two longitudinal clefts. The unilocular pistil is almost entirely divided by many placentary plates, which, emanating from the walls, advance nearly to the centre, bearing a number of anatropal ovules inserted over their whole surface. The Poppies are herbs with a white milky juice, with dentate leaves; the leaves emanating from the root are petiolate; those issuing from the stem are sessile, or embracing. Their peduncles, solitary, unifloral, droop before inflorescence.

1. The Corn Poppy (*Papaver Rhœas*) is very common in

corn fields, cultivated ground, and by roadsides, and forms, along with the Corn Centaury, their most graceful ornament, although it is the last plant the farmer likes to see in his fields. Its mucilaginous bitter petals are emollient and slightly narcotic.

2. The Poppy of the Levant, or Tournefort's Poppy, has scarlet or orange-coloured petals, with black and purple-coloured claws; it differs little from the Bracteated Poppy (*Papaver bracteatum*).

3. The Opium Poppy (*Papaver somniferum*, Fig. 406), of which there are two varieties: one, the White Poppy, so called because the seed is generally white, is the variety cultivated for the purpose of extracting opium; the other, the Black Poppy, because the seed is black, furnishes a sweet oil, known under the name of poppy oil. This seed is given to cage birds, under the name of maw seed. The oil is not narcotic, but perfectly wholesome, and is extensively used in India as a substitute for olive oil. Opium is the inspissated juice of the White Poppy, which flows out from incisions made in the capsule a little before ripening. The history of the Opium Poppy is very obscure as to the date of its first cultivation, but it was well known to the Greeks, and was cultivated for its seeds, as we learn from Theophrastus. It is also described by Arab authors. The White Poppy is supposed to be a native of Asia Minor, or of Persia, but it is nowhere found in a wild state. It is distinguished by obovoid or globular capsules, which, as well as the calyx, are smooth; the stem smooth and glaucous; leaves embracing the stem, and incised and obtusely dentate. The white-flowered variety Dr. Royle describes as cultivated in the plains of India, while the violet flowers he has only seen in the Himalaya mountains.

The Celandine (*Chelidonium majus*) belongs to the family of *Papaveraceæ*. It is a perennial plant with a reddish-yellow juice, which has a reputation for the cure of warts. The stems are straight, branched, and pubescent, being covered with long patulous hairs scattered over it. The leaves have from three to seven lobed oval segments, the lobes notched and crenulate, glaucous below. The flowers, which are yellow, are disposed in simple umbels, differing remarkably from those of the Poppy in the structure of the pistil. This organ, in reality, is composed of a unilocular ovary, having only two parietal placentas, surmounted by a short style with a bilobed stigma. The fruit is a linear capsule, and opens with two valves, which detach themselves from the base towards the summit, leaving a frame formed by the placentas. The seeds seem,

Fig. 406.—The Opium Poppy (*Papaver somniferum*).

besides other remarkable peculiarities, to be provided with a little white cellular excrescence like a crest.

Of this family we may also mention the *Eschscholtzia californica*, a perennial plant with a large solitary flower of a golden-yellow colour, closing itself up when it rains, the sepals of which, coherent at the base, detach themselves in one piece, after the fashion of a little pointed hat.

ALLIANCE XVI.—BERBERALES.

Flowers monodichlamydeous, unsymmetrical in the ovary; placentas sutural, parietal, or axile; stamens definite; embryo enclosed in a large quantity of hard fleshy albumen.

Flowers regular and symmetrical; placentas parietal; stamens alternate with the petals, or twice as many.	I. Droseraceæ.
Flowers irregular and unsymmetrical; placentas parietal; stamens opposite the petals.	II. Fumariaceæ.
Flowers regular, symmetrical; placentas sutural; stamens opposite the petals, with re-curved valves.	III. Berberidaceæ.
Flowers regular, symmetrical; placentas axile; stamens opposite the petals, anthers opening longitudinally.	IV. Vitaceæ.
Flowers regular, symmetrical; placentas axile and parietal; stamens alternate with the petals; ovules ascending or horizontal; corolla imbricated.	V. Pittosporaceæ.
Flowers regular, symmetrical; placentas axile; stamens alternate with the petals; ovules pendulous; corolla valvate.	VI. Olacaceæ.
Flowers regular, symmetrical; placentas axile; stamens alternate with the petals if equal to them in number; ovules pendulous; corolla imbricated.	VII. Cyrillaceæ.

The DROSERACEÆ, so called from δροσερός "dew," are mostly delicate herbaceous marsh-plants; some of them, as *Dionæa muscipula*, are distinguished by the irritability of the leaves when touched by a passing insect, which close upon it suddenly, and hold it a fast prisoner. The leaves of the order are usually covered with glands or glandular hairs, the flowers often arranged in circinate cymes. The calyx consists of five sepals; there are five petals, and five or more stamens; the fruit is capsular, one-celled, and contains minute seeds, having an embryo lying at the base of abundant albumen. The Sundews (*Drosera*) are remarkable for their singular red-coloured glandular hairs, which discharge a viscid acrid fluid in which insects are caught. Several of the foreign species have the reputation of being poisonous (*D. communis*) to sheep. *D. lunata* has viscid leaves with glandular fringes which close upon insects happening to touch them, and would probably yield a valuable dye. *D. rotundifolia*, the round-leaved Sundew, has the leaves close to the ground, nearly circular, and spreading, with a roundish limb abruptly tapering into a hairy petiole, the stem erect, springing from the centre of the leafy rosette. It is acrid and caustic; and in Italy the liquor called Rossoli is distilled from its juices. It curdles milk, and is said to cure corns and warts. There are about one hundred species of *Drosera* found in boggy places all over the world; most abundant in temperate Australia.

The FUMARIACEÆ, or Fumitories, are herbaceous plants, with slender brittle stems and twisting leaf-stalks, yielding a watery juice. The leaves of *Fumaria officinalis* are succulent, saline, and bitter; and the expressed juice is recommended in cutaneous and other diseases, to correct acidity. They are named from the word, *fumus*, smoke, because, according to Pliny, they had the same effect in producing a flow of tears from the eyes.

The BERBERIDACEÆ, or Barberries, are shrubs, or herbaceous perennials, with pale green, thin, deciduous leaves; flowers with faint pleasant odour; fruit and leaves with an agreeable acidity. The flowers are usually ternary, there being three or six sepals, and a like number of petals and stamens. The stamens are opposite the petals. They are characterised by the anthers opening by reflexed valves, the face of each anther-cell peeling off except at the point of attachment, where it adheres as if hinged; carpel solitary, free, and one-celled, style rather lateral, stigma orbicular; fruit, a berry, in some species a capsule. The peculiar arrangement and structure of the anthers occur in no other plants except the Laurels, and in the latter the flowers are without petals. This structure is well represented in Fig. 407, which exhibits an anther (*Berberis vulgaris*) in the act of dehiscence, which it accomplishes by the opening of two valves, *v v*, from which the pollen is shed in the form of fine dust, *p*. The bushes or herbs of which this order consists are extremely dissimilar in habits and appearance. They belong to the temperate parts of the world, being unknown in the tropics, except on mountain ranges. The Common Barberry (*Berberis vulgaris*) is well known from its pleasant acid flavour, its pinnate leaves, the leaflets being reduced to one, and the primary leaves to spines. The spiny leaves and stems are represented in Fig. 408, which shows the tripartite spiny leaves, in the axils of which buds are developed, bearing true leaves. It is also remarkable for the irritability of its stamens; these, when the filament is touched on the inside with the point of a pin in dry weather, bend forward towards

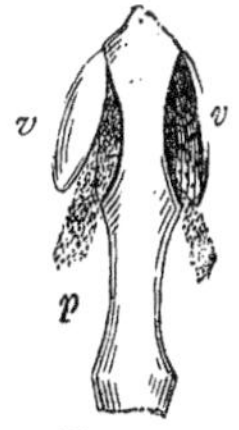

Fig. 407.

the pistil, touch the stigma with the anther, and remain curved for a short time, and then partially resume their erect position. In wet weather, when the filaments have lost their elasticity, the phenomenon is scarcely perceptible. The same result attended the experiment of applying corrosive sublimate to the filaments; they became rigid and brittle, and lost their irritability. On the other hand, on the application of narcotics, as prussic acid or belladonna, the irritability was destroyed by the filament becoming flaccid and relaxed.

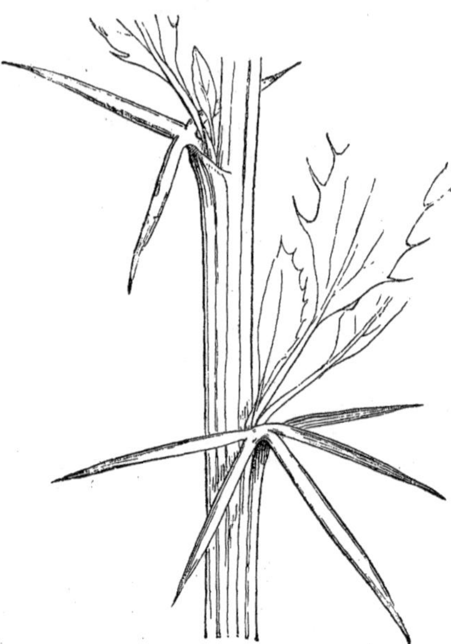

Fig. 408.—Stem and Foliage of the Barberry.

The Barberries are interesting from their graceful elegance, both in their racemes of yellow flowers and ovoid delicate berries. Their bark and root are used for dyeing leather and linen of a yellow colour.

Farmers in England have an idea that the proximity of the Barberry to wheat-fields interferes with their crops. It is difficult to imagine why they think so, unless the *Æcidium berberidis*, the fungus which is known to infest it, acts injuriously on the wheat crop, by infecting it with rust or mildew.

The VITACEÆ consist of sarmentose and climbing shrubs, with filiform twining tendrils; the Grape being the type of the order, which Kunth names *Ampelideæ*, from ἄμπελος, the vine; and Jussieu, *Vites*, from *vitis*, a vine. The order has acid properties, in common with Grossulaceæ and Berberidaceæ, and includes six genera, which are for the most part found in the temperate zone of both hemispheres. Like all extensively cultivated plants, the native country of the Vine (*Vitis vinifera*) is doubtful. It is among the plants spoken of in the Books of Moses, when the Vine appears to have been used as it is in the present day. There is little doubt of its being truly indigenous in Asia between the Black Sea and the Caspian, Mount Ararat and the Caucasus. In the forests of Mingrelia and Imeretia it flourishes in great magnificence, climbing to the tops of the highest trees, bearing bunches of fruit of delicious flavour. The Vine is found growing wild in many parts of France and Germany, and as far north as the 55th parallel, and also all over the South of Europe and Asia Minor, and southern latitudes as high as 40°, but it may be doubted if it is indigenous in any part of Europe.

According to Gibbon, "In the time of Homer, the Vine grew wild in the island of Sicily, and most probably in the adjacent district of the continent; but it was not improved by the skill, nor did it afford a liquor grateful to the taste of the savage inhabitants. A thousand years afterwards, Italy could boast that of the fourscore most generous and celebrated wines, more than two-thirds were produced from her soil. The blessing was soon communicated to the Narbonnese province of Gaul; but so intense was the cold to the north of the Cevennes, that, in the time of Strabo it was thought impossible to ripen the grape in those parts of Gaul. This difficulty, however, was gradually vanquished, and there is some reason to believe that the vineyards of Burgundy are as old as the age of the Antonines."—("Decline and Fall," vol. i., p. 69.)

The Sabine farm, of which Horace said,

> Angulus iste feret piper et thus ocyus uva.—Ep. i., 14, 23.

is altogether changed, for Vines hang in festoons from tree to tree over the site of his abode, supposing Mr. George Dennis is correct in placing the farm in the valley of the Digentia, and the Fons Bandusiæ in the narrow glen which opens just beyond it.

In Middle Germany the Vine ceases to grow at 1,500 feet above the level of the sea; south of the Alps it reaches 2,000 feet; on the Apennines and in Sicily it grows at 5,000 feet, and in the Himalayas it reaches 10,000 feet above the level of the Indian Ocean. In England the Vine was probably introduced by the Romans, for there are indications that vineyards were planted in the third century by the Emperor Probus which still existed in the eighth century, according to the Venerable Bede. William of Malmesbury, writing in the twelfth century, commends the vineyards of Gloucestershire; and we have read somewhere that the remains of vineyards are not unusual in the south and west of England. Although recent attempts have been made, on the system of cultivation introduced by Hoare, to establish vineyards of considerable magnitude in Wiltshire and elsewhere, we are not aware that they have prospered. This arises from the shortness of our summers. England has a mean temperature as high as that of many countries where the Vine flourishes in perfection, but there is a want of heat and sun in the months of September and October, at which time the Vine is ripening its fruit.

The actual limits of cultivation in Europe have been more exactly defined in the "Atlas de Physique Végétale" of M. H. Tricholet, who devotes a map to the limits of Vine culture on the globe. In this map a red line, drawn from Cape Finisterre along the coast of Spain, across France, leaving Paris a little to the north, across Germany, the northern shores of the Black Sea, to the Caucasus and Caspian Sea, forms the northern limits of cultivation for wine; the Mediterranean, the Sea of Marmora, and the southern shores of the Black Sea being its southern limits. A bold dash of green colour, which covers the whole of the western hemisphere, from 30° south latitude to the parallel of New York, and the west coast of Africa, indicates the range of the wild grape; while a delicate yellow tint, extending over a great part of Arabia, Asia Minor, and south and east of the Caspian as far as China and the seas of Japan, marks the range of country over which the grape is cultivated for its fruit only, either dried or in its green state.

The leaves of *Vitaceæ* are slightly heart-shaped, having two rounded lobes at the base; the Vine throws out tendrils, or branches, at the point of insertion of the leaves, by the aid of which it creeps along the object on which it is placed. The flowers of the Vine are disposed in a close panicle, very small and greenish, with a sweet odour (Fig. 409), which in spring perfumes the fields in the South of France. The calyx of the Vine is small and very short, composed of five teeth, scarcely visible. The corolla is composed of four or five petals, inserted on the summit of the disc, turning inwards at the edge in æstivation in a somewhat valvate manner; the apex of each petal is often inflexed and adherent to the others, so that the corolla is thrown off from below in one piece. To the petals are opposed four or five stamens, with free filaments, having bilocular anthers opening by two longitudinal clefts, and attached dorsally. From the centre of the flower rises a free ovary, surrounded at the base by a glandular disc, which is surmounted by a simple stigma, sessile and flat at the summit. This ovary is superior, and two-celled, each cell enclosing two collateral anatropal ovules, ascending from the base of the cell. The ovary ultimately becomes a globular berry, as in Fig. 410, which contains, when it does not prove abortive, four seeds or pips. These seeds, the shells of which are horny, enclose a very small embryo, erect in the axis of a very abundant fleshy albumen.

The fruit of the Vine is a grateful article of diet in various conditions. Muscatel grapes, dried—which is effected by cutting half through the footstalk while suspended from the tree—form one of our finest dried fruits; and currants, or corinths, are the dried fruit of a Vine which grows in Zante and Cephalonia.

Fig. 409.—Flower of the Vine.

Under the general name of wine is designated the juice of the grape when it has been subjected to fermentation, and we need not trouble ourselves to express here how important the grape harvest becomes to many countries. In France nearly 5,000,000 acres are planted with vines.

There are innumerable varieties of the Vine produced by cultivation and cross impregnation. Great differences are the result of soil and situation; gently-sloping hills, well isolated, with a south and westerly aspect, upon which the sun's rays rest the whole day, are the best localities; and the influence of temperature is such that it is quite usual to obtain upon the same hill grapes of the most opposite qualities, according to the different heights. As to the influence which the composition of the soil exercises, it appears to be more a question of bouquet than of quality in the wine. In short, excellent wines are produced from the soils of the most opposite quality. The best *crus* of Burgundy are the produce of an argillaceous chalky soil; those of Champagne proceed from a soil eminently calcareous. The vines of the Hermitage ripen their fruits upon the crumbling *débris* of granitic rocks; those of Châteauneuf upon a silicious soil. An unctuous sandy soil produces the wines of Graves and Medoc; while a schistose soil produces the wines of Lamalgue, near Toulon.

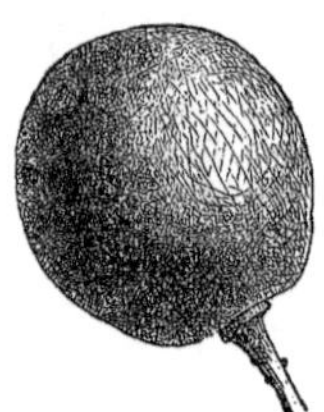

Fig. 410.—The Grape.

It is very important to choose the suitable manure on which to feed the Vine, which is essentially a coarse-feeding plant. Soils which are too energetic produce quantity at the expense of quality. Where the grape is high-flavoured, it alters the aroma of the wine. The nourishment most appropriate to the Vine is without smell, and of slow decomposition, such as woollen rags, clippings of horn, and such refuse. Its own ashes constitute an excellent manure, restoring to the vine the salts of potash and other mineral matters which were drawn from it in the previous years' growth.

The composition of the grape is sufficiently complex. It consists of the following substances, viz: water, lignine, glucose, pectine, tannin, albumen, essential oils, colouring matter (yellow, blue, and red, the first occurring only in white grapes), fatty matter, salts of lime and of potash, iron oxides and silica. Amongst these various substances the important and widely-distributed grape sugar, or *glucose*, which produces alcohol by its fermentation or chemical decomposition, plays the most important part in the act of converting the juice of the grape into wine.

It is only when the grape is thoroughly ripe that the vintage should take place, if it is desired to have wine of a good quality. Where the properties are enclosed, perfect ripening can be waited for; but in most vineyards of France the harvest is pushed on by the *ban de vendange*, which is fixed by the local authorities, acting under a council of vine-dressers.

The various operations which follow the vintage are reduced to four:—

1. Fullage of the grapes.
2. Fermentation of the *must*.
3. Decuivage.
4. Pressing.

Fullage is the process of dividing and crushing the fruit, exposing the juice momentarily to the action of the air, and bringing the fermenting principle into contact with the saccharine matters.

Formerly the operation of crushing the grapes was performed by the vine-dresser stamping upon them with the feet; now the grapes are crushed by passing them between two cylinders, channelled, and turning in reverse directions.

When the grapes are crushed effectually, they are given up to fermentation. In proportion as fermentation advances the temperature of the mass increases. The best fermenting temperature lies between 54° and 64° F. The mass originates much carbonic acid gas, which brings to the surface part of the stem and husk of the grape, and forms a thick sort of covering to the liquid mass, which is called the *chapeau*. The fermentation, which is well developed on the second day's Encuivage, is continued up to the eighth day, its state being indicated by the gas ceasing to be disengaged, and by the colour of the liquid (from red grapes), which takes a fine vinous tint, the alcohol, which is now present, having dissolved the colouring matter contained in the pellicles of the grape.

When Decuivage is about to take place, the liquor is drawn off, by means of a tap at the bottom of the vat, into casks, which are filled only to the fourth or fifth of their capacity, and which are then left open, in order that fermentation may continue slowly, and the disengagement of gas still proceed.

The mass remaining in the tun after the liquor is drawn off is subjected to pressure. The liquor which flows under this pressure, however, cannot be wine of equal quality. This is the process by which red champagne is produced.

White wine can always be obtained from red grapes. To effect this, in place of leaving the must to ferment with its residuum of husk, the liquor is drawn off as it is pressed, and fermented in separate tuns. As the colouring matter of the wine only exists in the pellicle of the grape, we can readily conceive that, being separated at once from the must, little or no colour is communicated to the liquid.

The wine preserved in the tuns, as we have said, ferments slowly by this second fermentary process, the liquid is clarified, the foreign matter is deposited, and forms the lees, which accumulate at the bottom of the tun. In order that the wine may retain its good quality, it becomes necessary to draw it off; that is, to separate it from the lees. In the months of March and April the wine is thus drawn off. If it is not quite limpid, recourse is had to clarification, or Collage. This operation is intended to make the wine pure and limpid, and divest it of the fermenting principle which is still held in suspension, and which might produce renewed fermentation. The fining for red wine is made of white of eggs, of blood, or of gelatine. The albumen or gelatine of these substances combining with the tannin dissolved in the wine, forms a precipitate, that is, a substance insoluble in the liquid, which is slowly deposited in the bottom of the tun, drawing with it all other foreign substances held in suspension in the liquor.

The sparkling wines of Champagne are prepared by special processes, which require more particular description. The greater part of these wines are made from the red or purple grape, the juice of which is generally richer in grape sugar than the white. A first pressure of the grape yields the liquor which produces the whitest wine. The residuum being subjected to further pressure, furnishes the juice which gives the rose-coloured wines. The must, white or rosy, is then put into great tuns, in which fermentation is established. After twenty-four hours the must or wort is drawn from the tun into another, which is filled and closed. This wine is drawn and fined three times, at intervals of a month, and in the month of April it is bottled. At this time three to five per cent. of crystallised or candied sugar is added to the liquid. At the end of a few months this sugar produces fermentation in the bottle, which increases the richness of the wine in alcohol and carbonic acid gas. In consequence of the expansive force of the gas thus evolved, the bottles ought to be well corked, and the corks strongly secured with iron wire. Sometimes a few grains of rice are put into the bottle with the wine, which induce also the second fermentation we have described above. The pressure of the carbonic acid has the effect of bursting about ten per cent. of the number bottled; the consequence has been a special manufacture of bottles for the wines of Champagne, many of the manufacturers supplying bottles under a warranty that they will support any pressure up to a certain fixed point. The bottles remain full, and are placed in horizontal beds for six months, without being disturbed. But during this period fermentation has produced in the midst of the liquor a deposit which proves clearly enough that fermentation has been going on. It is necessary to remove this deposit, which would otherwise destroy the transparency of the liquid. This new operation, which is called "disgorging" (*dégorgement*), is one of extreme delicacy. The operator shakes the bottle, so as to detach the deposit from it, and replaces it gently in a vertical position, the mouth downwards. The deposit thus descends to the neck of the bottle; if it is rapidly uncorked in that position the pressure of the liquor expels the deposit. The great delicacy of this operation lies in the necessity of losing the least possible quantity of gas and wine. The Champagne cellerman performs the difficult operation with great address.

The Liqueurs of the Vine are those which after fermentation preserve a great part of their sugar. It is chiefly in the countries of the South of Europe—in Italy, Spain, and the South of France—that these wines are prepared. To obtain Tokay, for example, the proportion of sugar is increased by leaving the most choice grapes to get thoroughly ripe, even to the extent of being slightly touched with frost; and even by placing them, after being cut, upon frames to dry, in order that the water may evaporate in part, thus increasing the saccharine richness of the must.

Various other substances besides alcohol are formed during the fermenting of grape juice. Of these the bouquet, most abundant in the Clarets of Bordeaux, but varying in character in different wines, gives to them their distinctive odours. The vinous odour, however, common to all wines, is due to œnanthic ether, perhaps the most volatile constituent of wine. The alcoholic richness of wines is very variable, as the following table of the natural strength of wines will show:—

Port	20 per cent.	of alcohol.
Sherry, or Lachryma-Christi . . .	17 ,,	,,
Malaga and White Sauterne . . .	15 ,,	,,
Vin de Baume	12·2 ,,	,,
Volney, Rhine Wine, and Frontignan	11 ,,	,,
Tokay	9·1 ,,	,,

It is alcohol which gives to wines their intoxicating quality. Tannin gives them astringency, which is corrected by many finings, which withdraw a portion of the tannin in combination with the albumen or gelatine employed. The acids are acetic and tartaric acid, the last in the form of potassium bitartrate, or cream of tartar, which latter gives wine its tartness. Potassium bitartrate is soluble in water, but not in alcohol; its solubility in a mixture of the two depends upon their proportions. In the fermentation of wine, therefore, as the alcohol increases, there arrives a point when the tartar begins to crystallise out. As the cream of tartar is gradually deposited in the tuns or in bottles, it follows that the wines in time lose much of their acidity. In *aging* they lose also much of their colouring matter, and take the tint which is called *pelure d'oignon* in France.

The PITTOSPORACEÆ are chiefly trees or shrubs of Australia, Africa, New Zealand, Norfolk Island, China, Japan, and the adjacent islands. The berries of *Billardiera mutabilis* are edible;

the fruit is green and cylindrical, becoming a pale amber colour when ripe.

The OLACACEÆ, trees or shrubs, often spiny, are a small order consisting of tropical or hardy tropical shrubs of the East Indies, New Holland, and Africa, with one of the West Indies, and a few of the Cape of Good Hope.

The CYRILLACEÆ are evergreen shrubs, with simple, non-stipulate leaves, all inhabitants of North America, and of little general interest.

ALLIANCE XVII.—ERICALES.

Flowers dichlamydeous; ovary symmetrical; placentas axile; stamens definite; embryo enclosed in copious fleshy albumen.

Flowers polypetalous; stamens all perfect, monadelphous; anthers two-celled, with a long membranous connective.	I. Humiriaceæ.
Flowers monopetalous; stamens all perfect, free; seeds with a firm skin; anthers one-celled, opening longitudinally.	II. Epacridaceæ.
Flowers half-monopetalous; stamens all perfect, free; seeds with a loose skin; embryo at the base of the albumen.	III. Pyrolaceæ.
Flowers polypetalous; stamenshalf-sterile and scale-like, free; seeds with a firm skin.	IV. Francoaceæ.
Flowers half-monopetalous; stamens all perfect, free; seeds with a loose skin or wing; embryo at the apex of the albumen.	V. Monotropaceæ.
Flowers monopetalous; stamens all perfect, free; seeds with a firm or loose skin; anthers two-celled opening by pores.	VI. Ericaceæ.

The HUMIRIACEÆ are trees or shrubs yielding a balsamic juice, all natives of the tropical parts of America. The balsam yielded by *Humirium floribundum* is a yellow liquid, called Balsam of Umiri, resembling the Copaiva and Balsam of Peru.

The EPACRIDACEÆ, are small trees and shrubs, remarkable for the great beauty of their flowers and the singular structure of their leaves, which are alternate, rarely opposite, stalked, and sometimes entire or serrated, dilated at their base, overlapping each other and half sheathing their stem, without a midrib, but with veins radiating from the base. The flowers, generally monopetalous, are white or purple, borne in spikes or terminal racemes. All the fruit-bearing section, such as the Australian Cranberries (*Lissanthe sapida*), are esculent, but the seeds are large, and the pulpy covering too thin to be very available for food. The Tasmanian Cranberry (*Astroloma humifusum*) is found all over that colony; the fruit, generally greenish-white in hue, is sometimes slightly red, and about the size of a Black Currant, consisting of a viscid pulp, apple-flavoured, and enclosing a large seed; it grows singly on a trailing stem. The native Currant (*Leucopogon Richei*) is a large, densely-foliaged shrub, growing on the sea-coast to the height of seven feet; the berries small and white. A French naturalist, named Riche, who accompanied the expedition in search of La Perouse, was lost for three days on the south coast of Australia, and supported himself chiefly upon the berries of this plant.

The *Epacridaceæ* scarcely differ from the small-leaved genera of *Ericaceæ* either in habit or character, except that in the former the anthers are one-celled. Dr. Brown was the founder of the order; and his reason was that the family of the Ericaceæ is now so vast that it seems to constitute a class rather than an order. "I may, therefore," he says, "be allowed to propose another order, *Epacridaceæ*, which is truly natural, although it depends upon the single character of the unusual simplicity of the anthers—a character, however, which is of the greater value as being opposed to the two-celled anthers of *Ericaceæ*, which are generally divided and furnished with appendages. The propriety of the measure is moreover confirmed, not only by the number of *Epacridæ*, large as it is, but also by their geographical position, for all, as far as we know them at present, are inhabitants of Australia and Polynesia—countries in which not more than one or two species of Ericaceæ are found."

The Ericaceæ, or Heaths (Fig. 411), are small trees, shrubs, or under-shrubs, with rigid evergreen leaves, whorled or opposite, and without stipules. Some few species are natives of Europe; several of the genera, such as the Heaths, Lings, *Azalea*, *Andromeda*, and *Arbutus*, having representatives in the British Flora. But the country of the Heaths is the Cape of Good Hope, where immense tracts of country are covered with them. The Tree Heath (*Erica arborea*) belongs to the region of the Mediterranean, overshading all other Heaths by its height, which sometimes reaches to sixteen or eighteen feet. Its flowers are numerous, and their sweet odour diffuses itself to a great distance. The Broom-Heath, or Ling (*Calluna vulgaris*) takes its name from the humble use to which it is applied; it grows in woods, and in sterile and uncultivated lands.

The Heaths have regular flowers; the calyx is monosepalous, and divided into four parts; the corolla varies in form from globular, or pitcher-shape, to tubular, bell, or rather patera-shaped, presenting four lobes alternating with the divisions of the calyx. The corolla does not carry the eight stamens which compose the androecium; these are inserted upon the receptacle. The pistil is composed of a superior ovary surmounted by a very erect style, and a dilated stigma. The ovary is four-celled, and in the internal angle of each cell the placenta is placed, charged with anatropal ovules. The fruit is capsular, opening loculicidally. The seeds are oval, reticulated, having an erect embryo in a fleshy albumen.

Fig. 411.
Heath (*Erica cinerea*).

Along with the Heaths are ranged the Azaleas, Kalmias, and Rhododendrons, producing flowers unequalled for their beauty in the whole vegetable world. For richness of colouring, elegance and variety of form, delicacy of texture, and minute perfection of corolla, the Heaths stand unrivalled. In the northern parts of the island, Heather or Ling (*Calluna vulgaris*), mingling with other species, covers vasts tracts of country; its hues of purple, red, and pink having the most brilliant effect. A cultivated variety of this species, with double flowers, is extremely beautiful; and the flowers of *Erica carnea* are the earliest harbingers of spring.

The Strawberry-tree (*Arbutus Unedo*), whose ripe red fruit of the past year is borne in the autumn on the same branch which supports the drooping panicled cluster of flowers of the present, is one of the most agreeable objects in Nature.

The trailing *Azalea procumbens* of the Scottish mountains, with leafy branches, tortuous stem, and small elliptical leaves with revolute margins, five-parted purple calyx, and bell-shaped corolla, is a pleasing object on the hill-side. In the Western hemisphere, ranging from Canada to Georgia, *A. viscosa* grows from three to eight feet in exquisite beauty, with deliciously fragrant whitish tubular flowers. The finest of the American species is *A. arborescens*, described by Pursh as rising from ten to twenty feet, on rivulets near the Blue Ridge, in Pennsylvania, forming, with its elegant foliage and large abundant rose-coloured flowers, the finest ornamental shrub with which he was acquainted. The Chinese Azalea (*A. sinensis*) has downy leaves, glaucous underneath, silky flowers, bell-shaped, with broadly ovate

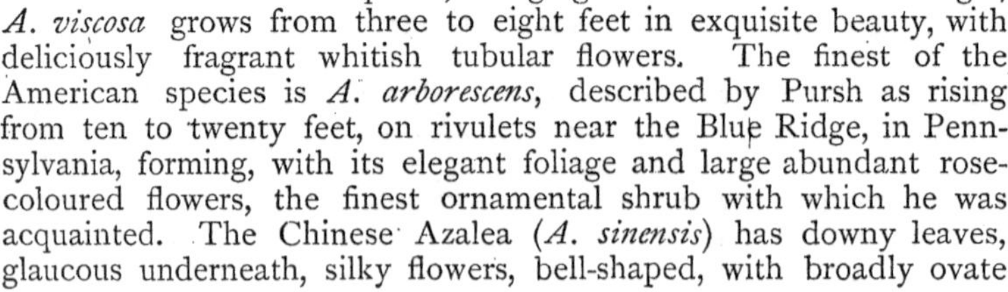

segments, wavy, and the upper one dotted after the manner of the Rhododendrons. *A. indica*, again, forms a bush two to six feet high, with drooping branches, leaves deep brownish green, and half evergreen, with large showy and brilliantly-marked flowers. But these sink into insignificance when compared with the gorgeous specimens which are produced in hundreds at our flower shows; pyramids, domes, and clusters, are there presented with flowers of every hue, in such profusion as to be almost oppressive, and one sighs for a glimpse of native heath as a relief from the gorgeous picture, though it should be but a hill-side covered with *Calluna vulgaris*.

Rhododendrons differ from Heaths chiefly in having ten instead of five stamens, and in the foliage being hard and evergreen. The first of the species introduced seems to have been from Asia Minor, whence *Rhododendron ponticum* was obtained; but the species introduced from the Sikkim Himalayas excel all others in beauty, and variety of form and colour.

The Kalmias are evergreen shrubs, with a small five-parted calyx, cup-shaped corolla with angular open limb, and a five-celled and many-seeded capsule; when in blossom their elegant flowers give them a beautiful appearance, but they are said to be poisonous. *Kalmia latifolia* yields a nectarous juice, eagerly sought after by bees and wasps, but the honey formed from it is said to be poisonous; and, swallowed by itself, it is said to have dangerously intoxicating properties.

Alliance XVIII.—Rutales.

Flowers monodichlamydeous, symmetrical; placentas axile; calyx and corolla imbricated; stamens definite; embryo with little or no albumen:—

Fruit consolidated, succulent, indehiscent; petals imbricated; stamens free, or nearly so; leaves dotted.	I. Aurantiaceæ.
Fruit consolidated, hard, dry, somewhat valvular; petals valvate; stamens free; leaves generally dotted.	II. Amyridaceæ.
Fruit consolidated, capsular; stamens deeply monadelphous or free; seeds numerous, winged.	III. Cedrelaceæ.
Fruit consolidated, berried, or capsular; stamens deeply monadelphous; seeds few, wingless; leaves dotless.	IV. Meliaceæ.
Fruit apocarpous; ovule single, suspended by a cord rising from the base of the carpel.	V. Anacardiaceæ.
Fruit apocarpous; ovules collateral, ascending, orthotropal, sessile	VI. Connaraceæ.

Fruit finally apocarpous, few-seeded, with the pericarp separating in two layers; ovules sessile, pendulous; flowers hermaphrodite.	VII. Rutaceæ.
Fruit finally apocarpous, few-seeded, with the pericarp separating in two layers; ovules sessile-pendulous; flowers polygamous.	VIII. Xanthoxylaceæ.
Fruit finally apocarpous, one-seeded, with the pericarp not laminating, and a succulent conical torus.	IX. Ochnaceæ.
Fruit finally apocarpous, one-seeded, with the pericarp not laminating, and a dry inconspicuous torus; albumen wanting; leaves alternate, without stipules.	X. Simarubaceæ.
Fruit finally apocarpous, few-seeded, with the pericarp not laminating, and a dry inconspicuous torus; albumen present; leaves opposite with stipules.	XI. Zygophyllaceæ.
Fruit finally apocarpous, many-seeded; flowers polypetalous. . .	XII. Elatinaceæ.
Fruit finally apocarpous, many-seeded; flowers apetalous, very imperfect.	XIII. Podostemaceæ.

The most important plants of the natural order AURANTIACEÆ are those belonging to the genus *Citrus*.

The Orange (*Citrus Aurantium*), so well known and highly appreciated as a fruit, comes from a fine evergreen tree, originally from Asia, and probably from Northern India. It is now largely cultivated in all the warm countries of the globe. Its glossy leaves are compound, with one leaflet, which is terminal on a winged leafy footstalk, and are oval or lanceolate, and entire. When examined through a strong light these leaves present little bright spots, which are minute cavities full of an odorous volatile oil. Its flowers, the elegance and delicate perfume of which are so well known, are composed of a somewhat bell-shaped calyx, a corolla with three to five petals, broad at the base, sometimes slightly combined, inserted on the outside of a hypogynous disc; the stamens double, or any multiple of the number, are united into several bundles. The fruit we need not describe; it is separated by membranous divisions into many cells (Fig. 412), containing seeds at their inner angle, it is filled with soft and juicy pulp, sweet and slightly acid.

There are numerous varieties of the Orange; the best are the China, the Maltese, the Lisbon, and St. Michael's, and, later in the season, an excellent variety comes from Valentia. The St. Michael's should be small, and flattened at the ends, with a thin smooth rind, the glands small, and the flesh a light coloured pulp without seeds; but many coarse varieties come to market. Those from Valentia resemble St. Michael's. Lisbon and Malta Oranges are larger, thicker in the rind, with glands; but the China Orange, when obtained of the best variety, excels all others for delicate flavour and aroma and sugary juice. The skin of this variety is always smooth and shining, and so thin that it is separated with difficulty from the flesh. The St. Michael's is probably a variety of this orange. The Bitter or Seville Orange is the fruit of *Citrus Bigaradia*, but there are many varieties. This is the plant from which the delicate condiment orange marmalade is prepared, and from which orange-flower water is produced.

The Orange is extensively cultivated in order to extract from its flowers and leaves the essential oil which they contain. In the

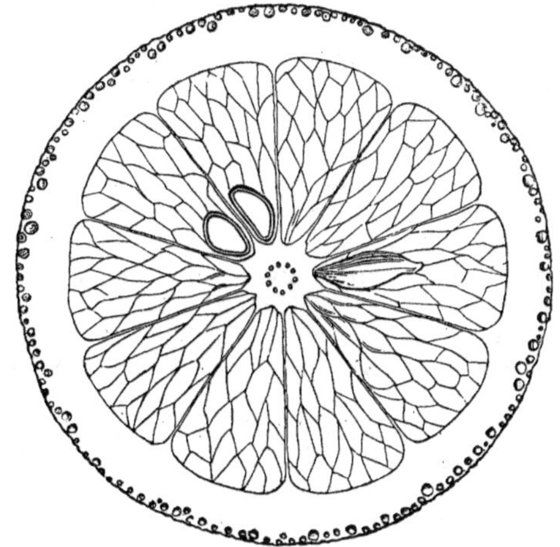

Fig. 412.—Section of an Orange.

South of France, but especially in Provence and Nice, the Orange is largely and successfully cultivated. In the South of Italy, about Sorrento, whole forests of Oranges exist, the fruit of which is carefully harvested. Lamartine sings—

> "On the sonorous shore, where the sea of Sorrento,
> At the foot of the Orange unrolls its blue wave."

The Orange sometimes attains great age and dimensions. In the orangery of Versailles a magnificent Bitter Orange (*C. Bigaradia*), familiarly called "the Great Constable," is known to be 450 years old; its trunk is thirty inches in circumference; it is growing with its roots in a large box. It was planted in 1421 by the gardener of the Queen of Navarre, and came from Chantilly. In 1532, Francis I., having confiscated all the property of the Constable of Bourbon, Lord of Chantilly, who had been driven into open rebellion, had the precious tree, which was quite unique in France, transported to his orangery at Versailles, where it remains in a highly flourishing state. The orange tree at the convent of S. Sabina, at Rome, dates from the year 1200. It is about thirty-three feet in height. At Nice there was in 1789, according to Risso, a tree which usually bore upwards of 5,000 oranges, and was more than fifty feet high, with a trunk which it took two men to grasp.

The substance to which odoriferous plants owe the qualities which render them so useful at the toilet is a volatile oil. It happens sometimes that distinct oils exist in the same plant. Let us take the Orange, for example. The essence drawn from the flowers of the Orange is very different from that furnished by the leaves. The essence furnished by the leaf differs again from that produced by the fruit. The volatile oil is contained in the vesicles or cells which pervade all its parts; and so completely are these enclosed, that the plants may be dried without divesting them of the odorous principle, which still remains in the cavities. In other cases, particularly in the flowers, the essence is volatilised at the rate at which it is produced.

The mode of extracting these essences varies according to their nature and condition. Some of them may be extracted by simple pressure. This may be done with the essential oils of the Citron and the Orange, which reside in the rind or envelope of the fruit. They are reduced to a pulp, adding water afterwards to the liquor produced by pressure, when the oil will swim on the surface. But the greater part of these essences are obtained by distillation. This process is performed by placing the leaves, flowers, or fruits of plants, with a sufficient quantity of water, in an alembic or still. The essential oils only enter into ebullition at a higher temperature than water, since their point of ebullition rises to 343° F. Nevertheless, the vapour of the essential oils diffuses itself in the steam which fills the alembic, which is condensed, and a new supply of steam succeeds, which in its turn is saturated with the vapour of the oil. In this manner we can explain the rapid and continued evaporation of oils which only enter into ebullition at 343° F. in steam, which has itself only a temperature of 212° F. In order to increase the temperature of the water, marine salt, a solution of which boils at 228° F., is added. This has the effect of increasing the evaporation of the oil. But the practice has its disadvantages, and in order to prevent the plants from being burnt by coming into contact with the bottom of the still, it is usual to place in it a diaphragm pierced with holes, which supports the bed of leaves or flowers being distilled. The steam which is thus condensed in the worm of the alembic is a mixture of water and essential oil, of which, however, the oil forms only a small part. In order to separate the two liquids, and secure the oil, a very ingeniously-conceived vase, known as the *Florentine receiver*, is employed. This vase separates the oil from the water on the simple principle of their respective specific gravities. Oil is lighter than water, consequently it floats on the surface; and if the mixed liquid is received in a vase or jar having a siphon tube rising from the bottom, but whose highest part is placed at a lower level than that of the neck of the vase, as long as the united liquid flows from the still, the water will sink to the bottom of the receiver and flow off, while the oil will accumulate.

The essential oils obtained by distillation from the Orange dissolve readily in fatty oils or alcohol, but very imperfectly in water. The condensed water, however, which passes over with the oil, is a true watery solution of the essences; in short, orange-flower water. It is very subject to putrefaction, which is indicated by the appearance of flocculent flaky matter accumulating at the bottom of the vessel in which it is kept. But a large proportion of the essences used by the perfumer are not made by distillation at all; they are extracted from plants through the agency of fat. At the season when the flowers are in bloom, clarified fat—generally lard—is melted in a water-bath, such as a double glue-pot, and as many flowers, such as those of Jasmine, Orange, or Rose, as it will saturate, are put into it. These are allowed to remain in contact with the flowers for twenty-four hours, at a temperature just sufficient to keep the fat liquid. The fat is then strained off from the flowers. It is strongly scented, and the flowers have nearly lost their perfume. The same fat is melted again, and an equal quantity of fresh flowers is added, and allowed to remain in contact with it. The process being repeated, in all seven times, the fat, in the end, becomes very highly scented, and is ready for exportation as a "pomade." The same process is applicable to all plants abounding in essential oil. To obtain the spirituous essence or tincture, suitable as a perfume for the handkerchief, this pomade is macerated in spirit of wine, which dissolves from it the greater portion of its perfume oil; what remains, however, is sufficient to render the grease a rich pomatum for the hair. This simple and perfect process is quite within the reach of the cottager, who may thus, even in England, make her own perfumes, and those, too, of a very superior quality—far better, in fact, than can be obtained by distillation. Fat has wonderful powers as an absorbent, as any lady having a greenhouse full of strongly-scented plants may prove. Let her suspend glass plates, smeared with inodorous fat, in the vicinity of her scented flowers, and they will be found to absorb no small portion of the perfume.

The AMYRIDACEÆ are trees and shrubs abounding in balsamic resins, and having all the appearance of Oranges, even to their dotted leaves, but the fruit forms a shell whose husk eventually splits into valve-like segments. The few known species are natives of tropical India, Africa, and America. The frankincense of Arabia is said to be the produce of *Boswellia serrata.* The *Balsamodendron Myrrha*, a dwarf shrub of Arabia, yields the myrrh of Mecca; and most plants of the order yield resins and balsams of great commercial and officinal value.

The ANACARDIACEÆ include the Cashews. *Pistachia vera*, a tree fifteen feet high, originally from Syria, yields the fruit so much esteemed as the Pistachio Nut. Gum mastic is drawn from *P. Lentiscus.* The leaves of *Rhus coriaria* (the Sumach) are used for tanning. The order generally yields resinous products of considerable commercial value.

The RUTACEÆ agree with the Aurantiaceæ in having dotted leaves, definite stamens, and a fleshy disc. The plants of this order emit an offensive odour from the glands which cover them. In the case of some of the genus *Dictamnus* the glands are filled with volatile oil, and in hot weather the surrounding atmosphere becomes so charged with it that a light coming in contact will inflame the air.

The XANTHOXYLACEÆ are tropical plants, mostly of America, and all possess in various degrees aromatic and pungent properties.

"The SIMARUBACEÆ, or Quassias, are known," says A. de Jussieu, "from all the rutaceous plants, by the co-existence of these characters —namely, ovaries with but one ovule; indehiscent drupes, exalbuminous seeds, a membranous integument to the embryo, and by the radicle being retracted within thick cotyledons." The plants of the order are intensely bitter—so bitter that the larva of a species of *Ptinus*, which attacked other dried specimens, refused to attack *Simaruba versicolor.*

ALLIANCE XIX.—GERANIALES.

Flowers monodichlamydeous, symmetrical; placentas axile; calyx imbricated; corolla twisted; stamens definite; embryo with little or no albumen.

Flowers symmetrical; styles distinct; carpels longer than the torus; seeds with little or no albumen.	I. Linaceæ.
Flowers regular, unsymmetrical, with a permanent cup-like involucre; stamens monadelphues; albumen abundant.	II. Chlænaceæ.
Flowers symmetrical; styles distinct; carpels longer than the torus; seeds with abundant albumen.	III. Oxalidaceæ.
Flowers very irregular and unsymmetrical, without an involucre; stamens distinct; seeds exalbuminous.	IV. Balsaminaceæ.
Flowers usually symmetrical; styles and carpels combined round a long beaked torus.	V. Geraniaceæ.

The LINACEÆ, or Flaxes, are a small order of annual plants, useful alike for their fibre, the tenacity of which renders it invaluable, and their seeds, which yield by expression the linseed-oil of commerce. Their leaves are alternate, free from all trace of volatile secretions, and destitute of stipules; the stem incapable of disarticulation. The plants of the order are distributed throughout the temperate regions of the earth, particularly along the shores of the Mediterranean, in Europe and Africa; but the flax-plant (*Linum usitatissimum*) is supposed to be a native of the great plateau of Upper Asia, whence it was introduced into Europe. It is this which furnishes man with his first and last clothing. From the time of Moses it was cultivated on a great scale in the plains of Egypt; under the Roman emperors the Egyptians were renowned for their linen fabrics. It soon spread over France, Germany, and other European countries, but in our times it has been most fully developed in Holland, Belgium, and the North of France. In Britain it seems to have become the speciality of the North of Ireland. It is the only species of the order cultivated, and is much more delicate in appearance than the *Cannabis*, or Hemp-plant. It branches out towards the summit, and carries its alternate leaves at an acute angle. Its flowers are of fine blue; the fruit a capsule containing ten small seeds. When the plant becomes yellowish, and its capsules begin to open and its leaves to fall, which usually happens at the end of June, it is at its maturity. It is gathered by tearing it up by the roots, and laid on the earth; after twenty-four hours it is bound up into small bundles, which are placed on end to dry. In preparing the fibre, it is placed under water for a sufficient time to destroy the non-textile part of the stem by a species of fermentation or decomposition.

The CHLÆNACEÆ are all natives of Madagascar, of whose uses little is known, but they present some curious anomalies to the botanist, along with remarkably showy flowers, usually red in colour, borne in racemes or panicles.

The OXALIDACEÆ, or Wood Sorrels, sometimes arranged among

the Crane's-bills, have regular flowers, beakless fruit, albuminous seeds, and a general tendency to form compound leaves. They are plentiful in tropical and temperate America, and the Cape of Good Hope, but thinly diffused in colder regions. Almost all the species are distinguished by acidity, owing to the presence of oxalic acid. Some are bitter and stimulating. The tubers of some contain a considerable quantity of starch. The species best known to us is the common Wood Sorrel (*Oxalis Acetosella*), abundant in the moist and shady woods in this and other European countries; it is one of the most elegant of our wild flowers, and the grateful acid of its leaves is well known. It was called of old, Allelujah, and Cuckoo's Meat, because, says old Gerarde, "when it springeth forth, the Cuckoo singeth most; at which time also Allelujahs were wont to be sung in our churches."

The BALSAMINACEÆ are best known by the *Impatiens Balsamina* of floriculture; they are natives of India especially, and are remarkable for their unsymmetrical flowers. The nature of the parts which constitute this irregularity has been much discussed by botanists. "According to Roper and others," says Dr. Lindley, "two membranous external scales and a spur alone belong to the calyx, of which the two other sepals are usually deficient on that side of the flower which is opposite the spur; on the other hand, the corolla consists of a large upper or back piece, and of two lateral inner wings, each of which consists of two petals." Kunth considers the large back piece of the flower to be composed of two sepals, and together with the spur and exterior scales to form a five-leaved calyx, while he finds in the innermost parts a corolla of four petals united in pairs, and he assumes the fifth petal to be abortive.

The GERANIACEÆ, Crane's-bills, Geraniums, and Pelargoniums, form perhaps the most popular group of plants of the whole vegetable world. They are herbaceous, soft, or tumid stemmed plants; the young stems jointed, and separable at the joints; the leaves opposite in the upper part, and often alternate.

The long beak-like torus, round which the carpels are arranged, and the membranous stipules at the joints are the true marks of the order. The Pelargoniums are chiefly natives of the Cape of Good Hope; the Geraniums and Erodiums, of Europe, North America, and Northern Asia.

In order to give the reader some idea of this interesting family, of which we meet with examples at every turn, whether we happen to be in town or country, in field or garden, let us examine in succession the genera, *Geranium*, *Erodium*, and *Pelargonium*.

The *Geraniums*, or Crane's-bills, have a calyx with five sepals, a hypogynous corolla with five free petals, and an andrœcium composed of ten stamens, five of which are large and five small. The latter are exterior and opposite to the petals; the larger stamens have a nectarous gland at their base. The pistil is composed of an ovary with five cells, surmounted by a style, dividing into five

Fig 413.—Herb Robert (*Geranium Robertianum*).

stigmatic branches. Each cell of the ovary contains two ovules. The fruit is a capsule with five cells, each containing by abortion only one seed, and separating, as it were by a spring detached from the central axis, from the base to the summit. Under its integuments the seed encloses an embryo without albumen, the flexuous cotyledons of which are fitted into one another. Herb Robert (*Geranium Robertianum*), Fig. 413, is frequently met with in hedgerows, shaded places, and on old walls. It flourishes in the months of April to August, and exhales a strong odour. This plant, which was once used in medicine, is an annual, with a diffused, branching, ascending, or straight stem, often of a reddish colour. It is coated with long, shaggy, patulous hairs, glandular at their summit. The leaves are divided into three to five petioled segments. The peduncles are longer than the leaves. The petals are purple in colour, veined with white.

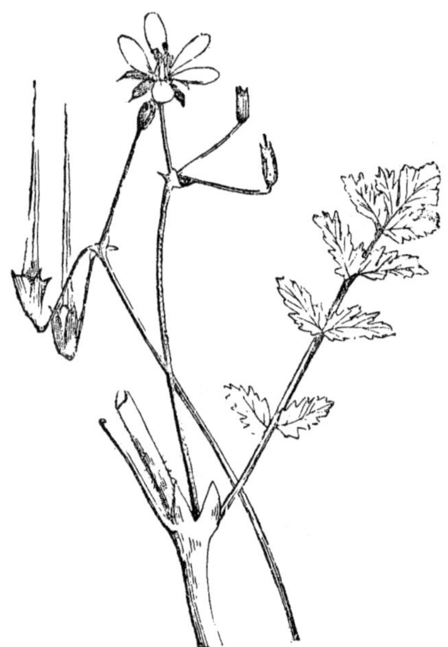

Fig. 414.—Stork's-bill (*Erodium cicutarium*).

The *Erodium*, or Stork's-bill, one species of which, the Hemlock Stork's-bill (*E. cicutarium*), Fig. 414, with pinnate leaves, is very common in all sandy places, has a corolla and calyx like the Geranium; but of the ten stamens five are abortive. These are small, with flattened filaments, destitute of anthers, and are opposite to the petals in the exterior whorl.

Pelargonium is particularly remarkable from the irregularity of the flowers. In the calyx the posterior sepal is prolonged at its base by a spur, which is straight, hollow, and adhering to the peduncle. The corolla generally bears unequal petals. The upper two are often largest, the other three differ from each other. As to the andrœcium, whilst in the *Erodium* the exterior verticil is completely abortive, in the Pelargonium, on the contrary, only three of the stamens of this verticil are sterile. The Pelargoniums are indigenous to the Cape of Good Hope. They contain a volatile oil, which gives them a strong odour, sometimes anything but agreeable, which is only redeemed by their beauty of form and colour. A great number of species are cultivated, to which horticulture has added innumerable varieties. Amongst others, we may mention *Pelargonium zonale*, the leaves of which are marked with a brownish band, and the petals of which are red or reddish-rose-coloured, or whitish; *Pelargonium inquinans*, the viscous leaves of which stain the fingers with brown, and the petals of which are scarlet or flesh-coloured; and *Pelargonium odoratissimum.*

ALLIANCE XX.—SILENALES.

Flowers monodichlamydeous; placenta free, central; embryo curved round a mealy albumen; pistil syncarpous.

Flowers dichlamydeous, symmetrical; corolla conspicuous; ovules amphitropal; leaves opposite, exstipulate.	I. Caryophyllaceæ.
Flowers dichlamydeous, symmetrical; corolla rudimentary; ovules amphitropal; stipules scarious.	II. Illecebraceæ.
Flowers dichlamydeous, unsymmetrical; sepals 2; petals 5, conspicuous; ovules amphitropal; leaves alternate, succulent.	III. Portulaceæ.
Flowers monochlamydeous; perianth often petaloid; ovules orthotropal.	IV. Polygonaceæ.

The CARYOPHYLLACEÆ as a group possess, with the exception of the Pinks and Carnations, little general interest; the beauty of form, rich colouring, and aromatic fragrance of the flowers, render these individuals of the family very general favourites. The order has been divided into—

Alsineæ, which have a polysepalous calyx; they are chiefly weeds, the best known being the common Chickweed (*Stellaria media*), of which birds are so fond.

Sileneæ, which have a gamosepalous calyx; they contain the Pinks, Carnations, and many popular plants.

The Pinks are herbaceous plants; the stem branching into forks, with tumid joints; the leaves simple, opposite, entire; the calyx tubular with five, sometimes four teeth, and supplied at its base with two or several bracts. The corolla is composed of five free hypogynous petals, with lengthened linear claws, and crenulate dentate limbs. The stamens are double the number of the petals, their anthers bilocular, and attached dorsally. The pistil is composed of a unilocular ovary, enclosing a great number of curved ovules, and surmounted by two very slight styles. The fruit is capsular, opening at the summit by double the number of valves there are styles.

Many hundred varieties, of great beauty are cultivated as florists' flowers. To grow them in perfection is quite a speciality in cultivation. The Carnation (*Dianthus Caryophyllus*) has red, rose-coloured, or white flowers, sometimes variegated or double. The qualities of the Clove Gillyflower, as it is sometimes also called from its spice-like odour, are sung by Chaucer :—

> "Ther springen herbes, grete and small,
> The licoris and the stetewall,
> And many a clove giliflore,
> ——— to put in ale,
> Whether it be moist or stale,"

whence another of its popular names, "Sops in Wine."

Fig. 415.—Carnation (*Dianthus Caryophyllus*).

The Sweet William (*Dianthus barbatus*) has the flowers in compact tufts, and protected by slight and pointed bracts, of the same length as the tube of the calyx. The Pink (*Dianthus plumarius*) has very straight barbate petals, which are sweet-scented, and of a pale rose colour, much varied by cultivation. Rousseau says in one of his letters : "Have you seen the *Dianthus superbus ?* At all events I will forward you one. It is really a most beautiful flower, with a sweet though somewhat faint odour. I can collect the seed very easily, for it grows in great abundance in a meadow which is just under my windows. Only the horses of the sun should be suffered to pasture on such plants as these."

Amongst other species belonging to the family of the *Caryophyllaceæ* we will mention the following :—the Soapwort (*Saponaria officinalis*), an indigenous plant, the roots of which contain a soft, gummy, resinous matter, which makes a lather like soap in water, and to which sudorific medicinal properties are attributed; many extremely beautiful species of *Lychnis*; *L. dioïca*, which the traveller frequently meets with by the wayside; *L. Flos-cuculi*, or Ragged Robin, the red petals of which are deeply cut, and which ornaments our fields in the spring-time; *Agrostemma coronaria*, the Rose Campion, a garden plant with purple flowers and whitish downy stem; the Corn Cockle (*Agrostemma Githago*), which abounds too often in our harvest-fields.

To the POLYGONACEÆ belong Buckwheat (*Polygonum Fagopyrum*), and the different species of Rhubarb (*Rheum*).

ALLIANCE XXI.—CHENOPODALES.

Flowers inconspicuous, monochlamydeous; placenta free, central; embryo external, curved or applied to the surface of a little mealy or horny albumen.

Perianth monophyllous; base of the tube finally indurated, and forming a spurious pericarp.	I. Nyctaginaceæ.
Perianth polyphyllous; stamens alternating with perianth segments, or indefinite; carpels, several or one.	II. Phytolaccaceæ.
Perianth polyphyllous, or nearly so; stamens opposite perianth segments; anthers often one-celled; ovary single, often several-seeded.	III. Amarantaceæ.
Perianth polyphyllous; stamens opposite perianth segments; anthers two-celled; ovary single, always one seeded.	IV. Chenopodiaceæ.

The group is of little general interest. NYCTAGINACEÆ are natives of the warmer parts of either hemisphere. PHYTOLACCACEÆ are natives of inter-tropical America, Africa, and India. AMARANTACEÆ are most frequently natives of the tropics; a few are natives of Europe, and a considerable number of Australia. CHENOPODIACEÆ include many of our vegetables, as Spanish Orach, the Spinach (*Spinacia oleracea*), and the well-known field Beet and Mangold-wurzel.

ALLIANCE XXII.—PIPERALES.

Flowers achlamydeous, embryo minute, either external, or just within the surface of a large quantity of mealy albumen.

Carpel solitary; ovule erect; leaves opposite or alternate	I. Piperaceæ.
Carpel solitary; ovule suspended; leaves opposite	II. Chloranthaceæ.
Carpels several, distinct; ovule erect; leaves alternate.	III. Saururaceæ.

The first are exclusively confined to tropical regions, common in America and the Indian Archipelago, and several species exist at the Cape of Good Hope. CHLORANTHACEÆ are also natives of the hottest parts of India, America, and the West Indies. The SAURURACEÆ agree in habit with the Peppers, but differ in the compound nature of the pistil. They are natives of North America, China, and the north of India, growing in marshy places.

SUB-CLASS III.—PERIGYNOUS EXOGENS.

Stamens adhering to calyx or corolla; pistil superior, or nearly so.

ALLIANCE XXIII.—FICOIDALES.

Flowers monodichlamydeous; placentas central or axile; corolla, if any, polypetalous; embryo external, curving round a mealy albumen.

Petals absent; sepals distinct; fruit enclosed in a membranous or succulent calyx; carpel single, solitary; seed erect.	I. Basellaceæ.
Petals numerous; carpels several, consolidated	II. Mesembryaceæ.
Petals absent; carpels several, consolidated	III. Tetragoniaceæ.
Petals absent; calyx gamosepalous; carpel single, finally surrounded by the hardened calyx.	IV. Scleranthaceæ.

The BASELLACEÆ are climbing, herbaceous, or shrubby tropical plants, with a double coloured perianth. The leaves of several are used as pot-herbs.

The MESEMBRYACEÆ are low-branching, half-shrubby herbaceous plants, chiefly of the hot and arid plains of Africa, the shores of the Mediterranean, Australia, and South America. The *Mesembryanthemum crystallinum*, the Ice-plant, has the leaves and stem densely covered with papillæ, resembling globules of water or granules of ice, which sparkle in the sun. This and other species are saline, somewhat nauseous to the taste, and supply alkali for glass-works in the South of Europe, where they abound. Medicinally they are diuretic. *M. edule*, the Hottentot Fig, is abundant on the sandy plains round the Cape of Good Hope, and is edible when ripe; the leaves, pickled, being substituted for the pickled cucumber, and its juice used medicinally. *M. Tripolium* has the singular property of gradually opening its seed-vessels on the approach of rain, contracting again when they become dry—a wonderful provision for the propagation of its species, for the seeds which have been closely shut up in the dry season are thus poured out of the open capsules when the growing season is at hand.

The TETRAGONIACEÆ are plants of no beauty and little interest. Like the Ice-plants, they are found on the shores of the Mediterranean, at the Cape of Good Hope, and also in the South Sea Islands; and the *Tetragonia expansa*, a New Zealand annual, is cultivated as a garden vegetable and substituted for spinach.

The SCLERANTHACEÆ are inconspicuous herbs of no known use.

ALLIANCE XXIV.—DAPHNALES.

Flowers monochlamydeous; carpel solitary; embryo amygdaloid, exalbuminous.

Anthers bursting lengthwise; ovule solitary, suspended; calyx imbricated.	I. Thymelaceæ.
Anthers bursting lengthwise; ovules erect; calyx valvate.	II. Proteaceæ.
Anthers with valvular dehiscence; leaves perfect.	III. Lauraceæ.
Anthers with valvular dehiscence; leaves rudimentary.	IV. Cassythaceæ.

The THYMELACEÆ occur in great abundance in the cooler parts of India, South America, South Africa, and Australia; they occur also in Europe. Their most common property is causticity, which resides in the bark, which is also very tough, and applicable to the purposes of cordage. *Lagetta lintearia*, the Lace-Bark tree of Jamaica, is remarkable for the facility with which it is separated into layers and meshes, which by lateral stretching become equal in delicacy to the finest lace. The bark of *Gnidia daphnoides* is manufactured in Madagascar into ropes; and a soft paper is made from the inner bark of *Daphne Bholua* in Nepaul, and from *Daphne cannabina* in China. Others, as *Passarina tinctoria* and *Daphne Gnidium*, yield a yellow dye, used to colour wool in the South of Europe. The Mezereon of gardens (*Daphne Mezereum*)—a deciduous plant, with dense clusters of white or purple flowers appearing on the plant before the leaves, which is found wild in the mountain forests of the middle and south of Europe—yields berries of a smooth, shining, and bright red, which are extremely acrid, and even poisonous. Linnæus speaks of a person having been poisoned by eating a dozen Mezereon berries. They are employed in Sweden to poison wild animals; and in Russia it is asserted that the Tartar women rub their cheeks with the berries to heighten their colour, instead of rouge. The Spurge Laurel (*Daphne Laureola*) of our hedgerows and woods is a handsome evergreen bush, with greenish flowers growing in clusters, concealed by the leaves.

The PROTEACEÆ are distinguished by the hard woody texture of their leaves, their irregular tubular calyx, with valvate æstivation. Brown considers that the radicle pointing towards the base of the fruit is a distinguishing feature. The order is named from the diversity of appearance presented by the several genera composing it. They are generally handsome evergreen shrubs, much prized by gardeners for the beauty and singularity of their flowers. Many of the genera are named after distinguished botanists, and their geographical distribution is extremely interesting. They are almost entirely confined to the larger continents of the southern hemisphere, being found at the Cape of Good Hope and adjacent regions, and New Holland; wherever the shores of Australia have been explored, the Proteaceous plants have been found; the great proportion of the order existing there in the same latitude as the Cape of Good Hope. On the south-east coast it forms the chief feature in the vegetation. *Grevillea*, named by Dr. Brown in honour of the Hon. Charles Francis Greville, is numerous in species there. Species of *Hakea* and *Banksia*—the latter genus named in honour of Sir Joseph Banks—are equally numerous.

The LAURACEÆ are trees sometimes of great size, distinguished from nearly all incomplete apetalous dicotyledons by the dehiscence of their anthers. Their habitat is cool places in the tropics of either hemisphere. *Laurus nobilis* is the only species found in a wild state in Europe. The species are all more or less aromatic and fragrant; some are valuable for their timber; others bear fruit like the nutmeg. Some yield fixed and essential oils, and camphor. Cinnamon and cassia are well-known products of the order from the hottest parts of Asia, the former being produced from *Cinnamomum zeylanicum*. Among the timber trees of the order is the Greenheart of Demerara (*Nectandra Rodiœi*).

The CASSYTHACEÆ, found in the hottest parts of the world only, are parasitic plants, resembling the Dodders. Their structure is nearly that of Laurels, the difference being in the fruit, which in *Cassytha* is enclosed in a berried calyx. Little is known of their properties or uses.

ALLIANCE XXV.—ROSALES.

Flowers monodichlamydeous; carpels more or less distinct, placentas, sutural; seeds, definite; corolla, if present, polypetalous; embryo amygdaloid, with little or no albumen.

Sepals and petals indefinite and similar; cotyledons convolute . .	I. Calycanthaceæ.
Flowers papilionaceous or regular; stamens usually definite; carpel single.	II. Leguminosæ.
Flowers regular; stamens usually indefinite; carpels one or many .	III. Rosaceæ.

The CALYCANTHACEÆ are well known in the garden for their delicious fragrance and for their chocolate-coloured flowers. These have an imbricated calyx and corolla which pass insensibly into each other, combining at the base into a thick fleshy tube, and a number of perigynous stamens with adnate anthers and projecting connective. Their wood presents four imperfect centres with concentric circles to each, lying at equal distances from the bark, which gives a square form to the circumference. *Chimonanthus fragrans* is called the

Japan Allspice; and the bark of *Calycanthus floridus* is sometimes used as a substitute for cinnamon.

The LEGUMINOSÆ form an extensive natural order, consisting of herbaceous plants, shrubs, and great trees, extremely variable in appearance. There are between six and seven thousand known species. Their most common feature is to have what are called papilionaceous flowers, which are readily recognisable, and leguminous fruit; although these characters do not always exist throughout the order, in some cases a kind of drupe taking its place, while the Mimoseæ have regular flowers and indefinitely hypogynous stamens.

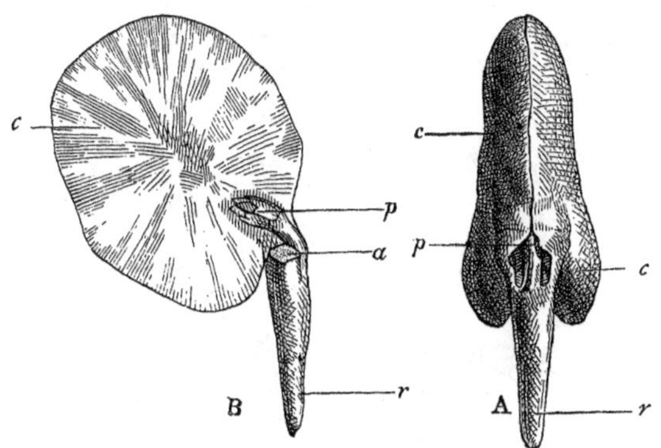

Fig. 416.—The Bean.
A, embryo, with the seed coat removed. B, embryo minus one cotyledon. *r*, radicle. *c*, cotyledon; *a*, point of separation of cotyledon; *p*, plumule.

The normal fruit of the Leguminosæ, however, may be considered a *Legume*, namely, a dry simple carpel, with a suture along both its margins, which opens at maturity into two valves.

This vast assemblage of plants has been subdivided into three sub-orders:—

Corolla papilionaceous, imbricate in the bud, odd petal exterior . . .	I. Papilionaceæ.
Corolla imbricate in the bud, odd petal interior	II. Cæsalpinieæ.
Corolla valvate	III. Mimoseæ.

The False Acacia, or Robinia (*Robinia pseud-Acacia*), which will serve us as a type of the *Papilionaceæ*, was derived originally from North America. It was first cultivated in France by Robin in the year 1601. It is a tree of great size, with a rounded crown and spreading branches; its russet bark is marked with deep longitudinal crevices; its branches are supplied with spines in the shape of strong prickles representing the stipules; its leaves are composed of numerous oblong leaflets; and its white and very odorous flowers are disposed in well-furnished hanging bunches (Fig. 417).

In the flower of the *Robinia* the calyx is nearly campanulate,

Fig. 417.—Branch and flower of *Robinia pseud-Acacia*.

almost bilobate, the upper lip truncate, or emarginate, the lower lip trifid. The corolla is composed of five petals. According to the expression used by botanists, it is said to be "papilionaceous." That part of the corolla called the "standard"—that is, the fifth petal—is orbicular; spreading backward, and scarcely extending beyond the "wings," which are free; and the "carina," formed by the two anterior petals, is pointed. The stamens are ten in number, of which nine form one bundle, leaving one free. The unilocular ovary encloses a score of ovules; the style is slender, and the stigma obtuse. The fruit, which forms an important character in this family, is a pod; the seeds are of a compressed ovoid shape, shiny, and of a dark colour, and enclose an embryo without albumen.

Amongst the *Papilionaceæ* a great number of species may be enumerated as plants of interest and usefulness to man; *Baptisia tinctoria*, the Wild Indigo, a bush of the United States, three feet high, yields a pale blue colouring matter, resembling indigo of an inferior kind; Bengal Hemp, the produce of *Crotalaria juncea;* Lupines; the beautiful yellow-flowering Gorse, Furze, or Whin (*Ulex europæus*, of which Linnæus preserved a plant in his greenhouse at Upsal; the Spanish Broom (*Spartium iunceum*), which yields a delicate fibre from which a fine linen-like cloth is made; the common yellow-flowered Broom (*Cytisus scoparius*), and the graceful Laburnum (*C. Laburnum*); the Lucern (*Medicago sativa*); the Trefoil (*M. Lupulina*); different species of Melilot (*Melilotus*); the Clovers (*Trifolium*), valuable in agriculture; the Indigo-plant (*Indigofera tinctoria*); the common Liquorice (*Glycyrrhiza glabra*); the ornamental Bastard Acacia or Locust-tree (*Robinia pseud-acacia*); the *Astragalus*, from which the gum tragacanth of commerce is obtained; the Bean (*Faba vulgaris*), from which the cultivated Beans have been derived (Fig. 416); the Peas (*Pisum sativum*); the Lentils (*Eroum Lens*); the Tares or Vetches (*Vicia sativa*); the Everlasting Pea (*Lathyrus latifolius*); the Sweet Pea, so fragrant and graceful; the Ground-nut of America, the seed of *Arachis hypogæa; Adesmia balsamifera*, better known as the Jarella-plant of Chili, which bear flowers of great beauty; the Moving-plant of Bengal (*Desmodium gyrans*); the French Honeysuckle (*Hedysarum coronarium*); Sainfoin, the well-known fodder plant; *Clitoria ternatea*, of which the root is said to be a cure for croup; *Soja hispida*, a plant of Japan, from which *soy* is prepared; the Ox-eye Bean (*Mucuna urens*); the Coral-tree (*Erythrina umbrosa*), grown in the Caraccas and Trinidad for shading the young chocolate plantations; the Dhak-tree of India (*Butea frondosa*), from which Bengal kino, or butea gum, is drawn (Indian kino is the inspissated juice of *Pterocarpus Marsupium*, a native of Coromandel); the Scarlet Runner (*Phaseolus multiflorus*), which is at once a beautiful ornament to our gardens and yields a useful vegetable for the table; the Dwarf Kidney Beans and Haricots; the Cabbage-tree (*Andira inermis*); the Tonquin Bean, which is the seed of *Dipterus odorata;* Balsam of Peru is supposed to be extracted from *Myrospermum Peruiferum*, and Balsam of Tolu from *M. Toluiferum; Virgilia capensis*, a handsome tree; the Judas-tree (*Cercis siliquastrum*).

The sub-order *Cæsalpinieæ* includes many medicinal plants, as the Nicker-tree (*Guilandina bonduc*), useful in intermittent fevers, the seeds of which are used as beads and marbles; Brazil-wood of commerce (*Cæsalpinia braziliensis*) is a tree of San Domingo, twenty feet high; *Hæmatoxylon campechianum* yields the logwood dye; *Tamarindus indica*, a large spreading tree of sixty feet, yields the well-known tamarind; the true Officinal Senna is the produce of *Cassia lanceolata; C. obovata* is the Alexandria Senna; *Cassia fistula*, a tree forty or fifty feet high, produces pods upwards of a foot in length, whose mucilage is cathartic; the wood of *Aloexylun agallochum* is much esteemed for its fragrant odour, this is the Aloe-wood of the East; Gum Animé is produced by *Hymenæa Courbaril*, a lofty tree of South America; the copal of Mexico is produced by some allied species; the bark of *Bauhinia racemosa* is used to make ropes, it is a climbing tree, known in India as the Maloo Creeper, which hangs in elegant festoons from the top of the loftiest trees, "which one is surprised," says Dr. Royle, "from the distance of its roots from the stems, how it could reach;" *Amherstia nobilis* is a Burmese tree thirty to forty feet high, which, "when in flower," as Dr. Wallich tells us, "is profusely ornamented with pendulous racemes of large vermilion-coloured blossoms, forming objects of beauty unequalled in the Indian Flora;" the cam-wood of commerce is the produce of *Baphia nitida*, an African tree of fifty or sixty feet high.

Amongst the *Mimoseæ* may be mentioned, *Erythrophleum guineense*, an immense tree of Guinea, growing a hundred feet high, called the Ordeal-tree, the juice being used by the natives as an ordeal of guilt or innocence; the Nitta-tree (*Parkia africana*), a large tree of Western Africa, where the seeds are roasted and used as a condiment, the pulp of the pods surrounding the seeds is sweet and farinaceous—a pleasant drink, and sometimes a sweetmeat, is made from it; *Adenanthera pavonina* is a gigantic tree one hundred feet high, its timber is valued for its solidity, and its scarlet highly polished seeds are used for personal decorations; several kinds of *Prosopis* yield edible fruits; Gum arabic is the produce of several species of *Acacia*, the most important being *A. vera* and *A. arabica; A. melanoxylon*, produces Black Wood, a hard, close-grained, dark, and richly-veined cabinet-wood of Australia, much used by the colonists; the pods of *Castanospermum australe* contain four seeds as large as a Spanish chestnut, which are eaten by the natives of Moreton Bay; *Brya ebenus*, a small tree, supplies Jamaica Ebony, its slender branches are flexible, and used as riding switches in the West Indies, where it was formerly used to punish refractory slaves.

The ROSACEÆ is a large family, of which the number of species is variously estimated at from 1,000 to 1,500 species. It is divided into the following well-marked sub-orders :—

Carpel single ; style basilar ; fruit drupaceous ; seed erect	I. Chrysobalaneæ.
Carpel single ; style terminal ; fruit drupaceous ; seeds suspended . .	II. Drupaceæ.
Carpels one—five, spuriously syncarpous, and adherent to calyx tube ; styles terminal.	III. Pomaceæ.
Carpels single or indefinite, not adherent to calyx tube ; styles lateral.	IV. Roseæ.
Flowers apetalous, often diclinous ; carpel single ; style terminal, often basilar.	V. Sanguisorbeæ.

The *Chrysobalaneæ* are exclusively natives of the tropics of Africa and America, and probably of the Indian forests. They are Stone-fruits, the drupes of many of which are edible.

The sub-order *Drupaceæ* includes the Almond, Peach, Nectarine, Apricot, Plum, and Cherry; fruits which are produced through the whole of Europe, Asia, and America.

Fig. 418.—Blossom of the Peach.

The Almond (*Amygdalus communis*) may be taken as the type. This tree, whose native country is unknown, although it is probably Central Asia, is now cultivated throughout the whole of Europe. Its branches are elongated, smooth, green, and slightly glaucous ; the leaves are alternate, lanceolate, and serrulate. The flowers appear before the leaves ; they are large and solitary, or in pairs along the branches. A hollowed receptacle in the shape of a cup, bears upon its edges five sepals, five petals, and from fifteen to thirty stamens, sheltering a sessile unilocular ovary, containing two collateral anatropal ovules suspended at the summit of its single cavity. It is surmounted by a terminal style. The fruit is a compressed oblong drupe, with fibrous coriaceous dry flesh, opening irregularly. Its stone is rugose, creviced, and hard ; it generally encloses only a single ovule, by reason of the abortion of the other. There are two varieties of the Almond ; the seeds of one are sweet, of the other bitter.

The Peach (*Amygdalus persica*) only differs essentially from the Almond in its fruit, the flesh of which is thick and succulent; and in the structure of its stone, which is furrowed with deep anfractuosities. This species, originally probably from Persia, presents three interesting varieties. In the first two the fruit is downy, in the third smooth. The first variety has firm flesh adhering to its nucleus; it comprehends the various kinds of Clingstone Peach. In the second variety the flesh is melting and easily detached from the stone. The third variety, distinguished from the two preceding by its skin, which is smooth and not downy, comprehends the different kinds of Nectarine.

Of the genus *Prunus*, the flowers present characteristics nearly identical with those of the genus *Amygdalus*, but it differs in the structure of the stone, which is not wrinkled or punctured, but is comparatively smooth. It comprehends the Apricot, Plum, and Cherry.

The Apricot (*Prunus armeniaca*) has a velvety drupe; it is of middle size, and has rotund leaves, nearly in the shape of a heart, terminating in a point, and dentate. The flowers are white, and disposed in little clusters very close together at the upper part of the branches. This tree is probably a native of Armenia. We may mention, among the varieties cultivated in France, the Early Apricot, the fruit of which is of a yellowish colour, small in size, with a firm and rather bitter saffron-coloured flesh. The Angoumois Apricot is of middle size, and the flesh is red and fragrant. The Peach-Apricot is the largest of all; the flesh is yellow, melting, and of a peculiar flavour.

The Common Plum is a fine branching tree, of from ten to twenty feet in height, with spreading branches, elliptical, sharp, crenulate, and dentate leaves. Its flowers are white, and appear before the leaves. It is often met with in hedges and on the borders of woods, but never in the interior of forests. This fact leads to the supposition that it is not indigenous. The Bullace (*Prunus insititia*) is a shrub of from six to ten feet in height, sometimes with prickly spines. It is found in the same situation as the last mentioned. The most esteemed varieties of the Plum appear to have come from the East, probably from Damascus. The number of the varieties is very considerable. Some have a round yellow fruit, as in the Golden Drop; in others the fruit is round, green, shaded with purple, like the Green Gage; in others, again, it is oval and globular, bluish or violet coloured, as in the Damson; in others it is nearly round, and the colour of wax, like the Yellow Magnum Bonum.

The Cherry-tree (*Prunus Cerasus*) bears a fruit (a drupe) with a smooth surface, without glaucous efflorescence. It is a rather tall tree, with straight cylindrical trunk, covered with a smooth and shiny bark. Its leaves are ovate, sharp, and dentate. The flowers are white, and in umbels. The fruit of the Wild or Bird Cherry-tree (*Prunus avium*) is used in the manufacture of cherry-water (*kirsch-wasser*) and of ratafia. The Bigarreau Cherry is very nearly allied to it, and has large red or yellow heart-shaped fruit (white-heart cherries), the flesh of which is with difficulty separated from the stone.

Fig. 419.—Blossom of Pear (*Pyrus communis*).

In the sub-order *Pomaceæ* we have the Apples, Pears, Quinces, Medlars, and other fleshy fruits whose flowers have indefinite stamens inserted in a ring, in the throat of the calyx (Fig. 419); the flowers, white or pink; the carpels adherent to the side of the ultimately fleshy calyx tube; the endocarp cartilaginous. The *Pomaceæ* are found plentifully in Northern Asia, Europe, and North America; they are rare in Mexico, unknown in Africa, except on the northern shore, and are entirely unknown in the southern hemisphere. Malic acid is contained in considerable quantities in the Apple and the Mountain Ash (*Pyrus Aucuparia*); prussic acid occurs in the seeds of most of the species, and even abundantly in some.

The Apple-blossom has a calyx with five lobes; a corolla with five nearly orbicular spreading petals; and a large number of stamens. The ovary is inferior, and generally presents five cells, with two collateral ascending and anatropal ovules; it has five free styles slightly coherent at their base. In the Pear (*Pyrus communis*) the fruit is nearly conical, not umbilicate at its base; the flesh is sweet, and towards the heart presents stony grains. In the Apple (*P. Malus*) the fruit is generally globular, always umbilicate at the base; the endocarp is cartilaginous, like that of the Pear; the flesh is acid, and never gritty. The Wild Apple grows spontaneously in

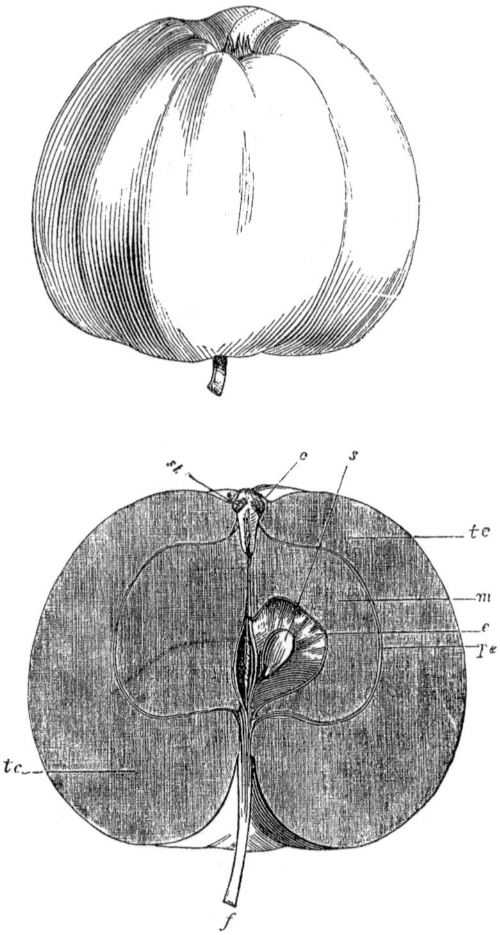

Fig. 420.—Exterior and Section of Fruit of *Pyrus Malus*.
c, remains of limb of calyx ; *st*, remains of stamens ; *tc*, calyx tube ; *pe*, epicarp ; *m*, mesocarp ; *e*, endocarp ; *s*, seed.

all European forests. Its rounded head is broader than it is high ; its leaves are ovate, dentate, acute, more or less downy on their lower face; its large red or white flowers form a cyme at the end of the young branches. The Apple is much modified by cultivation. From it we have the Pippins, Russets, Codlings, and many others. The Cider Apple takes the place of the Vine in many parts of Brittany, Normandy, and Picardy.

The Common Pear grows spontaneously in many forests of Europe; it is a tree with knotty branches, which attains from ten to twenty feet in height; its leaves, borne upon long petioles, are dentate, oval, and without hairs. The flowers are white, and disposed in cymes. From the wild state, the fruit, like that of the Apple, is ameliorated and much varied by culture; from it we have the Beurré, Bergamots, Saint Germain, Bon Chrétien, and hundreds of other varieties. Perhaps the most remarkable Pear is the Belle Angevine, grown in the neighbourhood of Angers, France, each of which weighs 3 lbs. avoirdupois.

The Medlar (*Mespilus germanica*) has an inferior ovary, like the other genera we have mentioned; and the fruit, crowned by five calycine segments, encloses five bony kernels. The genus *Cydonia*, or Quince, has an ovary, the cells of which enclose numerous ovules; and the fruit possesses a characteristic odour and flavour. The Medlar of Japan (*Eriobotrya japonica*) furnishes a yellow, melting, sweet and acid, comestible, fleshy fruit.

The sub-order *Roseæ* admits of division into several tribes, of which we may take the Rose, Bramble, and Meadow Sweet as types.

The Roses have a calyx with five foliaceous segments, which alternate with five petals; its perigynous stamens are numerous, and their filaments free, bearing anthers with two cells; all these organs are inserted upon the upper edge of an oval or spherical receptacle. At the bottom of this receptacle, which resembles a bladder or small bottle, a large number of free carpels stand erect; each forms a unilocular ovary, with a single anatropal ovule and an elongated style, surmounted by an obtuse stigma. When arrived at maturity these pistils become achænes, which are enveloped by the now fleshy receptacle; the seeds enclose a straight embryo, destitute of albumen. Roses are often supplied with prickles or spines, they have alternate leaves, with stipules adnate to the petiole, and beautiful terminal flowers, either solitary or in clusters, which have a sweet and unequalled odour. The Rose long ago gained the sceptre for beauty over all the most beautiful flowers of our gardens and hedgerows. There are numerous species of the genus *Rosa*, from which innumerable varieties have been produced. We must content ourselves here with briefly describing a few well-known species. The

Dog Rose (*Rosa canina*) is an indigenous species, common in our hedgerows and upon the borders of woods, the fruit of which is of a coral red, with a yellowish, acid, and astringent pulp, enclosing a

Fig. 421.—Branch of the Red Rose (*Rosa gallica*).

number of hard hairy pips or seeds. The Red Rose (*Rosa gallica*) is represented in Fig. 421; the petals were formerly employed in medicine as astringents. The Red Rose was brought from Syria to France in the times of the Crusades. The Cabbage Rose (*Rosa centifolia*), whose admirable flowers are the ornament of our gardens, came originally from the Caucasus. The Damask Rose (*Rosa damascena*) still preserves some stamens not changed into petals; its odour is very sweet, and it is used in the preparation of rose-water. From the Musk Rose (*Rosa moschata*), like the two preceding species, a volatile oil is extracted, called otto of roses.

Brambles (*Rubus*, Fig. 422), like the Rose, have five sepals five petals, and numerous stamens and pistils; but in this case the receptacle, instead of being hollowed out in the shape of a bottle, rises like a disc or cone, and upon this the carpels are disposed. When at maturity these become changed into little drupes, grouped together upon a spongy and persistent receptacle. Brambles are sarmentose, and provided with prickles, with simple, alternate, digitate leaves, with stipules adnate to the petioles, and terminal or axillary flowers, which are really solitary, but are disposed in a panicle. We often meet with the Bramble (some form of *Rubus fruticosus*), the Dewberry (*Rubus cæsius*), represented in Fig. 422, and the Raspberry (*Rubus idæus*).

Fig. 422.—The Dewberry (*Rubus cæsius*).

In the Strawberry the calyx is composed of five sepals joined at the base, and furnished with an involucre of five bracteoles. The stamens, which are numerous, are inserted upon the edge of a receptacle in the shape of a cup, which rises again in the middle like the bottom of a bottle. The numerous carpels are inserted upon the lower part of the receptacle, and have a lateral style. They are changed at the time of maturity into achænes, which are implanted upon the fleshy and succulent receptacle. Strawberry plants are long-lived, creeping perennials, with alternate trifoliate leaves, sometimes simple by abortion, with stipules adnate to the petioles. *Fragaria vesca* furnishes several wild varieties known under the name of Wood Strawberries. *Fragaria chilensis* is the Chili Strawberry, the fruit of which is erect, rose-coloured, white within, and sometimes as large as a pigeon's egg. *Fragaria collina* is not very common; its fruit is of a lively red, ovoid, contracted at the base, almost destitute of carpels, and shiny in its lower parts, and is with difficulty detached from the bottom of the calyx.

The Meadow Sweet (*Spiræa Ulmaria*) displays its corymbs of delicate white flowers at the edges of water, or in damp fields. It has a five-parted calyx, corolla with four to five petals, numerous stamens, generally also five carpels, which are sessile at the bottom of a receptacle hollowed like a rather deep cup, and enclosing in a single cavity two series of ovules, generally suspended. When at maturity these become follicles, which open at the sutures. Meadow Sweets are herbs, shrubs, or under-shrubs, with simple or compound alternate leaves, with stipules adnate to the petioles, having flowers, disposed in white or red bunches, in corymbs, panicles, or fascicles. The Dropwort (*Spiræa Filipendula*) is frequently to be met with in chalky places; its flowers white, in terminal corymbs. Another species, the *Spiræa Aruncus*, is not British; the root was highly extolled in olden times as a tonic and febrifuge. These three species are perennial herbs. Amongst the ligneous species which belong to ornamental gardening, many have showy flowers, generally white or red.

The *Sanguisorbeæ* are mostly herbaceous; *Poterium muricatum* is used as a fodder plant. They are natives of extra-tropical regions.

ALLIANCE XXVI.—SAXIFRAGALES.

Flowers monodichlamydeous; carpels consolidated; corolla, if any, polypetalous; placentas sutural or axile; seeds indefinite; embryo slender with a long radicle; albumen little or none.

Styles distinct; leaves alternate	I. Saxifragaceæ.
Styles distinct; leaves opposite, exstipulate	II. Hydrangeaceæ.
Styles distinct; leaves opposite, with large interpetiolar stipules . .	III. Cunoniaceæ.
Styles connate; calyx polysepalous; leaves alternate	IV. Brexiaceæ.
Styles connate; calyx gamosepalous; leaves opposite	V. Lythraceæ.

The SAXIFRAGACEÆ are so called from *saxum*, "a stone," and *frango*, "I break," from the supposed virtues of the genus *Saxifraga* as a specific in calculus. Of the species, many are British, most of them rock plants, and many of them old favourites, as *S. umbrosa* (London Pride, None so Pretty), often used as an edging plant in old cottage gardens. They also abound in South America; and Professor Martins found them with other old favourites on the rocky declivities of North

XVI.—The Common Elm (*Ulmus campestris*).

Cape. The CUNONIACEÆ are of frequent occurrence in Australia, the Tropics, and at the Cape of Good Hope. *Cunonia capensis* is in great request for its timber, which is called Red Alder, or Rood Elze. The LYTHRACEÆ are represented in the British Flora by *Lythrum Salicaria*, the Purple Loosestrife.

ALLIANCE XXVII.—RHAMNALES.

Flowers monodichlamydeous; carpels consolidated; placentas axile; fruit capsular, berried, or drupaceous; seeds definite; embryo amygdaloid with little or no albumen.

Flowers apetalous; ovary composed of four carpels; calyx tubular; cotyledons consolidated.	I. Penæaceæ.
Flowers apetalous; ovary composed of two carpels; calyx tubular; cotyledons amygdaloid.	II. Aquilariaceæ.
Flowers apetalous; ovary composed of two carpels; calyx campanulate, irregularly divided; cotyledons thin and leafy.	III. Ulmaceæ.
Flowers polypetalous; calyx valvate; stamens opposite petals; seeds erect.	IV. Rhamnaceæ.
Flowers polypetalous; calyx valvate; stamens alternate with petals; seeds pendulous.	V. Chailletiaceæ.
Flowers polypetalous; calyx imbricated; stamens monadelphous . .	VI. Hippocrateaceæ.
Flowers polypetalous; calyx imbricated; stamens distinct	VII. Celastraceæ.
Flowers monopetalous; stamens episepalous	VIII. Stackhousiaceæ.
Flowers monopetalous; stamens epipetalous; ovules ascending; radicle short; cotyledons amygdaloid.	IX. Sapotaceæ.
Flowers monopetalous; stamens epipetalous; ovules sometimes suspended; radicle long; cotyledons leafy.	X. Styracaceæ.

These orders consist of evergreen shrubs, trees, and herbaceous plants, which are, with few exceptions, confined to the tropics of both hemispheres and the warmer parts of Europe. The exceptions are the ULMACEÆ, or Elms, which have usually been classed among the *Urticaceæ*, from which they differ in having a two-celled fruit and hermaphrodite flowers.

The Elm (*Ulmus campestris*), PLATE XVI., is generally found in mountain woods, and it is not uncommon to find it planted by road-sides and in places of public resort. It is a large tree, with a lofty head, formed of strong ascending branches, abundantly furnished with close compact twigs. Its leaves are alternate, furnished with two caducous stipules, ovate, acute, and oblique at the base; doubly dentate, and generally pubescent and rough. They only appear after the flowers, which are reddish, and arranged in sessile fascicles, or bundles; each flower, always without a corolla, consisting of a calyx of four or five lobes, with four or five stamens opposite to these lobes; a free two-celled ovary, containing in each cell a single anatropal ovule. The fruit, or samara, of the Elm (Fig. 423) is dry, compressed, winged, membranous in its circumference, indehiscent, and unilocular.

The various species of Elm are found wild in most parts of the world. They are trees or shrubs of the north part of Asia, of the mountains of India, of North America, China, and Europe. The seeds of the Elm produce trees often so unlike the parent tree, that some botanists think more than one of the so-called species have originated in this manner.

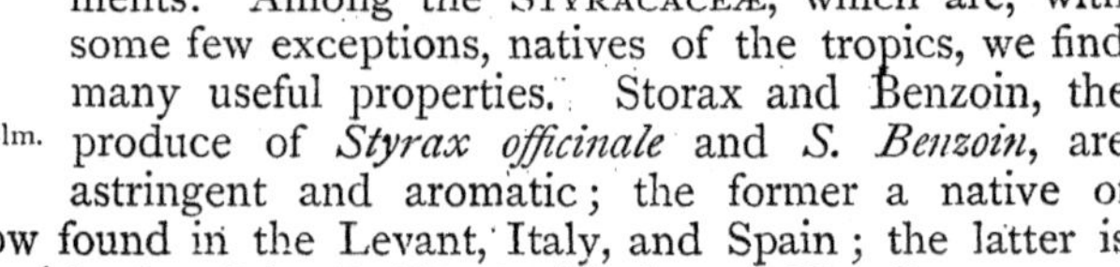

Fig. 423.
Samara of the Elm.

The RHAMNACEÆ produce medicinal plants of considerable importance. The berries of various species of *Rhamnus* are violent purgatives; some of them yield excellent dyes, others are grateful condiments. Among the STYRACACEÆ, which are, with some few exceptions, natives of the tropics, we find many useful properties. Storax and Benzoin, the produce of *Styrax officinale* and *S. Benzoin*, are astringent and aromatic; the former a native of Syria, but now found in the Levant, Italy, and Spain; the latter is found in the islands of the Indian Archipelago. The SAPOTACEÆ, chiefly natives of the tropics, produce several dessert fruits.

ALLIANCE XXVIII.—GENTIANALES.

Flowers dichlamydeous, monopetalous; placentas axile or parietal; embryo minute; albumen copious.

Exstipulate; stigmas simple, sessile, radiating	I. Ebenaceæ.
Exstipulate; stigmas on a style; placentas axile; seeds definite, pendulous; corolla imbricate.	II. Aquifoliaceæ.
Exstipulate; stigmas capitate, with a membranous ring below, and contracted in the middle.	III. Apocynaceæ.
Leaves opposite, with interpetiolar stipules	IV. Loganiaceæ.
Exstipulate; stigmas on a style; placentas axile; seeds indefinite, peltate.	V. Diapensiaceæ.
Exstipulate; stigmas on a style; placentas axile; seeds definite, erect; flowers unsymmetrical.	VI. Stilbaceæ.
Exstipulate; stigmas on a style; placentas parietal; flowers didynamous.	VII. Orobanchaceæ.
Exstipulate; stigmas on a style; placentas parietal; flowers regular.	VIII. Gentianaceæ.

The EBENACEÆ are Indian and tropical, although a few species are found as far north as Switzerland in the Old World, and New York in the New. They are chiefly remarkable for their hard black wood, sometimes variegated, and known as Ebony and Ironwood; also for the extreme acerbity of their unripe fruit.

The AQUIFOLIACEÆ, of which the Common Holly is a type, are found sparingly in the West Indies and South America, where *Ilex paraguayensis* yields the Paraguay tea, in the leaves of which Mr. Stenhouse detected theine. The evergreen shrub, *Prinos glabra*, is employed as a substitute for tea in many parts of North America. In Europe the Common Holly (*Ilex Aquifolium*), with its numerous varieties of gold and silver-blotched, broad, narrow, and thick serrated leaves, is a beautiful object in ornamental clumps of shrubberies, especially when clothed with its profusion of red or yellow berries.

The APOCYNACEÆ, or Dogbanes, are trees or shrubs, chiefly tropical, *Vinca* and *Apocynum* alone belonging to northern countries. For the most part they are handsome plants, with large, showy, symmetrical flowers. They all yield a milky juice by incision in the stem, which is generally poisonous. The Tangin poison-tree of Madagascar, which was at one time used as an ordeal of guilt or innocence, is remarkable for its poisonous properties. In the Periwinkle (*Vinca minor*), the calyx (Fig. 424) is five-parted, while the root, like that of the Gentians, is bitter, acrid, and astringent. Others are not only harmless, but nourishing; the *Hya-Hya*, or Milk-tree of Demerara, and Cream-fruit of Sierra-Leone, are of this description. Caoutchouc is yielded in abundance by *Vahea gummifera*, *Urceola elastica*, and *Willoughbeia edulis*. Many of them yield valuable medicines, but from the great prevalence of the poisonous properties in the order, they require to be administered with caution. Even the Oleander (*Nerium*) is a formidable poison, as well as a destroyer of cutaneous vermin. It is related that while the French troops occupied Madrid, a marauding soldier cut some branches of *Nerium Oleander* to employ as spits on which to roast his plunder, and of the twelve comrades who partook of the feast seven died, and the other five were dangerously ill.

The LOGANIACEÆ, separated from the former order, are either tropical or inhabit countries near the tropics, a few species belonging to Australia and America. There is no order more venomous than this, which now includes *Strychnos Nux-vomica*, an Indian tree, with small greenish-white flowers, ribbed leaves, and a beautiful round orange-coloured fruit, the size of a small apple, having a brittle stalk, and white, gelatinous pulp. The pulp of the fruit, according to Roxburgh, is perfectly harmless, and is greedily devoured by birds, but the seeds are extremely dangerous.

The other members of the group have few properties worth mentioning. The OROBANCHACEÆ are parasites upon the roots of other plants. The seeds, according to Vaucher, will be dormant until they come into contact with the roots of the plants on which they grow parasitically, when they immediately begin to germinate.

The GENTIANACEÆ range over the entire globe. They bloom on the verge of eternal snow in the Alps, in the chinks of the rocky steeps of North Cape, in the Himalayas, on Mexican mountains, and in the hottest sandy plains of India and South America. As

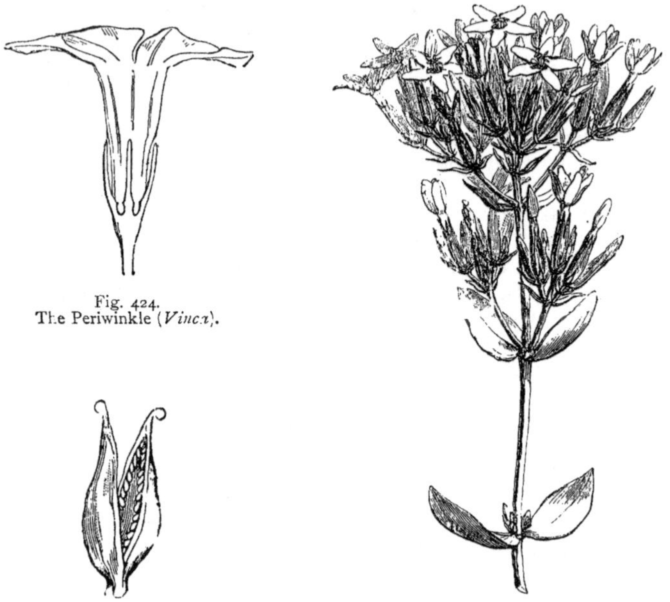

Fig. 424.
The Periwinkle (*Vinca*).

Fig. 425.

Fig. 426.—*Erythræa Centaurium.*

ornamental plants, they are remarkable for the brilliant colours and beautiful form of their flowers, whose prevailing colours are either an intense blue or a clear bright yellow.

As an example of the order we shall limit ourselves to describing the Centaury (*Erythræa Centaurium*), Fig. 426.

It is a little plant, common in woods, fields, and glades; its opposed leaves are entirely sessile, and disposed like the Thyme. The flowers are regular and hermaphrodite; the calyx tubular, with five linear divisions. The corolla is in the shape of a funnel, with very long tube and limbs, with five divisions. Five stamens are inserted upon the tube of the corolla. Before the expansion of the flower the anthers are straight, but become spirally twisted after the emission of the pollen. The pistil is composed of a superior ovary, surmounted by a filiform style, divided into two branches, which are rounded at the summit. The ovary is unilocular, with parietal placentas, bearing a great number of anatropal ovules. The fruit is a capsule opening with two valves, which bear the seeds upon their sides. These enclose a very small embryo in a fleshy albumen.

The Gentians (*Gentiana*) only differ from the preceding in having straight anthers. The root of the Yellow Gentian (*G. lutea*) is employed in medicine, being the plant which furnishes the gentian root of the druggists. It is a native of the Alps and other mountains of Central Europe, growing vigorously in calcareous soils, where its numerous fascicled bright yellow flowers surprise the traveller by their unexpected appearance. Most other species of Gentians are now abandoned in medicine; but botanists cherish them for their elegance of form and the brilliant colouring of their flowers. We may cite as belonging to this family two of the most graceful ornaments of our rivers and ponds: the Bog-bean (*Menyanthes trifoliata*), whose trifoliate leaves, and flowers disposed in spikes, snowy white, rose-coloured, or purple in their tints, are bearded on the inside of their corolla with delicate filaments. Again, there is the Fringed Buckbean (*Villarsia nymphæoides*), the elegant rival of the yellow Water Lily, which is being introduced into our ornamental waters, where its round green leaves and yellow flowers form a very graceful ornament.

ALLIANCE XXIX.—SOLANALES.

Flowers dichlamydeous, monopetalous, symmetrical; placentas axile; fruit, two-three-celled; embryo large, surrounded by a small albumen. Anomalous genera, having no corolla or separate petals, occur.

Stamens free, two or four	I. Oleaceæ.
Stamens free, five ; placentas axile ; embryo terete	II. Solanaceæ.
Anthers and stigma consolidated into a column	III. Asclepiadaceæ.
Stamens free, five ; placentas axile ; cotyledons leafy, folded longitudinally.	IV. Cordiaceæ.
Stamens free, five ; placentas basal ; cotyledons leafy, doubled up .	IV. Convolvulaceæ.
Stamens free, five ; placentas basal ; embryo filiform, spiral . . .	VI. Cuscutaceæ.
Stamens free, five ; placentas axile ; cotyledons straight, plano-convex	VII. Polemoniaceæ.

Among the OLEACEÆ we have the Common Olive (*Olea europæa*), whose fruit and oil are alike valuable. The Ash (*Fraxinus*) is extremely abundant in this country and over the whole of Europe. The bark of *Fraxinus rotundifolia* yields the manna of the druggists in the warm climates of Southern Europe. This substance, which contains a distinct principle called mannite, is also produced by the Common Ash (*F. excelsior*).

The Lilacs were originally from Asia. Their leaves are opposite and simple. Two species are more especially cultivated in our gardens: the Common Lilac (*Syringa vulgaris*) and the Persian Lilac (*Syringa persica*). They have regular and hermaphrodite flowers; the monosepalous calyx is four-lobed; the corolla four-cleft, monopetalous; the tube much elongated, and surmounted by a spreading four-lobed limb; two stamens are inserted upon the tube of the corolla. The pistil is composed of a superior ovary, surmounted by a style with two stigmatic branches. This ovary is two-celled, the fruit a capsule opening loculicidally into two valves, having a partition in the centre, and containing two narrow-winged seeds, provided with a fleshy albumen and with a straight embryo.

The Olive (*Olea europæa*) is a sober greyish-green tree of middle size, and from twelve to twenty feet high, without beauty, having a rugged stunted appearance. Its leaves are oblong, or entirely lanceolate; the upper surface smooth and whitish green, the lower scaly. The flowers of this tree form axillary bunches, straight during efflorescence, drooping when at maturity. Its fruit is a drupe, with a two-celled stone, one of the cells being frequently abortive, and the other ripening only one seed. The pericarp of this drupe contains a fixed oil, obtained by pressure, holding the highest rank among oils used for alimentary purposes. In November, when the fruit is quite ripe, and assuming a black colour, it is gathered, taken to the mill, and passed between two grinding-stones, placed at such a distance as to crush the fleshy part without breaking the stone; the mass is put into bags made of rushes, and moderately pressed. What is called "virgin oil" is thus obtained; the pulp is moistened with water, and again submitted to pressure, and oil of second quality is obtained; the process is repeated more than once.

The Privet (*Ligustrum vulgare*) is nearly allied to the Lilac. Its leaves are astringent, and the berries, which the birds eat, furnish a black colour used in dyeing. Country-people in France manufacture a writing ink with the crushed fruit.

To understand the structure of the Ash (*Fraxinus*), let us examine two species, namely, the Flowering Ash (*F. Ornus*) and the

XVII.—The Common Ash (*Fraxinus excelsior*).

Common Ash (*F. excelsior*). The former is a very ornamental tree from thirty to forty feet in height. The compound leaves of the Flowering Ash have from seven to nine sessile, lanceolate, and dentate leaflets; they are green and smooth on the upper surface, the lower is a little paler, and hairy the whole length of the midrib. The flowers appear with the leaves; they are regular and hermaphrodite, and of a greenish white. The calyx is four-toothed, and the white corolla consists of four very long linear petals. There are two stamens and one pistil with two cells, each containing two suspended anatropal ovules, as in the Lilac; the fruit is a winged samara.

From this and some other species of the Ash—notably, as already observed, from *F. rotundifolia*—a liquid is drawn by incisions made in the bark, which solidifies when exposed to the air. When fresh, it is sweet and nutritious; with age it becomes slightly purgative, and serves a useful purpose in medicine. The most highly-esteemed manna is obtained from Sicily.

The Common Ash (*Fraxinus excelsior*), PLATE XVII., is a large tree, which when in good condition reaches the height of seventy or eighty feet, with a trunk of eight or ten feet in circumference. It grows in the woods, in clumps, or by itself, blossoming in April and May. Its leaves present from nine to fifteen nearly sessile, lanceolate, opposite leaflets, smooth on the upper part, velvety below at the base of each side of the midrib. The flowers of the Common Ash, contrary to those of the Flowering Ash, are completely destitute of floral envelopes; they are composed of two stamens and a pistil. The Common Ash is found in many parts of Northern Asia. Its rapid growth and tough, hard wood make it one of the most useful British trees, besides being singularly graceful as an ornament on lawns and parks; but more especially does it seem the natural ornament of architectural ruins, such as Melrose and Netley Abbeys, where it may be seen in great perfection, blending its slender branches and airy foliage with the venerable walls and tracery of the windows. The Ash, however, is an enemy to all interloping vegetation; its rapid growth, and the extension of its roots may be traced in the languor of its vegetable neighbours. The Weeping Ash (*F. excelsior* var. *pendula*) has many of the characters of the Common Ash, but the tendency of its branches is to bend downwards, so that the arching boughs, when grafted on a stem of suitable height, will soon reach the ground, and form a natural arbour. It was first, says Loudon, discovered in a field at Gamlingay, in Cambridgeshire.

The SOLANACEÆ, or Nightshades, are natives of all parts of the world without the polar circle, but they culminate in number and energy in the tropics. At first sight it would seem an anomalous system which places the Potato and the Love Apple in the same order with the Deadly Nightshade and Henbane. The arrangement is founded on the structure of the flower, the resemblance of which is obvious in most species. Most plants of the order contain the narcotic principle solanine in greater or less proportion.

If we examine the Potato (*S. tuberosum*) as a type of the family, we find that the calyx (Fig. 427) is monosepalous, with five divisions; the corolla monopetalous, and shaped like a wheel or cup, having lobes alternating with the division of the calyx, five stamens with short filaments, and bilocular anthers, opening at the summit by two pores. The pistil consists of a superior ovary, surmounted by an elongated style, which terminates in an obtuse stigma. This ovary has two cells, and for each cell a large placenta with numerous ovules inserted upon the partition which separates them. The fruit is a berry containing a number of compressed seeds, supplied with a fleshy albumen and an embryo bent round upon itself. *Solanum tuberosum* came originally from the Cordilleras of Peru and Chili. We have already said that its tubers are not the roots, but the subterranean branches of the plant: P (Fig. 429) being the parent tuber, from the evolution of an eye of which the axis of underground stem A has issued; *a* is a branch given off at a node of the primary axis A, at which also the root-fibres *r* are developed; T, a young tuber, is the thickened extremity of the branch.

Fig. 427.—Potato Blossom.

As other examples of the genus *Solanum* we may also cite the Black Nightshade (*S. nigrum*), and the woody Nightshade (*S. Dulcamara*). *Solanum nigrum* is a herbaceous shrubby plant, which grows under hedges, in woods, by river sides, on the walls of cottages, and in cultivated places. It has white flowers, producing a black berry, while *S. Dulcamara* has purple flowers and scarlet berries. The Egg-plant (*Solanum esculentum*), is a herb originally from tropical Asia, which culture has spread along the shores of the Mediterranean and southern regions of Europe. It is now naturalised in America. Its large, shiny, ovoid, generally violet, but sometimes yellow-coloured fruit contains a white flesh, which becomes comestible by cooking. The Apple of Sodom (*S. sodomeum*), is a native of

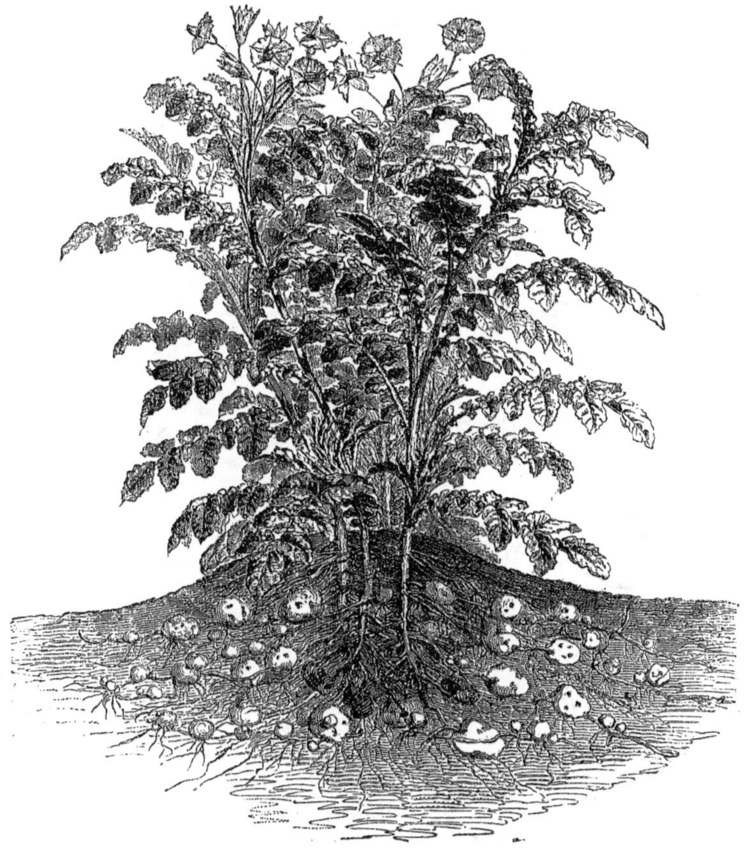

Fig. 428.—Potato Plant and Tubers.

South Europe and North Africa. Of this plant Sir John Mandeville, the old English traveller, speaking of the Dead Sea, tells us that, "Besyden growen trees that baren fulle faire apples and faire of colour to beholden, but whosoe breakethe them or cuttethe them in two, he shall find them within coles and cyndres." Milton alludes to this tradition :

> "They, fondly thinking to allay
> Their appetite with gust, instead of fruit
> Chewed bitter ashes."

The Tomato (*Lycopersicum esculentum*) is cultivated in most gardens, and produces a fruit called the Love Apple, of a lively red

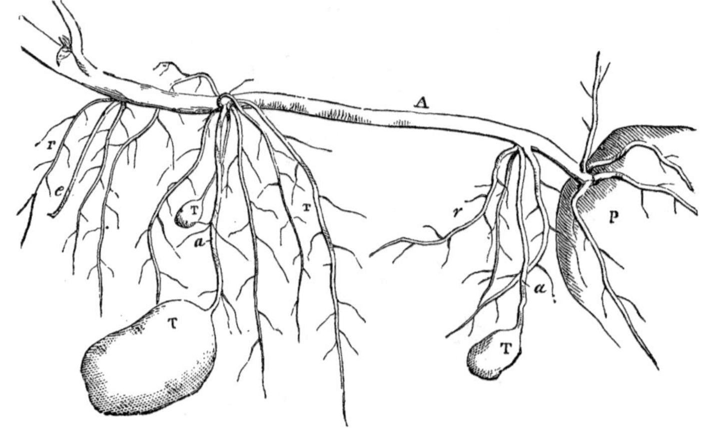

Fig. 429.—Roots and Tubers of the Potato.

colour, with round lobes, filled with an orange, sourish pulp, having a very agreeable perfume.

The Belladonna, or Deadly Nightshade (*Atropa belladonna*), is a perennial, herbaceous, elegant plant, with sombre leaves, livid flowers, and fruit resembling little black cherries. The sweet flavour of these fruits is deceitful, for their juices constitute a deadly poison. Yet as a medicine the berry (Fig. 430) has powerful soothing properties. The expressed juice of the leaves produces a remarkable dilatation of the pupil of the eyes—a singular property, which has been utilised in operations for cataract to facilitate the extraction of the crystalline lens. The odour of the whole plant produces headache and sickness.

The Mandrake (*Mandragora officinalis*), which has properties analogous to those of Belladona, but less violent, in olden times was employed by magicians and sorcerers to produce mental hallucinations and disturb the reason. The Winter Cherry (*Physalis Alkekengi*), a slightly acid succulent berry, enclosed in an accrescent calyx, sometimes used at dessert, is recommended as a diuretic. The Capsicum, the shiny berries of which are green at first, and red when arrived at maturity, contains a resinous, balsamic, but very acrid principle, which makes these fruits to be much esteemed as a condiment in all countries.

Fig. 430.—Berry of Belladonna.

Tobacco (*Nicotiana*) belongs to another section of the family of the *Solanaceæ.* Their anthers open by two longitudinal clefts. Their fruit is dry; it is a capsule which opens in two valves, leaving the placentary partition filled with seeds in the centre (Fig. 431). The Tobacco is a renowned plant, which has made the conquest of the world. In all parts of the globe it is consumed.

Henbane (*Hyoscyamus niger*) is distinguished from *Solanum* and its congeners, as well as from Tobacco, by its capsular fruit, which opens circularly, like a little box. Lastly, we must mention the Thorn Apple (*Datura Stramonium*), the capsule of which is covered with prickly tubercles.

Fig. 431.—Flowers of Tobacco-plant.

The ASCLEPIADACEÆ are succulent plants, chiefly of South Africa, where they flourish in the dry and sterile soil. In tropical India, America, and Australia, they also abound; only two species being found in northern regions. Of *Asclepias* there are many North American species, and *Cynanchum* is found between 59° north and 32° south latitude. The roots of most of the species are acrid, and yield a milky juice. Many of them possess useful medicinal properties.

The CORDIACEÆ are trees, native of the tropics of both hemi spheres, the flesh of their fruit being succulent, mucilaginous, and emollient.

The CONVOLVULACEÆ, or Bindweeds, are familiar to most readers, and universal favourites, from their elegant twining stem, which accounts for their name—derived from *volvo*, "to wind round"—heart-shaped or lobed leaves, and the innumerably varied colours of their bell-shaped flowers.

The CUSCUTACEÆ are climbing colourless parasites of both hemispheres. Mr. Griffiths speaks of a gigantic species which even preys upon itself in Affghanistan, where he saw one mass covering a willow-tree from twenty to thirty feet high.

ALLIANCE XXX.—CORTUSALES.

Flowers monodichlamydeous, monopetalous; placenta free, central; embryo lying within a large quantity of hard albumen.

Stamens alternate with the petals; styles two; inflorescence circinate.	I. Hydrophyllaceæ.
Stamens opposite the petals; fruit membranous, one-seeded; styles five; stem herbaceous.	II. Plumbaginaceæ.
Stamens alternate with the petals; style one; inflorescence straight.	III. Plantaginaceæ.
Stamens opposite the petals; fruit capsular, many-seeded; style one; stem herbaceous.	IV. Primulaceæ.
Stamens opposite the petals; fruit indehiscent, drupaceous; style one; stem woody.	V. Myrsinaceæ.

HYDROPHYLLACEÆ are little known out of the American continent. *Nama* and *Hydrolea* occur in India; some species of *Nemophila* are garden favourites, cherished for their elegant flowers.

PLUMBAGINACEÆ grow in salt marshes and sea-coasts of the temperate parts of the world, along the Mediterranean basin; others in Greenland and the mountain ranges of Europe, and a few within the tropics; *Plumbago zeylanica* from Ceylon to Port Jackson; *Ægialitis* grows among the Mangroves of Australia; *Vogelia* at the Cape of Good Hope.

PLANTAGINACEÆ, often stemless herbs, are scattered over the world, but they prevail chiefly in temperate latitudes; their foliage is slightly bitter and astringent; their seeds are covered with mucus, which renders many of them emulcent.

The PRIMULACEÆ are herbaceous plants, with simple alternate leaves, and without stipules; their stems are chiefly subterranean, and their leaves form a rosette at the surface of the soil, from whence springs the stalk bearing the flowers. They are mostly natives of temperate regions, rare between the tropics, but abounding in mountainous parts of Europe and Asia. The Cowslip (*Primula veris*), Fig. 432, grows in our woods and fields. The Bird's-eye Primrose

Fig. 432.—The Cowslip (*Primula veris*).

is not uncommon in boggy places as far north as Yorkshire. The different varieties of Auricula are derived from the Yellow, *P. Auricula*, a native of the Swiss Alps.

The calyx of the Primrose is monosepalous, and forms a tube terminated at the summit by five lobes or teeth. The corolla is monopetalous and hypocrateriform; its limb presents five lobes alternating with the teeth of the calyx; five stamens are inserted upon the tube of the corolla opposite the lobes. The pistil presents a superior ovary, surmounted by a more or less elongated style. The ovary is unilocular, and has in its interior a large free central placenta: it is filled with a great number of ovules. The fruit is a capsule, with five valves, opening at the summit, through which the seeds, which are supplied with a fleshy albumen enveloping a straight embryo, are suffered to escape.

Next to the Primrose we will place *Cyclamen*, so easily recognised from its elegant corolla with reflexed lobes, and by its subterranean stem, sometimes much enlarged; it is a favourite food with swine, whence the name of Sow's Bread, not uncommonly given to it. We have also *Lysimachia*, with wheel-shaped corolla, and bitter, astringent roots; one of them, the Yellow Loosestrife (*L. vulgaris*), unfolds its large handsome flowers by the sides of rivers and in shady, watery places.

Slightly removed from the Primroses are the Pimpernels (*Anagallis*); they have a round ovary, with a thread-like style; the fruit is a capsule, opening, like a box, with a lid.

ALLIANCE XXXI.—ECHIALES.

Flowers dichlamydeous, monopetalous, symmetrical or unsymmetrical; fruit consisting of several one-seeded nuts; embryo large, with little or no albumen.

* Flowers regular.

Flowers unsymmetrical; stamens two	I. Jasminaceæ.
Flowers symmetrical; stamens four; stigma naked	II. Salvadoraceæ.
Flowers symmetrical; stamens five; ovary not lobed; stigma naked	III. Ehretiaceæ.
Flowers symmetrical; stamens five; ovary more than four-lobed	IV. Nolanaceæ.
Flowers symmetrical; stamens five; ovary four-lobed	V. Boraginaceæ.
Flowers symmetrical; stamens hypogynous; ovary not lobed; stigma indusiate.	VI. Brunoniaceæ.

* * Flowers irregular, unsymmetrical.

Ovary of two deeply-lobed carpels; ovules erect	VII. Labiatæ.
Ovary not lobed; ovules erect	VIII. Verbenaceæ.
Ovary not lobed; ovules pendulous; anthers two-celled	IX. Myoporaceæ.
Ovary not lobed; ovules pendulous; anthers one-celled	X. Selaginaceæ.

The BORAGINACEÆ are mostly natives of temperate regions. These plants are abundant in Southern Europe; their properties are unimportant; the Borage (*Borago officinalis*), and the pretty Forget-me-not (*Myosotis palustris*) being, perhaps the best-known species.

The Common Comfrey (*Symphytum officinale*), Fig. 433, which we will take for a type of this family, is a herb, angular, rough, with simple alternate leaves, without stipules; the ample radical leaves

Fig. 433.—The Comfrey (*Symphytum officinale*).

are ovate pointed, petiolated; the cauline leaves decurrent, lanceolate. The flowers, disposed in scorpioid cymes, are white or violet coloured; they are regular and hermaphrodite. The calyx is five-toothed. The corolla is tubular, with an urceolate limb, and short triangular lobes. Underneath these five lobes the neck is furnished with five lanceolate ciliate scales, forming a white cone. Five stamens are inserted upon the tube of the corolla, and alternate with its lobes. The fruit is composed of four achænes. The Comfrey is common in England, growing in meadows, near rivers and ditches.

By the side of *Symphytum* are grouped other genera, such as the Borage (*Borago officinalis*), the rotate corolla of which, purple in the bud, becomes blue when fully expanded; the Viper's Bugloss (*Echium vulgare*), a common plant in waste places, on chalk or lime stone; *Pulmonaria officinalis*, employed in olden times in medicine; *Myosotis*, which from its pretty blue colour and from its beauty and freshness, has obtained the name of Forget-me-not.

The LABIATÆ, from *labium*, a lip, describes the peculiar form of the divisions of the corolla in this important order. They are herbaceous or half-shrubby plants, usually yielding more or less aromatic essential oil. Their stems are four-sided, with opposite branches and leaves; flowers in axillary opposite clusters, sessile, or on short stalks; and the upper lip of the corolla consists of two united petals opposite to the three united sepals of the bilabiate calyx, while the two united sepals are opposed to the three united petals of the corolla. The *Labiatæ* are spread over the whole world, but are most abundant in temperate regions, diminishing in number towards the poles and the tropics.

The White Dead Nettle (*Lamium album*), Fig. 434, is a herbaceous plant, frequently met with in grassy places and by road-sides. It will serve as a type of the family. The stems and branches are four-sided; the leaves simple, ovate, and opposite, long and acuminate, unequally dentate, and slightly rugose. The inflorescence is in small contracted cymes, with sessile flowers, thus forming what botanists call *glomerules* which spring from the axils of the upper leaves. The flowers are hermaphrodite and irregular. The calyx is monosepalous; the corolla rather large, white, tinged with yellow inside, monopetalous, and bilabiate. The stamens are four in number, of unequal length, two large and two small, inserted upon the corolla. The pistil consists of a superior ovary, the external and upper surface of which presents four protuberances, and a style, which is inserted in the midst of them, terminating in two branches covered with stigmatic papillæ; each protuberance is a cell of the ovary, and each cell contains an anatropal ovule; when at maturity, each cell becomes an achæne. The seed encloses a straight embryo surrounded by a fleshy albumen, slightly developed.

All the *Labiatæ* possess similar characteristic organs of vegetation

Fig. 434.—The White Dead Nettle (*Lamium album*).

to those described, with some slight differences, depending on the shape of the calyx, that of the corolla, or the number and relative dimensions of the stamens. For instance, the Sages have two stamens. These have an anther of a very remarkable structure; the connective is very long, and is placed perpendicular to the filament, like the beam of a balance. At one extremity of this beam is a cell filled with pollen; at the other an appendage which represents the other cell, which is abortive. Most species of the extremely natural family of the Labiatæ are endowed with stimulating properties, due to an essential aromatic oil, which resides in the glands placed under the epidermis. The Common Sage and many other species of the genus are thus provided; so also is the Rosemary (*Rosmarinus officinalis*), the Wild Thyme (*Thymus Serpyllum*), Peppermint (*Mentha Piperita*), and the Common Balm (*Melissa officinalis*), from all of which useful medicines are obtained by distillation. The ground Ivy (*Glechoma hederacea*), the Hyssop (*Hyssopus officinalis*) are efficacious as bitters and aromatics; *Teucrium Chamædrys* is also sometimes employed in medicine.

Alliance XXXII.—Bignoniales.

Flowers dichlamydeous, monopetalous, unsymmetrical; fruit capsular or berried with consolidated carpels, embryo with little or no albumen.

Character	Order
Placentas parietal; fruit bony, or capsular; embryo amygdaloid; radicle short.	I. Pedaliaceæ.
Placentas parietal; fruit capsular, or baccate; embryo with minute cotyledons; radicle long.	II. Gesneraceæ.
Placentas parietal; fruit succulent, hard-shelled; embryo amygdaloid; radicle short.	III. Crescentiaceæ.
Placentas axile; seeds winged, sessile, exalbuminous; cotyledons large, leafy.	IV. Bignoniaceæ.
Placentas axile; seeds wingless, attached to hard placental processes, exalbuminous; cotyledons large, fleshy.	V. Acanthaceæ.
Placentas axile; seeds albuminous	VI. Scrophulariaceæ.
Placentas free, central	VII. Lentibulariaceæ.

Pedaliaceæ occur in small numbers in the tropics, chiefly in Africa. *Sesamum*, yielding an oil which is substituted for olive oil, and several other species have useful medicinal properties.

Gesneraceæ are small bushes, frequently mere herbs. The species of *Achimenes* and *Gloxinia*, well-known in our gardens, are natives of tropical America. Other genera are found in all parts of the world, from the cooler parts of Asia, the Cape of Good Hope, the warm valleys of the Himalayas, and Australia.

Crescentiaceæ and Bignoniaceæ, are natives of the tropics of both hemispheres. The Calabash-tree (*Crescentia Cujete*) bears a great gourd-like fruit, filled with a subacid pulp, much eaten by the negroes. The chief station of *Bignonia*, whose trumpet-shaped flowers are the glory of the places they inhabit, extends from Pennsylvania to the southern provinces of Chili.

The ACANTHACEÆ, distinguished by their large leafy bracts, which nearly conceal their flowers, are almost all tropical, although the tropical genus, *Acanthus*, is found as far north as Greece, where it became the model for a graceful architectural ornament.

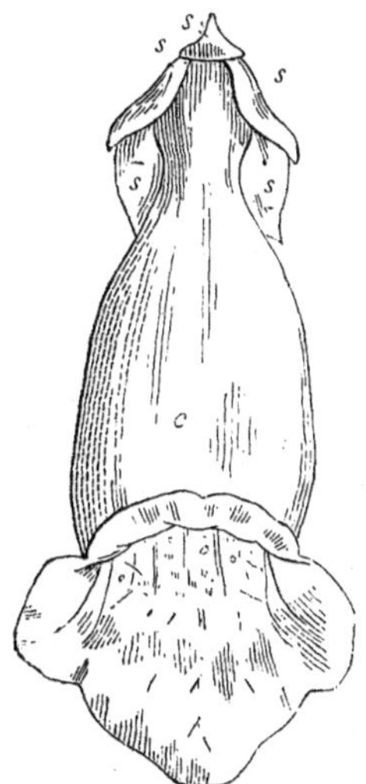

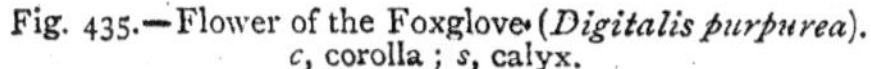
Fig. 435.—Flower of the Foxglove (*Digitalis purpurea*). *c*, corolla; *s*, calyx.

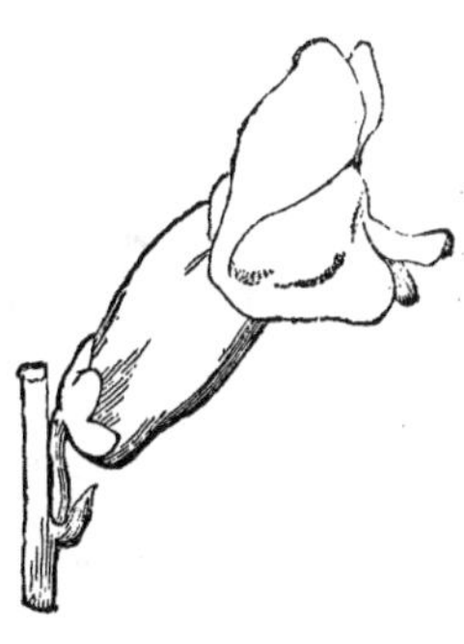
Fig. 436.—Flower of Snapdragon (*Antirrhinum*).

The SCROPHULARIACEÆ include a large number of well-known plants, of apparently anomalous structure, and forms which at first glance would seem to belong to other orders. They are distinguished from *Solanaceæ* by the absence of a fifth stamen, and the imbricated æstivation of the corolla. The tendency in the flowers in many cases is to form pouches, or spurs. The species are generally acrid, bitter, and suspected. In the Foxglove (*Digitalis purpurea*), Fig. 435, and several other species, these qualities become dangerous. *Schizanthus*, *Calceolaria*, *Alonsoa*, *Antirrhinum*, *Maurandia*, *Lophospermum*, *Rhodochiton*, *Collinsia*, *Pentstemon*,

Russellia, *Mimulus*, *Hemianthus*, present us with galaxy of green-house and garden flowers such as scarcely any other order can produce.

SUB-CLASS IV.—EPIGYNOUS EXOGENS.

Stamens adherent to the calyx or corolla; ovary inferior or nearly so.

ALLIANCE XXXIII.—CAMPANALES.

Flowers dichlamydeous, monopetalous; embryo with little or no albumen.

Ovary two or more celled; anthers free or half-united; stigma naked; corolla valvate, regular.	I. Campanulaceæ.
Ovary two or more celled; anthers syngenesious; stigma surrounded by hairs; corolla valvate, irregular.	II. Lobeliaceæ.
Ovary two or more celled; anthers syngenesious or free; stigma indusiate; corolla induplicate.	III. Goodeniaceæ.
Ovary two or more celled; stamens and style combined; corolla imbricate.	IV. Stylidiaceæ.
Ovary one-celled; corolla imbricate; anthers free; ovule pendulous; seeds exalbuminous.	V. Valerianaceæ.
Ovary one-celled; corolla imbricate; anthers free; ovule pendulous; seeds albuminous.	VI. Dipsacaceæ.
Ovary one-celled; corolla valvate; anthers syngenesious; ovule pendulous; seeds albuminous.	VII. Calyceraceæ.
Ovary one-celled; corolla valvate; anthers syngenesious; ovule erect; seeds exalbuminous.	VIII. Compositæ.

The CAMPANULACEÆ inhabit the temperate parts of the world, being rare in the tropics, and chiefly found, according to De Candolle, in our hemisphere, between the 36th and 37th parallels of north latitude; the chain of the Alps, Italy, Greece, the Caucasus, and the Altai range; the Cape of Good Hope is another centre. The whole order is more or less ornamental, and most of the plants yield a white milky juice, which is somewhat bitter and acrid.

The Bell-flower, of which we give an example (*Campanula Medium*, Fig. 437), derives its name from its large, full-blown, bell-shaped corolla, opening in great numbers at the same time. Its stem is erect, branching towards the top; the leaves are sessile, ovately lanceolate, irregularly crenate and dentate, with the flowers slightly inclined, and disposed in loose bunches. The flowers are regular and hermaphrodite; the calyx gamosepalous with five lobes; the corolla, campanulate or bell-shaped, is divided in its upper part into five teeth, alternating with the calycine lobes. The stamens, five in number, are free, and are not inserted in the tube of the corolla; the anthers are bilocular, and the filaments flattened and enlarged in their lower part in order that they may embrace the ovary. The pistil is composed of an inferior ovary, surmounted by a style, divided into five stigmatic branches. The ovary is five-celled; the fruit a capsule, with five cells which open at the base.

Many species of *Campanula* are cultivated. We may cite, as worthy of attention, the Peach-leaved Bell-flower (*C. persicifolia*), with flowers of pale blue, erect and in long compound panicles; it is not indigenous to this country, but is naturalised in one or two places. This species grows double, and forms a fine garden flower. *C. pyramidalis* is another stately Bell-flower, which reaches the height of three or four feet, forming a pyramidal mass of flower. The pretty Harebell (*C. rotundifolia*), with slender stem and nodding cluster of "heavenly blue," is not the least worthy of those that deserve notice.

Fig. 437.—The Bell-flower (*Campanula Medium*).

The LOBELIACEÆ are frequently found within or upon the borders of the tropics of both hemispheres. They are extremely beautiful when in blossom, and great favourites in the greenhouse, but the milky juice with which they are charged is powerfully acrid and narcotic, corrodes the skin, and is fatal taken internally. Burton says that if horses eat *Isotoma longiflora*, inflammation is produced, so that they swell until they burst. Like most of the poisonous orders, it includes species possessing valuable medicinal properties. *Lobelia inflata* is a North American species, which, when dried, is formed into a cake, when it has a slightly irritating, acrid taste, somewhat like tobacco, causing a flow of saliva and a feeling of nausea.

The VALERIANACEÆ are natives of temperate climates; they are generally strongly-scented aromatic plants. The DIPSACACEÆ include the Teazel (*Dipsacus*) used by fullers.

The COMPOSITÆ include an immense number of herbaceous and shrubby plants, sometimes of small trees, amounting to ten thousand species. The flowers of this family have an arrangement quite

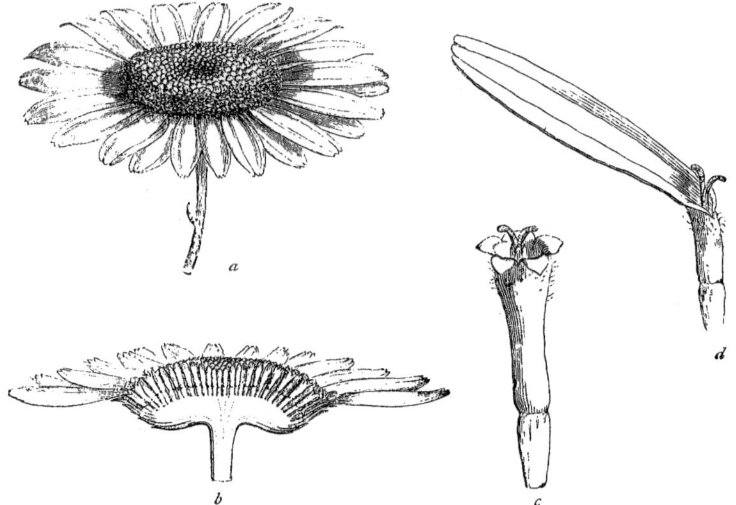

Fig. 438.—Disc, section, and detached florets of Ox-eye Daisy (*Chrysanthemum leucanthemum*).

characteristic. They are all disposed on a head or disc, so as to have the appearance of being a single flower, though they are really a union of many flowers; hence the name of COMPOSITÆ, which has been given to them. This arrangement is easily understood if we examine the representation of the *capitulum* as it appears in the Ox-eye Daisy (*Chrysanthemum leucanthemum*), of which we give in Fig. 438 the whole of the capitulum in *a*; a section of the head is seen at *b*; finally, the isolated flowers of the centre, and the circumference of the same capitulum, are represented in *c* and *d*. The flowers of the same capitulum may be all hermaphrodite; or they may also be of two kinds—the exterior, female; the interior, male. The calyx of these flowers may be of various shapes. Sometimes it is so reduced that it seems as if it were absent; in other cases it forms a sort of cup or crown; it even degenerates into a kind of silky tuft, or pappus. The corolla is either regular or irregular. In the former case it is tubular, and five-lobed. In the latter it is split in its greatest extent, and forms a tongue, dentate at the summit, or it separates into two lips. The stamens are inserted upon the tube of the corolla, and alternate with its divisions. The filaments are generally free, but the two-celled anthers are united at their edges into a tube which sheathes the style. The pistil is composed of a unilocular ovary, containing a single, straight, anatropal ovule; it is surmounted by a minute style, which is divided into two branches, both in the hermaphrodite and female flowers, but it is undivided in the male. The branches of the style are furnished with stigmatic papillæ and collecting hairs. Before expansion the style is shorter than the stamens; but at the time of fecundation it increases rapidly, and rises into the hollow cylinder formed by the anthers. As they rise, the hairy collectors sweep off the pollen which the gaping anthers contain, and soon appear charged with its precious dust. It is observed that the female flowers are destitute of hairy collectors; that the male flowers are destitute of the stigmatic papillæ; that the neuter flowers have neither stigmatic papillæ nor collecting hairs. The fruit is an achæne, often furnished with a plume, to favour its dissemination. The solitary seed encloses a straight embryo without albumen.

This enormous family of flowers has recently been systematised by Mr. Bentham; his arrangement is not as yet published, and that of De Candolle, which has long been in use, is given here instead.

SUB-ORDER I.—*LIGULIFLORÆ;* florets all hermaphrodite, and slit or ligulate.

1. *Cichoraceæ.*

SUB-ORDER II.—*LABIATIFLORÆ;* hermaphrodite florets, or at least the unisexual ones, divided into two lips.

2. *Mutisiaceæ.*—Style cylindrical or somewhat tumid; its arms usually blunt or truncate, very convex at the outside, and covered at the upper part by a uniform downiness, or absolutely naked.

3. *Nassauviaceæ.*—Style never tumid; the branches long, linear, truncate, fringed only at the point.

SUB-ORDER III.—*TUBULIFLORÆ*; hermaphrodite florets, tubular, with five or rarely four, equal teeth.

4. *Vernoniaceæ.*—Style cylindrical; its arms generally long and subulate, occasionally short and blunt, always hispid all over.

5. *Eupatoriaceæ.*—Style cylindrical; its arms long and clavate, with a downy papillose surface on the outside near the end.

Fig. 439.—Chicory. Fig. 440.—Isolated Flower of Chicory.

6. *Asteroideæ.*—Style cylindrical; its arms linear, flat on the outside, and equally and finely downy.

7. *Senecionideæ.*—Style cylindrical; its arms linear, fringed at the point, generally truncate, but sometimes extended beyond the fringe into a cone or appendage of some sort.

8. *Cynareæ.*—Style thickened upwards, and often fringed at the tumour; its arms downy externally.

The *Cichoraceæ* possess a milky juice, contained in a system of latiferous vessels, which has bitter, resinous, narcotic principles. The properties and virtues of these species vary according to the relative proportion of these principles, and according to the age of the plants and the development of their different organs.

The Chicory (*Cichorium Intybus*) is an indigenous species in France, and probably with us; the roots are used in medicine. Fig. 439 represents the Wild Chicory; Fig. 440 an isolated flower from the capitulum of this plant. The roots of the Cultivated Chicory, dried, roasted, and ground, are sometimes mixed with coffee. The young leaves, dressed as a salad, are eaten by the lower classes in France. The Endive (*C. Endivia*) is a Mediterranean plant, less bitter, and is appropriated to alimentary use as a salad.

The Wild Lettuce (*Lactuca sativa*) has a bitter juice, of an offensive odour. The Cultivated Lettuce yields a pharmaceutical extract, called *lactucarium*, which possesses narcotic properties. It is sometimes employed in medicine as a substitute for opium. The young leaves of the Lettuce, of which many varieties are cultivated in our gardens, are used as salads.

To this sub-order belong also *Scorzonera*, Salsify (*Tragopogon*), and the Dandelion (*Taraxacum*).

The Tubulifloræ include plants some of which were held in great favour for medical uses in olden times, but are now abandoned. Such are the Holy Thistle (*Cnicus benedictus*), Milk Thistle (*Silybum Marianum*), Star Thistle (*Centaurea Calcitrapa*), Corn Bluebottle (*Centaurea Cyanus*), common in corn-fields. One of this family, the Safflower (*Carthamus tinctorius*), furnishes a dye soluble in alcohol. It came originally from India, but is now cultivated in Asia, America, and nearly over the whole of Europe. The colour drawn from the Safflower is not very strong, but its shades are very delicate and varied. Some of the species are comestible. Such are the Artichoke (*Cynara scolymus*), of which the base of the bracts of the involucre, and the receptacle are eaten; of the Cardoon (*Cynara Cardunculus*), the midrib of the leaves is eaten, being whitened and rendered fleshy by blanching.

The sub-order also comprises plants in which a bitter principle is generally combined with a volatile oil or resin, whilst the roots contain a matter more or less analogous to fecula, somewhat resembling starch, known under the name of *inuline*. We may mention Wormwood (*Artemisia absinthium*), from which the well-known liqueur *absinthe* is obtained; Tansy (*Tanacetum vulgare*); Yarrow or Milfoil (*Achillea Millefolium*); several Alpine species of *Achillea* are also used by the Swiss as tea; the Chamomile (*Anthemis nobilis*), of which we give a representation in Fig. 441, an example in which all the flowers have become ligulate by culture; the *Arnica* and Elecampane (*Inula Helenium*), &c.

Fig. 441.—Common Chamomile.

It is to this sub-order that most of the cultivated *Compositæ* belong. Such are the Chrysanthemums, of which we have so many beautiful varieties in our town gardens, for which they are especially adapted; the Daisy; the species of *Aster*, autumnal plants, originally from North America; *Cineraria cruenta*, with innumerable garden forms; *Helichrysum orientale*, whose flowers are known as *Immortelles; Zinnia;* lastly *Dahlia*, of which the species, originally from Mexico, with simple single flowers, has yielded varieties—the glory of our gardens in the autumn—vying even with the Chrysanthemums.

ALLIANCE XXXIV.—MYRTALES.

Flowers dichlamydeous, polypetalous; placentas axile; embryo with little or no albumen.

Characters	Order
Ovary one-celled; ovules pendulous; leaves dotless; seeds exalbuminous; cotyledons convolute.	I. Combretaceæ.
Ovary one-celled; ovules pendulous; leaves dotless; seeds albuminous; cotyledons flat.	II. Alangiaceæ.
Ovary one-celled; ovules ascending; leaves dotted; embryo fused into a solid mass.	III. Chamælauciaceæ.
Ovary usually several-celled; flowers polypetalous or apetalous; calyx open, minute; stamens definite; ovules pendulous; cotyledons minute.	VI. Haloragaceæ.
Ovary several-celled; flowers polypetalous or apetalous; calyx valvate; stamens definite; ovules horizontal or ascending; cotyledons flat, much larger than the radicle.	V. Onagraceæ.
Ovary several-celled; flowers polypetalous; calyx valvate; stamens indefinite; cotyledons flat, much shorter than the radicle.	VI. Rhizophoraceæ.
Ovary several-celled; flowers monopetalous; calyx valvate; stamens indefinite, monadelphous; cotyledons amygdaloid.	VII. Belvisiaceæ.
Ovary several-celled; flowers polypetalous; calyx imbricate; stamens definite; anthers rostrate; leaves usually dotless.	VIII. Melastomaceæ.
Ovary several-celled; flowers polypetalous or apetalous (or valvate); calyx imbricate; stamens indefinite; anthers oblong; leaves usually dotted.	IX. Myrtaceæ.
Ovary several-celled; flowers polypetalous; calyx valvate or imbricate; stamens indefinite, part collected into a fleshy hood; anthers oblong; leaves dotless.	X. Lecythidaceæ.

The COMBRETACEÆ are natives of the tropics, mostly astringents; the bark of several is useful in tanning; others yield gum; the galls of several species are useful dyes. ALANGIACEÆ are chiefly Indian, with aromatic roots; fruit edible but insipid; timber valuable. CHAMÆLAUCIACEÆ, are small, beautiful flowering bushes, resembling the Heaths; they abound in Australia; of their uses and properties nothing is known. HALORAGACEÆ, are found in ditches, sluggish streams, mostly in temperate parts of the world. Species of *Trapa* supply an article of food in their kernels.

The ONAGRACEÆ include *Fuchsia*, *Clarkia*, and some other garden favourites of great beauty, natives of the northern hemisphere, and abundant in the New World. Some of these, as *Jussiæa* and *Fuchsia*, are used in Brazil and Chili as dyes; others are astringents. The species of *Œnothera* expand their flowers in the evening, whence their name of Evening Primrose. RHIZOPHORACEÆ, or Mangroves, grow in muddy waters on the coast, where they soon form dense thickets which the sun's rays fail to penetrate; hence the putrid exhalations which render many tropical regions near the coast and large rivers so unhealthy; the seeds have the curious habit of germinating while still attached to the branch that bears the fruit. BELVISIACEÆ are smooth-leaved Camellia-like bushes, wholly African and tropical, of whose uses little is known. The MELASTOMACEÆ are natives of both hemispheres, but mostly American; they are all slightly astringent. Many produce edible fruit, and some are useful in medicine. The MYRTACEÆ are natives of hot climates, within and without the tropics; *Myrtus communis*, the Myrtle, the species with which we are most familiar, is a native of Persia. The order includes the *Eucalyptus*, the Gum-trees of Australian travels; the Pomegranate (*Punica Granatum*); the Guavas (*Psidium*), and a number of other exotics equally well known from their fruits or flowers. LECYTHIDACEÆ are natives of the hottest parts of South America. Brazil and Sapucaya "nuts" are seeds of plants belonging to this order.

ALLIANCE XXXV.—CACTALES.

Flowers dichlamydeous, polypetalous; placentas parietal; embryo with little or no albumen.

Sepals and petals distinct; stamens opposite the petals; styles separate; ovules pendulous.	I. Homaliaceæ.
Sepals and petals distinct: stamens scattered; styles confluent; ovules pendulous; seeds albuminous.	II. Loasaceæ.
Sepals and petals numerous, similar; stamens scattered; styles confluent; ovules horizontal; seeds exalbuminous.	III. Cactaceæ.

HOMALIACEÆ are all tropical.

LOASACEÆ are American plants found over the whole continent; their most noted peculiarity being the secretion of an acrid juice with which the hairs on the stem are charged.

The CACTACEÆ came originally from the American continent. They are at the same time fleshy and ligneous. Their branching stems present the most varied, often the most grotesque forms. Sometimes they are erect, like a tall fluted column; at others they are massed together like a solid sphere, tapering off into cylindrical branches, or flattened after the manner of the Indian Fig. In short, nothing is more varied than the aspect of the numerous species which grow naturally in strange profusion in America, and which art has brought together in great quantities in our gardens for the purposes of study or gratification. The stem is generally destitute of leaves, the position of which is only indicated by a small cushion situated under the bud. Nevertheless, *Pereskia* has true petioled leaves, which are large and oblong, caducous, or deciduous. The buds, situated in the axils of the leaves, are of two kinds; the lower are furnished with spines, the upper are developed into branches and

Fig. 442.—*Mamillaria elephantidens.*

flowers. Fig. 442 represents *Mamillaria elephantidens,* one of the genus cultivated in greenhouses.

The flowers in the genus *Cactus* are regular and hermaphrodite; their envelopes are composed of a great number of divisions, the exterior of which are sepals, whilst the internal ones are petals, it is not always possible to find the precise limit between the corolla and calyx. The stamens are very numerous. The ovary is inferior, and surmounted by a lengthened style, divided into several stigmatic branches; it is unilocular, and has as many parietal placentas on its interior as there are stigmatic branches; upon each of these placentas are a number of anatropal ovules. The fruit is a pulpy berry; the seeds have a straight or curved embryo, and little or no albumen.

Opuntia has the stem more or less flattened, with oval or oblong articulations bearing bunches of spines or hairs. The flowers are large and magnificent. Nothing is more curious than these large corollas clothed in the most vivid colours, and which seem to be nailed upon the strong, prickly, and succulent ragged stem of the plants. The flowers seem to spring from the bunches of hair, or from the edges of the articulations; they are white, red, or yellow, according to the species. The fruit, of various size and colour, is edible, but very insipid in taste. The Prickly Pear, or Indian Fig (*Opuntia Tuna*), is a plant originally from America. It has long been naturalised in the South of Europe, in Spain, Italy, Sicily, Greece, &c., where it is cultivated to make hedges and enclosures, its fruit being, to some extent, the food of the inhabitants of these countries. *O. cochinellifera* is the plant on which the cochineal insect feeds and breeds. This produces the rich-coloured pigment employed in the manufacture of carmine.

The species of *Cereus*, or Torch Thistles, have continuous angular stems, the angles with bunches of hairy prickles. The flowers are large and beautiful. Those of the Torch Thistle of Peru are solitary, about six inches in length, white within, the tube greenish and the limb rose colour. *Cereus giganteus*, indigenous to Mexico, resembles an immense candelabrum, sixty feet high. The genus *Echinocactus* is frequently cultivated in this country. It resembles a rolled-up hedgehog, but its sides present longitudinal furrows, furnished with white cottony excrescences with short and spreading spines. It is from these thorny tubercles that the flowers spring; they are always large and beautiful, and last for many days. The *Melocactus* has a globular, ovoid, or pyramidal stem, with the sides separated by straight furrows. This stem is surmounted by a dense mass of bristly wool mixed with slender compact spines, from amongst which the flowers spring; they are very small, and ephemeral in their duration. *Melocactus communis* is cultivated in greenhouses under the name of Turk's-cap Cactus.

Lastly, we must mention the *Mamillaria*, of which we have previously given a representation. The thorny tubercles in this genus are spirally disposed round the stem. The flowers, which last a long time, often surmount it, forming a kind of zone or crown.

Alliance XXXVI.—Grossales.

Flowers dichlamydeous, polypetalous; seeds numerous, minute; embryo small, in a copious albumen.

Fruit pulpy; placentas parietal	I. Grossulaceæ.
Fruit capsular; placentas axile; style one; stamens definite; calyx imbricate.	II. Escalloniaceæ.
Fruit capsular; placentas axile; stamens indefinite; calyx valvate.	III. Philadelphaceæ.
Fruit pulpy or fibrous; placentas axile; style one; stamens indefinite; calyx imbricate.	IV. Barringtoniaceæ.

The Grossulaceæ are chiefly natives of the temperate and

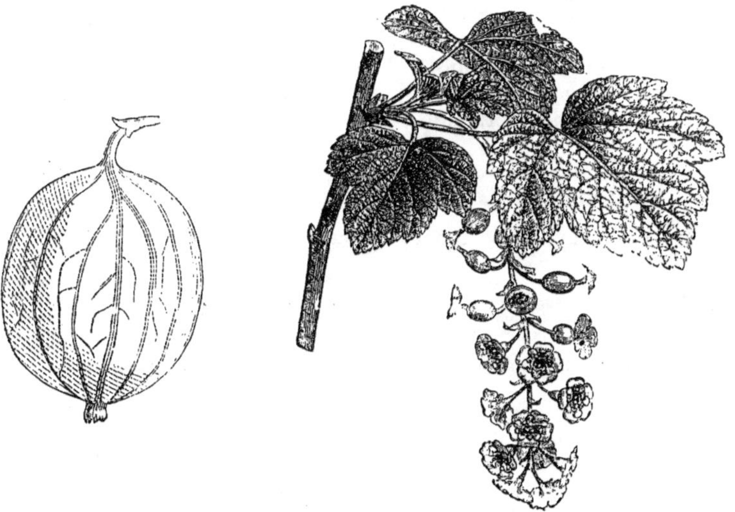

Fig. 443.—Gooseberry.

Fig. 444.—Branch and Flowers of the White Currant.

colder regions of the northern hemisphere. The berry is mucilaginous and agreeable, from the presence of malic acid. The Common Gooseberry (*Ribes Grossularia*) is a shrubby plant, often armed with spines, placed under the leaves, which are alternate or fasciculate, and palmately-lobed with a dilated petiole at the base. The flowers are arranged in bunches or clusters, axillary in the species destitute of spines; solitary, or at least less numerous, in species thus armed. The calyx monosepalous, with five divisions; corolla with five free petals, alternating with the lobes of the calyx; stamens five, and perigynous, and opposite to the sepals; the pistil consists of an inferior ovary, surmounted by two short styles with obtuse stigmas. In the interior of the ovary, which is unilocular, are two placentas bearing numerous ovules. The fruit is a berry (Fig. 443), crowned by the persistent limb of the calyx and the dried-up petals. The integuments of the seeds become gelatinous externally, internally crustaceous, and contain an albumen, hard and nearly horny, and very abundant, at the base of which is found a very small, straight embryo.

Many species of Ribes are cultivated in gardens as ornamental plants, such as *R. aureum* and *R. sanguineum*. Others are cultivated for their fruit, such as the Gooseberry, which we have described, the Currants, Red and White (both varieties of *R. rubrum*), and the Black Currant (*R. nigrum*), the fruit of which contains, as does its leaf, an aromatic resinous principle.

ALLIANCE XXXVII.—CINCHONALES.

Flowers dichlamydeous, monopetalous; embryo minute, in a copious albumen.

Stamens epigynous, dehiscence porous	I. Vacciniaceæ.
Stamens epipetalous; anthers sinuous; flowers unsymmetrical . . .	II. Columelliaceæ.
Stamens epipetalous; anthers straight; leaves with interpetiolar stipules.	III. Rubiaceæ.
Stamens epipetalous; anthers straight; leaves exstipulate	IV. Caprifoliaceæ.

The VACCINIACEÆ, or Cranberries, are found in mountainous and marshy places in temperate regions of the Old and New World, chiefly in the northern hemisphere. The Bilberry or Whortleberry (*Vaccinium Myrtillus*) is a well-known example. The Bleaberry (*V. uliginosum*), and the Cranberry (*Oxycoccus palustris*), are equally well known in many parts of the British Isles.

The COLUMELLIACEÆ occupy an uncertain anomalous position among the surrounding orders. The only known species are from Mexico and Peru, and their uses still unknown.

The RUBIACEÆ form one of the largest orders, and includes plants with a considerable diversity of properties. It may be divided into three sub-orders:—

Stipules leaf-like	I. Stellatæ.
Stipules small; ovary with one-two seeds in each cell	II. Coffeæ.
Stipules small; ovary many-seeded	III. Cinchoneæ.

Fig. 445.—*Sherardia arvensis.*

As a type of the *Stellatæ* we may take the Field Madder

(*Sherardia arvensis*), Fig. 445, a small annual with lilac flowers, nearly sessile, and disposed in a dense head. They are hermaphrodite and regular; the calyx gamosepalous, with six teeth; the corolla monopetalous, and four-lobed. There are four stamens alternating with these lobes, inserted upon the tube of the corolla. The pistil consists of an inferior ovary, surmounted by a style divided into two stigmatic branches. Each of the cells contains an ascending anatropal ovule. The fruit forms two *achænes*, crowned by the calyx; the embryo is slightly curved, in a horny albumen. The leaves are simple, opposite, and accompanied by two lateral stipules, which resemble the leaves sufficiently to make it appear that there are six verticillate leaves without stipules. Besides *Sherardia*, several other species, such as the Woodruff (*Asperula*), the Bedstraws (*Galium*), the Madder (*Rubia*), are all common indigenous plants. *Rubia tinctoria* is cultivated in the South of France for the sake of its roots, which contain a colouring matter of a beautiful red, much used in dyeing.

The *Coffeæ* are an important sub-order of the *Rubiaceæ*. The Coffee-tree (*Coffea*) is an evergreen shrub, having lanceolate, wavy, and smooth leaves, resembling those of the Laurel. They are opposite, and accompanied by two interpetiolar stipules; the flowers are white and fragrant, forming clusters in the axils of the leaves; the calyx is four or five-lobed; the corolla funnel-shaped and also five-lobed; stamens equal in number; the fruit, a red berry, about the size of a cherry, consists of a thick and rather sweetish pulp, which encloses two seeds, in a parchment-like shell. Each of these is convex and smooth on the outside, and furrowed on the inside. The short and straight embryo is at the base of a hard albumen, which constitutes nearly the whole of the seed. The Coffee-plant, which was brought originally from Abyssinia, was in the fifteenth century transported into Arabia, which has since become as a second home to this shrub, no coffee being equal to that produced in the neighbourhood of Mocha.

Cephaëlis belongs to the same sub-order. The species are small shrubs, natives of the forests of Brazil. They are chiefly distinguished by the properties of their roots, which yield Ipecacuanha, a drug having bitter, acrid, and nauseous properties, but a valuable medicine; that of commerce is the produce of *Cephaëlis Ipecacuanha* and some other species.

The sub-order *Cinchoneæ* is represented by the genus *Cinchona* (Fig. 446). This consists of evergreen trees or shrubs, which grow in the tropical Andes, between ten degrees of north latitude and nineteen degrees south, at a height of from 2,000 to 3,000 feet above the level of the sea. They have regular hermaphrodite flowers; calyx monosepalous with five teeth; corolla monopetalous, cup-shaped, and five-lobed; stamens five, alternate with these lobes, inserted upon the tube of the corolla; ovary inferior, surmounted by a style, divided into two stigmatic branches. The fruit is two-celled, and contains numerous winged seeds. The bark of most of the ligneous species contains astringent and bitter principles, which, though existing in other genera, are more abundant in *Cinchona*, and especially in *C. Calisaya*, the bark of which appears to be the richest in quinine of all the known species.

Fig. 446.—Flowers of *Cinchona*.

The mode of procuring this invaluable febrifuge is interesting, and has been recorded in the following notes:—"About the end of June, 1847," says Mr. Weddell, "I set out to the province of Casabaya. This province is divided by the Cordillera into two distinct regions; the one forming table-lands, the other comprehending a long series of parallel valleys. . . . . . These valleys furnish the greater part of the Peruvian bark. It would be difficult to give an idea of all the treasures of vegetation buried in these vast solitudes. The thirst for gold formerly peopled them, but the wilderness has resumed its empire, and the axe of the *cascarillero* alone breaks its silence now. The name of "cascarillero" is given to those men who cut the Peruvian bark in the woods; they are brought up to this occupation from their childhood, and instinctively, as one might say, they find their way to the centre of the forest, through almost inextricable labyrinths, as if the horizon were open before them.

"These *cascarilleros* do not gather the Peruvian bark for their own profit; generally they are enrolled in the service of some tradesman or small company, who send a sort of overseer to superintend their labour. Having fixed upon a portion of the forest favourable to their purpose, the party proceed to make roads to the point which is to be the centre of their operations. From this time, every part of the forest—a view of which is commanded by the new pathway—becomes provisionally the property of the party, and no other *cascarillero* dare work it. The overseer, having established his camp, proceeds to build a wooden hut, in which he can shelter himself and store his provisions; and if their stay is likely to be prolonged, he proceeds to sow maize and vegetables for the use of the party; the *cascarilleros*, in the meantime, wandering over the forest one by one, or in small bands, each enveloped in his poncho, with provisions for several days, and the blankets which constitute their beds. They range the forest, axe or knife in hand, to clear away the innumerable obstacles which arrest their progress at every step; for the *cascarillero* is exposed to perils which often endanger his life. The forests are rarely composed entirely of *Cinchona*; they form groups more or less numerous, scattered here and there in the depths of the forest; sometimes—and this is commonly the case—they are completely isolated. If the position be favourable, a glance at the branches—a slight play of colour peculiar to the leaves—the aspect produced by a large mass of inflorescence, reveals the tree at a great distance. In other circumstances he must content himself with an inspection of the trunk, in which the outer layer of bark—the fallen leaves, even—are sufficient to make known the neighbourhood of the object of their search. Having marked the group, they begin operations by felling the tree with the axe, a little above the root, taking care, in order to lose none of the bark, to bare it at the place where the axe is to be laid; and as the thickest part is surrounded by the largest quantity of bark, and is consequently the most profitable, it is usual to dig out the earth at the foot of the trunk, so that the barking should be complete. The Cinchona is sometimes completely surrounded with *lianes*, which shoot from tree to tree. I remember having cut down a large tree, hoping to get the flowers, but after having knocked down three neighbouring trees it still remained standing, supported in that position by the *lianes*, which were wound round its branches, supporting it as if wrapped in a shroud. When at last the tree falls, the outer bark is removed by means of a wooden mallet, or the back of an axe. The part thus stripped is then brushed, and divided throughout by uniform incisions. The bark is separated from the trunk by means of a knife, with the point of which the surface of the wood is as far as possible shaved. The bark of the branches is separated much as that of the trunk. The details of dressing the bark vary a little in the two cases; in fact, the thinner plates of the bark of the branches, which make the rolled bark, called *canuto*, are merely exposed to the sun, when they take of themselves the desired form, which is that of a hollow cylinder; but those which are the produce of the trunk, and constitute the ordinary bark, which is called *tabla*, are subjected during the drying process to great pressure, without which they would take the shape of the others. After their first

XIX.—Gathering Cinchona Bark in a Peruvian forest.

exposure to the sun, the squares are disposed one on the top of the other, just like the planks of deal in a timber-yard, and are kept level by means of heavy weights laid on the pile. The next day the squares of bark are put back again in the sun for a short while, then back again into the press, and so on. But the work of the *cascarillero* is not nearly finished, even when the preparation of the bark is over; his spoil has to be conveyed to the camp. With a heavy load upon his shoulder, he has to retrace the intricate paths that he traversed with difficulty without his burden. I have seen more than one district where the bark had to be carried through the wood during fifteen or twenty days—it is difficult to conceive how such labour can be properly remunerated. The care of packing the bark, which devolves upon the overseer, is no unimportant part of the labour. He arranges the different loads, as the cutters bring them into the camp, in parcels, which are sewn up in woollen canvas packing." In this condition the bales are transported on the backs of men, asses, or mules, to the town depôts, where they are packed in fresh leather, in which they acquire a great solidity. When dry they are called *surons*, and in this condition they reach Europe. PLATE XIX., copied from Mr. Weddell's work, represents the harvesting of the bark of the Cinchona in the manner described in a Peruvian forest.

Many ornamental species of *Rubiaceæ* embellish our hot-houses. Such are *Ixora coccinea*, a beautiful shrub of the island of Ceylon, with persistent slightly succulent leaves, and bright red flowers, disposed in tufts, which long preserve their brilliancy; *Ixora odorata*, a native of Madagascar, whose flowers exhale a delicious odour; *Rondeletia speciosa*, from the Havanna, has tubular flowers, of a brilliant scarlet outside, with yellowish orange inside the throat; *Rogiera*, from Guatemala; *Bouvardia*, from Mexico; *Luculia gratissima*, from Nepaul, whose rose-coloured corollas exhale a delicious perfume; *Gardenia florida*, commonly called the Cape Jasmine.

The CAPRIFOLIACEÆ are natives of the northern parts of Europe, Asia, and America, but are rare in Northern Africa, and still less known in the southern hemisphere. Many of the family are climbing plants, of which the Honeysuckle is an example.

The Elders are familiar inhabitants of our hedgerows, and about cottages and farmhouses, generally near ponds or ditches with stagnant water. The dwarf species (*S. Ebulus*) is fœtid, and somewhat nauseous. The common Elder (*S. nigra*) is a small bushy tree, with delicate cream-coloured flowers in cymes, which are in full blossom in June, and its dark purple clusters of berries are equally beautiful in September and October. Large orchards of Elders are cultivated in Kent for the purpose of making wine from their fruit. The flowers are also distilled with water and alcohol, and yield a perfumed liquid known as *elder-flower water*, much approved for the toilet.

The balls of white blossom of the Guelder Rose (*Viburnum Opulus*) have a fine effect in a well-arranged shrubbery.

The Honeysuckle (*Lonicera Periclymenum*) is the Woodbine of the poets, the "twisted Eglantine" of Milton. Shakespeare writes:—

> "So doth the woodbine, the sweet honeysuckle,
> Gently entwist the maple."

In many green lanes in Britain this sweet-scented climber may be observed encircling the stem of some young tree, which bears indelible marks of its embrace as it winds round the stem from left to right. The bright red, or rather crimson berries, succeed the fragrant clusters of flowers, equalling them in beauty.

Linnæa is a lowly plant, the name of which was changed from *Nummularia*, at the request of the great botanist, to commemorate his own name. "Its lonely, depressed growth," he said, "was a fitting emblem of his own early condition." It is found in Fir woods in the North of England and Scotland, and also in the northern regions of Europe, distinguished by its slender, trailing stem, and drooping flowers of pale purplish-rose colour, with something between a bell and funnel-shaped corolla.

ALLIANCE XXXVIII.—UMBELLALES.

Flowers dichlamydeous, polypetalous; seeds large, solitary; embryo small, lying in a copious albumen.

Ovary two-celled, crowned with a double fleshy disc; styles distinct .	I. Umbelliferæ.
Ovary more than two-celled; corolla valvate; leaves alternate . . .	II. Araliaceæ.
Ovary two-celled; style filiform; corolla valvate; leaves opposite . .	III. Cornaceæ.
Ovary two-celled; styles two; corolla imbricate; leaves alternate . .	IV. Hamamelidaceæ.
Ovary one to three-celled; style simple or bifid; anthers extrorse . .	V. Bruniaceæ.

This important group of Exogens, which is familiarly represented by the Hemlocks, Wild Celery, Parsleys and Fennels, takes the most singular forms in *Astrantia*, *Eryngium*, and *Leucolæna*, where, instead of the hollow and fistular stem of the Hemlock, they become solid, branching bushes, with panicled flowers, and the inconspicuous involucre of *Œnanthe* becomes great white three-lobed plates surrounding the flower in *Leucolæna rotundifolia*.

The order includes about 1,300 species; it has been divided by Bentham and Hooker into the following sub-orders:

I. HETEROSCIADIÆ . . .	Umbel simple ; vittæ none	Including Astrantia. Eryngium. Hydrocotyle.
II. HAPLOZYGIÆ . . .	Umbels compound ; primary ridges of fruit only conspicuous ; vittæ rarely absent.	Including Œthusa. Anethum. Angelica. Carum. Cicuta. Conium. Fœniculum.
III. DIPLOZYGIÆ . . .	Umbels compound ; primary and secondary ridges of fruit both conspicuous.	Including Coriandrum. Cuminum. Daucus.

The family of the *Umbelliferæ* is one of the most important of the vegetable world, as well for the number of the species which compose it, as for the medicinal and economic properties which belong to the different species. One of the characteristic traits in their organisation consists in the presence of reservoirs or canals (*vittæ*) within the fruit, which contain aromatic volatile oils.

Some of the species are valuable contributions to the table, others deadly poisons when improperly used. We shall take a cursory glance at the more interesting, taking them at random as they present themselves. The Garden Angelica (*Angelica Archangelica*), formerly largely cultivated on account of its aromatic pungent leaf-stalks (Fig. 447), is a pretty herbacous plant, indigenous in the mountains of the south and east of France, and naturalised in England. Its tap-root is rather voluminous; the bluish-green stem attains the height of three feet and upwards. This stem is tumid or hollow, as are also the petioles of the leaves, which are large, doubly compound, and serrate. The flowers form little umbellules, disposed again in umbels ; they are small, and of a greenish colour. The calyx presents a limb formed of five very small teeth. The corolla is composed of five petals, free, elliptic, entire, curved ; there are five stamens, dorsally attached to the filaments, and alternating with the projecting petals. The pistil is composed of an inferior ovary, surmounted by two spreading styles, each terminated by a small ovoid stigma. This ovary is two-celled, each cell enclosing a suspended anatropal ovule. When at maturity, the fruit, which is winged, constitutes two achænes, one for each cell, which finally separate, and remain suspended at the extremity of two filaments, which are prolongations of the receptacle. Each achæne encloses one seed, formed almost entirely of horny albumen, towards the upper extremity of which a small cylindrical embryo is enclosed.

The Wood Archangel (*A. sylvestris*), which grows wild on banks

Fig. 447.—Angelica.

of rivers, and in wet and marshy places, contains analogous properties, but in a less degree. The same is the case with the Alexanders and the Masterwort.

A greater number of the species cultivated in Europe furnish fruits with a hot and aromatic flavour, which have been employed from time immemorial as condiments. Such are the Aniseed (*Pimpinella Anisum*), Cumin (*Cuminum cyminum*), Dill (*Anethum graveolens*), Coriander (*Coriandrum sativum*), Caraway (*Carum Carui*), Fennel (*Fœniculum vulgare*), &c.

Several of the Umbelliferæ occupy important places in our kitchen gardens. The root of the Wild Carrot (*Daucus Carota*), so common in our fields, is small, heavy, fibrous, and of an acrid flavour. In its wild state this root cannot be eaten, but under the influence of culture it becomes fleshy, voluminous, nutritious, and sweet, while retaining its aromatic flavour. The Parsnip (*Pastinaca sativa*) grows spontaneously in chalky districts. Like the Carrot it is tap-rooted, and culture has rendered it alimentary. Smallage (*Apium graveolens*), when cultivated, takes the name of Celery. Its petioles in the wild state are acrid and strongly odorous, but under the influence of culture they acquire a sweeter flavour. The colour is taken from them by blanching, that is, by keeping them in the dark covered with earth. Parsley (*Petroselinum sativum*), indigenous to the South of Europe is now cultivated chiefly for the sake of its leaves; the same is the case with the Chervil (*Anthiscus Cerefolium*).

Some of the *Umbelliferæ* have poisonous or narcotic properties. The first in this list is the Hemlock (*Conium maculatum*). It is a common plant on the road-side, on rubbish heaps, in burial grounds, and in damp shaded places in the neighbourhood of habitations. Its root is white and spindle-shaped. Its straight branching herbaceous stem is from a yard to six feet high; it is smooth, cylindrical, glaucous, slightly fluted, and spotted with marks of a deep purple colour. It has very large, alternate, deeply-cut, compound leaves, with small serrate segments; its flowers are small, white, and disposed in terminal umbels, consisting of from ten to twelve rays. Its petals are almost equal, sessile, and somewhat heart-shaped. Upon each of the two lateral portions of the fruit are five projecting crenulate ribs, which give it the appearance of being covered with small asperities or tubercles. Any part of the Hemlock, when crushed between the fingers, exhales a fœtid and disagreeable odour. It is well known that this plant constitutes a violent poison to man, and still more so to animals. The poisonous properties of Hemlock have been known from the most ancient times; Socrates and Phocion were recompensed for the services they had rendered to the Greeks by having to drink the juice of this plant.

The Water-Hemlock, or Cowbane (*Cicuta virosa*), is a still more active and violent poison than the Common Hemlock. It is happily very rare; it grows on the banks of ponds and ditches, and in turfy marshes.

Lastly, we must mention the *Æthusa Cynapium*, or Fool's Parsley, which is commonly found in cultivated places. In kitchen gardens this plant may possibly be mistaken for Parsley, which it somewhat resembles when young and imperfectly developed. It may be distinguished from this pot-herb by the following characteristics:—The leaves of the Parsley are divided twice; their segments are broad and divided into three cuneiform and dentate lobes: the leaves of the Fool's Parsley are divided three times; their segments are more numerous, straighter, sharp-pointed, deeply cut, and dentate. Besides, the odour of the Parsley is agreeable, refreshing, and aromatic; whilst that of the Fool's Parsley is nauseous and fœtid. If the two plants are in flower they will be distinguished at the first glance, for the flowers of the Parsley are yellowish, whilst those of the Fool's Parsley are white. The stems of these plants also present different characteristics: that of the Fool's Parsley is almost smooth, the lower part reddish, and the whole slightly tinged with red; the stem of our aromatic vegetable, on the contrary, is channelled and green.

The ARALIACEÆ are trees and shrubs of the tropics, and of their borders in both hemispheres. *Adoxa moschatellina*, and the Ivy (*Hedera Helix*), are the only plants of the order indigenous to the British Islands; the former is a small herbaceous plant distinguished by its slightly musky odour, and its greenish-yellow flowers, which grow in woods and shady places. The Ivy is universally diffused in woods, hedges, on old buildings and rocks, or trunks of trees, on which its coriaceous evergreen leaves and clinging and trailing branches form a prominent object. Some strange confusion has arisen between the names of the Ivy and the Yew, which Dr. Prior explains thus: "The Chamæpitys of Pliny, as we learn from Parkinson, was called in English, Ground Pine and Ground Ivie, after the latin word *Iva*. But the name Ground Ivy had been assigned to another plant, which was called in Latin, *Hedera terrestris* (*Nepeta Glechoma*), and thus Ivy, and Hedera came to be regarded as equivalent terms. But there was again another plant which was also called *Hedera terrestris*, viz., the creeping form of *Hedera Helix*, and as Ivy had become equivalent to Hedera in the former case, so it did in this too, and eventually was appropriated to the full-grown evergreen shrub so well known. The botanical names of the Yew are so completely confused by the older botanists with those of the Ivy, that, dissimilar as are the trees, there can be no doubt that the origin of their names is identical."

The root of *Panax quinquefolium*, a species belonging to this order, furnishes a drug much used by the Chinese under the name of Gingseng; and *P. fruticosum*, *P. cochleatus*, natives of the Moluccas, are used as aromatic medicines by native practitioners in the East.

CORNACEÆ are found all over the temperate parts of Europe and America. Some of them, as *Cornus florida*, *sericea*, and *circinata*, are said to possess tonic properties of a high order. The Cornel, or Dogwood, is a tree sometimes seen in our hedges, and cultivated in our plantations, being useful for making butchers' skewers, and the Cornelian Cherry (*Cornus mascula*) is common on the Continent, where its little clusters of starry yellow flowers are the earliest harbingers of spring.

HAMAMELIDACEÆ, or Witch Hazels, are found in North America, Japan, China, Central Asia, and South Africa; its most attractive member being the genus *Rhodoleia*, whose great red involucral leaves, says Dr. Lindley, give quite a new aspect to the order, and point at an affinity of some kind with *Liquidambar*.

BRUNIACEÆ are, with the exception of one species, all natives of the Cape of Good Hope.

ALLIANCE XXXIX.—ASARALES.

Flowers monochlamydeous; embryo small, lying in a copious albumen.

Ovary one-celled; ovules definite; perianth 4-5-cleft; occasionally root-parasites.	I. Santalaceæ.
Ovary one-celled; ovules definite; perianth 4-8-cleft; stem-parasites .	II. Loranthaceæ.
Ovary three to six-celled; ovules indefinite.	III. Aristolochiaceæ.

The SANTALACEÆ, or Sandal-woods, are found in Europe and North America as humble plants, but in Australia, the East Indies, and the South Sea Islands, they expand into large shrubs or small trees. Sandal-wood, the produce of *Santalum album* is hard, heavy, admits of high polish, and yields a fine perfume; qualities which recommend it for all kinds of fancy furniture and boxes. It is also burnt in temples, as incense, its fragrant odour being due to an essential oil said to be heavier than water.

The LORANTHACEÆ are natives of the tropics, both of Asia and America, but rare in Africa. Their economy presents some very curious phenomena. In *Viscum*, the Mistletoe, according to Decaisne, the ovule does not appear till three months after the pollen has taken effect. Griffith, who has also minutely studied *Loranthus*, states that the ripe seeds adhere firmly to the substance on which they are applied, by means of their viscid coating, which hardens into a transparent glue, and in two or three days after application the radicle curves towards its support, becoming enlarged and flattened as soon as it reaches it. By degrees a union is formed between the woody system of the parasite and stock, the fibres of the sucker-like root of the former expanding on the wood of the latter in the form of a bird's foot. Up to this time the parasite has been nourished by its own albumen, but as soon as it has acquired the height of one or two inches a lateral shoot is sent out, which adheres to the stock by means of sucker-like productions, which frequently run to a considerable distance, covering the tree with parasites.

Mr. Miers, who has carefully studied the order, draws a distinction between *Loranthus* and *Viscum:* the former distinguished by its large, showy, dichlamydeous, crimson flowers, with lengthened stamens, and an ovary containing a solitary ovule, suspended from the summit of a cell, with a large fleshy cotyledon; *Viscum*, on the contrary, having small, pale, diœcious, monochlamydeous flowers, with stamens sessile, or nearly so, different in structure, with dissimilar pollen; a unilocular, turbinate ovary, with three ovules, of which only one arrives at maturity, attached to a free central placenta, with almost obsolete cotyledons. On these grounds he founds the new order *Viscaceæ*.

The Mistletoe is supposed to be propagated by birds, especially by the Fieldfare and Misselthrush, which feed on the berries. The mode in which the propagation of *Myzodendron* is effected is also clearly demonstrated by Dr. Hooker. Here the fruit is provided with long, feathery processes analogous to the pappus of the Compositæ, which float them in the air, and afterwards assist to hold them on to the branches while the radicle insinuates itself into the plant.

The ARISTOLOCHIACEÆ, or Birthworts, are common in tropical America, sparingly in North America, Europe, and Siberia, and in small numbers in India. Two species are said to be British plants, but among the rarest of our reputed species, and probably an accidental importation. The *Aristolochia clematitis* is recorded as being found growing on old walls, near Spittal, in Lincolnshire, &c.

The distinguishing characteristic of the order resides in the flowers, which have the perianth constantly divided into three segments. The stamens have the same ternary characters, and the cells of the fruit are three or six, always adherent to the perianth. The arrangement of the wood is also peculiar, their stems being composed of longitudinal plates, surrounded by a central pith, with an exterior bark; but these plates are not placed in concentric circles, as in other exogenous plants, but continue to grow, uniformly and uninterruptedly, as long as the plant lives. The most remarkable species are found in tropical America, where the gigantic size and grotesque appearance of their flowers excite the wonder of the traveller; of these, *A. cymbifera*, the borders of whose calyx resembles one of the lappels of a Norman woman's cap, measures seven or eight inches in length; while the flowers of *A. cordiflora* and *A. gigantea* are from fifteen to sixteen inches across. They are generally tonic and stimulating, and several of them are used in medicine.

PART IV.

GEOGRAPHICAL DISTRIBUTION OF PLANTS.

LINNÆUS, whose singular genius foresaw most of the conquests reserved for his favourite science—the study of botany—laid the foundations of Botanical Geography. In the prolegomena of his "Flora Lapponia," the immortal botanist of Upsal says, in the poetical and concise style which is peculiar to him: "The dynasty of the Palms reigns in the warm regions of the globe; the tropical zones are inhabited by whole races of trees and shrubs; a rich crown of plants adorns the plains of Southern Europe; troops of green *Graminaceæ* occupy Holland and Denmark; numerous tribes of Mosses are settled in Sweden; but the brownish-coloured *Algæ*, and the white and grey Lichens, alone vegetate in cold and frozen Lapland, the most remote habitable spot of earth: the lowest of the vegetables alone live on the confines of the earth."

The modifications in the distribution of plants which Linnæus had observed journeying from south to north, Tournefort had already observed during his travels in Armenia, upon the slopes of Mount Ararat. At the foot of this mountain he saw the plants of Armenia; higher up he found the plants resembling those of Italy; higher up still he found representatives of those of the environs of Paris; above these were allies of the plants of Sweden; finally, on the borders of eternal snow, near the summit of the mountain, he found plants which recalled those of Lapland.

Buffon had also a glimpse of the laws which apply to the distribution of plants. "The vegetation which covers the earth," he says, "and which is still more closely attached to it than the animals which browse on it, are even more interested than they in the nature of climate. Each country, each changing degree of temperature, has its particular plants. We find at the foot of the Alps the plants of France and Italy; at their summits we find the plants of the frozen North; and the same northern plants we find again at the summit of the mountains of Africa. Upon the range of the hills which separate the Mogul empire from the kingdom of Cashmere, we find on the southern slopes many of the plants of India, and it is not without surprise that we find on the north flanks many of those of Europe. It is also from the extremes of climate that we draw our drugs, perfumes, and poisons, and all the plants whose properties are in excess. Temperate climates, on the contrary, only produce temperate things; the mildest of herbs, the most wholesome of vegetables, the most refreshing of fruits, the quietest of animals, the most polished of men, are the heritage of the mildest climates."

At the commencement of the eighteenth century Geographical Botany was in a manner created by Alexander von Humboldt, whose genius is so universal that his traces are found in connection with every modern science. On his return from a voyage to the equinoctial regions of America, Von Humboldt, in one of his finest memoirs, demonstrated that it is the predominance of certain forms of vegetation which enables us to recognise a country immediately. A forest of Firs and Pines transports us at once to the northern or to the high mountain ranges of Europe; the Oaks and Beeches to the temperate zone; the Olives to the south, and the Palms into intertropical regions; the Cape of Good Hope is the country of the Heaths, and Mexico is perhaps the country most typical of the Orchids. In another memoir Humboldt attempted to estimate the total number of plants diffused over the surface of the globe, and the influence of climate upon their distribution. For the first time he established clearly that localities, each equally distant from the equator, and at an equal elevation above the level of the sea, might nevertheless have climates very little resembling each other, while countries situated under parallels very remote one from the other might have analogous climates.

The travels of naturalists of our own day in all parts of the globe have established, to the satisfaction of botanists, that certain characteristics belong to the vegetation of each climate, an interesting fact of which we shall endeavour to convey to the reader some succinct idea. The researches of explorers, combined with the labours of descriptive botanists, enable us to give some precision to the principles of botanical geography.

Theophrastus, who was born B.C. 370, enumerated 500 kinds of plants, and Pliny (died A.D. 79), in his "Historia Naturalis," treats of 800. The Greek, Roman, and Arab naturalists mention 1,400 species, but in the sixteenth century, according to G. Bauhin, this number was raised to 6,000. Linnæus's "Species Plantarum" (1764) contains descriptions of 8,551 species. Wildenow's edition (published from 1797 to 1810) described 17,457 species of flowering plants, which, with the Cryptogamia, made 20,000 of known plants. Persoon, in his "Synopsis Plantarum" (1805-1807), brings the number up to 26,000. Since that time the number of known plants has increased much more rapidly. R. Brown, in his "General Remarks on the Botany of Terra Australis," speaks of 37,000 Phanerogamia, and Humboldt ("De Distributione Geographica Plantarum") of 44,000 Phanerogamous and Cryptogamous plants. De Candolle, in his "Essai Élémentaire de Géographie Botanique" (1820), thought that the writings of botanists and the various European collections of dried specimens might contain 56,000 species; at which number also, in 1820, the species in the herbarium of the Jardin des Plantes were estimated. The collection of M. Benjamin Delessert, of Paris, was supposed to contain, in 1847, 95,000 species, a number which, about the same time, Lindley conjectured to comprise all known species. It has been estimated as probable that the existing species of Phanerogams amount to from 150,000 to 200,000. The number of Cryptogams is probably not less.

The numerical proportion of species belonging to the Phanerogams or Cryptogams varies according to the latitudes of the globe. As we advance towards the north, the number of Cryptogams increases; the proportion of Phanerogams, on the other hand, rises as we approach the equator. In the frozen or temperate zones the Cryptogamia are humble vegetables which scarcely raise themselves above the surface of the soil; but in tropical Brazil, the arborescent ferns rise to the height of the loftiest Palm-trees.

The vegetation of each species corresponds with a determinate interval in the scale of the thermometer, and this interval is not the same for all plants. Barley occupies the northernmost points of grain culture—about 70° north latitude in Lapland; 67° to 68° in western Russia, and 66° in the eastern parts; on the north-western parts of Europe it does not extend so far, since the mean temperature of the summer months falls, and the climate is made less favourable by excessive moisture. It extends from the north of Scotland to the Shetlands and the Faroës, but seldom ripens properly in the latter, so that the seed-corn is chiefly brought from Denmark. Barley will not grow in Iceland. In warmer countries it occupies a corresponding region on the hills, and is grown as high as 3,500 feet on the Alps; on the Himalayas up to 12,000 feet of altitude.

The northern limit of Wheat in Great Britain is about 58° north latitude, in Norway 64°, and in Sweden 62°, but it is not much cultivated beyond 60°. In the west of Russia it is at 61° 15′, and falls off gradually in the east. The Polar limits in North America are not yet ascertained, as its culture has not perhaps been pushed there to its utmost extent; it is known to be cultivated at 54° in the middle of the continent. Its equatorial limit in the plains of the northern hemisphere seems to be not far distant from the tropic, but it grows luxuriantly on the plateaux of the intertropical regions of America at 8,000 feet or more above the level of the sea. The possible limits of the culture of Spelt (*Triticum Spelta*) are not ascertained, since the existing diffusion of this grain depends more on the habits of men, and their systems of cultivation, than on climatal conditions.

The Vine is a native of the warmer temperate zones, and is said to attain a diameter of three to six inches, and to climb to the top of the highest trees in the forests of Mingrelia. Humboldt estimates that the cultivation of the Vine succeeds only in those climates where the annual mean temperature is about 60° Fahr., if the mean summer heat is not below 68° Fahr. In the Old World these conditions are found to exist as far north as latitude 50°, in the New World not beyond 40°. In both hemispheres the profitable culture ceases within 30° of the equator, unless in elevated regions, or in such as Teneriffe, where the heat is moderated by the sea-breeze. Thus the region of the Vine occupies a band of about 20° in breadth in the Old World, and not more than half that breadth in America.

We can now comprehend why certain vegetables live in some countries without flowering, and others without bearing fruit. The short summers and short days in such countries fail to yield the aggregate amount of heat, and that supplied is just sufficient to develop their leaves, but not enough to expand their flowers; and their fruits are abortive. The influence of heat on vegetation is so marked that we can scarcely name a single species which is truly cosmopolitan. Most vegetables occupy a determinate zone of their own, which they rarely pass. The cold prevents them from passing its limits towards the north; the heat exercises the same influence towards the south.

Humidity of the atmosphere and the solar influence have, also, a notable influence on the geographical distribution of plants. It is still more necessary to consider the influence of altitude. In proportion as we rise in the atmosphere the temperature decreases, and this lowering of the temperature is so rapid, that in ascending a mountain we pass through a wide range of temperature in the course of a few hours. From this it follows that a high mountain under the equator may be clothed at its base with the richest vegetation, while its summit is covered with eternal snow, and the space between is clothed with the same kind of diversity of vegetation (on a limited scale) which the traveller meets with in his journey from the equator to the pole.

We will now consider, in order to give a general idea of their vegetable products, the five geographical divisions of the whole world—Europe, Asia, Africa, America, and Australia.

EUROPE.

We can distinguish in Europe three great botanical regions. 1. The region of the North; 2. The Middle region; and 3. The region of the South, or Mediterranean.

The Northern region comprehends Lapland, Iceland, Sweden, Norway, and the northern provinces of Russia. The vegetation is monotonous; the ligneous species form only the one-hundredth part of the plants; the cryptogams predominate. The trees are principally coniferous and amentaceous. The Oak, the Hazel, and Poplar are arrested at 60° N. lat.; the Beech, the Ash, and the Lime at 63°; the Conifers at 67°; Barley and Oats can be cultivated up to 70°. Spitzbergen, the most northerly island of Europe, situated between 76° 30′ and 81° contains only ninety-three species of phanerogamous plants, belonging principally to the families of *Graminaceæ*, *Cruciferæ*, *Caryophyllaceæ*, *Saxifragaceæ*, *Ranunculaceæ*, and *Compositæ*. Among these plants there is scarcely a single tree or shrub, but only an under-shrub, *Empetrum nigrum*, and two small creeping Willows.

Some idea of the vegetation of Norway, with its deeply-intersecting fjords, may be gathered from the engraving, Fig. 448. Martins, to whom botanical geography is indebted for many valuable observations, made a voyage along the western coast of Norway, from Drontheim to North Cape, in recording which he has traced with a vigorous hand the picturesque vegetation of that country. "On the 28th of June," he says, "we arrived at Drontheim. While disembarking I was much surprised to see Cherry-trees bearing fruit about the size of peas. Lilac, Mountain Ash, Black Currant, and *Iris germanica* were covered with expanding flowers. My astonishment ceased, however, when I learnt that the spring had been a very fine one. The most common tree in the gardens and streets of the town is the Mountain Ash. I remarked also four Oaks (*Quercus Robur*), which appeared to suffer from the cold; in fact, upon the west coast of Norway the northern limit of the Oak lies half a degree south of Drontheim. The Ash is a more hardy tree, but it never attains the dimensions of the Oak in Sweden, and in latitude 61° 18′ I noted the last of them. The Lime lives at Drontheim, as do the Poplar (*P. balsamifera*), and

Fig. 448.—A Norwegian Fjord.

the Horse Chestnut; the Lilac blooms in every garden. All fruit-trees can only be cultivated as espaliers. Even in the most favoured situations, the Apple, Pear, and Plum do not ripen every year. In the environs of Drontheim, groups of Elder, Birch, Fir, intermingled with Ash, Maple, Aspen, Bird-cherry, Hazel, Juniper, and Willow, crown the heights. The fields are dry and well exposed, while the meadows occupy the lower ground. This fine fresh landscape has something exceedingly pleasing about it, although severe and cold.

"Towards the north I pushed on to Cape Ladehamer, which is crowned with light-foliaged Birches. On the east is the cascade of Leerfes, where the foaming waters of the Nidelven precipitate themselves over the rocks in the middle of a black forest of Pines. I arrived there at midnight. The dawn and the twilight which mingled together on the horizon, projected a hazy, doubtful light upon the landscape, for at this period of the year the sun scarcely sinks beneath the horizon in this latitude, and the bright lights which burn in the heavens towards the north already announce that it will soon reappear.

"In the fields and by the roadsides I found a great many plants which occupy similar situations in France." "Nevertheless," he continues farther on, "the eye of the botanist was rejoiced by the sight of a vegetation belonging at once to the Flora of the Boreal regions of the Alps and of the sea-shore." In the thickets grow *Geranium sylvaticum*, *Aquilegia vulgaris*, *Aconitum septentrionale*, *Pedicularis lapponica*, *Trientalis europæa*, *Paris quadrifolia;* in the less sheltered places, *Cornus suecica*, *Vaccinium Vitis-idæa*, *Polygonum viviparum;* in the marshes, the Bleaberry, and *Geum rivale;* upon the sandy sea-shore, *Plantago maritima*, *Glaux maritima*, *Elymus arenarius*, *Triglochin maritimum*, and many others equally interesting to the botanist.

In the first days of July the traveller reached Heldringen, a post-town situated on the borders of Nordland and the Government of Drontheim under latitude 65° 15′. He scaled a mountain whose denuded summit was 2,100 feet above the level of the sea. Its vegetation resembled that of the summit of the Alps. *Salix lapponica* and *Diapensia lapponica* alone reminded him that he was in Norway.

"At Bodoë, in 67° 16′," he continues, "I saw for the first time houses covered with turf, upon which grew many tufts of grass. According to my custom, I first examined the cultivated vegetables, but I saw only a few Potatoes, Peas, Radishes, a few Gooseberry-trees without fruit, and some fields of Barley and Rye. In the meadows just above the sea-level I found some plants which would have demonstrated to me, in the absence of other proofs, how much the climate of this country approaches that of the most elevated Alpine regions. There was *Dryas octopetala*, *Silene acaulis*, *Arctostaphylos alpina*, *Alchemilla alpina*, *Bartsia alpina*; and besides them, species which are unknown in Alpine regions, namely, such as *Aconitum septentrionale*, and *Tofieldia borealis*. Besides these, notwithstanding the difference of climate, some of the plants which are most common in the neighbourhood of Paris are found here, as the Dandelion, the Coltsfoot (*Tussilago*), Ladies' Smock (*Cardamine pratensis*), *Viola sylvatica;* they seemed a souvenir of France thrown at random in the midst of this northern vegetation."

At Hammerfest, which is under 70° 48′ north latitude, all attempts at cultivation had disappeared. The energies of the place are turned to commerce; it is from curiosity rather than for profit or utility that a few vegetables are cultivated.

"Near the city," adds the Professor, "I observed rich meadows, that were cut once a year, and some herds of half-wild reindeer, which grazed and roamed about freely. We shall deceive ourselves, however, if we consider Hammerfest a dull or melancholy city. Its principal streets, on the contrary, consists of very fair new wooden houses, well ordered, and in all respects comfortable. These are the habitations of the better class of inhabitants. The houses of the lower classes are poorer and older; borrowing, however, a particular charm from the flowery turf with which they are covered. The roofs are formed of great squares of turf, on which a number of plants have germinated and grow vigorously. In seeing these aërial gardens I have for the first time been able to comprehend the phrase '*in textis*,' which often occurs in the writings of Linnæus, indicative of the locality. In short, it was upon the roofs of houses that the learned botanist of Upsal herborised at Hammerfest; indeed, I frequently borrowed a ladder myself from the proprietor in order to gather the plants which grew round the chimney of one of these picturesque old houses. What I often found there were *Cochlearia anglica*, *Lychnis diurna*, *Chrysanthemum inodorum*, Shepherd's Purse, *Poa pratensis*, and *P. trivialis*. In autumn, when the flowers of *Chrysanthemum inodorum* are in full bloom, these hanging meadows rival in beauty those of our own more genial climate, and give the city a smiling physiognomy which contrasts most happily with the severe aspect of surrounding Nature. *Ranunculus glacialis*, *Arabis alpina*, *Silene acaulis*, *Saxifraga nivalis*, Bilberries, *Diapensia lapponica*, *Salix reticulata*, *S. herbacea*, &c., grow in the neighbourhood."

"How great was my surprise on landing at the North Cape, in latitude 71°," he continues, "to find myself in the middle of the richest subalpine meadows that can be imagined! High and tufted grass, which reached my knees. I found here, in short, at the northern extremity of Europe, the flowers which had so often attracted my admiration at the foot of the Swiss Alps; there they were, as vigorous, as brilliant, and much larger than among the mountains. The globe flower (*Trollius europæus*), *Alchemilla alpina*, *Geranium sylvaticum*, *Hieracium alpinum*, *Phleum alpinum*, and *Poa alpina*. On the right rises the imposing mass of North Cape, steep and inaccessible; before us a steep and sloping but verdant path, which permitted us to attain the summit by winding round the side of the mountain. In the ascent I gathered with enthusiasm all the plants which presented themselves; to me they possessed a peculiar interest as being, so to speak, the most robust and adventurous of all their European congeners. They seemed, like myself, to be expatriated, and exposed on this black rock to be battered by the waves. I was tempted to ask them why they had quitted the skirts of the cultivated fields and peaceful shades of the woods of Meudon, where they could receive the homage of Parisian botanists, in order to lead this exposed life among strangers. They were the *Spiræa Ulmaria*, *Cerastium arvense*, Shepherd's Purse, *Dandelion*, Golden Rod, &c. Nevertheless the Boreal or Alpine plants were in the majority on these slopes. I found there *Thalictrum alpinum*, *Pedicularis lapponica*, *Salix reticulata*, the Snowy Gentian, *Cornus suecica*, &c. The loftiest summit of North Cape is 1,020 feet above the level of the sea; it is surmounted by a small rock, on which many visitors have engraved their names. But even this was not destitute of all vegetation; small circular patches of Lichens (*Parmelia saxatilis*, *Umbilicaria erosa*), black as the rock, were attached to it, and a small microscopic moss hid itself in the clefts. There were a few miserable-looking plants, which had been destroyed by the winds, scattered on the ground, and seeking shelter behind such elevations of the soil as would protect them from the continuous squalls which swept the North Cape. Among the shrubs I found the Dwarf Birch and *Azalea procumbens*. The herbaceous plants were much less numerous; *Silene acaulis*, the *Diapensia lapponica*, and *Saxifraga oppositifolia*.

The mid-European region includes southern Russia, Germany, Holland, Belgium, Switzerland, the Tyrol, and the British Isles, Upper Italy, and the greater part of France. This region, whose exact limits it would be difficult to trace, is very different from the preceding. It is milder, more temperate; its woods and forests consist essentially of Oak (*Quercus Robur*), to which we may add Chestnut, Beech, Birch, Elm, Hornbeam, Alder, &c.; but the Oak predominates. These trees, all of which lose their leaves during winter, give to the landscape a very peculiar feature, varying with the season. This region is especially favourable to the cultivation of the Cereals. An oblique line, drawn from east to west, with certain inflections of

XX.—Vegetation of the Banks of the Loire.

its course, but ranging between the forty-seventh and forty-eighth parallel, and inclining a little towards the north, would divide it into two zones—one, the Northern, in which the Vine and the Mulberry yield to the rigour of winter, whose forests are chiefly composed of Conifers, where the culture of the Apple and Pear takes their place, and which includes more *Cyperaceæ*, *Rosaceæ*, and *Cruciferæ*; the other, the Southern, characterised by the culture of the Vine, the Mulberry, and the Maize, and in which *Labiatæ* begin to predominate. Some idea of the vegetation of this region will be gathered from PLATE XX., which represents the banks of the Loire in the glory of its summer vegetation.

In the Southern region the Mediterranean forms the centre. It is a vast basin, whose shores present a vegetation which, if not identical, is at least analogous in its whole extent. *Labiatæ* abound there, and in certain seasons the air is filled with their sweet perfume. To this extensive family we may add a large number of *Caryophyllaceæ*, *Cistaceæ*, *Liliaceæ*, and *Boraginaceæ*. The Mediterranean draws its distinctive character, however, from the vast extent of uncultivated country, where the Kermes Oak, *Phillyrea*, the Evergreen Oak, and various half frutescent Labiatæ, reign supreme. These plants more especially abound in Italy, Spain, Greece, Algeria, and in the northern portion of Asia Minor. Nevertheless, a new vegetation makes its appearance at Rhodes and Jaffa, which becomes closely connected with that of Egypt. The vegetation of the Mediterranean often presents itself with a smiling and agreeable aspect. Clumps of odorous Myrtles, *Arbutus*, and *Vitex Agnus-castus*, frequently occur on its shores; magnificent Oleanders, whose praises have been sung by the poets, occupy the edges of the brooks. In Italy, Sicily, and Spain, the Orange-trees bear without cessation flowers and fruit. The Prickly Pear (*Opuntia vulgaris*), and the American *Agave*, naturalised here, form impenetrable hedges in the southern parts of these countries, to which they give a marked and very characteristic landscape. The forests consist essentially of the Evergreen Oak (*Quercus Ilex*), whose persistent leaves remain until after their third year, and whose acorns, which have a very agreeable taste, form a considerable portion of the people's food, and of the Cork-tree (*Quercus Suber*), mixed with other characteristic trees and shrubs, such as *Erica arborea*, numerous species of *Cistus*, with ephemeral flowers, often large and of dazzling brilliance, and of *Cytisus*, *Genista*, &c.

Among the other species characteristic of these happy regions we may cite the Cypress (*Cupressus*), the Aleppo Pine, the Stone Pine, Planes, the Olive, which we scarcely meet with elsewhere;

Mastic-tree (*Pistacia lentiscus*), and the Pomegranate (*Ceratonia Siliqua*), &c.

Over a great part of the south coast of Sicily, a Palm, the *Chamærops humilis*, with fan-like foliage, waves sometimes beside the Date, from the bosom of a clump of Oranges and Citrons, its tall stipe crowned with an elegant panicle of drooping and feather-like leaves.

ASIA.

It would require a volume to give even an idea of the rich and varied vegetation of Asia. We must limit ourselves to a rapid glance of the features most characteristic of its Northern, Central, and Southern divisions.

The Northern region, or Siberia, forms a botanical region in close connection with the Northern region of Europe in the one direction, and with its own middle region on the other. It has its own peculiar character, nevertheless, from the predominance of certain families, such as *Leguminosæ*, *Ranunculaceæ*, *Cruciferæ*, *Liliaceæ*, and *Umbelliferæ*. Some genera are remarkable for the number of their species; we may quote *Astragalus* among the *Leguminosæ; Spiræa* among the *Rosaceæ;* and *Artemisia* among the *Compositæ*. Considering that the mean temperature varies from 29° to 46° Fah., we cannot reckon on a condition of vegetation very varied. Forests are formed by Larch, Spruce, *Pinus Cembra*, *P. sibirica*, *P. sylvestris*, &c.; White and Balsam Poplars and isolated Balsamic plants, dwarf Birches, Service-trees, Alder Buckthorn, Alders, Willows, accompany them, while Whortleberries and Rhododendrons form the under-shrubs. The Flora of the Steppes of Kamtschatka does not differ materially from that of the pasturages of Central Europe. According as the spectator expects these to be rich or sterile, he is the more or less surprised to find stately Tulips and graceful Irises mingling with the grassy turf in spring, but the Wormwood (*Artemisia*), and other monotonous forms of vegetation, succeed them.

Humboldt assigns to the forests of the Oural the vegetation characteristic of a park. "They present," he says, "an alternation consisting of a mixture of needle-leaved and round-leaved trees, and lawns; an assemblage which is completed by masses of brushwood, formed by Wild Roses, Honeysuckles, and Junipers, whilst *Hesperis*, *Polemonium*, *Cortusa Mathioli*, magnificent Primroses, and Larkspurs, form a perfect carpet of flowers; while the Water Buckbean, with white blossoms, is the grace of the marshes." He saw also "on the banks of the Irtisch great spaces entirely coloured

red by *Epilobium*, with which were associated tall-stemmed Larkspurs (*Delphinium*), with blue flowers, and the fiery-scarlet *Lychnis chalcedonica*."

The central region consists of Northern China and Japan. The Magnolias—those grand-leaved trees, with magnificent flowers and delicate aroma, which give such an attractive feature to gardens where they can be cultivated—are natives of this vast region. So is the Camellia, which has been, as it were, naturalised in the greenhouses of Europe, whose evergreen, glossy, and persistent foliage is the admiration of travellers, and of which we may reckon upwards of 700 varieties; and the Tea-plant (*Camellia Thea*), of whose leaves so many millions of pounds are annually imported into Europe. Also the *Aucuba*, with coriaceous leaves and clustered flowers, so ornamental in our gardens and shrubberies; *Celastrus*, Hollies, Spindle-tree, *Lagerströmia*, *Spiræa*, *Elæagnus*, &c.

The most remarkable trees and shrubs besides these are the Palm, *Raphis flabelliformis;* the Paper Mulberry (*Broussonetia papyrifera*); *Osmanthus*, whose flowers are employed to give flavour to Tea leaves; the Ebony-tree (*Diospyros Kaki*), with white flowers, and berries of a cherry-red, and of a delicious flavour; the Loquat (*Eriobotrya japonica*); *Salisburia adiantifolia*, which is planted round the temples; Yews (*Taxus nucifera* and *verticillata*); Cypress (*Cupressus japonica*); Junipers; Thujas; Oaks (*Quercus glabra* and *glauca*); *Alnus japonica*, *Juglans nigra*, and several species of Laurels and Maples.

Among the cultivated plants we find Rice; Wheat; Barley; Oats; *Sorghum vulgare;* Sago (*Cycas revoluta*); Taro (*Caladium esculentum*); *Convolvulus Batatas;* Apple; Pear; Quince; Plum; Apricot; Peach; Orange; Radish; Cucumber; Gourds; Water-Melons; Anise (*Pimpinella Anisum*); Peas; Beans; Hemp; and Cotton (*Gossypium herbaceum*)—a remarkable mingling of vegetable productions, which transports us at one moment from Asia to Europe, and at the next from America to Asia. We might dwell upon a crowd of ornamental plants, many of which are now well known in Europe, as the *Glycine*, the Lily of Japan, Tiger Lily, and Chinese Primrose.

The Southern region of Asia comprehends the two Indian peninsulas. Here non-tropical species disappear, or only present themselves very rarely. Tropical families become more numerous; the trees cease to lose their leaves; ligneous species are more numerous than without the tropics; the flowers are larger, more magnificent; climbing, creeping, and parasitic plants increase in number and size. India may be considered the true country of aromatic plants. Nor is the rich soil less fruitful in the production of suitable timber for constructive purposes.

Among the most abundant arborescent plants in this botanical region are *Bombax; Sapindus; Mimosa; Acacia; Cassia; Jambosa; Gardenia;* Ebony (*Diospyros Ebenus*) has been celebrated for its black-coloured, solid wood from the most ancient times; *Bignonia;* Teak (*Tectona grandis*) is a magnificent tree, which furnishes timber well adapted for building purposes from its great endurance; *Isonandra Gutta* produces *gutta-percha;* Laurels have an aromatic bark; the Nutmeg-tree (*Myristica*) produces seeds which are employed as spice; Figs (*Ficus religiosa, indica, elastica*); Palms, such as the Borassus (*Borassus flabelliformis*), with magnificent large fan-like leaves; *Sagus*, whose soft pulp yields sago, a farinaceous product very rich in starch; *Calamus*, whose twining and creeping stem is sometimes upwards of 500 feet in length, of one uniform thickness, and of which the canes used in Europe are made; Areca (*Areca Catechu*), the nut of which is a favourite masticatory with the natives; *Corypha umbraculifera*, the trunk of which, sometimes reaching the height of sixty or seventy feet, is crowned with an ample tuft of leaves spread out in umbrella form, covering a space of eighteen feet; *Dracæna*; Screw-pines (*Pandanus*); last but not least the Bamboo.

If we throw a glance, moreover, at the plants under cultivation, we find them equally important:—Rice; Earth-nut; *Sorghum;* Indian Corn; the Cocoa-nut, the elegant and useful tree which gives to man almost all the necessaries of life, supplying him at once with shelter, food, light, heat, and clothing; the Clove-tree (*Caryophyllus aromaticus*), the unopened flower of which is the well-known clove; Pepper (*Piper nigrum*), the fruit of which, gathered before maturity, has been constantly brought to Europe since the expedition of Alexander the Great; and the Betel (*Chavica Betel*), with bitter and aromatic leaves, in which the Southern Asiatics enclose a few slices of the Areca-nut, which they chew; the Tamarind (*Tamarindus indica*), a magnificent tree, the fruit of which encloses a pulp of acid flavour; the Mango (*Mangifera indica*), whose much-vaunted fruit has a sweet and richly-perfumed flavour accompanied with a grateful acidity; the Mangosteen (*Garcinia Mangostana*), whose berry encloses, under a bitter and astringent epicarp, a delicious pulp; the Banana, whose yellow-clustered fruit, each six or eight inches long, furnishes a very nourishing food; the Rose Apple (*Jambosa vulgaris*); the Guava (*Psidium pomiferum*), with yellow fruit of the size of a Pear; Oranges; Water-Melons; Sugar-cane; and Coffee.

XXI.—Landscape of an Indian Forest.

We have attempted in PLATE XXI. to give an ideal representation of the principal species of vegetables belonging to the botanical region which we have just described. Some rustic species are figured in the foreground. On the left a *Corypha*, surmounted by *Arenga saccharifera* and a group of Bamboos. Towards the centre, but still on the left, and near the trunk of a great Sandal-wood tree, is a Sago Palm (*Sagus*). In the middle distance is the Areca Palm, its stem inclined and surrounded by embracing Lianes. On the right is the *Borassus*; close by is the Banana; both are in the shade of a large Mango-tree. To the left of this is a Cinnamon and Gutta-percha tree backed by a lofty Cocoa-nut.

The plants cultivated in the background are Pepper (*Piper*); the Camphor-tree (*Camphora officinarum*), behind the Cocoa-nut, and in the distance the Nutmeg and the Clove-tree, near a row of Bamboos and Rotangs.

AFRICA.

Africa, like Asia, presents three very distinct regions:—1st, the Northern, which comprehends the Mediterranean littoral, and the Sahara; 2nd, the Central, which is tropical; 3rd, the Southern, which includes the Cape of Good Hope.

The Mediterranean region, by which we mean the African littoral bathed by the Mediterranean, includes Algeria from the northern slopes of the Atlas to the sea, and the Delta of the Nile. This part of Africa represents, in many respects, a vegetation analogous to that of South Europe. In the mountain region of North Africa all the plants of Central Europe may be cultivated with advantage. The Vine prospers in the neighbourhood of Tlemcen, Milianah, Mascara, and Medeah, where the colonists and even the natives have undertaken its cultivation. The Olive, so generally spread over North Africa, constitutes one of the chief sources of wealth to the Kabyle tribes. The Cork-tree forms immense forests in the lower mountain region of the littoral: in the province of Constantine, gathering the cork has become an important trade since its conquest by France. With respect to the Sahara, M. Cosson, a traveller and botanist, thus expresses himself:—

"Northern Africa is especially characterised by the extreme rarity of rains, the dryness of the atmosphere, and the extremes of temperature; the absence of great ranges of mountains and of permanent water-courses gives an aspect quite special to this desert-like vegetation. The number of species growing spontaneously does not exceed 500. The greater number of these are perennials, which grow in tufts, and have a dry and sterile aspect, giving them a characteristic rugged and hard appearance. The families represented in the Algerian Sahara in greatest number are *Compositæ*, *Graminaceæ*, *Leguminosæ*, *Cruciferæ*, and *Chenopodiaceæ*. Among the ligneous species are Tamarisks, a genus of elegant flowering shrubs, and the *Pistacia atlantica*. The Date-tree is, however, the chief source of wealth in the gardens of the oases. This tree is cultivated, not alone for the abundance and variety of its products, but also for its shade, which secures other cultivated plants from the violence of the winds, and maintains in the soil the moisture required for the cultivation of other crops.

"Besides the Date, an oasis generally presents an abundant crop of Figs, Pomegranates, Apricots, frequently the Vine. The Peach, the Quince, the Pear, and the Apple, are planted in gardens, and in the oases the Citron, the Orange-tree, Olives, Barley, more rarely still, Wheat, are cultivated in the irrigated lands of the neighbourhood, and in the intervals between the Date plantations. Onions, Beans, Carrots, Turnips, and Cabbages, occupy a large place among the plants cultivated. Pimento is also largely cultivated for the stimulating properties of its fruit, which render it a favourite condiment with the Arabs. The Egg-plant, and the Tomato, are cultivated in some gardens for their fruit. Numberless species of *Cucurbitaceæ* are also sown in the gardens in summer, and sometimes attain a great size. The Gombo (*Hibiscus esculentus*) is cultivated here and there by the negroes for its mucilaginous fruit. The industrial and fodder plants are principally Hemp, represented by a dwarf variety (Haschich), which is not employed as a textile plant, but its extremities are smoked by some of the less fervent Mussulmen. Tobacco is also cultivated. Henna (*Lawsonia inermis*), the leaves of which have been employed in dyeing a black colour, scarcely exists except in the oasis of Ziban."

The central region is only very imperfectly known, in consequence of the terribly insalubrious nature of its coast. The same forms of vegetation, however, prevail there which are found in other tropical regions. We may remark here that the plants, which are usually herbaceous in countries without the tropics, become ligneous in these regions. This is the case with plants of the families *Rubiaceæ*, and *Malvaceæ*. We note here also the almost entire disappearance of *Cruciferæ* and *Caryophyllaceæ*. The prevailing families are *Leguminosæ*, *Terebinthaceæ*, *Malvaceæ*, *Rubiaceæ*, *Acanthaceæ*, *Capparidaceæ*, and *Anonaceæ*. If we take a glance at prevailing vegetation proper to this region of Africa, we find upon the humid coasts

XXII.—Vegetation of White's River, Abyssinia.

impenetrable forests formed of Mangroves (*Rhizophora Mangle*), and *Avicennia tomentosa*, *Musa*, *Canna*, *Amomum*, *Pandanaceæ*, gigantic *Malvaceæ* (such as the Baobab), *Bromeliaceæ*, *Aroideæ*. Aloes (*Aloe socotrina*) furnishes the aloes of medicine; and several fleshy Euphorbias impress their strange characteristics upon the vigorous vegetation of this region.

It would be depriving African vegetation of its richest ornament not to mention its admirable Palms. At their head stands the Oil Palm (*Elæis guineensis*), the fruit of which, of the size of an Olive, contains so much oil that the liquid flows out when it is pressed between the fingers. The seed contains a sort of butter. The sap of this precious tree yields an excellent wine; its leaves prove excellent food for sheep and goats. But the true palm wine is produced from *Raphia vinifera*. Another remarkable member of this elegant family is *Lodoicea Seychellarum*, the fruit of which is larger than a man's head, and weighs upwards of twenty pounds; it sometimes floats as far as the coast of India. It is a fact worthy of remark that in this region very few Ferns or Orchids are observed, and yet these groups of plants are extremely numerous in other tropical countries.

Among the exotic vegetables which are successfully cultivated in Central Africa we may reckon Maize; Rice; *Sorghum*; Indian Corn; Manioc; *Caladium esculentum*, belonging to the family of the *Araceæ*, the rhizome and leaves of which are alimentary; the Banana; the Mango; the Papaw-tree (*Carica Papaya*), the fruit of which, about the size of a small melon, is eaten either raw or cooked, and the pulp mixed with sugar, forms a delicious marmalade; the Pine Apple; Figs; Coffee; Sugar-cane; Ginger; various species of *Dolichos*; the Earth-nut; Cotton; Tobacco; and the Tamarind. PLATE XXII., which represents an Abyssinian village, will give some idea of the vegetation of equatorial Africa. Alongside the lofty Palms and Baobabs (*Adansonia*) we see the arrangements for cultivating Rice.

The Southern region of the Cape of Good Hope is the country of the species of *Protea*, *Pelargonium*, *Epacridaceæ*, *Oxalis*, and *Ixia*, which decorate our hothouses and parterres. No other country can compare with this region for the prodigious abundance and dimensions of its Heaths. While the plains of Europe, the Alps included, scarcely yield a dozen species, at the Cape there are many hundreds. They attain sometimes the height of fifteen or sixteen feet. Their leaves are small, inconspicuous, and acicular; but their flowers are large, and the colours which decorate them brilliant in the extreme, varying from the softest shades to dazzling.

The Flora of this region is rich in vegetable forms, but it is by no means smiling in its aspect. We find no true forests, grand and sombre, in the whole region; there are few creeping plants, but, on the other hand, there are many succulents. The most characteristic families are the *Restiaceæ*, *Iridaceæ*, *Proteaceæ*, *Ericaceæ*, *Mesembryanthaceæ*, *Rutaceæ*, *Geraniaceæ*, *Oxalidaceæ*, and *Polygalaceæ*. Among the characteristic genera we may mention the *Ixia*; *Gladiolus*, with their sword-shaped leaves and parti-coloured flowers; *Strelitzia*, so remarkable for their inflorescence, and for their blue and yellow flowers; *Protea*, so named for their diversity of appearance; *Leucadendron*, of which one species, *L. argenteum* (the Silver-tree), rises to the height of from thirty to forty feet, its branches bearing lanceolate leaves, silky and silvery; *Helichrysum* and *Gnaphalium*, corymbiferous composites, better known as *Immortelles*; *Mesembryanthemum*, or Ice-plants; *Stapelia*, leafless Asclepiads, with angular fleshy stem and showy flowers but somewhat fœtid odour; *Phylica*, a genus of Rhamnads somewhat resembling Heaths, with abundant evergreen foliage, and small cottony heads of white flowers; *Pelargonium*, of which an infinite variety of forms, the result of culture, are known; *Oxalis*, the evergreen *Sparmannia*, whose white flowers, stamens with purple filaments and irritable anthers, are so ornamental in orangeries. It is upon the sandy coast of this curious botanical region that the species of *Stapelia*, *Iridaceæ*, *Mesembryanthemum*, and *Diosma* abound. The Heaths and Crassulas grow upon the slopes of the mountains.

The cultivated plants are the Cereals, most of the fruits and vegetables of Europe, the Sorghum of Caffreland, Yam, Banana, Tamarind, and Guava.

AMERICA.

Vegetation is richer and more varied in America than in any other part of the globe. Beginning with North America, we find its Polar vegetation quite analogous to that of Europe and Asia under the same latitudes. The Willow, Birch, and Poplar, exposed to the persistent action of the cold, become stunted bushes; and Saxifrages, Mosses, and Lichens prevail.

Without dwelling on the Arctic regions, then, we may divide this immense country into two regions; one of which, descending as far as 36°, may be called the Northern region; the other, comprehended between 36° and 30° of latitude, will constitute the Southern region.

The Northern region well deserves to be called the region of

Agave americana. *Cereus.* *Melocactus.*

XXIII.—Mexican Plants.

Aster and *Solidago;* those beautiful composites abound there with *Liatris*, *Rudbeckia*, and *Galardia*, of the same family. *Œnothera*, *Clarkia*, *Andromeda*, and *Kalmia*, charming ornamental plants, well known in our flower gardens, likewise characterise this vegetable zone. Amongst the most abundant arborescent species, we may mention numerous species of Pine, Fir, Larch, *Thuja*, Juniper; no less than twenty-seven species of Willow; twenty-five of Oak, Beeches, Chestnuts, Elms, Hornbeams, Alders, Birches, Poplars, and Ashes. With these are mingled the American Plane; *Liquidambar*, the trunk and branches of which furnish juices used in medicine; the Tulip-tree, with singularly truncate leaves, and large, spreading, solitary, yellowish flowers; different species of Maple, Lime, *Robinia*, and Walnut. Together with these numerous and varied arborescent species, which attain considerable dimensions, grow the *Myrica cerifera*, which furnishes an abundant wax drawn from the fruit by boiling; the Currant (*Ribes*), with coloured and ornamental flowers in great varieties of red, yellow, and white; the elegant *Andromeda*, *Azalea*, *Rhododendron*, and *Spiræa*, present themselves in endless varieties; Sumacs, a species of which (*Rhus toxicodendron*), with greenish yellow flowers, contains a juice so acrid that contact with it produces blisters and erysipelas, and is a dangerous poison; *Ceanothus*, Hollies, and Buckthorns.

In the Southern region the vegetation somewhat resembles that of the tropics, being a transition between that of the temperate and torrid zones. Walnuts, Elms, Chestnuts, and Oaks are found there, and with them three species of Palms, one of which is *Chamærops Palmetto;* species of *Yucca;* of *Zamia*, among the *Cycadeaceæ; Passiflora;* of woody twining plants, such as *Bignonia sapindus;* Cactuses, and Laurels. Lastly, by the side of Tulip-trees, *Pavia*, and *Robinia*, grow magnificent species of *Magnolia*, of which this is the true domain. The vegetation of this region is thus remarkable in its variety. The Sugar-cane, Indigo, Cotton, and Tobacco cover the cultivated plains. In Missouri, Texas, Arkansas, and Mexico, the great colony of the Cactuses raise their lofty stems. In this region *Cactus*, *Opuntia*, *Cereus*, *Echinocactus*, and *Melocactus*, raise their oddly-branching stems and clustering flowers, the most remarkable of all doubtless being *Cereus giganteus.* It inhabits the wildest and most inaccessible regions, requiring little or no soil to attain a prodigious development. It has at first the appearance of an enormous tomahawk. Thence rises a column, three yards high, which branches off and assumes the shape of an immense candelabrum, the height of which may be twelve or thirteen yards. PLATE XXIII. is a representation of several Cactuses belonging to this region, from an original drawing by M. Bende, a French traveller in this country. Mexico, according to the reports of botanists, may be divided into three regions of altitude. The first extends from the valleys as far as the Oak forests—this is the region of Palms, Cotton, Indigo, Sugar-cane, Coffee, and tropical fruits. The second, situated at an elevation of from 3,500 to 9,000 feet above the sea, is the temperate region. It stretches from the Oak forests to the forests of *Coniferæ*. At this height the temperature is still sufficient to ripen some tropical fruits. The third, or cold region, occupies a space comprehended between the Conifers and perpetual snow. In many places it possesses a climate under which Pear, Apple, and Cherry trees, and the Potato, can still grow. In ascending from the foot of Orizaba, one sees successively appear and disappear *Mimosa*, *Acacia*, Cotton, *Convolvulus*, *Bignonia*, Oaks, Palms, Bananas, Myrtles, Laurels, *Terebinthaceæ*, Tree-ferns, *Magnolia*, Arborescent Composites, Plane, *Storax*, Apples, Pears, Cherries, Apricots, Pomegranates, Lemon and Orange trees, Orchids, *Fuchsia*, and *Cactus*.

In the plains of Venezuela, known under the name of Llanos, over which we propose to conduct the reader, we shall find in Alexander von Humboldt a faithful and eloquent guide to the vegetation. "We entered," he says, "into the basin of the Llanos, in the Mesa de Paja. The sun was nearly at its zenith; the earth, wherever it was sterile and destitute of vegetation, had a temperature of 118° to 122° Fahrenheit; not a breath of wind was felt as we rode upon our mules. Nevertheless, in the midst of this apparent calm, whirlwinds of dust rose unceasingly, driven by little currents of air, which only skimmed the surface of the soil, and were produced by the difference of temperature between the naked sandy places and those covered with vegetation; they rendered the heat of the air still more suffocating." Through this atmosphere of quartz grains and banks of vapour, rendering the horizon sometimes waving and sinuous, sometimes striate, continues the learned traveller, "I saw naked trunks of Palm-trees, destitute even of their crowning tuft of verdure. The trunks appeared in the distance like the masts of ships on the horizon. There is something imposing, but sad and melancholy, in the uniform appearance of these steppes. Everything appears immovable; only the shadow of a little cloud, which traverses the zenith, announcing the approach of the rainy season, is projected upon the savannah. The steppes are principally covered with graminaceous plants, such as *Killengia*, *Cenchrus*, and *Paspalum*. With these we find a few dicotyledonous plants, such as *Turnera*; some *Malvaceæ*, and, what is very remarkable, species of *Mimosa*, with leaves quite sensitive to the touch, which the Spaniards call *Dornuderas*. The same race of cows which in Spain fatten upon sainfoin and clover, here find excellent nourishment in the herbaceous sensitive plants. The pasturage is richest, not only near rivers subject to inundations, but also where the trunks of the Palm-trees are the most crowded, which cannot be attributable to the shelter and protection which they have from the sun's rays, since the Palm of the Llanos (*Corypha tectorum*) has only a very few corrugated and palmate leaves, like those of *Chamærops*, and the lower are always parched and dried up. Besides the isolated trunks of Palms we also find, here and there, in the Llanos, groups of Palms, in which the *Corypha* mingles with a tree of the family of *Proteaceæ*—a new species of *Rhopala*, with hard and resonant leaves. In the Llanos of Caraccas, the *Corypha* extends from the Mesa de Paja to Guayaval. More to the north and north-west it is replaced by another species of the same genus, with leaves equally palmate, but much larger. To the south of Guayaval other Palms predominate, chiefly the pinnate-leaved *Piritu* (*Guilielma speciosa*) and the *Mauritia flexuosa*, the Sago-tree of America, which supplies farinaceous food, good wine, thread to weave into hammocks, clothes, and baskets; its fruit, in shape resembling pine-cones, being covered with scales, like those of *Calamus* (Rotang), with something of the taste of an apple. The Guaranes, whose very existence, so to speak, depends on the Murichi Palm, obtain an acid and very refreshing fermented liquor from it. This Palm has large, shiny, corrugated, and fan-like leaves, maintaining a most beautiful verdure in times of the greatest drought. The sight of it alone in the Llanos produces an agreeable and refreshing sensation; and the Murichi, laden with its scaly fruit, contrasts singularly with the sad aspect of the Palm of Cobija, the leaves of which are always grey and covered with dust."

As we ascend from the low country of Central America towards the high craters of the Cordilleras, whirlwinds of snow and hail succeed, each day, and for several hours, to the hot rays of the sun. If we ascend the Andes, between 20° south latitude and 5° north, at a height of from 5,000 to 10,000 feet above the sea level, we shall find extra-tropical forms of vegetation become more abundant: *Gramínaceæ*; some *Amentaceæ*—such as the Oaks, Willows; *Labiatæ; Ericaceæ;* numerous *Compositæ; Caprifoliaceæ; Umbelliferæ; Rosaceæ; Cruciferæ;* and *Ranunculaceæ*. Tropical plants, on the contrary, disappear, or become very rare; but still, isolated species of Palms, Pepper-plants, *Cactaceæ*, Passion-flowers, and *Melastomaceæ* are found at considerable heights. Among the most abundant ligneous species are the *Ceroxylon andicola*, the highest of all the Palms, which reaches the height of 200 feet, and produces a wax which exudes from its leaves, and from the base of their petioles; Willow and Humboldt's Oak; several species of *Cinchona*, which here reign supreme; a few Hollies, and species of *Andromeda*. Vegetables cultivated between the tropics, in Mexico, and as far south as the river Amazon, disappear almost entirely here; but Maize and Coffee, the Cereals and European fruits, are cultivated in these regions; Potatoes; *Chenopodium Quinoa*, the seeds of which, when boiled, serve as food for the inhabitants of the mountains.

If we ascend to the height of 10,000 feet above the sea on the Andes, and in the same latitude, tropical forms of vegetation almost entirely disappear. Those, on the contrary, which characterise temperate climates, and even the Polar regions, become abundant. Large trees are no longer seen. Alders; Bilberries; Currants; *Escallonia*, with bitter and tonic leaves, of which this is the home; Hollies and *Drymis*, are bushes belonging to these regions, as well as the curious Calceolarias, with shoe-shaped corolla, the seeds of which have supplied horticulture with an infinite number of varieties. Amongst the characteristic families we also find *Umbelliferæ*, *Caryophyllaceæ*, *Cruciferæ*, *Cyperaceæ*, Mosses, and Lichens. Returning to more circumscribed botanical districts, the climate of Caraccas has often been called one of perpetual spring. A more delicious temperature cannot be conceived. During the day it ranges between 60° and 68° Fahr., and in the night between 60° and 64°, at once favourable to the growth of the Banana, the Orange, Coffee, the Apple, Apricot, and Wheat.

We must not quit these regions without mentioning two beneficent trees—the *Theobroma Cacao* and the Cow-tree, *Brosimum Galactodendron*. The roasted and crushed seeds of *Theobroma Cacao*, with the addition of sugar, make chocolate. Humboldt gives the following account of the Cow-tree, which has the habit of *Chrysophyllum Cainito*:—"The fruit is rather fleshy, consisting of one, sometimes two nuts. When incisions are made in the trunk an abundance of thick glutinous milk flows, which is without any acidity. This substance exhales a very agreeable balsam-like odour. It was presented to us in the fruit of the Calabash-tree. We drank considerable quantities of it in the evening before going to bed, and again early in the morning, without experiencing any injurious effects. Negroes and free people who work in the plantations drink of it, and soak their maize or manioc bread in it. The master of the farm assured us that the slaves fattened visibly during the season when the *Palo de Vacca* furnishes them with most milk. Upon the arid flank of a rock," adds Von Humboldt, "there grows a tree whose leaves are dry and coriaceous, its great ligneous roots almost piercing the stone. During many months of the year not a shower waters its foliage, the branches appear dry and dead; but when the trunk is pierced a sweet and nourishing milk follows the incision."

In order to penetrate to the heart of the vegetation of Brazil, the region of Palms and *Melastomaceæ*, the land of promise to the naturalists, we shall take as our guide Martins and Auguste de Saint-Hilaire, who have written with much exactness on the vegetable wonders displayed in the Brazilian forests. Their aspect varies according to the nature of the soil, and the distribution of water traversing them. If these forests are not the seat of a constant supply of moisture, or if the moisture is only renewed by periodical rains, the drought stops the vegetation, and it becomes intermittent, as in European climates. This is the case in the Catingas. The vegetation of the untrodden forests, on the contrary, of which Saint-Hilaire gives an eloquent picture, is the reverse of this; excited by the ceaseless action of the two agents, humidity and heat, the vegetation of the virgin forests remains in a state of continual activity. The winter is only distinguished from the summer by a shade of colour in the verdure of the foliage; and if some of the trees lose their leaves, it is to assume immediately a new appearance. "When a European arrives in America, and sees from a distance the untrodden forests for the first time, he is astonished not to see the singular forms which he admired in European hothouses, but which are here mingled in masses and lost. And he is astonished at the little difference in the outline of the forests between those of his own country and those of the New World, and he is only struck with the proportions and the deep green colour of the leaves, which, under the most brilliant sky imaginable, impart a grave and severe aspect to the landscape. In order to appreciate all the beauties of a tropical forest we must plunge into retreats as old as the world. Nothing there reminds us of the fatiguing monotony of our Oak and Fir forests; each tree has a bearing peculiar to itself. Each has its own foliage, and often its own peculiar shade of verdure. Gigantic specimens of vegetation, each belonging to different, sometimes to remote families, mingle their branches and blend their foliage. Five-leaved *Bignoniaceæ* grow beside *Cæsalpinia*, and the golden leaves of *Cassia* spread themselves in falling upon Arborescent Ferns. Myrtles and *Eugenia*, with their thousand-times-divided branches, are finely contrasted with the elegant simplicity of the Palms; *Cecropia* spreads its broad leaves and branches, which resemble immense candelabra, among the delicate foliage of *Mimosa*. There are trees with perfectly smooth bark, others are defended by prickly spines; and the enormous trunk of a species of Wild Fig spreads itself out with sloping plates, which seem to support it like so many arched buttresses. The obscure flowers of our Beeches and Oaks only attract the attention of naturalists; but in the forests of South America gigantic trees often display the most brilliant colours in their corolla. Long golden clusters hang from the branches of the *Cassia*. *Vochysia* erect a thyrsus of odd-shaped flowers. Yellow and sometimes purple corollas, longer than those of our *Digitalis*, cover in profusion the species of Trumpet-flowered *Bignonia*; and *Chorisia* is decked with flowers which resemble our Lily in shape, and remind us of *Alströmeria* from the mixture of colours they present. Certain vegetable forms, which assume at home very humble proportions, present themselves with a floral pomp unknown in temperate climates; some *Boraginaceæ* become shrubs; many *Euphorbiaceæ* assume the proportions of majestic trees, offering an agreeable shelter under their thick umbrageous foliage."

But it is principally among the *Graminaceæ* that the greatest difference is observable. Of these there are a great number which attain no larger dimensions than our *Bromus*, forming masses of grass only distinguished from European species by their stems being more branchy, and the leaves larger. Others shoot up to the height of the forest tree, with a graceful habit. At first they are as upright as a lance, terminating in a point, with only one leaf, resembling a large scale, at each internode; when these fall, a crown of short branches spring from their axils, bearing the true leaves. The stems of the Bamboos are thus decorated with verticils at regular intervals. It is to the *Lianes* principally that tropical forests are indebted for their picturesque beauty, and these are the source of the most varied effects. Our own Honeysuckle and the Ivy give but a faint idea of the appearance presented by the crowd of climbing and creeping plants belonging to many different families. These are *Bignoniaceæ*, *Bauhinia*, *Cissus*, and *Hippocrateaceæ*, and while they all require a support, they each have notwithstanding a bearing peculiar to themselves. One of those climbing parasites will encircle the trunk of the largest trees to a prodigious height, the marks left by the old leaves seeming in their lozenge-shaped design to resemble the skin of a serpent. From this parasitic stem spring large leaves of a glossy green, while its lower parts give birth to slender roots, which

XXIV.—A Brazilian Forest.

descend again to the earth straight as a plumb-line. The tree which bears the Spanish name of *Cipo-Matador*, "the murderous Liane," has a trunk so slight that it cannot support itself alone, but must find support on a neighbouring tree more robust than itself. It presses against its stem, aided by its aërial roots, which embrace it at intervals like so many flexible osiers, by which it secures itself and defies the most terrible hurricanes. Some *Lianes* resemble waving ribbons, others are twisted in large spirals, or hang in festoons, spreading between the trees, and darting from one to another, twining round them, and forming masses of stem, leaves, and flowers, where the observer often finds it difficult to assign to each species what belongs to it.

Thousands of different species of shrubs, *Melastomaceæ*, *Boraginaceæ*, Peppers, and *Acanthaceæ*, springing up round the roots of large trees, fill up the intervals left between them. Species of *Tillandsia* and Orchids, with flowers of strange and whimsical shape, make their appearance, and these often serve as supports to other parasites. Numerous brooks generally run through these forests, communicating their own freshness to the forest vegetation, presenting to the tired traveller a delicious and limpid water, while the banks of the stream are carpeted with Mosses, Lycopodiums, and Ferns, from the midst of which spring Begonias, with delicate and succulent stems, unequal leaves, and flesh-coloured flowers. PLATE XXIV. is a reproduction of a celebrated engraving published about the year 1825. It represents the untrodden depths of a Brazilian forest, from a picture by the Count de Forbin. Let us glance at the vegetation of the countries of the great American continent situated below the tropic of Capricorn, which comprehends Chili, La Plata, and Patagonia. We find two Palms in Chili, *Jubæa spectabilis* and *Ceroxylon australis*. A magnificent tree, *Araucaria imbricata*, which rises to the height of 150 feet, its verticillate branches lying almost horizontally, and covered with spiny leaves, here forms immense forests. A few *Graminaceæ*, Heaths, *Labiatæ*, *Umbelliferæ*, Fuchsias, *Loasaceæ*, Myrtles and Laurel-bushes, but particularly ligneous Composites, form the chief part of the vegetation.

The forests of Paraguay, still little known, situated along the coast of the Atlantic, consist of ligneous *Compositæ* and *Ilex paraguayensis*, the Paraguay tea, of which a large quantity is annually exported.

In the Argentine Republic Auguste de Saint-Hilaire found only 500 species of plants, amongst which only fifteen belonged to families which are not European.

When we reach the south coast of Patagonia and the Falkland

Islands, a few brown and coriaceous *Graminaceæ* and *Cyperaceæ*, such as *Dactylis cæspitosa*, *Carex trifida*, *Bolax glebaria*, *Cardamine glacialis*, *Veronica*, *Calceolaria*, *Aster*, *Opuntia Darwinii*, *Lomaria magellanica* among the Tree Ferns, a few Brambles, thickets of Bilberries and *Arbutus*, include nearly the whole of the vegetation of these desert lands, where Mosses, Hepaticas, and Lichens reign supreme. We now reach the southern part of South America. In the stormy region of Terra del Fuego thick forests cover the mountains, where they are sheltered from the wind, to the height of 1,500 feet above the level of the sea. *Fagus betuloides* predominates there; then comes *F. antarctica*, accompanied by Barberry and Currant Bushes.

At the Island of Hermite, the most southerly point of the American continent, there is still some arborescent vegetation. Hooker there observed eighty-four flowering plants and many Cryptogams. A Fungus parasitic on the Beech (*Cyttaria Gunnii*) constitutes there a principal aliment of the miserable inhabitants of these gloomy regions.

Australia.

The Australian Flora and Fauna are so different from those of any other part of the world, that from the state of our geological knowledge it does not appear possible that this part of the world can be considered contemporary with any of the other divisions. The study of animals and plants of Oceania leads naturalists to the conclusion that these countries belong to an earlier creation than those of the rest of the earth; it seems to belong to the tertiary or secondary epochs. In fact, all the Marsupial animals belong to a type of mammals similar to those found in the fossil state in Jurassic rocks, and the vegetation presents such anomalies as might be expected in the tertiary period more than in that of our days. It presents forms more ancient than any other contemporary vegetation. More than nine-tenths of the species found between 33° and 35° south latitude, in Australia, are absolutely limited to these regions. Many constitute completely distinct families; others form families which are scarcely represented in any other part of the globe. Those even which belong to groups more generally diffused disguise their natural affinities under forms isolated and unlike their congeners. The different species of two genera, namely *Eucalyptus* among *Myrtaceæ*, and *Acacia* among *Leguminosæ*, form perhaps, from their number and dimensions, one-half of the vegetation which covers the country (Fig. 449). Their leaves are reduced to phyllodes. Neither these phyllodes nor the limb of the real leaves are placed horizontally, like those of Europe and other parts of the world, represented in the engraving, they are perpendicular to the su: the soil, so that the light shining between these vertical blade: arrested, as in the case with our trees and bushes, in which the leaves are placed transversely one above the other. The effect produced by masses of Australian verdure is thus entirely different from that to which we are accustomed in Europe. The aspects of these forests particularly struck the first travellers who visited them, from the singular sensation communicated to the eye by this mode of distributing light and shade.

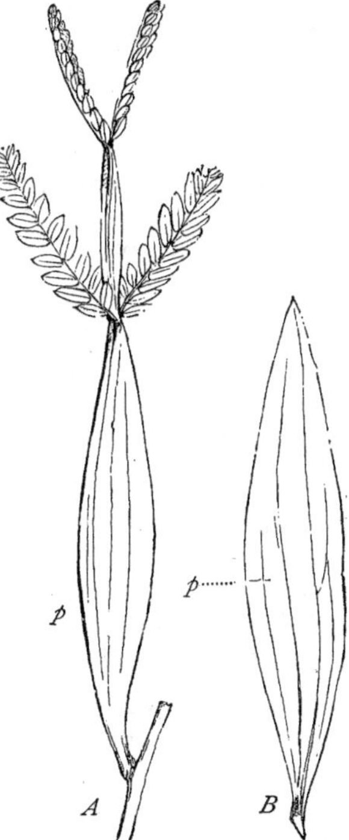

Fig. 449.—Australian *Acacia*. *p*, the petiole, forming a flat, leaf-like phyllode. *A*, with leaflets; *B*, without any leaflets.

Eucalyptus, which occupies such a large place in Australian vegetation, may be said to be the sacred tree with the natives; it shadows the tombs of the savage inhabitants of these countries. Sir Thomas Mitchell, the traveller to whom we owe the first scientific description of Australia, has given a remarkable picture of "these groves of death," which are daily becoming more and more rare, and will disappear under the influence of European colonisation. He relates that these groves mark the centre of the patrimonial land of each great Australian tribe. Little *tumuli* of grass, and sandy footpaths, surround the clumps of these funereal squares, over which spreads the shadow of the *Eucalyptus* and *Xanthorrhœa*. If to the magnificent *Eucalyptus* and simple-leaved *Acacia*, which predominate in the forests and give quite a special character to the vegetation, we add the *Xanthorrhœa*, with its thick stem, long, narrow, linear leaves, curved and spreading at the summit, from the centre of which rises an elongated stem, terminated by a spike of robust flowers; the *Casuarina*, with long, pendent, and drooping boughs, most delicately articulated; *Araucaria excelsa*, whose column-like trunk and verticillate branches rise to the height of ninety or a hundred feet; the elegant *Epacridaceæ*, with flowers so varied; a vast number of pretty Leguminosæ, which now add to the riches of our hothouses; more than 120 terrestrial *Orchidaceæ*, nearly all belonging to genera peculiar to Australia, we shall have an idea of the vegetation which covers and decorates in so original a way the shores of New Holland.

The large islands of New Zealand almost correspond in latitude with the zone which we have been examining. These islands are the nearest land (considering Van Diemen's Land as part of Australia), and are interesting as being the exact antipodes of Western Europe, and because they repeat as it were our Mediterranean region on the other side of the globe. While resembling it in climate, however, the native vegetation has its own characteristics. It has some features in common with Australia and the tropics, as will appear from the account given of them by Messrs. Richard and Lesson, whose account we chiefly follow.

In the large island of Ika-na-Nawi there are immense forests of Lianes and interlacing shrubs, which render them impenetrable. In these forests there exist, no doubt, trees of gigantic dimensions, for the canoes of the natives are sometimes as much as sixty feet long, and from three to four broad, all hollowed out of one trunk. At from two to four miles from the coast Messrs. Richard and Lesson saw large spaces, very low and probably marshy, covered with great masses of green trees, of which the *Dacrydium cupressinum* and *Podocarpus dacrydioides* and some others, form the principal species. The European is surprised to meet there many familiar plants, or species closely allied to them, such as *Senecio*, *Veronica*, Rushes, *Ranunculus acris*, &c. On the other hand, several plants peculiar to New Zealand grow abundantly in these localities, such, amongst others, as the *Phormium tenax*, called by Europeans New Zealand Flax, because its fibres furnish a very strong thread, much used in the manufacture of certain fabrics.

Ferns form a tenth of the number of species in the whole vegetation of this country; among Monocotyledons, are *Graminaceæ* and *Cyperaceæ;* among Dicotyledons, *Umbelliferæ*, *Cruciferæ*, and *Onagrariaceæ.* New Zealand only furnishes a small number of alimentary plants. The aboriginal inhabitants of this archipelago, for the most part icthyophagous, were long reduced to the feculent root of a Fern, the *Pteris esculenta*, for food, when they could not obtain fish. None of their trees produce large fruit. The Taro (*Caladium esculentum*) and the Sweet Potato (*Convolvulus Batatas*) also serve as nourishment to the inhabitants of these countries. It is to be remarked that

European vegetables, introduced into New Zealand by sailors, are propagated there with such facility that the aspect of the ground, as well as conditions of life, are greatly modified. Amongst the vegetables proper to the archipelago in question we may note the *Corypha australis* amongst the Palms; arborescent species of *Dracæna*, forests of Coniferæ, with large leaves, such as *Dammara*, and *Metrosideros* amongst the *Myrtaceæ.*

Mountain Vegetation.

We have briefly traversed the principal botanical regions of the globe; and in the course of our survey, we have seen that vegetation changes with the latitude; that is to say, according to the distance from the equator. As we advance from the equator towards the poles we meet in succession with the equatorial, tropical, temperate, and polar zones—vegetation gradually losing its power, a fact which is proved most satisfactorily by the decreasing number of species, and by their dwarfed appearance, until vegetation altogether ceases in the region where snow reigns eternal. When heat disappears, organic life is extinguished, and vegetable organisation experiences loss of power and vigour proportionable to the degradation of atmospheric heat.

But a natural reflection presents itself immediately, as a corollary upon these remarks.

When we ascend a mountain, or, in fact, when we ascend by any means whatever—in a balloon, for instance, as Mr. Glaisher's experiments seem to show—the temperature decreases by something like one degree for every hundred yards above the surface. It follows from these premisses that every stage in the ascent of a mountain should exhibit different forms of vegetation, each forming a zone or botanic region similar to those we have passed in tracing their geographical latitudes.

"Let us imagine a spectator at the foot of the Alps," says Jussieu, "opposite to one of those grand rocky masses crowned with eternal snow. As his eye ranges along the side of the mountain, he observes that the vegetation which immediately surrounds him, and which is that which characterises Central and Northern France, disappears at a certain height, giving place to another, which, in turn, disappears at a higher range. Beyond a certain distance the eye can only seize the masses indicated by large trees, the humbler plants being concealed behind them, so that they look like a series of bands superposed one over the other on the slopes of the mountain. At first these belts are composed of deciduous-leaved plants, which drop their leaves early, and are readily distinguishable by their more tender verdure; then Conifers of deeper green, which in the mass appear nearly black. Another belt succeeds of an undecided green, interrupted here and there by clumps of another colour, which goes straggling up to the sinuous line where the snow commences. This is owing to the circumstance that the trees whose branches are too closely intermingled have died out, making room for shrubs or herbaceous plants, more dwarfed in their growth, and more on a level with the soil. If the spectator approaches the mountain, and scales it, he will find other plants very different from the masses he looked at in the distance, which we call Alpine plants, such as the Aconites, *Astrantia*, certain species of *Artemisia*, *Senecia*, *Prenanthes*, *Achillea*, *Saxifraga*, and *Potentilla*. After having skirted the Walnut-trees, and traversed the woods formed of Chestnut, these will be observed to cease, and forests of Oak, Beech, and Birch take their place. Of these the Oaks disappear first, at the height of about 2,600 feet above the level of the sea, the Beeches about 3,000. Beyond this the trees consist entirely of evergreen trees, as Firs, Larches, and the common Pine, which stop also at certain successive stages, about 6,000 feet. The Birch ascends a little higher, but disappears at about 6,500 feet of elevation. A Conifer (*Pinus Cembra*) continues for another hundred yards. Beyond this limit the trees become dwarfed in size; for example, a species of Alder (*Alnus viridis*) becomes a low shrub. Near to this the botanist will find himself surrounded by shrubs very characteristic of the Alps, sometimes called the Alpine Rose (*Rhododendron*), which ceases in its turn only a little higher, giving place to plants much more lowly, which scarcely rise above the soil. These are specially known as Alpine plants. They belong to families which he observed at his point of departure, but of different species; *Cruciferæ*, *Caryophyllaceæ*, *Rosaceæ*, *Leguminosæ*, *Compositæ*, *Cyperaceæ*, *Graminaceæ*, but of different species. With them representatives of other families, which rarely show themselves in the plains, such as Saxifrages and Gentians, are numerous. Annuals cease almost entirely, as might be foreseen, since an unfavourable season, in which the ripening of their seeds was checked, would be sufficient to destroy their race." The roots of perennial or woody plants bury themselves under the soil, where a higher temperature is preserved. They submit themselves to the influence of the atmosphere, and develop themselves when it is milder and sufficiently warm. But this can only be done during a short season, and on some places only once in many years. It follows that the stems are short, and scarcely rise out of the soil, while those that are frutescent usually hug the ground, sometimes creeping, more often producing short, hardy, intertwining stems, forming thick stunted patches, as would result in ordinary cases from pruning shrubs every year very near the ground. The general appearance proper to the plant is thus effaced in some respects, and replaced by the physiognomy belonging to Alpine vegetation. These plants are generally of the arborescent kind, like the Willows, whose roots creep along the ground. The more elevated they are, the more scattered and impoverished is the vegetation, until, towards the summit of the rocks, it only appears in the form of Lichens, whose crust differs from the monotonous tint of their own surface. When the limit of eternal snow is reached, organised life can no longer exist.

Mont Ventoux, in Provence, presents us with an interesting application of the same facts. This mountain rises abruptly from a plain, the temperature of which may be compared with that of Sienna, Brescia, or Venice, while the summit of the mountain approaches the climate of northern Sweden. To ascend its sides and reach the summit is as if we had actually traversed nineteen degrees of latitude, or from 44° to 63°. Martins has published an interesting account of the vegetation of this mountain : "It presents a succession of well-defined botanical regions, each characterised by the presence of plants which are wanting on the others. These regions are six in number upon the southern slopes, and five on its northern side.

Ascending the southern slope, its base, in respect to its vegetation, is like that of the valley of the Rhone. All the plants of the plains are found in the region at the foot of the mountain, and they are well characterised by two trees—the Aleppo Pine and the Olive. Both belong to the basin of the Mediterranean, round which they form a girdle, only interrupted by the delta of the Nile. The Aleppo Pine is found upon all the hills which lie at the southern foot of Mont Ventoux, but ceases at the height of 1,400 feet above the level of the sea. The Olive ascends a little higher, but ceases also at 1,600. Under these trees we meet with all the species which characterise the vegetation of Provence, the Kermes Oak, the Rosemary, the Spanish Broom, and *Dorycinium suffruticosum*. A narrow zone, scarcely exceeding 180 feet, succeeds to this, which is characterised by the Evergreen Oak. Among the under-shrubs we find *Plumbago*, *Juniperus oxycedrus*, *Euphorbia Characias*, *europea*, and *Psoralea bituminosa*.

"A region altogether destitute of arborescent vegetation follows.

The soil is here naked, stony, and generally uncultivated; nevertheless, here and there fields of Chick Peas, Oats, and Barley appear, the last of which disappears at 3,500 feet above the Mediterranean; but a shrub, the Box, two under-shrubs, Thyme and Lavender, another herbaceous Labiate (*Nepeta geaveolens*), and *Vincetoxicum officinale*, predominate as to size and number. It is at this point that the attempts to replant woods of Oak and Pine are pursued with success. It is necessary to ascend to 3,800 feet before again reaching arborescent vegetation. It is composed of Beeches; at first sparse and undersized, they get larger 300 feet higher, especially in the deep ravines and valleys, where they are sheltered from the wind. This region extends as high as 5,500 feet. At this height the depressions are slight; valleys and ravines almost cease, and the trees exposed to the action of the winds become humble bushes, with short, hard, and crowded branches. One of these bushes, like a large ball or mattress extending on the earth, is often as old as the great Beeches which elevate their proud heads to the heavens in the valleys below. Numerous species of plants occupy the region of Beeches, many of them belonging to the sub-alpine zone of the mountains of Central Europe, never descending into the plains, unless transplanted. Such are the Buckthorn, the Gooseberry, the Wallflower, the Mountain Sorrel, *Anthyllis montana*, *Cacalia alpina*, &c.

"At the height of 5,600 feet the cold is too great, the summer too brief, and the wind too violent for the Beech to exist any longer. As upon Mont Ventoux, so it is on the Alps and Pyrenees—on all, a tree of the family of Conifers is the last representative of arborescent vegetation. It is a humble species of Pine, called the Mountain Pine (*Pinus uncinata*), because the scales of its cone are curved into a sort of claw. These Pines are found many feet in height in sheltered places, but become mere bushy shrubs when exposed to the sweep of the winds. They ascend as high as 6,000 feet, the extreme limit of arborescent vegetation. The herbaceous plants of this region are the same as in the region of Beeches; they nearly all attain the limit of the Pines. In addition to the common Juniper—resting on the soil, as it always does on high mountains, where the weight of the snow crushes it all the winter—we find the Mountain Germander (*Veronica montana*) and the Tufted Saxifrage (*S. cæspitosa*), which is found on the loftiest ridges of the Alps.

"Its flora thus teaches us, in the absence of the barometer, that we have reached the Alpine region of Mont Ventoux, and that the region of arborescent vegetation has disappeared. But here the botanist will be delighted to find the flora of Lapland, of Iceland, and of Spitzbergen. In the Alps this region extends to the line of perpetual snow, the home of eternal winter. But as Mont Ventoux is only about 6,300 feet high, the summit only extends to the lower zone of the Alpine regions in the Alps and Pyrenees. At this point all trees have disappeared, but a crowd of small plants expand their corollas on the stony surface. Among them the orange-flowering Poppy, *Viola cenisia*, the blue-flowered *Astragalus*, and, quite at the summit, *Poa alpina*, *Euphorbia Gerardiana*, and the Common Nettle, which is generally found wherever man fixes his dwelling. A chapel has been built on the summit of the mountain since the ascent of Petrarch. But it is not on the south of the summit that the botanist will seek for the Alpine plants characteristic of the loftier regions. It is on the northern declivities, on the rocks exposed to the glacial north winds, nearly deprived of the sun during long months, and covered with snow from June. These I have surveyed as I would survey an old friend. The Purple Saxifrage (*S. oppositifolia*) was the first plant I recognised; I had gathered it on the summit of the Reculet, the loftiest ridge of the Jura, and upon all the summits of the Alps which reached or passed the limits of perpetual snow. When I put foot for the first time on the icy shores of Spitzbergen, the Purple Saxifrage was among the first plants which attracted my attention; for here are found, on the shore of the sea, the cold summers and the melting snow of the summits which crown the Alps and the Pyrenees. Upon Mont Ventoux other Saxifrages, equally Alpine, accompany it. The blue bell-shaped flowers of *Campanula Allioni* raised their head from a heap of stones and dwarf plants, which covered all these heights; the round-headed *Phyteuma*, the hairy *Androsace*, the *Ononis* of Mont Cenis, and three species of *Arenaria*, clung to the rocks or peeped through the stones."

For the sake of comparison let us leave Provence and Europe, and glance at the ranges, in the heart of Asia, of the lofty Himalayas, or "abode of snow," as the word means in the figurative language of the Asiatics. Dr. Hooker passed the rainy season of 1848 in the sanitary establishment of Dorjiling, the farthest English possession in Sikkim, 7,200 feet above the sea, having in sight the loftiest peaks of the range. Twelve of these are more than 24,000 feet high, and one of them, Kinchinjunga, attains the height of 28,177 feet. Mount Chumulari, another giant of the Himalayas of Thibet, was visible from a neighbouring peak, the Sinchul, during the ascent of which the traveller made his first acquaintance with some of the beautiful Rhododendrons with which he afterwards enriched the gardens of Europe. "In the month of May," says the Doctor," "when the Magnolias and Rhododendrons are in flower, the magnificent vegetation of the Sinchul yields nothing in certain respects to that of the tropics, the beauty of the effect being, however, much diminished by the constant gloom of the season. The white-flowered Magnolia (*M. excelsa*) is one of the trees which predominate at the elevation of 7,000 to 8,000 feet, and in 1848 it had flowered so abundantly that it seemed as if the broad sides of the Sinchul and other mountains at the same elevation were covered with snow. The purplish-flowered species (*M. Campbellii*) does not appear under the elevation of 8,000 feet. It is a large but unsightly tree, with dark—almost black—bark, and few branches, destitute of leaves in winter and while in blossom, but throwing out at the extremity of the branches great bell-shaped flowers of a purplish-rose colour, the fleshy petals of which cover all the surrounding soil.

"Upon its branches, and upon the Oaks and Laurels, the *Rhododendron Dalhousiæ*, a slender creeping shrub, grows as an epiphyte, bearing at the extremity of its branches from three to six white bell-shaped flowers, citronous in odour, and four and a half inches in length. The scarlet-flowered Rhododendron is rare in these woods, but is much surpassed by *R. argentum*, which here becomes a tree forty feet high, with leaves twelve or fifteen inches in length, of a deep green above and silvery green on the lower surface, and with flowers large as those of *R. Dalhousiæ*. Oaks, Laurels, Maples, Birches, *Hydrangea*, a species of Fig which grows on the very summit of the mountain, and three Chinese and Japan genera constitute the chief woodland vegetation of this part of the Sinchul.

"Below Dorjiling, the zones of vegetation are well marked. Above 6,500 feet, Oaks, Chestnuts, and Magnolias characterise the vegetation; immediately below a Tree Fern appears (*Alsophila gigantea*); next, a species of Palm, of the genus *Calamus*, and a *Plectocomia*; this last shoots up the branches of the loftiest trees, extending itself over the forest to the distance sometimes of 120 feet from its stem; finally, a last characteristic trait of the region is a Wild Plantain, which attains nearly the same height as the preceding species."

With some difficulty Dr. Hooker obtained permission of the native authorities to go beyond Dorjiling, and, in particular, to visit the higher passes of the Himalaya in Tibet, and especially the principal pass of the Kinchinjunga. Following his steps in this ascent, he found at 8,100 feet *Abies Brunoniana*, a fine species, which assumes the form of an obtuse pyramid, with spreading branches like the Cedar; it is unknown in the exterior chain, and occupies in the interior a zone less elevated by 1,000 feet than the Silver Firs (*Abies Webbiana*). We meet also at this level with a great number of sub-Alpine plants belonging to *Leycesteria*, *Thalictrum*, *Rosa*, *Gnaphalium*, *Alnus*, *Betula*, *Ilex*, *Berberis*, *Rubus*, and some Ferns, *Anemone*, Strawberries, Alpine Bamboos, and Oaks.

On the higher level our traveller saw Junipers mingling with Silver Firs, which were even superseded by evergreen Rhododendrons, spreading along the slopes in immense profusion; *Spiræa*, dwarf Junipers, and small Birch-trees, Willows, Honeysuckles, Berberries, and a species of Service-tree. At 12,200 feet the vegetation was almost limited to numerous species of *Rhododendron*, which formed a continuous zone of 1,200 feet broad on the steep slopes of the mountain. A little *Andromeda* made itself quite remarkable there; and by the roadside the botanist saw two plants which reminded him of his far-distant home—the Meadow Grass (*Poa annua*) and the Shepherd's Purse (*Capsella*). At 13,200 feet the soil becomes hard and frozen, and at 22,000 feet perpetual snow covers the mountain side.

The traveller attained the summit of the pass into Tibet at 16,700 feet above the level of the sea, where he still found many species of *Compositæ*, *Graminaceæ*, and an *Arenaria*, with great masses of the curious *Saussurea gossypina*, covered with a white down, which felt soft to the touch, and about ten inches high. The species of covering given to this plant is almost exceptional among the plants of the Himalayas; the Alpine species which are scattered about, such as *Arenaria*, Primroses, Saxifrages, *Ranunculus*, Gentians, Grasses, and Cyperaceæ, having their foliage perfectly naked.

The following year Dr. Hooker in one of his ascents towards Tibet collected upwards of 200 species upon one of the crests of the Himalayas, among which he found ten *Cruciferæ*, twenty *Compositæ*, ten *Ranunculaceæ*, nine *Caryophyllaceæ*, ten species of *Astragalus*, eight species of *Potentilla*, twelve *Graminaceæ*, fifteen species of *Pedicularis*, and seven *Boraginaceæ*. Finally, on the 7th of September, 1849, he reached the culminating point of the Himalayan flora on Mount Donkia, at an elevation of 18,400 feet. The *Arenaria rupifraga* is the only Phanerogam which he met with at this elevation; *Festuca ovina*, a *Saussurea*, and a little Fern,

(*Woodsia*,) were, however, found very near the summit, where he observed many Lichens and some Mosses. The Lichens and Mosses are thus the last plants which disappear on the confines of organic life.

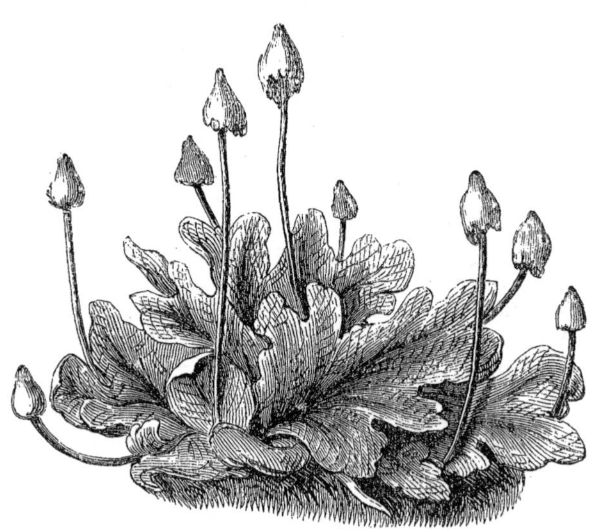

GLOSSARY

OF

BOTANICAL TERMS IN COMMON USE.

ABBREVIATE (*abbreviare*, to shorten), used to indicate that one part is shorter than another.
ABERRANT, deviating from the natural form.
ABORTION, suppression of an organ, depending on non-development.
ABRADED, rubbed off.
ABRUPT, ending in an abrupt manner, as the truncated leaf of the Tulip-tree; *abruptly pinnate*, ending in two pinnæ — in other words, paripinnate; *abruptly acuminate*, a leaf with a broad extremity, from which a point arises.
ACAULESCENT, without an evident stem.
ACCESSORY, an addition to a usual number.
ACCRESCENT, when parts continue to grow and increase after flowering, as the calyx of *Physalis* and the styles of *Anemone pulsatilla.*
ACCRETION, growing of one part to another.
ACCUMBENT, applied to the embryo of *Cruciferæ* when the cotyledons have their edges applied to the folded radicle.
ACEROSE, needle-like, narrow and slender, with a sharp point.
ACHÆNE, or ACHÆNIUM, a monospermous seed-vessel which does not open, but the pericarp of which is separable from the seed.
ACHLAMYDEOUS, having no floral envelope.
ACHROMATIC, applied to lenses which prevent chromatic aberration, *i.e.*, show objects without any prismatic colours.
ACICULAR, like a needle in form.
ACICULUS, a strong bristle.
ACINACIFORM, shaped like a sabre or scimitar.
ACOTYLEDONOUS, having no cotyledons.
ACROCARPI, Mosses having their fructification terminating the axis.
ACROGENOUS, having a stem increasing by its summit.
ACULEATE, furnished with prickles.
ACULEUS, a prickle, a process of the bark, not of the wood, as in the Rose.
ACUMINATE, drawn out into a long point.
ACUTE, terminating in a sharp point.
ADHERENT, adhesion of parts that are normally separate, as when the calyx is united to the ovary.
ADNATE, when an organ is united to another throughout its whole length: as the stipules to the petiole in Roses, and the filament and anther in *Ranunculus.*
ADPRESSED, or APPRESSED, closely applied to a surface.
ADULT, full grown.
ADVENTITIOUS, organs produced in abnormal positions, as roots arising from aërial stems.
ÆRUGINOUS, having the colour of verdigris.
ÆSTIVATION, the arrangements of the parts of the flower in the flower-bud.
AGGLOMERATED, collected in a heap or head.
AGGREGATE, gathered together.
ALA, a wing, applied to the lateral petals of papilionaceous flowers, and to membranous appendages of the fruit, as in the Elm, or of the seed, as in Pines.
ALBUMEN, the nutritious matter stored up with the embryo within the seed, called also Perisperm and Endosperm.
ALBURNUM the outer young wood of a dicotyledonous stem.
ALEXIPHARMIC, that which counteracts poisons.
ALGOLOGY, the study of sea-weeds.
ALTERNATE, arranged at different heights on the same axis, and towards different sides.
ALVEOLÆ, regular cavities on a surface, as in the receptacle of the Sunflower, and in that of *Nelumbium.*
ALVEOLATE, like a honeycomb.
AMENTUM, a catkin, or deciduous unisexual spike; plants having catkins are *Amentiferous.*
AMNIOS, the fluid or semi-fluid matter in the embryo-sac.
AMORPHOUS, without definite form.
AMPHISARCA, an indehiscent, multilocular fruit, with a hard exterior, and pulpy round the seeds, as seen in the Baobab.
AMPHITROPAL, an ovule, curved on itself, with the hilum in the middle.
AMPLEXICAUL, embracing the stem over a large part of its circumference.
AMPULLA, a hollow leaf, as in *Utricularia.*
AMYLACEOUS, starch-like.
ANASTOMOSING, inosculation of vessels.

ANASTOMOSIS, union of vessels ; union of the final ramifications of the veins of a leaf.
ANATROPAL, an inverted ovule, the hilum and micropyle being near each other, and the chalaza at the opposite end.
ANCEPS, two-edged.
ANDRŒCIUM, the male organs of the flower.
ANDROGYNOUS, male and female flowers on the same peduncle, as in some species of *Carex*.
ANDROPHORE, a stalk supporting the stamens, often formed by a union of the filaments.
ANFRACTUOSE, wavy or sinuous, as the anthers of *Cucurbitaceæ*.
ANGIOSPERMOUS, having seeds contained in a seed-vessel.
ANISOSTEMONOUS, stamens not equal in number to the floral envelopes, nor a multiple of them.
ANNOTINUS, a year old.
ANNULUS, applied to the elastic rim surrounding the sporangia of some Ferns, also to a cellular rim on the stalk of the Mushroom, being the remains of the veil.
ANTERIOR, same as inferior when applied to the parts of the flower in their relation to the axis.
ANTHELMINTIC, a vermifuge.
ANTHER, the part of the stamen containing pollen.
ANTHERIDIUM, the male organ in cryptogamic plants, frequently containing moving filaments.
ANTHERIFEROUS, bearing anthers.
ANTHEROZOIDS, moving filaments in an antheridium.
ANTHESIS, the opening of the flower.
ANTHOCARPOUS, applied to fruits, formed by the ovaries of several flowers.
ANTHODIUM, the capitulum or head of flowers of the Composite plants.
ANTHOPHORE, a stalk supporting the inner floral envelopes, and separating them from the calyx.
ANTHOS, a flower; in composition, *Antho;* in Latin, *Flos.*
ANTHOTAXIS, the arrangement of the flowers on the axis.
APETALOUS, without petals ; in other words, monochlamydeous.
APHYLLOUS, without leaves.
APICULATE, having an apiculus.
APICULUS, or APICULUM, a terminal soft point, springing abruptly.
APOCARPOUS, ovary and fruit composed of numerous distinct carpels.
APOPHYSIS, a swelling at the base of the theca in some Mosses.
APOTHECIUM, the rounded, shield-like fructification of Lichens.
APTEROUS, without wings or membranous margins.
ARACHNOID, applied to fine hairs so entangled as to resemble a cobweb.
ARBOREOUS, tree-like.
ARCHEGONIUM, the female organ in cryptogamic plants.
ARCUATE, curved in an arched manner.
AREOLÆ, little spaces on a surface.
AREOLATE, divided into distinct angular spaces, or areolæ.
ARILLATE, having an arillus.
ARILLUS and ARILLODE, an extra covering on the seed ; the former proceeding from the placenta, the latter from the exostome, as in Mace.
ARISTA, an awn, a long pointed process.
ARMATURE, the hairs, prickles, &c., covering an organ.
ARTICULATED, jointed, separated easily and cleanly at some point.
ASCENDING, applied to a procumbent stem which rises gradually from its base ; to ovules attached a little above the base of the ovary ; and to hairs directed towards the upper part of their support.
ASCI, tubes containing the sporidia of the Cryptogamia.
ASCIDIUM, a pitcher-like leaf, as in *Nepenthes.*
ASPERITY, roughness, as on the leaves of *Boraginaceæ.*
ATROPAL, the same as Orthotropous.
ATTENUATE, thin and slender.
AURICULATE, having appendages ; applied to leaves having lobes (ear-shaped) or leaflets at their base.
AWN and AWNED. See *Arista.*
AXIL, the upper angle, where the leaf joins the stem.
AXILE, or AXIAL, belonging to the axis.
AXIL-FLOWERING, flowering in the axilla.
AXILLARY, arising from the axil of a leaf.
AXIS is applied collectively to the stem and root—the ascending and descending axis, respectively.

BACCA, berry, a unilocular fruit, having a soft outer covering and seeds immersed in pulp.
BACCATE, resembling a berry.
BALAUSTA, the fruit of the Pomegranate.
BARBATE, bearded, having tufts of hair.
BARK (*cortex*), the outer cellular and fibrous covering of the stem ; separate from the wood in dicotyledons.
BARREN, not fruitful ; applied to male flowers, and to the non-fructifying fronds of Ferns.
BASAL, or BASILAR, attached to the base of an organ.
BASIDIUM, a cell bearing on its exterior one or more spores in some Fungi, which are hence called *Basidiosporous.*
BAST, or BASS, the inner fibrous bark of dicotyledonous trees.
BEAKED, like the sharp-pointed beak of a bird in form.
BEDEGUAR, a hairy excrescence on the branches and leaves of Roses, caused by an attack of a cynips.
BIDENTATE, having two tooth-like processes.
BIFARIOUS, in two rows, one on each side of an axis.
BIFID, two-cleft, cut down to near the middle into two parts.

BIFORINE, a raphidian cell with an opening at each end.
BILABIATE, having two lips.
BILOBED, divided into two lobes.
BILOCULAR, having two cells.
BINATE, applied to a leaf composed of two leaflets at the extremity of a petiole.
BIPARTITE, cut down to near the base into two parts.
BIPINNATE, a compound leaf, divided twice in a pinnate manner.
BIPINNATIFID, a simple leaf, with lateral divisions extending to near the middle, and which are also similarly divided.
BIPINNATIPARTITE, differing from bipinnatifid in the divisions extending to near the midrib.
BIPLICATE, doubly folded in a transverse manner.
BISERRATE, when the serratures are themselves serrate.
BITERNATE, a compound leaf divided into three, and each division again divided into three.
BLADE, the lamina or broad part of a leaf, as distinguished from the petiole or stalk.
BLANCHING. See *Etiolation*.
BLETTING, a peculiar change in an austere fruit, by which, after being pulled, it becomes soft and edible, as in the Medlar.
BLISTERED, applied to raised spots in leaves.
BOLE, the trunk of a tree.
BOTHRENCHYMA, dotted or pitted vessels.
BRACT, a leaf more or less changed in form, from which a flower or flowers proceed; flowers having bracts are called *bracteated*.
BRACTEOLE, a small bract at the base of a separate flower in a multifloral inflorescence.
BRANCHLETS, little branches.
BRYOLOGY, the study of Mosses; same as Muscology.
BULB, an underground stem covered with scales.
BULBIL, or BULBLET, separate buds in the axil of leaves, as in some Lilies.
BYSSOID, very slender, like a cobweb.

CADUCOUS, falling off very early, as the calyx of a Poppy.
CÆSIOUS, grey.
CÆSPITOSE, growing in tufts.
CALCAR, a spur, projecting hollow or solid process from the base of an organ, as in the flower of Larkspur or Snapdragon; such flowers are called *calcarate*, or spurred.
CALCEOLATE, slipper-like, applied to the hollow petals of some Orchids; also to the corolla of *Calceolaria*.
CALLOSITY, or CALLOUS, a leathery or hardened thickening on a limited portion of an organ.
CALYCIFLORÆ, a sub-class of polypetalous Exogens, having the stamens attached to the calyx.
CALYCINE, belonging to the calyx.
CALYPTRATE, in form resembling an extinguisher.
CALYX, the outer envelope of a flower.
CAMBIUM, the young active cells between the bark and the young wood.
CAMPANULATE, shaped like a bell, as the flower of Harebell.
CAMPYLOTROPAL, a curved ovule, with the hilum, micropyle, and chalaza near each other.
CANALICULATE, channelled, having a longitudinal groove or furrow.
CANCELLATE, latticed, composed of veins alone.
CANESCENT, hoary.
CAPILLARY, filiform, thread-like, or hair-like.
CAPITATE, pin-like, having a rounded summit, as some hairs.
CAPITULUM, head of flowers in *Compositæ*.
CAPREOLATE, having tendrils.
CAPSULE, a dry seed-vessel, opening by valves, teeth, pores, or a lid.
CARINA, keel, the two partially united lower petals of papilionaceous flowers.
CARINATE, keel-shaped.
CARPEL, the leaf which contains the ovules. Several carpels may enter into the composition of one pistil.
CARPOLOGY, the study of fruits.
CARPOPHORE, a stalk bearing the pistil, and raising it above the whorl of the stamens, as in *Lychnis* and *Capparis*.
CARUNCLE, a fleshy or thickened appendage of the raphe of the seed.
CARYOPSIS, the monospermal seed-vessel of a grass, the pericarp being adherent with the seed.
CATKIN, same as Amentum.
CAUDATE, having a tail or feathery appendage.
CAUDEX, the stem of Palms and of Tree Ferns.
CAUDICLE, the process supporting a pollen mass in Orchids.
CAULESCENT, having an evident stem.
CAULICLE, the rudimentary axis of the embryo.
CAULINE, produced on the stem.
CAUSTICITY, having a burning quality.
CELLULAR, composed of cells.
CELLULOSE, the chemical substance of which the cell wall is composed.
CENTIMÈTRE, a French measure, equal to 0·3937079 British inch.
CENTRIFUGAL, applied to that kind of inflorescence in which the central flower opens first.
CENTRIPETAL, applied to that kind of inflorescence in which the flowers at the circumference or base open first.
CERAMIDIUM, an ovate conceptacle, having a terminal opening, and with a tuft of spores arising from the base; seen in Algæ.
CEREAL, a general term applied to wheat, oats, barley, and rye.
CHALAZA, the place where the nourishing vessels enter the nucleus of the ovule.
CHLOROPHYLL, the green colouring matter of leaves.
CHORISIS, separation of a lamina from one part of an organ, so as to form a scale or a doubling of the organ; it may be either transverse or collateral.

CHROMULE, the colouring matter of the cells of flowers; also of the lower Algæ.
CILIA (*cilium*), short stiff hairs fringing the margin of a leaf; also the delicate vibratile hairs of zoospores.
CILIATO-DENTATE, toothed and fringed with hairs.
CIRCINATE, rolled up like a crozier, as the young fronds of Ferns.
CIRCUMSCISSILE, cut round in a circular manner, such as seed-vessels opening by a lid.
CIRCUMSCRIPTION, the periphery or margin of a leaf.
CIRRHUS, a modified leaf in the form of a tendril.
CLATHRATE, latticed, like a grating.
CLAVATE, club-shaped, becoming gradually thicker towards the top.
CLAW, the narrow base of some petals, corresponding with the petiole of leaves.
CLEFT, divided to about the middle.
CLOVES, applied to young bulbs, as in the Onion.
CLYPEATE, having the shape of a buckler.
COCCIDIUM, a rounded conceptacle in Algæ without pores, and containing a tuft of spores.
COCHLEAR, a kind of æstivation, in which a helmet-shaped part covers all the others in the bud.
COCHLEARIFORM, shaped like a spoon.
COCHLEATE, shaped like a snail shell.
COLEORHIZA, a sheath, surrounding the radicles of a monocotyledonous embryo.
COLLATERAL, placed side by side, as in the case of some ovules.
COLLUM, neck, the part where the plumule and radicle of the embryo unite.
COLUMELLA, central column in the sporangia of Mosses.
COLUMN, a part of a flower of an Orchid, supporting the anthers and stigma, and formed by the union of the styles and filaments.
COMA, a tuft of hair on a seed.
COMMISSURE, union of the faces of the two achænes in the fruit of *Umbelliferæ*.
COMOSE, furnished with hairs, as the seeds of the Willow.
COMPOUND, composed of several parts, as a leaf formed by several leaflets.
COMPRESSED, flattened laterally or lengthwise.
CONCENTRIC, curves with common centre.
CONCEPTACLE, a hollow sac containing a tuft or cluster of spores.
CONCRETE, hardened into a mass.
CONDUCTING TISSUE, applied to the loose cellular tissue in the interior of the style.
CONDUPLICATE, followed upon itself, applied to leaves and cotyledons.
CONE, a dry multiple fruit, formed by bracts covering naked seeds.
CONFERRUMINATE, indistinguishably united together.
CONFERVOID, formed of a single row of cells, or having articulations like a *Conferva*.
CONFLUENT, when parts unite together in the progress of growth.
CONJUGATION, union of two cells, so as to develop a spore.
CONNATE, when parts are united, even in the early state of development; applied to two leaves united by their bases.
CONNECTIVE, the part which connects the anther-lobes.
CONNIVENT, when two organs, as petals, arch over so as to meet above.
CONSTRICTED, contracted in some particular place.
CONTORTED, when the parts in a bud are imbricated and regularly twisted in one direction.
CONVOLUTE, when a leaf in the bud is rolled upon itself.
CORDATE, of leaves heart-shaped at the base.
CORDIFORM, having the shape of a heart.
CORIACEOUS, having a leathery consistence.
CORM, thickened underground stem, as in *Arum* and *Colchicum*.
CORNUTE, horned.
COROLLA, the inner envelope of the flower.
COROLLIFLORÆ, gamopetalous exogens.
CORONA, a coralline appendage, as the crown of the Daffodil.
CORPUSCLE, a small body or particle.
CORRUGATED, wrinkled or shrivelled.
CORTEX, the bark.
CORTICAL, belonging to the bark.
CORYMB, a raceme, in which the lower stalks are longest, and all the flowers come nearly to a level above.
COSTATE, provided with ribs; primary.
COTYLEDON, the temporary leaf of the embryo.
CREMOCARP, the fruit of *Umbelliferæ*, composed of two separable achænes or mericarps.
CRENATE, having superficial, rounded, marginal notches.
CRENATURES, divisions of the margin of a crenate leaf.
CREST, an appendage to fruits or seeds.
CRIBRIFORM, riddled with holes.
CRISP, having an undulated margin.
CRUCIFORM, arranged like the parts of a cross, as the flowers of *Cruciferæ*.
CRUSTACEOUS, hard, thin, and brittle.
CRYPTOGAMOUS, with the organs of reproduction obscure.
CUCULLATE, formed like a hood or cowl.
CULM, stem or stalk of Grasses.
CUNEIFORM, or CUNEATE, shaped like a wedge.
CUPULA, the cup of the acorn, formed by aggregate bracts.
CUSPIDATE, prolonged into an attenuated point.
CUTICLE, the thin membrane that covers the epidermis.
CYCLOSIS, movement of the latex in laticiferous vessels, and of the fluid-cell contents within the cell.
CYMBIFORM, shaped like a boat.
CYME, a kind of definite inflorescence, in which the flowers are in racemes, corymbs, or umbels, the successive central flowers expanding first.

CYPSELA, monospermal fruit of Compositæ.
CYTOBLAST, the nucleus of a cell.
CYTOGENESIS, cell development.

DECIDUOUS, falling off after performing its functions for a limited time, as the calyx of *Ranunculus*.
DECIDUOUS TREES, those which lose their leaves annually.
DECIMÈTRE, the tenth part of a mètre, or ten centimètres.
DECLINATE, directed downwards from its base.
DECOMPOUND, a leaf cut into numerous compound divisions.
DECORTICATED, deprived of bark.
DECUMBENT, lying flat along the ground, and rising from it at the apex.
DECURRENT, leaves which are attached along the side of a stem below their point of insertion; such stems are often called Winged.
DECUSSATE, opposite leaves crossing each other in pairs at right angles.
DEDUPLICATION, same as Chorisis.
DEFINITE, applied to inflorescence when it ends in a single flower, and the expansion of the flower is centrifugal; also when the number of the parts of an organ is limited, as when the stamens are under twenty.
DEFLEXED, bent downwards in a continuous curve.
DEFOLIATION, the fall of the leaves.
DEGENERATION, when an organ is changed from its usual appearance, and becomes less highly developed, as when scales take the place of leaves.
DEHISCENCE, mode of opening of an organ, as of the seed-vessels and anthers.
DELTOID, like the Greek Δ in form.
DEMULCENT, an emollient.
DENTATE, toothed, having short triangular divisions of the margin.
DENTICULATE, finely toothed, having small toothed-like projections along the margin.
DENTIFORM, tooth-shaped.
DEPENDENT, hanging down.
DEPRESSED, flattening of a solid organ from above downwards.
DETERGENT, having a cleansing power.
DIADELPHOUS, stamens in two bundles, united by their filaments.
DIANDROUS, having two stamens.
DIAPHANOUS, transparent.
DICHLAMYDEOUS, having calyx and corolla.
DICHOTOMOUS, stem dividing by twos.
DICLINOUS, unisexual flower either monœcious or diœcious.
DICOTYLEDONOUS, embryo having two cotyledons.
DICTYOGENOUS, applied to monocotyledons having netted veins.
DIDYNAMOUS, two long and two short stamens.
DIFFUSE, scattered.
DIGITATE, compound leaf, composed of several leaflets attached to one point.
DIGYNOUS, having two styles.
DIMEROUS, when the parts of a flower are in twos.
DIMIDIATE, when one half of an organ is smaller than the other half.
DIŒCIOUS, staminiferous and pistilliferous flowers on separate plants.
DIPLOSTEMONOUS, stamens double the number of the petals or sepals.
DIPTEROUS, having two wings.
DISCOID, in the form of a disc or flattened sphere; *discoid pith*, divided into cavities by discs.
DISCS, the peculiar rounded and dotted markings on the fibres of coniferous wood.
DISK, a part intervening between the stamens and the pistils in the form of scales, a ring, &c.
DISSECTED, cut into a number of narrow divisions.
DISSEPIMENT, a division in the ovary; true when formed by the edges of the carpels, false when formed otherwise.
DISTICHOUS, in two rows on opposite sides of a stem.
DIVARICATING, branches coming off from the stem at a very wide or obtuse angle.
DODECANDROUS, having twelve stamens.
DOLABRIFORM, shaped like an axe.
DORSAL, applied to the suture of the carpel which is farthest from the axis.
DOUBLE FLOWER, when the organs of reproduction are converted into petals.
DRUPE, a fleshy fruit like the Cherry, having a stony endocarp.
DRUPELS, small drupes aggregated to form a fruit, as in the Raspberry.
DURAMEN, heart-wood of dicotyledonous trees.

ELATERS, spiral fibres in the spore-cases of Hepaticæ.
ELLIPTICAL, having the form of an ellipse.
EMARGINATE, with a notch at the end.
EMBRACING. This is said to be the case when a leaf clasps the stem.
EMBRYO, the young plant contained in the seed.
EMBRYO-SAC, the cell in which the embryo is formed.
ENDOCARP, the inner layer of the pericarp, next the seed.
ENDOCHROME, the colouring matter within the cells of the lower plants.
ENDOGEN, a monocotyledon.
ENDOPHLŒUM, the inner bark or liber.
ENDOPLEURA, the inner covering of the seed.
ENDORHIZAL, numerous rootlets arising from *within* a common radicle, and passing through sheaths, as in endogenous germination.
ENDOSMOSE, movement of fluids inwards through a membrane.
ENDOSPERM, albumen formed within the embryo-sac.
ENDOSTOME, the inner faramen of the ovule.
ENDOTHECIUM, the inner coat of the anther.
ENSIFORM, in the form of a sword, as the leaves of *Iris*.
ENTIRE (*integer*), without marginal divisions
ENVELOPES, FLORAL, the calyx and corolla.

EPICALYX, outer calyx formed either of sepals or bracts, as in Mallow and *Potentilla.*
EPICARP, the outer covering of the fruit.
EPICHILIUM, the terminal portion of the lip (*labellum*) in Orchids.
EPIDERMIS, the cellular layer covering the external surface of plants.
EPIGYNOUS, above the ovary by adhesion to it.
EPIPETALOUS, inserted on the petals.
EPIPHYLLOUS, growing upon a leaf.
EPIPHYTES, attached to another plant, and growing suspended in the air.
EPISPERM, the external covering of the seed.
EQUITANT, applied to leaves folded longitudinally, and overlapping each other without any involution.
ERECT, applied to an ovule which rises from the base of the ovary.
ERODED, gnawed or bitten.
EROSE, irregularly toothed, as if gnawed.
ERUMPENT, as if bursting through the epidermis.
ESCHAROTIC, having the power to scar or burn the skin.
ETÆRIO, the aggregate drupes forming the fruit of Rubus.
ETIOLATION, blanching ; losing colour through growth in the dark.
EXALBUMINOUS, without a separate store of albumen or perisperm.
EXANNULATE, without a ring ; applied to some Ferns.
EXCENTRIC, removed from the centre or axis ; applied to a lateral embryo.
EXCIPULUS, a receptacle containing fructification in Lichens.
EXCORIATED, stripped of skin or bark.
EXCURRENT, running out beyond the edge or point.
EXOGEN, Dicotyledon.
EXORHIZAL, radicle proceeding directly from the axis, and afterwards branching, as in exogens.
EXOSMOSE, the passing outwards of a fluid through a membrane.
EXOSTOME, the outer opening of the foramen of the ovule.
EXOTHECIUM, the outer coat of the anther.
EXSERTED, extended beyond an organ, as stamens beyond the corolla.
EXSICCATED, dried up.
EXSTIPULATE, without stipules.
EXTINE, the outer covering of the pollen grain.
EXTRA-AXILLARY, removed from the axil of the leaf, as in the case of some buds.
EXTRORSE, applied to anthers which dehisce on the side farthest removed from the pistil.

FÆCULA, starchy matter.
FALCATE, or FALCIFORM, bent like a sickle.
FARINACEOUS, mealy, containing much starch.
FASCIATION, union of branches of stems so as to present a flattened ribbon-like form.
FASCICLE, a shortened umbellate cyme, as in some species of *Dianthus.*
FASCICULATE, arranged in bundles.
FASTIDIATE, having a pyramidal form, from the branches being parallel and erect, as in Lombardy Poplar.
FAUCES, the gaping part of a monopetalous corolla.
FEATHER-VEINED, a leaf having the veins passing from the midrib at a more or less acute angle, and extending to the margin.
FECUNDATION, fertilisation.
FENESTRATE, applied to a leaf with perforations.
FERRUGINOUS, rusty.
FERTILE, applied to pistillate flowers, and to the fruit-bearing fronds of Ferns.
FIBROUS, composed of numerous fibres, as some roots.
FIBRO-VASCULAR TISSUE, containing vessels and fibres.
FILAMENT, stalk supporting the anther.
FILAMENTOUS, a string of cells placed end to end.
FILIFORM, like a thread.
FIMBRIATED, fringed at the margin.
FISSIPAROUS, dividing spontaneously into two parts by means of a septum.
FISSURE, a straight slit in an organ for the discharge of its contents.
FISTULOUS, hollow, like stems of Grasses.
FLABELLIFORM, fan-shaped, as the leaves of some Palms.
FLACCID, feeble, weak.
FLAGELLUM, a runner, a weak creeping stem, bearing rooting buds at different points, as in the Strawberry.
FLEXUOSE, having alternate curvations in opposite directions.
FLOCCOSE, covered with wool-like tufts.
FLORETS, little florets forming a compound flower.
FOLIACEOUS, having the form of leaves.
FOLLICLE, a fruit formed by a single carpel dehiscing by one suture, which is usually the ventral.
FOVEOLATE, having pits or depressions, called foveæ or foveolæ.
FOVILLA, minute granular matter in the pollen grain.
FROND, the leaf-like organ of Ferns, bearing the fructification.
FRONDOSE, applied to Cryptogams with foliaceous or leaf-like expansions.
FRUCTIFICATION, the seed or fruit of plants.
FRUSTULES, the parts or fragments into which Diatomaceæ separate.
FRUTICOSE, shrubby.
FUGACIOUS, evanescent, falling off early, as the petals of *Cistus.*
FULVOUS, tawny, yellow.
FUNGOUS, having the substance of Fungi or Mushrooms.
FUNICULUS, the cord connecting the hilum of the ovule to the placenta.
FURCATE, divided into two branches, like a two-pronged fork.
FURFURACEOUS, scaly or scurfy.
FUSCOUS, blackish brown.
FUSIFORM, shaped like a spindle.

GALBULUS, the polygynœcial fruit of Juniper.
GAMOPETALOUS, same as monopetalous, petals united.
GAMOPHYLLOUS and GAMOSEPALOUS, same as Monophyllous and Monosepalous, sepals united.
GEMINATE, twin organs combined in pairs; same as Binate.
GEMMATION, the development of leaf-buds.
GEMMULE, same as plumule, the first bud of the embryo.
GENICULATE, bent like a knee.
GERMEN, or GERM, a name for the ovary.
GERMINAL VESICLE, a germ contained in the embryo-sac, from which the embryo is developed.
GERMINATION, the sprouting of the young plant.
GIBBOSITY, a swelling at the base of an organ, such as the calyx or corolla.
GIBBOUS, swollen at the base, or having a distinct swelling at some part of the surface.
GLABROUS, smooth, without hairs.
GLAND, an organ of secretion consisting of cells, and generally occurring on the epidermis of plants.
GLANDULAR HAIRS, hairs tipped with a gland, as in *Drosera* and Chinese Primrose.
GLANS, nut, applied to the acorn and hazel-nut, which are enclosed in an involucre formed of consolidated bracts.
GLAUCOUS, covered with a pale green bloom.
GLOBOSE, round-shaped.
GLOBULE, male organ of Chara.
GLOCHIDIATE, barbed; applied to hairs with two reflexed points at their summits.
GLOMERULE, a rounded cymose inflorescence, as in *Urtica*.
GLUMACEOUS, chaffy.
GLUME, a bract covering the organs of reproduction in the spikelets of Grasses.
GLUTEN, a highly nitrogenous substance found in seeds.
GONIDIA, green cells in the thallus of Lichens.
GRAIN, caryopsis, the fruit of Grasses.
GRUMOUS, collected into granular masses.
GYMNOGEN, a plant with naked seeds, *i.e.*, seed not in a true ovary.
GYMNOSPERMOUS, plants with naked seeds, *i.e.*, seeds not in a true ovary; such as Conifers.
GYNANDROUS, stamen and pistil united in a common column, as in Orchids.
GYNOBASE, a central axis, to the base of which the carpels are attached.
GYNŒCIUM, the female organs of the flower.
GYNOPHORE, a stalk supporting the ovary.
GYRATE, same as Circinate.

HABIT, general external appearance.
HASTATE, halbert-shaped, applied to a leaf with two portions at the base projecting more or less completely at right angles to the blade.
HAULM, dead stems of herbs, as of the Potato.
HAUSTORIUM, the sucker at the extremity of the parasitic root of Dodder.
HEART-WOOD, same as Duramen.
HELICOIDAL, having a coiled appearance like the shell of a snail; applied to inflorescence.
HERB, a plant with an annual stem, opposed to a woody plant.
HERBACEOUS, green succulent plants which die down to the ground in winter; annual shoots, with green-coloured cellular parts.
HERMAPHRODITE, stamens and pistils in the same flower.
HESPERIDIUM, the fruit of the Orange and other *Aurantiaceæ*.
HETEROCYSTS, peculiar large cells in *Nostochineæ*.
HETEROGAMOUS, composite plants having hermaphrodite and unisexual flowers on the same head.
HETEROPHYLLOUS, presenting two different forms of leaves.
HILUM, the base of the seed to which the placenta is attached either directly or by means of a cord. The term is also applied to the mark at one end of some grains of starch.
HIRSUTE, covered with long stiff hairs.
HISPID, covered with long, very stiff hairs.
HISTOLOGY, the study of microscopic tissues.
HOMOGENEOUS, having a uniform structure or substance.
HYALINE, transparent or colourless.
HYBRID, a plant resulting from the fecundation of one species by another.
HYMENIUM, the part which bears the fructification in Agarics.
HYPANTHODIUM, the receptacle of *Dorstenia*, bearing many flowers.
HYPOCHILUM, the lower part of the labellum of Orchids.
HYPOCRATERIFORM, shaped like a salver, as the corolla of *Primula*.
HYPOGEOUS, under the surface of the soil; applied to cotyledons.
HYPOGYNOUS, inserted below the ovary or pistil.

IMBRICATE, parts overlying each other like tiles on a house. *Imbricated æstivation*, the parts of the flower-bud alternately overlapping each other, and arranged in a spiral manner.
IMPARI-PINNATE, unequally pinnate; pinnate leaf ending in an odd leaflet.
INARCHING, a mode of grafting by bending two growing plants towards each other, and causing a branch of the one to unite to the other.
INARTICULATE, without joints or interruption to continuity.
INCISED, cut down deeply.
INCLUDED, applied to the stamens when enclosed within the corolla, and not pushed out beyond its tube.
INCUMBENT, cotyledons with the radicle on their back.
INCURVED, bending inwards.
INDEFINITE, applied to inflorescence with centripetal expansion; also to stamens above twenty, and to ovules and seeds when very numerous.

INDEHISCENT, not opening, having no regular line of suture.
INDIGENOUS, an aboriginal native in a country.
INDUPLICATE, edges of the sepals or petals turned slightly inwards in æstivation.
INDUSIUM, epidermal covering of the fructification in some Ferns.
INFERIOR, applied to the ovary where it seems to be situated below the calyx, and to the part of the flower farthest from the axis.
INFLEXED, bending inwards.
INFLORESCENCE, the mode in which the flowers are arranged on the axis.
INFUNDIBULIFORM, in shape like a funnel, as seen in some gamopetalous corollas.
INNATE, applied to anthers when attached to the top of the filament.
INSPISSATED, thickened or dried-up juice or sap.
INTERNODE, the portion of the stem between two nodes or leaf-buds.
INTERPETIOLAR, between the petioles.
INTERRUPTEDLY-PINNATE, a pinnate leaf in which pairs of small pinnæ occur between the larger pairs.
INTINE, the inner covering of the pollen grains.
INTRAMARGINAL, within the margin.
INTRORSE, applied to anthers which open on the side next the pistil.
INVERSE, inverted.
INVOLUCEL, bracts surrounding the partial umbel of *Umbelliferæ*.
INVOLUCRE, bracts surrounding the general umbel in *Umbelliferæ*, the heads of flowers in Compositæ, and in general any verticillate bracts surrounding numerous flowers.
INVOLUTE, edges of leaves rolled inwards spirally on each side in æstivation.
IRREGULAR, a flower in which the parts of any of the verticils differ in size.
ISOMEROUS, when the whorls of a flower are composed each of an equal number of parts.
ISOSTEMONOUS, when stamens and floral envelopes have the same number of parts or multiples.
ISOTHERMAL, lines passing through places which have the same mean annual temperature.

JUGATE, applied to the pairs of leaflets in compound leaves; *Unijugate*, having one pair; *Bijugate*, two pairs, and so on.

KEEL, same as Carina.
KNOTTED, when a cylindrical stem is swollen at intervals into a knob.

LABELLUM, lip, one of the divisions of the inner whorl of the flower in Orchids. This part is in reality superior, but becomes inferior by the twisting of the ovary.
LABIATE, lipped; applied to irregular gamopetalous flowers, with an upper and under portion separated more or less by an hiatus or gap.
LACINIATE, irregularly cut into narrow segments.
LACTESCENT, yielding milky juice.
LACUNA, a large space in the midst of a group of cells.
LAMELLÆ, gills of an Agaric; also applied to flat divisions of the stigma.
LAMINA, the blade of the leaf; the broad part of the petal or sepal.
LANCEOLATE, tapering to each end, but broadest *below* the middle.
LATERAL, arising from the side of the axis, not terminal.
LATEX, granular fluid contained in laticiferous vessels.
LATICIFEROUS, vessels containing latex, which anastomose.
LAX, not compact.
LEAFLETS, the small portions of compound leaves.
LEGUME, a pod composed of one carpel, opening usually by a ventral and dorsal suture, as in the Pea.
LEGUMINOUS, plants bearing pods.
LENTICEL, a small cellular process on the bark of the Willow and other plants.
LENTICULAR, in the form of a doubly-convex lens.
LEPIDOTE, covered with scales or scurf.
LIANES, twining woody plants.
LIBER, the fibrous inner bark of Endophlœum.
LID, the calyx which falls from the flower in one piece.
LIGNINE, woody matter which thickens the cell walls.
LIGULATE, strap-shaped.
LIGULE, a process arising from the petiole of Grasses, where it joins the blade.
LIGULIFLORÆ, composite plants having ligulate florets.
LIMB, the blade of the leaf; the broad part of a petal or sepal. When sepals or petals are united, the combined broad parts are denominated collectively the limb.
LINE, the twelfth part of an inch.
LINEAR, very narrow when the length greatly exceeds the breadth.
LINGUIFORM, strap-shaped.
LIPPED, having a distinct lip or labellum.
LOBE, large division of a leaf or any other organ, applied often to the divisions of the anther.
LOCULAMENTS, divisions of the cells of a seed-vessel.
LOCULICIDAL, fruit dehiscing through the back of the carpels.
LOCULUS, a cavity in an ovary. The terms are also applied to the anther.
LOCUSTA, a spikelet of Grasses.
LODICULE, a scale at the base of the ovary of Grapes.
LOMENTUM, an indehiscent legume or pod with transverse partitions, each division containing one seed.
LURID, a colour combining yellow, purple, and grey.
LYRATE, a pinnatifid leaf with a large terminal lobe, and smaller ones as we approach the petiole.

MACROPODOUS, applied to the thickened radicle of a monocotyledonous embryo.
MARCESCENT, withering, but not falling off until the part bearing it is perfected.
MEDULLA, the pith.
MEDULLARY RAYS, cellular prolongation uniting the pith and the bark.
MEDULLARY SHEATH, sheath containing spiral vessels, surrounding the pith in exogens.
MEMBRANOUS, having the consistence, aspect, and structure of a membrane.
MERICARP, carpel forming one-half of the fruit of *Umbelliferæ*.
MERITHAL, a term used in place of Internode; applied by Gaudichaud to the different parts of the leaf.
MESOCARP, middle covering of the fruit.
MESOCHILUM, middle portion of the labellum of Orchids.
MESOPHLŒUM, middle layer of the bark.
METRE, equal to 39·3707 inches British.
MICROMETER, instrument for measuring microscopic objects.
MICROPYLE, the opening or foramen of the seed.
MILLIMETRE, equal to 0·0393707 English inch.
MONADELPHOUS, stamens united into one bundle by union of their filaments.
MONILIFORM, beaded; cells united with interruptions, so as to resemble a string of beads.
MONOCARPIC, producing flowers and fruit once during life, and then dying.
MONOCHLAMYDEOUS, flowers having a single envelope.
MONOCLINOUS, stamens and pistils in the same flower.
MONOCOTYLEDONOUS, having one cotyledon in the embryo.
MONŒCIOUS, stamens and pistils in different flowers on the same plant.
MONOPETALOUS, same as Gamopetalous.
MONOPHYLLOUS, same as Gamophyllous.
MONOSEPALOUS, having one sepal or division in the calyx. Same as Gamosepalous.
MONSTROSITY, an abnormal development; applied more especially to double flowers.
MORPHOLOGY, the study of the forms which the different organs assume, and the laws that regulate their metamorphoses.
MUCILAGE, a thick viscid fluid.
MUCRO, a stiff point abruptly terminating an organ.
MUCRONATE, having a mucro.
MUCRONULATE, having a little hard point.
MURICATE, covered with firm sharp points or excrescences.
MURIFORM, like bricks in a wall; applied to cells.
MYCELIUM, the cellular spawn of Fungi.

NAKED, applied to seeds not contained in a true ovary; also to flowers without any floral envelopes.
NAPIFORM, shaped like a turnip.
NATURALISED, originally introduced by artificial means, but become apparently wild.
NAVICULAR, hollowed like a boat.
NECTARY, any abnormal part of a flower. It ought to be restricted to organs secreting a honey-like matter, as in the Crown Imperial.
NERVATION, same as Nevation.
NERVES, the veins of leaves.
NETTED, applied to reticulated venation.
NODDING, drooping.
NODE, the part of a stem from which the leaf-bud proceeds.
NODOSE, having swollen nodes or articulations.
NUCLEUS, the body which gives origin to new cells; also applied to the central cellular portion of the ovule and seed.
NUCULE, female part of fructification in the Characeæ.
NUT, any dry one-celled indehiscent fruit with hard pericarp.

OBCORDATE, inversely heart-shaped, with the divisions of the heart at the opposite end from the stalk.
OBLONG, about three-fourths as long as broad.
OBOVATE, reversely ovate, the broad part of the egg being uppermost.
OBSOLETE, imperfectly developed or abortive; applied to the calyx when it is in the form of a rim.
OBTUSE, not pointed, with a rounded or blunt termination.
OCHRACEOUS, clay or ochre colour.
OCHREA, the sheathing stipule of *Polygonaceæ*.
OFFICINAL, sold in the shops.
OLERACEOUS, used as an esculent potherb.
OLIVACEOUS, having the colour of olives.
OOPHORIDIUM, organ in Lycopodiaceæ containing large spores.
OPAQUE, dull, not shining.
OPERCULAR, covered with a lid.
OPERCULUM, lid; applied to the separable part of the theca of Mosses; also applied to the lid of certain seed-vessels.
OPPOSITE, applied to leaves placed on opposite sides of the same stem at the same level.
ORBICULAR, rounded leaf with petiole attached to the centre of it.
ORGANOGRAPHY, the description of the organs of plants.
ORTHOTROPAL, ovule with foramen opposite to the hilum; embryo with radicle next the hilum.
OSMOSE, the force with which fluids pass through membranes in experiments on exosmose and endosmose.
OVAL, elliptical, blunt at each end.
OVARY, the part of the pistil which contains the ovules.
OVATE, shaped like an egg; applied to the broader end of the egg next the petiole or axis.
OVOID, egg-shaped.
OVULE, the young seed contained in the ovary.

PALE, the part of the flower of Grasses within the glume; also applied to the small scaly laminæ which occur in the receptacle of some Compositæ.

PALÆOPHYTOLOGY, the study of fossil plants.

PALEACEOUS, chaffy, covered with small, erect, membranous scales.

PALMATE and PALMATIFID, applied to a leaf with radiating venation, divided into lobes to about the middle.

PALMATIPARTITE, applied to a leaf with radiating venation, cut nearly to the base in a palmate manner.

PANDURIFORM, shaped like a fiddle.

PANICLE, inflorescence of Grasses, consisting of spikelets on long peduncles coming off in a racemose manner.

PANICULATE, forming a panicle.

PAPILIONACEOUS, corolla composed of vexillum, two alæ, and carina, as in the Pea.

PAPILLOSE, covered with small nipple-like prominences.

PAPPUS, the hairs at the summit of the ovary in Compositæ. They consist of the altered calycine limb. *Pappose*, provided with pappus.

PARAPHYSES, filaments, sometimes articulated, occurring in the fructification of Mosses and other Cryptogams.

PARASITE, attached to another plant, and deriving nourishment from it.

PARENCHYMA, cellular tissue.

PARIETAL, applied to placentas on the wall of the ovary.

PARI-PINNATE, a compound of pinnate leaf ending in two leaflets.

PARTHENOGENESIS, production of perfect seed with embryo, without the application of pollen.

PATENT, spreading widely.

PATULUS, spreading less than when patent.

PECTINATE, divided laterally into narrow segments like the teeth of a comb.

PEDATE and PEDATIFID, a palmate leaf of three lobes, the lateral lobes bearing other equally large lobes on the edges next the middle lobe.

PEDICEL, the stalk supporting a single flower.

PEDUNCLE, the general flower-stalk or floral axis; sometimes it bears one flower, at other times it bears several sessile or pedicellate flowers.

PELAGIC, growing in the ocean.

PELLUCID, transparent.

PELORIA, a name given to a teratological phenomenon. which consists in a flower that is usually irregular becoming regular; for instance, when *Linaria*, in place of one spur, produces five.

PELTATE, shield-like. fixed to the stalk by a point within the margin; peltate hairs, attached to their middle.

PENDULOUS, applied to ovules which are hung from the upper part of the ovary.

PENICILLATE, resembling a camel's-hair pencil.

PENNI-NERVED and PENNI-VEINED, the veins disposed like a feather, running from the middle of the leaf to the margin.

PENTAMEROUS, composed of different whorls in five, or multiples of that number.

PEPO, the fruit of the Melon, Cucumber, and other *Cucurbitaceæ*.

PERENNIAL, living, or rather flowering, for several years.

PERFOLIATE, a leaf with the lobes at the base, united on the side of the stem opposite the blade, so that the stalk appears to pass through the leaf.

PERIANTH, a general name for the floral envelopes; applied in cases where there is only a calyx, or where the calyx and corolla are alike.

PERICARP, the covering of the fruit.

PERICHÆTIAL, applied to the leaves surrounding the fruit-stalk or seta of Mosses.

PERICLADIUM, the large sheathing petiole of *Umbelliferæ*.

PERIDERM, a name applied to the outer layer of the barks.

PERIDIUM, the envelope of the fructification in gasteromycetous Fungi.

PERIGONE, same as Perianth. Some restrict the term to cases in which the flower is female, or pistilliferous. It has also been applied to the involucre of *Jungermannieæ*.

PERIGYNOUS, applied to the corolla and stamens when attached to the calyx.

PERIGYNUM, applied to the pistil in the genus *Carex*.

PERIPHERICAL, applied to an embryo curved so as to surround the albumen, following the inner part of the covering of the seed.

PERISPERM, the albumen or nourishing matter stored up with the embryo in the seed.

PERISTOME, the opening of the sporangium of Mosses after the removal of the calyptra and operculum.

PERITHECIUM, a conceptable in Cryptogams, containing spores, and having an opening at one end.

PERSISTENT, not falling off, remaining attached to the axis until the part which bears it is matured.

PERSONATE, a gamopetalous irregular corolla, having the lower lip pushed upwards, so as to close the hiatus between the two lips.

PERTUSE, having slits or holes.

PERULÆ, the scales of the leaf-bud.

PETALOID, like a petal.

PETALS, the leaves forming the coralline whorl.

PETIOLATE, having a stalk or petiole.

PETIOLE, a leaf-stalk; *Petiolule*, the stalk of a leaflet in a compound leaf.

PHÆNOGAMOUS, same as Phanerogamous.

PHANEROGAMOUS, having conspicuous flowers.

PHYCOLOGY, the study of Algæ, or Seaweeds.

PHYLLARIES, the leaflets forming the involucre of composite flowers.

PHYLLODIUM, the leaf-stalk, enlarged so as to have the appearance of a leaf.

PHYLLOTAXIS, the arrangement of the leaves on the axis.

PHYSIOGNOMY, general appearance, without reference to botanical characters.

PHYSIOLOGY, vegetable, the study of the functions of plants.

PHYTOLOGY, the study of plants; same as Botany.
PHYTOZOA, moving filaments in the antheridia of Cryptogams.
PILEATE, having a cup or lid like the cup of a Mushroom.
PILEORHIZA, a covering of the root, as in *Lemna*.
PILEUS, the cap-like portion of the Mushroom, bearing the hymenium on its under side.
PILOSE, provided with hairs; applied to pappus composed of simple hairs.
PINNA, the leaflet of a pinnate leaf.
PINNATE, a compound leaf having leaflets arranged on each side of a central rib.
PINNATIFID, a simple leaf cut into lateral segments to about the middle.
PINNATIPARTITE, a simple leaf cut into lateral segments, the divisions extending nearly to the central rib.
PINNULE, the small pinnæ of a bipinnate or tripinnate leaf.
PISTIL, the female organ of the flower, composed of one or more carpels; each carpel being composed of ovary, style, and stigma.
PISTILLATE and PISTILLIFEROUS, applied to a female flower or a female plant.
PISTILLIDIUM, the female organ in Cryptogams.
PITCHERS, vessels of this form at the end of the leaves of *Nepenthes*, &c.
PITH, same as Medulla.
PLACENTA, the cellular part of the carpel, bearing the ovule.
PLACENTATION, the formation and arrangement of the placentas.
PLEURENCHYMA, woody tissue.
PLEUROCARPI, Mosses with the fructification proceeding laterally from the axils of the leaves.
PLICATE, folded like a fan.
PLUMOSE, feathery; applied to hairs having two longitudinal rows of minute cellular processes.
PLUMULE, the first bud of the embryo, usually enclosed by the cotyledons.
PLURILOCULAR, having many loculaments.
PODETIUM, a stalk bearing the fructification in some Lichens.
PODOSPERM, the cord attaching the seed to the placenta.
POLLARD-TREES, cut down so as to leave only the lower part of the trunk, which gives off numerous buds and branches.
POLLEN, the powdery matter contained in the anther.
POLLEN-TUBE, the tube emitted by the pollen grain after it is applied to the stigma.
POLLINIA, masses of pollen found in Orchids and Asclepiads.
POLYADELPHOUS, stamens united by their filaments so as to form more than two bundles.
POLYANDROUS, stamens above twenty.
POLYCARPIC, plants which flower and fruit many times in the course of their life.
POLYCOTYLEDONOUS, an embryo having many cotyledons, as in Firs.
POLYGAMOUS, plants bearing hermaphrodite as well as male and female flowers
POLYMORPHOUS, assuming many shapes.
POLYPETALOUS, a corolla composed of separate petals.
POLYPHYLLOUS, a calyx or involucre composed of separate leaflets.
POLYSEPALOUS, a calyx composed of separate sepals.
POME, a fruit like the Apple and Pear.
POROUS VESSELS, same as Pitted or Dotted Vessels.
POSTERIOR, applied to the part of the flower placed next the axis; same as Superior.
POUCH, the short pod or silicle of some *Cruciferæ*.
PREMORSE, bitten; applied to a root terminating abruptly, as if bitten off.
PRICKLES, hardened epidermal appendages of a nature similar to hairs.
PRIMINE, the outer coat of the ovule.
PRIMORDIAL UTRICLE, the lining membrane of cells in their early state.
PROCESS, any prominence or projecting part, or small lobe.
PROCUMBENT, lying on the ground.
PROEMBRYO, cellular body in an ovary, from which the embryo and its suspensor are formed. Sometimes Proembryo is used for Prothallus.
PROLIFEROUS, bearing abnormal buds.
PRONE, prostrate, lying flat on the earth.
PROPAGULUM, an offshoot or germinating bud attached by a thickish stalk to the parent plant.
PROSENCHYMA, fusiform tissue forming wood.
PROTHALLIUM, or PROTHALLUS, names given to the first part produced by the spore of an acrogen in germinating.
PROTOPLASM, the nitrogenous gelatinous matter in which the vital activity of cells resides.
PSEUDO-BULB, the peculiar aërial stem of many epiphytic Orchids.
PUBESCENCE, short and soft hairs covering a surface.
PULULATING, budding.
PULVERULENT, covered with fine powdery matter.
PULVINATE, shaped like a cushion or pillow.
PULVINUS, cellular swelling at the point where the leaf-stalk joins the axis.
PUNCTATED, applied to the peculiar dotted woody fibres of *Coniferæ*.
PUTAMEN, the hard endocarp of some fruits.
PYCNIDES, cysts containing stylospores found in some Lichens.
PYXIS, a capsule opening by a lid.

QUATENARY, composed of parts in fours.
QUINARY, composed of parts in fives.
QUINATE, five leaves coming off from one point.
QUINCUNX, when the leaves in the bud are five, of which two are exterior, two interior, and the fifth covers the interior with one margin, and has its other margin covered by the exterior. *Quincuncial*, arranged in a quincunx.

RACE, a permanent variety.
RACEME, an indefinite inflorescence, in which there is a primary axis bearing stalked flowers.
RACEMOSE, flowering in racemes.
RACHIS, the axis of inflorescence; also applied to the stalk of the frond in Ferns, and to the common stalk bearing the alternate spikelets in some Grasses.
RADICAL, belonging to the root; applied to leaves close to the ground, clustered at the base of a flower-stalk.
RADICLE, the young root of the embryo.
RAMENTA, little brown withered scales with which the stems of some plants are covered.
RAMIFICATIONS, sub-divisions of roots or branches.
RAPHE, the line which connects the hilum and the chalaza in anatropal ovules.
RAPHIDES, crystals found in cells, which are hence called *Raphidian*.
RECEPTACLE, the flattened end of the peduncle rachis, bearing numerous flowers in a head; applied also generally to the extremity of the peduncle or pedicel.
RECLINATE, curved downwards from the horizontal, bent back up.
RECURVED, bent backwards.
REDUPLICATE, edges of the petals or sepals turned outward in æstivation.
REGMA, seed-vessels composed of elastic cocci, as in *Euphorbia*.
REGULAR, applied to an organ, the parts of which are of similar form and size.
RELIQUIÆ, remains of withered leaves attached to the plant.
RENIFORM, in shape like a kidney.
REPAND, having a slightly undulated or sinuous margin.
REPLUM, a longitudinal division in a pod formed by the placenta, as in *Cruciferæ*.
RESUPINATE, inverted by a twisting of the stalk.
RETICULATE, netted, applied to leaves having a network of anastomosing veins.
RETINACULUM, the glandular viscid portion at the extremity of the caudicle in some Pollinia.
RETRORSE, turned backwards.
RETUSE, when the extremity is broad, blunt, and slightly depressed.
REVOLUTE, leaf with its edges rolled backwards in vernation.
RHIZOME, a stem creeping horizontally, more or less covered by the soil, giving off buds above and roots below.
RHIZOTAXIS, the arrangement of the roots.
RHOMBOID, quadrangular form, not square with equal sides.
RIB, the projecting vein of a leaf.
RINGENT, a labiate flower in which the upper lip is much arched.
ROOT-STOCK, same as Rhizome.
ROSETTE, leaves disposed in close circles forming a cluster.
ROSTELLUM, a prolongation of the upper edge of the stigmas in some Orchids.
ROSTRATE, beaked.
ROTATE, a regular gamopetalous corolla, with a short tube, the limbs spreading out more or less t right angles.
RUBEFACIENT, that which reddens the surface.
RUDIMENTARY, an organ in an abortive state arrested in its development.
RUFOUS, rust-red.
RUGOSE, wrinkled.
RUMINATE, applied to mottled albumen.
RUNCINATE, a pinnatifid leaf with a triangular termination, and sharp divisions pointing downwards, as in Dandelion.
RUNNERS, procumbent shoots which root at their extremity.
RUSTY, rust-coloured.

SAGITTATE, like an arrow; a leaf having two prolonged sharp-pointed lobes projecting downwards beyond the insertion of the petiole.
SAMARA, a winged dried fruit, as in the Elm.
SAPONACEOUS, soap-like.
SARMENTOSE, yielding runners.
SARMENTUM, sometimes meaning the same as Flagellum, or runner; at other times applied to a twining stem which supports itself by means of others.
SCABROUS, rough, covered with very stiff short hair.
SCALARIFORM, vessels having bars like a ladder, seen in Ferns.
SCALES, small processes resembling minute leaves.
SCANDENT, climbing by means of supports, as on a wall or rock.
SCAPE, a naked flower-stalk, bearing one or more flowers arising from a short axis, and usually with radical leaves at its base.
SCARIOUS, or SCARIOSE, having the consistence of a dry scale, membranous, dry, and shrivelled.
SCION, the young twig used as a graft.
SCLEROGEN, the thickening matter of woody cells.
SCORPIOIDAL, like the tail of a scorpion; a peculiar twisted cymose inflorescence, as in Boraginaceæ.
SCURFY, applied to stems and leaves covered with loose scales.
SECUND, turned to one side.
SECUNDINE, the second coat of the ovule, within the primine.
SEGMENTS, divisions.
SEGREGATE, separated from each other.
SEMINAL, applied to the cotyledons, or seed-leaves.
SEPAL, one of the leaflets forming the calyx.
SEPTATE, divided by septa or partitions.
SEPTICIDAL, dehiscence of a seed-vessel through the septa or edges of the carpels.
SEPTIFRAGAL, dehiscence of a seed-vessel through the back of the loculaments, the valves also separating from the septa.
SEPTUM, a division in an ovary formed by the sides of the carpels.
SERICEOUS, silky; covered with fine, close-pressed hairs.

SERRATE, having sharp processes arranged like the teeth of a saw; *Biserrate*, when these are alternately large and small, or where the teeth are themselves serrated.
SERRULATE, with very fine serratures.
SESSILE, without a stalk, as a leaf without a petiole.
SETA, a bristle or sharp hair; also applied to the gland-tipped hairs of *Rosaceæ* and *Hieracium*, and to the stalk bearing the thecca of Mosses.
SETACEOUS and SETIFORM, in the form of bristles.
SETIFORM, bristle-shaped.
SETOSE, covered with setæ and bristles.
SHEATH, the lower part of the leaf surrounding the stem.
SILICULA, a short pod with a double placenta and replum, as in some *Cruciferæ*.
SILIQUA, a long pod, similar in construction to the silicle.
SIMPLE, not branching, not divided into separate parts. Simple fruits are those formed by one flower.
SINUOUS, with a wavy or flexuous margin.
SINUS, the base or recesses formed by the lobes of leaves.
SLASHED, divided by deep and very acute incisions.
SOCIAL PLANTS, such as grow naturally in groups or masses.
SOREDIA, powdery cells on the surface of the thallus of some Lichens.
SPADIX, a succulent spike bearing male and female flowers, as in Arum.
SPATHE, large membranous bract covering numerous flowers.
SPAWN, same as Mycelium.
SPECIFIC CHARACTER, the essential character of a species.
SPERMAGONE, a microscopic conceptacle in Lichens, containing reproductive bodies called spermatia; also a conceptacle containing fructification in Fungi.
SPERMATIA, motionless spermatozoids in the spermagones of Lichens and Fungi.
SPERMODERM, the general covering of the seed, sometimes applied to the episperm or outer covering.
SPHEROIDAL, nearly spherical.
SPIKE, inflorescence consisting of numerous flowers sessile on an axis.
SPINE, or THORN, an abortive branch with a hard, sharp point.
SPIRAL VESSELS, having a spiral fibre coiled up inside a tube.
SPONGIOLE, the cellular extremity of a young root.
SPORANGIUM, a case containing spores.
SPORE, a cellular germinating body in cryptogamic plants.
SPORIDIUM, a cellular germinating body in Cryptogamia, containing two or more cells in its interior.
SPORULES, the small spores in Cryptogamia.
SQUAMIFORM, like scales.
SQUAMOSE, covered with scales.
SQUARROSE, covered with processes spreading at right angles, or in a greater degree.
STAMEN, the male organ of the flower formed by a stalk or filament, and the anther containing pollen.
STAMINATE, applied to a male flower, or to plants bearing male flowers.
STAMINODIUM, an abortive stamen.
STANDARD, same as Vexillum.
STELLATE, like a star.
STERIGMATA, cells bearing naked spores; also cellular filaments bearing spermata and stylospores in the spermogones and pycnides.
STERILE, male flowers not bearing fruit.
STICHIDIA, pod-like receptacles, containing spores.
STIGMA, the upper cellular secreting portion of the pistil uncovered with epidermis.
STIGMATIC, belonging to the stigma.
STIPE, the stalk of Fern fronds; the stalk bearing the pileus in Agarics.
STIPEL, appendage at the base of a leaflet.
STIPITATE, supported on a stalk.
STIPULATE, furnished with stipules.
STIPULE, appendage at the base of leaves.
STOLON, a sucker at first aërial, and then rooting.
STOLONIFEROUS, having creeping runners, which root at the joints.
STOMATA, openings in the epidermis of plants, especially in the leaves.
STOOL, a plant from which layers are propagated by bending down the branches so as to root in the soil.
STRAP-SHAPED, same as Ligulate; linear, or about six times as long as broad.
STRIATED, marked by streaks or striæ.
STRIGOSE, covered with rough, strong, adpressed hairs.
STROBILUS, a cone, applied to the fruit of Firs, as well as to that of the Hop.
STROPHIOLE, a swelling on the surface of a seed.
STRUMA, a cellular swelling at the point where a leaflet joins the midrib; also a swelling below the sporangium of Mosses.
STYLE, the stalk interposed between the ovary and the stigma.
STYLOPOD, an epigynous disc seen at the base of the styles of *Umbelliferæ*.
STYLOSPORE, a spore-like body, borne on a sterigma, or cellular stalk, in the pycnides of Lichens.
SUBEROUS, having a corky texture.
SUBTERRANEAN, underground; same as Hypogeal.
SUBULATE, shaped like a cobbler's awl.
SUCCULENT, soft and juicy.
SUFFRUTICOSE, having the characters of an under-shrub.
SULCATE, furrowed or grooved.
SUPERIOR, applied to the ovary when free, or not adherent to the calyx; to the calyx, when it is adherent to the ovary; to the part of a flower placed next the axis.
SUPERNATANT, floating on the surface.
SUPRA-DECOMPOUND, doubly compounded.

SUSPENDED, applied to an ovule which hangs from a point a little below the apex of the ovary.
SUSPENSOR, the cord which suspends the embryo, and is attached to the radicle in the young state.
SUTURAL, applied to that kind of dehiscence which takes place at the sutures of the fruit.
SUTURE, the part where separate organs unite, or where the edges of a folded organ adhere; the ventral suture of the ovary is that next the centre of the flower; the dorsal suture corresponds with the midrib.
SYMMETRY, applied to the flower, has reference to the parts being of the same number, or multiples of each other.
SYNANTHEROUS, anthers united together.
SYNCARPOUS, carpels united so as to form one ovary or pistil.
SYNGENESIOUS, same as Synantherous.

TAP-ROOT, root descending deeply in a tapering, undivided manner.
TEGMEN, the second covering of the seed; called also Endopleura.
TEGMENTA, scales protecting buds.
TENDRILS, curling, twining organs, with which plants grasp supports.
TERATOLOGY, study of monstrosities and morphological changes.
TERCINE, the third coat of the ovule, forming the covering of the central nucleus.
TERETE, nearly cylindrical.
TERMINAL, at the top or end.
TERNARY, parts arranged in threes.
TERNATE, compound leaves composed of three leaflets.
TESTA, the outer covering of the seed; some apply it to the coverings taken collectively.
TETRADYNAMOUS, four long stamens and two short, as in *Cruciferæ*.
TETRAGONOUS, having four angles.
TETRAMEROUS: a flower is tetramerous when its envelopes are in fours.
TETRASPORE, a germinating body in Algæ, composed of spore like cells, but also applied to those of three cells.
THALAMIFLORAL, parts of the floral envelope inserted separately into the receptacle of the thalamus.
THALAMUS, the receptacle of the flower, or the part of the peduncle into which the floral organs are inserted.
THALLOGENS, or THALLOPHYTES, plants producing a thallus.
THALLUS, cellular expansion in Lichens and other Cryptogams, bearing the fructification.
THECA, sporangium or spore case, containing spores.
THROAT, the orifice of a gamopetalous corolla.
THYRSUS, a sort of panicle, in form like a bunch of grapes, the inflorescence being mixed.
TIGELLUS, the young embryonic axis.
TOMENTOSE, covered with cottony, entangled pubescence, called tomentum.
TOMENTUM, dense, close hair.
TOOTHED, dentated.
TORUS, another name for Thalamus; sometimes applied to a much-developed thalamus, as in *Nelumbium*.
TRANSPIRATION, the exhalation of fluids by leaves, &c.
TRIADELPHOUS, stamens united in three bundles by their filaments.
TRIANGULAR, having three angles, the faces being flat.
TRICHOTOMOUS, divided successively into three branches.
TRIFOLIATE, or TRIFOLIOLATE, same as Ternate. When the three leaves come off at one point the leaf is *ternately trifoliate;* when there are a terminal stalked leaflet and two lateral ones, it is *pinnately trifoliate*.
TRIGONOUS, having three angles, the faces being convex.
TRIMEROUS; a trimerous flower has its envelopes in three or multiples of three.
TRIPARTITE, deeply divided into three.
TRIPINNATE, a compound leaf three times divided in a pinnate manner.
TRIPINNATIFID, a pinnatifid leaf with the segments twice divided in a pinnatifid manner.
TRIQUETROUS, having three angles, the faces being concave.
TRITERNATE, three times divided in a ternate manner.
TRUNCATE, terminating abruptly, as if cut off at the end.
TRYMA, drupaceous fruit, like the Walnut.
TUBER, a thickened underground stem, as the Potato.
TUBERCLE, the swollen root of some terrestrial Orchids.
TUBERCULATE, covered with knobs or tubercles.
TUBEROUS, applied to roots in the form of tubercles.
TUBULAR, bell-shaped; applied to a campanulate corolla, which is somewhat tubular in its form.
TUMID, swelling.
TUNIC, a coat or envelope.
TUNICATED, applied to a bulb covered by thin external scales, as the Onion.
TURBINATE, in the form of a top.
TURGID, swollen.
TYPICAL, applied to a specimen which has eminently the characteristics of the species, or to a species or genus characteristic of an order.

UMBEL, inflorescence in which numerous stalked flowers arise from one point.
UMBELLULE, a small umbel, seen in the compound umbellate flowers of many *Umbelliferæ*.
UMBILICATE, fixed to a stalk by a point in the centre.
UMBILICUS, the hilum or base of a seed.
UNARMED, without prickles or spines.
UNCINATE, provided with an uncus, or hooked process.
UNCTUOUS, oily.

UNDULATE, waved.

UNGUICULATE, furnished with a short unguis.

UNGUIS, claw, the narrow part of a petal; such a petal is called *Unguiculate.*

UNICELLULAR, composed of a single cell, as some Algæ.

UNILATERAL, arranged on one side, or turned to one side.

UNISEXUAL, of a single sex; applied to plants having separate male and female flowers.

URCEOLATE, urn-shaped; applied to a gamopetalous globular corolla with a narrow opening.

VALVATE, opening by valves, like the parts of certain seed-vessels, which separate at the edges of the carpels.

VALVATE ÆSTIVATION and VERNATION, when leaves in the flower-bud and leaf-bud are applied to each other by the margins only.

VALVES, the portions which separate in some dehiscent capsules.

VASCULAR TISSUE, composed of vessels.

VEINS, fibro-vascular skeleton of leaves.

VELUM, veil; the cellular covering of the gills of an Agaric in its early state.

VENATION, the arrangement of the veins.

VENTRAL, applied to the part of the carpel which is next the axis.

VERNATION, the arrangement of the leaves in the bud.

VERRUCOSE, covered with wart-like excrescences.

VERSATILE, applied to an anther which is attached by one point of its back to the filament, and hence is very easily turned about.

VERTEX, the uppermost point.

VERTICAL, perpendicular.

VERTICIL, a whorl; parts arranged opposite to each other at the same level, or, in other words, in a circle round an axis. The parts are said to be *Verticillate.*

VERTICILLASTER, a false whorl, formed of two nearly sessile cymes, placed in the axils of opposite leaves, as in Dead Nettles.

VESICLE, another name for a cell or utricle.

VEXILLARY, applied to æstivation when the vexillum is folded over the other parts of the flower.

VEXILLUM, standard, the upper or posterior petal of a papilionaceous flower.

VILLOUS, covered with long soft hairs, and having a woolly appearance.

VIRESCENT, green.

VIRGATE, long and straight, like a wand.

VISCOUS, or VISCID, clammy, like bird-lime.

VITELLUS, the embryo-sac when persistent in the seed.

VITTÆ, cells or clavate tubes containing oil in the pericarp of *Umbelliferæ.*

VIVIPAROUS, plants producing leaf-buds instead of fruit.

VOLUBILE, twining; a stem or tendril twining round other plants.

VOLVA, wrapper; the organ which encloses the parts of fructification in some Fungi in their young state.

VULNERARY, having a healing power.

WATTLED, having processes like the wattles of a cock.

WHORLED, same as Verticillate.

WINGS, the two lateral petals of a papilionaceous flower, or the broad flat edge of any organ.

XANTHOPHYLL, yellow colouring matter in plants.

ZONES, stripes or belts.

ZOOSPORE, a moving spore provided with cilia; called also Zoosperm and Sporozoid.

INDEX.

*** ITALICS ARE WOODCUT ILLUSTRATIONS.

www.ingramcontent.com/pod-product-compliance
Lightning Source LLC
LaVergne TN
LVHW021223110826
845150LV00002B/227